W9-BFC-765

APPENDIX 4 AREAS FOR t DISTRIBUTIONS

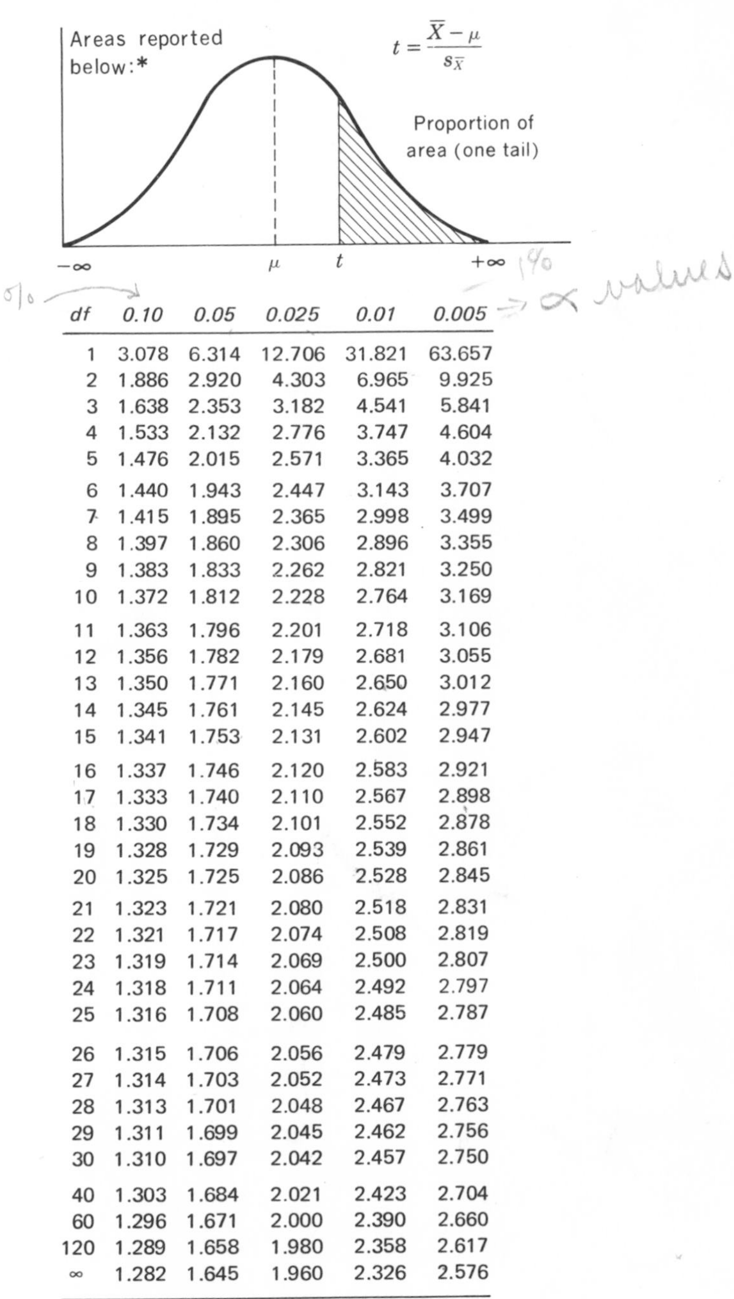

df	*0.10*	*0.05*	*0.025*	*0.01*	*0.005*
1	3.078	6.314	12.706	31.821	63.657
2	1.886	2.920	4.303	6.965	9.925
3	1.638	2.353	3.182	4.541	5.841
4	1.533	2.132	2.776	3.747	4.604
5	1.476	2.015	2.571	3.365	4.032
6	1.440	1.943	2.447	3.143	3.707
7	1.415	1.895	2.365	2.998	3.499
8	1.397	1.860	2.306	2.896	3.355
9	1.383	1.833	2.262	2.821	3.250
10	1.372	1.812	2.228	2.764	3.169
11	1.363	1.796	2.201	2.718	3.106
12	1.356	1.782	2.179	2.681	3.055
13	1.350	1.771	2.160	2.650	3.012
14	1.345	1.761	2.145	2.624	2.977
15	1.341	1.753	2.131	2.602	2.947
16	1.337	1.746	2.120	2.583	2.921
17	1.333	1.740	2.110	2.567	2.898
18	1.330	1.734	2.101	2.552	2.878
19	1.328	1.729	2.093	2.539	2.861
20	1.325	1.725	2.086	2.528	2.845
21	1.323	1.721	2.080	2.518	2.831
22	1.321	1.717	2.074	2.508	2.819
23	1.319	1.714	2.069	2.500	2.807
24	1.318	1.711	2.064	2.492	2.797
25	1.316	1.708	2.060	2.485	2.787
26	1.315	1.706	2.056	2.479	2.779
27	1.314	1.703	2.052	2.473	2.771
28	1.313	1.701	2.048	2.467	2.763
29	1.311	1.699	2.045	2.462	2.756
30	1.310	1.697	2.042	2.457	2.750
40	1.303	1.684	2.021	2.423	2.704
60	1.296	1.671	2.000	2.390	2.660
120	1.289	1.658	1.980	2.358	2.617
∞	1.282	1.645	1.960	2.326	2.576

*Example: For the shaded area to represent 0.05 of the total area of 1.0, the value of t with 10 degrees of freedom is 1.812.

Source: Reprinted by Hafner Press, a division of Macmillan Publishing Company, from *Statistical Methods for Research Workers*, 14th ed., abridged Table IV, by R. A. Fisher. Copyright © 1970 by University of Adelaide.

STATISTICS A First Course

FIFTH EDITION

Statistics: A First Course

Donald H. Sanders

Educational Consultant
Fort Worth, Texas

Elizabeth Farber

CONSULTING EDITOR
Bucks County Community College

McGRAW-HILL, INC.

New York St. Louis San Francisco Auckland Bogotá Caracas
Lisbon London Madrid Mexico City Milan Montreal
New Delhi San Juan Singapore Sydney Tokyo Toronto

STATISTICS: A First Course

Copyright © 1995 by McGraw-Hill, Inc. All rights reserved. Previously published under the title of *Statistics: A Fresh Approach.* Copyright © 1990, 1985, 1980, 1976 by McGraw-Hill, Inc. All rights reserved. Printed in the United States of America. Except as permitted under the United States Copyright Act of 1976, no part of this publication may be reproduced or distributed in any form or by any means, or stored in a data base or retrieval system, without the prior written permission of the publisher.

This book is printed on acid-free paper.

2 3 4 5 6 7 8 9 0 VNH VNH 9 0 9 8 7 6 5

ISBN 0-07-054900-1

This book was set in Minion by York Graphic Services, Inc.
The editors were Jack Shira and Margery Luhrs;
the designer was Armen Kojoyian;
the production supervisor was Friederich W. Schulte.
Von Hoffmann Press, Inc., was printer and binder.

Library of Congress Cataloging-in-Publication Data

Sanders, Donald H.
Statistics: a first course / Donald H. Sanders.—5th ed.
p. cm.
Includes index.
ISBN 0-07-054900-1
1. Statistics. I. Title.
QA276.12.S26 1995
519.5—dc20 94-32146

INTERNATIONAL EDITION

Copyright 1995. Exclusive rights by McGraw-Hill, Inc. for manufacture and export. This book cannot be re-exported from the country to which it is consigned by McGraw-Hill. The International Edition is not available in North America.

When ordering this title, use ISBN 0-07-113564-2

About the Author

DONALD H. SANDERS is the author of eight books about computers and statistics. Over 20 editions of these texts have been published in English, and several have been released in French, German, Spanish, Chinese, and other languages. Well over a million copies of these books have been used in college courses and in industry and government training programs.

Dr. Sanders has 20 years of teaching experience. After receiving degrees from Texas A & M University and the University of Arkansas, he was a professor at the University of Texas at Arlington and at Memphis State University. He was a tenured full professor at Texas Christian University for 14 years.

In addition to his books, Dr. Sanders has contributed articles to journals such as *Data Management, Automation, Banking, Journal of Small Business Management, Journal of Retailing,* and *Advanced Management Journal.* He has also encouraged his graduate students to contribute articles to national periodicals, and more than 70 of these articles have been published. Dr. Sanders has chaired the "Computers and Data Processing" Subject Examination Committee, CLEP Program, College Entrance Examination Board, Princeton, New Jersey.

Special Acknowledgment

Professor Elizabeth Farber of Bucks County Community College served as consulting editor for this edition and was instrumental in preparing the complete first drafts of Chapters 4, 5, and 10. In addition, she supplied the many hundreds of new problems using real data that are found throughout the text. Her expertise, talents, and tireless efforts in helping to prepare this revision are gratefully acknowledged.

To Those Who Open This Book with Dismay

Contents

CHAPTER 6 Sampling Concepts

CHAPTER 7 Estimating Parameters

CHAPTER 8
Testing Hypotheses: One-Sample Procedures

CHAPTER 9
Testing Hypotheses: Two-Sample Procedures

CHAPTER 10
Quality Control

CHAPTER 11 Analysis of Variance

CHAPTER 12 Chi-Square Tests: Goodness-of-Fit and Contingency Table Methods

CHAPTER 13 Linear Regression and Correlation

CHAPTER 14

Nonparametric Statistical Methods

APPENDIXES

WARNING!
Preface
Ahead

Preface

If I had only one day left to live, I would live it in my statistics class . . . it would seem so much longer.

—***Quote in a university student calendar***

It's that time again—time to attempt once more to present the subject of statistics in an interesting, timely (and occasionally humorous) way so that a period spent on the subject doesn't seem to students to represent the eternity suggested by the above quote.

Actually, most readers of this book accept the fact that an educated citizen must have an understanding of basic statistical tools to function in a world that's becoming increasingly dependent on quantitative information. But most who read this text have never placed the solving of mathematical problems at the top of their list of favorite things to do. In fact, many probably don't care much for math and have heard numerous disturbing rumors about statistics courses.

A motivating force behind the preparation of this text is the distinct possibility that the misgivings and apathy implicit in the introductory quote are related in some way to the unfortunate fact that many existing statistics books are rigorously written, mathematically profound, precisely detailed—and excruciatingly dull!

Philosophy of This Book

The *main difference between this text and many others* is that an attempt is made here to (1) present material in a rather relaxed and informal way without omitting important concepts, (2) demonstrate the wide range of relevant issues and questions that can be addressed with the help of statistical analysis techniques by presenting more than *2,000* realistic problems of the type that are dealt with all the time in health care, business and economics, the social and physical sciences, engineering, education, and leisure activities, and (3) use an intuitive and/or commonsense approach (and an occasional humorous situation or name) to develop concepts whenever possible. In short, this book is written to communicate with students rather than to lecture to them, and its intent is to convince readers that the study of statistics can be a lively, interesting, and rewarding experience.

More specifically, the *purpose of this book* is to introduce students at an early stage in a college program to many of the important concepts and procedures they'll need to (1) evaluate such daily inputs as organizational reports, newspaper and magazine articles, and radio and television commentaries, (2) improve their ability to make better decisions over a wide range of topics, and (3) improve their ability to measure and cope with changing conditions both at home and on the job. And since users of this text may frequently be consumers rather than producers of statistical information, the emphasis here is on explaining statistical procedures and interpreting the resulting conclusions. However, the *mathematical demands are modest*—no college-level math background is required or assumed.

Features of This Edition

An obvious change in this fifth edition is that it is now printed in an attractive *four-color format* and is organized into 14 chapters rather than 16. Three chapters dealing with index numbers, time-series analysis, and an overview of other topics that appeared in the fourth edition have been deleted. The chapter on probability and probability distributions has been rewritten and expanded into two new chapters (Chapters 4 and 5). And a new chapter (Chapter 10) focusing on statistical quality control concepts has been added.

The following feature introduces each chapter:

➤ A LOOKING AHEAD section previews upcoming content and learning objectives while helping students focus their attention on the important ideas.

In the body of the chapters you'll find:

➤ STATISTICS IN ACTION minicases are included in each chapter to show students how statistics is used throughout various aspects of our daily lives. Each is keyed with an icon that relates it to one of the following categories:

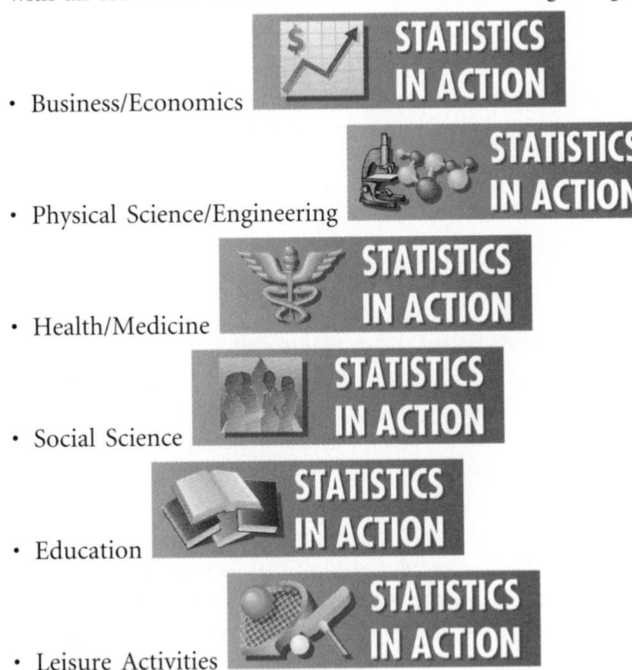

- ➤ KEY TERMS and FORMULAS are introduced in boldface type and some are highlighted in color-shaded boxes for easy reference.
- ➤ SELF-TESTING REVIEW PROBLEMS are presented after each section to support understanding of major topics and help students master the material.
- ➤ LOGIC FLOWCHARTS, with procedural steps outlined in color, were designed to help guide students through the material and give them a visual representation of the concepts.
- ➤ MINITAB is used in numerous chapter example problems to demonstrate the ways in which statistical software packages can be used to support computing, analysis, and decision-making activities in statistics. *Minitab* was selected for illustration purposes mainly because of its wide popularity, but the use of it or any other statistical package is not a requirement.

And at the end of each chapter, you'll find:

- ➤ A LOOKING BACK section summarizes the main points and helps students to recap the objectives of the chapter.
- ➤ The REVIEW EXERCISES sections use actual data in hundreds of new and real-world problems to help students gain additional practice and confidence. More than 2,000 problems dealing with a wide variety of realistic topics are included in the Self-Testing Review and in the end-of-chapter exercises. At least 1,800 of these exercises are new for this edition.

- ➤ A TOPICS FOR REVIEW AND DISCUSSION section presents questions that are designed to actively engage students in discussion and/or writing activities regarding what they have learned. There are more than 140 such questions throughout the text.

- ➤ A PROJECTS/ISSUES TO CONSIDER section offers suggested topics for extended research and student exploration of ideas presented in the chapter.

- ➤ A COMPUTER EXERCISES section presents problems that are keyed to relevant chapter topics. By using the appropriate hardware/software resources, students can see that tedious procedures can be carried out with a few keystrokes. There are more than 120 such exercises throughout the text.

Changes in This Edition

The first three chapters of this text focus on descriptive statistics. Chapter 1 includes a new six-step numbered format for statistical problem-solving. A new discussion of samples and sample selection is introduced, and the section on the role of the computer in statistics has been condensed and updated. Chapter 2 still introduces examples of how statistical methods are often used improperly, and it contains new tables and charts for emphasis. A discussion of index numbers has also been added to this chapter. Chapter 3, which covers measures of central tendency and dispersion, has been substantially reworked to replace two chapters from the fourth edition.

Chapters 4 and 5 focus on probability concepts and probability distributions, replacing a single earlier chapter, and are completely new for this edition. Coverage of these topics has been greatly expanded. A new section on estimating the population variance has been added to Chapter 7.

CHAPTER 1

Let's Get Started

LOOKING AHEAD

This chapter gives you an understanding of the purpose of this book (and this course). You'll see that the word "statistics" is used in several ways, and you'll also learn the meaning of other important terms. The need for statistics is outlined here, and a series of steps that are used to solve statistics problems is presented. Finally, the role of the computer in statistics is briefly discussed.

Thus, after studying this chapter, you should be able to:

- Define the meaning of the terms in boldface type such as **statistics, population, census, sample, parameter, statistic, descriptive statistics,** and **statistical inference.**
- Understand and explain why a knowledge of statistics is needed.
- Outline the basic steps in the statistical problem-solving methodology.
- Identify various methods of obtaining samples.
- Discuss the role of computers and data-analysis software in statistical work.

1.1 What to Expect

In O. Henry's *The Handbook of Hymen,* Mr. Pratt is wooing the wealthy Mrs. Sampson. Unfortunately for Pratt, he has a poet for a rival. To offset this romantic disadvantage, Pratt selects a book of quantitative facts to dazzle Mrs. Sampson.

> "Let us sit on this log at the roadside," says I, "and forget the inhumanity and ribaldry of the poets. It is in the glorious columns of ascertained facts and legalized measures that beauty is to be found. In this very log we sit upon, Mrs. Sampson," says I, "is statistics more wonderful than any poem. The rings show it was sixty years old. At the depth of two thousand feet it would become coal in three thousand years. The deepest coal mine in the world is at Killingworth, near Newcastle. A box four feet long, three feet wide, and two feet eight inches deep will hold one ton of coal. If an artery is cut, compress it above the wound. A man's leg contains thirty bones. The Tower of London was burned in 1841."
>
> "Go on, Mr. Pratt," says Mrs. Sampson. "Them ideas is so original and soothing. I think statistics are just as lovely as they can be."

It's possible (even likely) that you don't yet share the view of statistics expressed by Mrs. Sampson. Oh, you may agree that an understanding of statistical tools is needed in a modern world. But you've never placed the solving of mathematical problems at the top of your list of favorite things to do, you've possibly heard disturbing rumors about statistics courses, and you've not been eagerly awaiting this day when you must crack open a statistics book. If the comments made thus far in this paragraph apply to you, you needn't apologize for your possible misgivings. After all, many statistics books are rigorously written, mathematically profound, precisely detailed—and excruciatingly dull!

It isn't possible in this book to avoid the use of formulas to solve statistical problems and demonstrate important statistical theories. But a knowledge of advanced mathematics certainly *isn't* required to grasp the material presented here. In fact, you'll be relieved to know that a beginning-level high school algebra course prepares you for all the math required.

You'll find in the pages and chapters that follow that the intent is to use an intuitive, commonsense approach to develop concepts. The goal is to communicate with you rather than lecture to you. Thus, important concepts are presented in a rather relaxed and informal way. (Occasional quotes, ridiculous names, and unlikely situations are sometimes used to recapture your attention.) In short, this book is written for beginning students rather than statisticians, and its intent is to convince you that the study of statistics is a lively and rewarding experience. (If Mr. Pratt could convince Mrs. Sampson, then maybe you too can be converted.)

1.2 Purpose and Organization of the Text

To do is to be—*J.-P. Sartre*

To be is to do—*I. Kant*

Do be do be do—*F. Sinatra**

Purpose of This Book

The purpose of this book is to acquaint you with the statistical concepts and techniques needed to organize, measure, and evaluate data that may then be used to support informed decisions. Thus, the emphasis here is placed on explaining statistical procedures and interpreting the resulting conclusions. In short, the following dialog from K. A. C. Manderville's *The Undoing of Lamia Gurdleneck* concludes with an important message that's kept in mind throughout this text.†

> *"You haven't told me yet," said Lady Nuttal, "what it is your fiancé does for a living."*
>
> *"He's a statistician," replied Lamia, with an annoying sense of being on the defensive.*
>
> *Lady Nuttal was obviously taken aback. It had not occurred to her that statisticians entered into normal social relationships. The species, she would have surmised, was perpetuated in some collateral manner, like mules.*
>
> *"But Aunt Sara, it's a very interesting profession," said Lamia warmly.*
>
> *"I don't doubt it," said her aunt, who obviously doubted it very much. "To express anything important in mere figures is so plainly impossible that there must be*

---

* Have you ever noticed that chapters and sections of chapters in learned books and academic treatises are often preceded by quotations such as these that are selected by the author for some reason? In some cases a quotation is intended to emphasize a point to be presented; in other cases (often in the more erudite sources) there appears to be no discernible reason for the message, and it forever remains a mystery to the reader. In this particular case, the quotations from the above philosophers unfortunately fall into the *latter category!* However, we will from time to time throughout the book attempt to use quotations to emphasize a point.

† Frontispiece from Maurice G. Kendall and Alan Stuart, *The Advanced Theory of Statistics*, vol. 2: *Inference and Relationships*, Hafner Publishing Company, Inc., New York, 1967.

endless scope for well-paid advice on how to do it. But don't you think that life with a statistician would be rather, shall we say, humdrum?"

Lamia was silent. She felt reluctant to discuss the surprising depth of emotional possibility which she had discovered below Edward's numerical veneer.

"It's not the figures themselves," she said finally, "it's what you do with them that matters."

Swashbuckling Statistician

The black-and-gold flag looks like it should be flying from the mast of a pirate ship. But this flag isn't the banner of a swashbuckling buccaneer. Rather, it's the pennant of NM Direct, the mail-order division of the exclusive Neiman Marcus Group of retail stores. The chief executive of NM Direct is B.D. "Bernie" Feiwus, a convention-flouting captain who uses his training as a statistician to get the most out of Neiman Marcus' customer data. These data are fine-tuned to produce valuable mailing lists that include only those households that are likely to buy the type of merchandise found in the NM catalogs. Feiwus' statistical training has also been used to arrive at decisions on how to (1) time the speed of conveyor belts that are used in the order-filling process, and (2) make more money with few additional costs by putting out more catalogs and thus increasing business from January through August.

Definitions and Organization

Let's pause here just long enough to define a few terms. The word "statistics" is commonly used in two ways. In the first context, "statistics" is a *plural* term meaning numerical facts or data. In the previous quote, Lamia begins the last sentence with "It's not the figures themselves . . ." An identical expression would be "It's not the statistics themselves . . ."

But the word "statistics" can refer to much more than just numerical facts. When used in a broader and *singular* sense, "statistics" refers to a subject of study (in Lamia's words, "it's what you do with them that matters.") Or, more formally,

> **Statistics** is the science of designing studies, gathering data, and then classifying, summarizing, interpreting, and presenting these data to support the decisions that are needed.

Thus, "statistics" refers to a subject of study in the same way that "mathematics" refers to such a subject.

Let's briefly look here at some important terms used in the study of statistics. Two such terms are *population* and *sample.*

> A **population** is the complete collection of measurements, objects, or individuals under study.

You'll notice in this definition that "population" isn't limited to a group of people. Rather, the term refers to *all* of the measures, counts, or qualities that are of interest in a study. Thus, a population may be all Boffo batteries built in a day's production run, the weights of all packages in a shipment, the mileage on all the police cars in Los Angeles, the outstanding balances on all credit accounts in a Nordstrom's department store, or the length of stay of all patients currently in Mercy Hospital. To gain information about this characteristic of interest, a researcher may occasionally try to survey all the elements in a population. Such a survey is called a **census.**

But it's more common to first select a sample from the population and then analyze the sample data.

> A **sample** is a portion or subset taken from a population.

Since a sample is only a subset of a population, the data of interest that it supplies are necessarily incomplete. But if sampling is done scientifically, it's usually possible to obtain sample results that are sufficiently accurate for the researcher's needs.

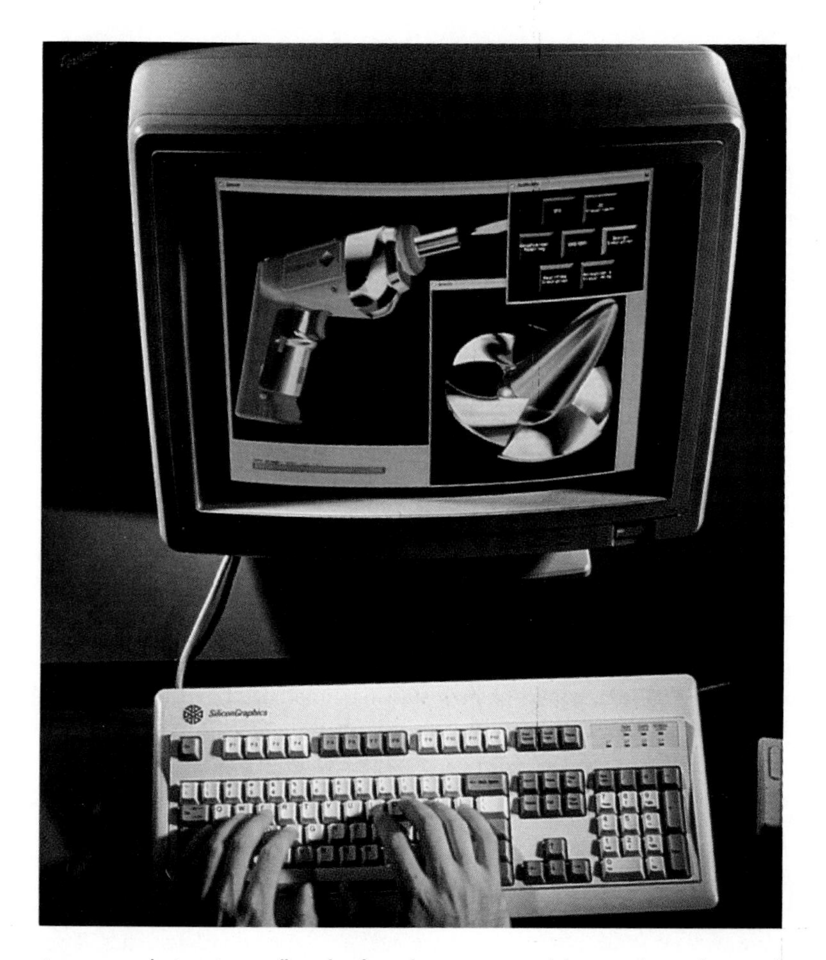

Computers make it easier to collect, classify, and present statistical data to aid in quality control efforts. (*Tim Davis/Science/Photo Researchers*)

Two other statistical terms are closely related to the concepts we've just examined.

A **parameter** is a number that describes a *population* characteristic.

The average (arithmetic mean) weight of all packages in a population is a single number, and this number is a parameter. Other parameters are (1) the percentage of all hospital patients receiving a treatment who are responding favorably to the treatment and (2) the average lifetime (in hours) of all Boffo batteries. Thus, if a percentage figure or an average value describes a population, it is a parameter.

Often, however, a parameter is unknown and a *statistic* must be used.

A **statistic** is a number that describes a sample characteristic.

A statistic is to a sample what a parameter is to a population—it's a single value that summarizes some characteristic of interest. Thus, if we receive a shipment of 1,000 packages, select a sample of 25 of them, weigh each of the 25, and compute the average weight of the 25 packages, the result is a statistic. But if we compute the average weight of all 1,000 packages in the shipment, the result is a parameter.

Back now to the general subject of study known as "statistics." This subject can be further broken down into two essential parts: (1) descriptive statistics and (2) inferential statistics. Figure 1.1 presents an overview of these categories that we'll now consider.

Descriptive statistics includes the procedures for collecting, classifying, summarizing, and presenting data. Charts, tables, and summary measures such as averages are used to describe the basic structure of the study subject. Obviously, statistical description remains an important part of the study of statistics. Sales or government agency data, for example, may be classified or grouped by (1) volume, size, or quantity, (2) geographic location, or (3) type of product or service. To be of value, masses of data are often condensed or sifted—that is, summarized—so that resulting information is concise and effective. A general sales manager, for example, may be interested only in the average monthly total sales of particular stores. Although she could be given a report that breaks sales down by department, product, and sales clerk, such a report is more likely to be of interest to a department manager. Once the facts have been classified and summarized, it's then often necessary that they be presented or communicated in a usable form—perhaps through the use of tables and charts—to the final user.

Descriptive statistics is the focus of the next two chapters. With a knowledge of descriptive procedures, you can evaluate information presented in reports, articles, and broadcasts and improve your ability to measure and thus cope with changing conditions. In Chapter 2, for example, you'll see how statistical methods have been *improperly used* to confuse or deliberately mislead people. Many of the invalid uses presented

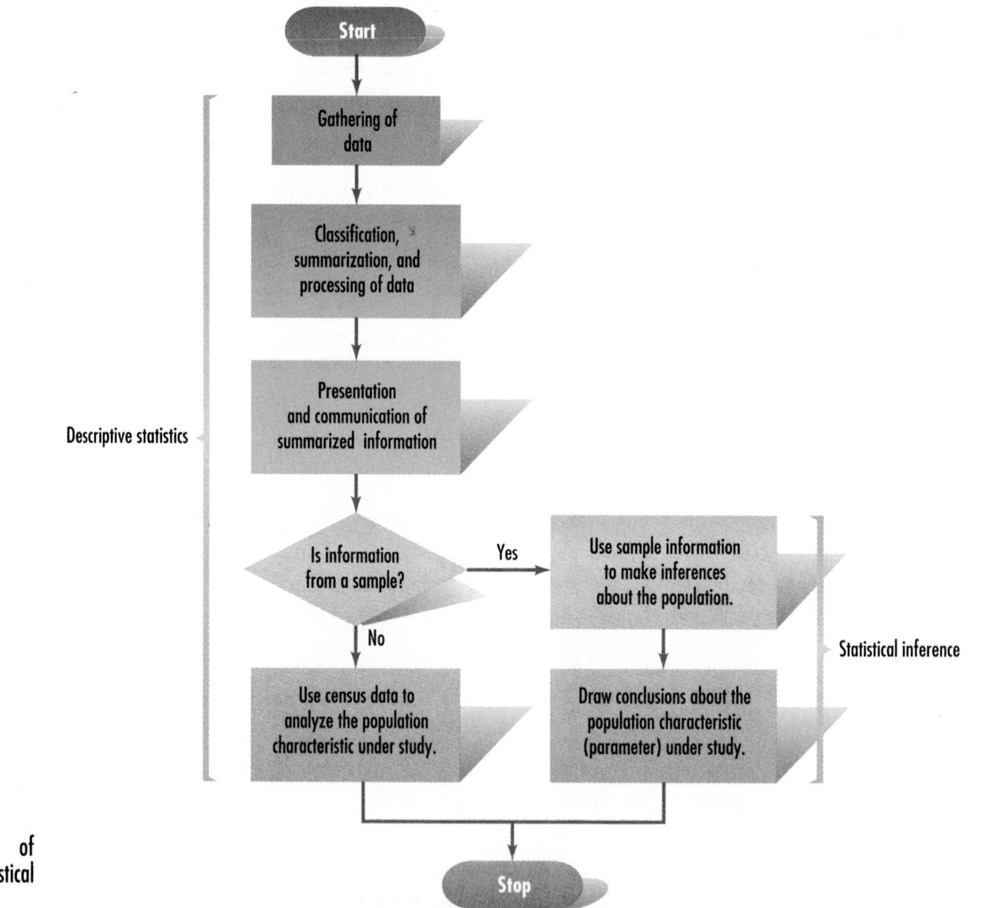

FIGURE 1.1 An overview of descriptive statistics and statistical inference.

involve descriptive procedures. And in Chapter 3 we'll consider data collection and classification, *measures of central tendency* (or averages), and *measures of dispersion* that are frequently used by decision makers to describe data sets.

A second topic that must receive extensive treatment in any statistics text is inferential statistics. **Statistical inference** is the process of arriving at a conclusion about a population parameter (which is usually an unknown quantity) on the basis of information obtained from a sample statistic (a known value). Statistical inference concepts allow an analyst to use sample data to *estimate* an unknown population parameter of interest—for example, the average weight of all 1,000 packages in the shipment described earlier. If the average weight of the sample of 25 packages is 7.1 pounds, the average weight of all 1,000 packages may be estimated to probably be between 6.9 and 7.3 pounds. Or perhaps the analyst will use statistical inference to conduct a *test* to see if the average weight of all 1,000 packages is 7.5 pounds. On the basis of the sample statistic of 7.1, the test results may indicate that the population mean weight falls below the specified 7.5-pound parameter.

Figure 1.2 graphically presents the statistical inference process of using a sample statistic to make a judgment about an unknown population parameter. In Chapters 4 and 5, you'll learn the conceptual foundations of probability that support sampling applications. Then, in the remaining chapters you'll see how statistical inference concepts are applied.

Most of the chapters in this book contain *self-testing review sections* following the presentation of important material. You're encouraged to test your understanding of the reviewed subjects presented in these sections before moving to the next topic. Answers to odd-numbered self-testing review exercises are given at the end of the chapters. (Answers to even-numbered exercises are found in the *Study Guide* that is available for this text.) You'll also find at the end of a chapter (1) a *looking back* chapter summary, (2) *additional review exercises* (with answers to odd-numbered exercises found at the back of the book), (3) *topics for review and discussion*, (4) *projects/issues to consider*, and (5) *computer exercises*.

1.3 Need for Statistics

At the beginning of this chapter the following sentence appeared: "Oh, you may agree that an understanding of statistical tools is necessary in a modern world." Per-

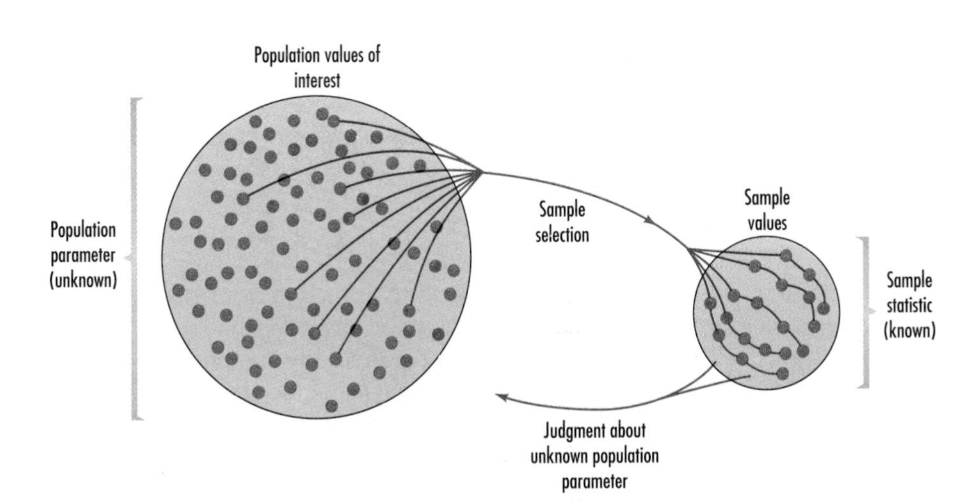

FIGURE 1.2 The statistical inference process involves the use of a known sample statistic to arrive at a judgment about an unknown population parameter.

haps that statement was premature; perhaps you don't agree at all. It *is* a fact, though, that you need a knowledge of statistics to help you (1) describe and understand numerical relationships and (2) make better decisions.

Describing Relationships between Variables

The amount of data that's gathered, processed, and presented to people is overwhelming. Thus, it's necessary that you be able to sift through this huge mass to identify and describe those sometimes obscure relationships between variables that are often so important in decision making. Consider these examples:

- A management consultant wants to compare a client's investment return for this year with related figures from last year. He summarizes masses of revenue and cost data from both periods and, based on his findings, presents his recommendations to his client.
- A public health official decides to see if there is any connection between inhaling the smoke produced by cigarette smokers (second-hand smoking) and the incidence of asthma in young children. She applies statistical techniques to large amounts of data and reaches conclusions that will affect the health of large numbers of people.
- A marketing manager wants to analyze the relationships that may exist between the demand for her product and such demographic factors as income, age, and ethnic background. On the basis of the relationships that may exist, she can advise her staff to design an appropriate advertising campaign.
- A college admissions director needs to find an effective way of selecting student applicants. He designs a statistical study to see if there's a significant relationship between SAT scores and the grade point average achieved by freshmen at his school. If there is a strong relationship, high SAT scores will become an important criteria for acceptance.

Aiding in Decision Making

Statistical methods allow people to make better decisions in the face of uncertainty. Consider these examples:

- The purchasing manager for a processing plant that supplies fresh poultry to eight states is responsible for buying chickens in shipments of 10,000 birds. Standards specify that the birds should weigh 6 pounds on the average. (Birds over that weight tend to be too tough; lighter birds are too scrawny.) The truck of a new supplier rolls up to the unloading dock with a shipment. The supplier's salesperson assures the manager that the shipment meets standards. Should the shipment be accepted on the basis of this claim? Probably not. Rather, statistical inference techniques are used to select a sample of, say, 50 birds from the population of 10,000. Each bird in the sample is weighed, the average weight of the 50 birds is computed, some other statistical calculations are made, and a conclusion about the average weight of the population of 10,000 birds is then reached. Given this information, the manager can now make a more informed decision about accepting the shipment.

- A quality control engineer at a plant that produces shotgun shells knows that certain variations will exist in the shells produced. These variations in shot patterns and in shot velocity can be tolerated if they don't exceed specified limits more than 1 percent of the time. By using a carefully designed sampling plan, the engineer arrives at reliable conclusions or inferences about the quality of a production run. These conclusions are based on tests conducted by firing a relatively small number of shells randomly selected from those produced during the run.
- Suppose that the manager of Big-Wig Executive Hair Stylists, Hugo Bald, has advertised that 90 percent of the firm's customers are satisfied with the company's services. If Polly Tician, a consumer activist, feels that this is an exaggerated statement that might require legal action, she can use statistical inference techniques to decide whether or not to sue Hugo.
- A personnel manager has noted that job applicants who score high on a manual dexterity test later tend to perform well in the assembling of a product, while those with lower test scores tend to be less productive. By applying a statistical technique known as "regression analysis" (the subject of Chapter 13), the manager can forecast how productive a new applicant will be on the job on the basis of how well he or she performs on the test.

1.4 Statistical Problem-Solving Methodology

Mark Twain used a simple approach to "problem solving" in *Sketches Old and New*:

"If it would take a cannon ball 3⅓ seconds to travel four miles, and 3⅜ seconds to travel the next four, and 3⅝ to travel the next four, and if its rate of progress continued to diminish in the same ratio, how long would it take to go fifteen hundred million miles?"—***Arithmeticus, Virginia, Nevada***

"I don't know."—***Mark Twain***

In most statistical problem-solving situations, though, it's best to follow a more enlightened approach. Several steps are followed to get rational answers to statistical problems, and when one of these steps is ignored, the final results may be invalid, inaccurate, or needlessly expensive. The basic steps in statistical problem solving are outlined below.

Step 1—Identifying the Problem or Opportunity

The researcher must clearly understand and correctly define exactly what it is that the study is to accomplish. For example, is the goal to study some population and then estimate an unknown average or percentage, or is it to impose some treatment on the group and then gauge the response? That is, can the study goal be achieved through mere *counts or measurements* of the group, or must an *experiment* be performed on the group? If samples are needed (and they usually are), how large should they be and how should they be taken? Time and effort are wasted if the answers to such questions are too blurred or too broad.

Psst . . . Hey, Mac, Need Some Data . . . ?

Each year the Census Bureau publishes its latest edition of the *Statistical Abstract of the United States.* With about 1,000 pages and 1,500 tables, the *Abstract* can overwhelm browsers with facts about such things as who they are, what they do, how they travel, and the nature of the world they occupy. For example, did you know that since 1975 the average American has been eating less beef but that chicken consumption has increased sharply? Or do you need to know the number of recent graduates in chemical engineering, economics, nursing, and psychology? The *Abstract* has the figures. What about the percentage of Americans over 45 who wear glasses? The *Abstract*'s got it.

Step 2—Gathering Available Facts

Data must be gathered that are accurate, timely, as complete as possible, and relevant to the problem. Sources of available data may be classified into internal and external categories.

Internal data are found in the departments of an organization. Business data are produced in accounting, production, and marketing departments; in a college or hospital, the registrar's office or admitting section generates internal facts. *External data* are facts produced by outside sources such as professional associations and government agencies. External data are supplied by nongovernment publications such as the *American Economic Review, The American Statistician, BioScience, Business Week, Chemical Engineering, Forbes, New England Journal of Medicine, Science, Teaching of Psychology*—this list could go on and on. A wealth of external data are also supplied by government publications such as the *Census of the United States*, the *Census of Business*, the *Survey of Current Business*, the *Statistical Abstract of the United States*, the *Monthly Labor Review*, and the *Federal Reserve Bulletin.*

It's generally preferable to gather available data from **primary sources**—those that initially gather the data and first publish them—rather than from **secondary sources**—those that republish the data. This is true because secondary sources may introduce reproduction errors and may not explain how the facts were gathered or what limitations exist to their use. Secondary sources may also fail to show how the primary source defined the variables. During the depression month of November, 1935, for example, the National Industrial Conference Board estimated the number unemployed at about 9 million; the National Research League estimated 14 million; and the Labor Research Association topped them all with a figure of 17 million. These estimates varied primarily because of differences in the way unemployment was defined. Sherlock Holmes summarized the importance of data gathering in *The Adventure of the Copper Beeches* when he said: "Data! Data! Data! I can't make bricks without clay."

Step 3—Gathering New Data: Tools and Samples

In many cases the data needed by decision makers simply aren't available elsewhere, and so there's no alternative but to gather them. There are advantages to gathering new data, for one who's aware of the problem can define the variables so that the resulting facts possess the properties needed to solve the problem. Instruments are used to gather data that are measured, weighed, or timed. When the data are supplied by people, personal interviews and mail questionnaires are commonly used tools.

In a *personal interview*, an interviewer asks a respondent the prepared questions that appear on a form and then records the answers in the spaces provided. This data-gathering approach allows the interviewer to clarify any terms that aren't understood by the respondent, and it results in a high percentage of usable returns. But it's an expensive approach and is subject to possible errors introduced by the interviewer's manner in asking questions. Interviews are often conducted over the telephone. This is less expensive, but, of course, some households don't have telephones or have unlisted numbers, and this may bias the survey results.

When *mail questionnaires* are used, the questions are printed on forms, and these queries are designed so that they can be answered by the respondent with check marks or with a few words. The use of questionnaires is often less expensive than personal interviews. But the percentage of usable returns is generally lower. And those who do answer may not always be the ones to whom the questionnaire was addressed, and/or they may respond because of a nonrepresentative interest in the survey subject.

Instruments, interviews, and mail questionnaires can be used to gather new data from entire populations. Usually, though, the new data gathered with these tools come from samples. A sample should be representative of the population, but there are many ways that samples can be selected.

Judgment Samples. Sample selection is sometimes based on the opinion of one or more persons who feel sufficiently qualified to identify items for a sample as being characteristic of the population. Any sample based on someone's expertise about the population is known as a **judgment sample**. Let's assume that a political campaign manager intuitively picks certain voting districts as reliable places to measure the public's opinion of her candidate. The poll that is then taken in these districts is a judgment sample based on the campaign manager's expertise.

A judgment sample is convenient, but it's difficult to assess how closely it measures reality. This difficulty of objective assessment leaves an uncomfortable uncertainty in any estimate based on sample results. This doesn't mean, though, that a judgment sample should never be used. The quality of such a sample depends on the researcher's expertise, but experience may serve as a valuable tool in surveys.

Probability Samples. Unlike a judgment sample, a probability sample produces results that *can* be objectively assessed. A **probability sample** is one in which the chance of selection of each item in the population is known before the sample is picked. The simple random, systematic, stratified, and cluster samples discussed below are all types of probability samples that are used to gather new data.

Simple Random Samples. If a probability sample is chosen in such a way that all possible groupings of a given size have an equal chance of being picked and if each item in the population has an equal chance of being selected, then the sample is called a **simple random sample**. Let's assume that every item in a population is numbered and each number is written on a slip of paper. Now if each numbered slip of paper is placed in a bowl and mixed and if a group of slips is then picked, the items represented by the selected slips constitute a random sample.

A more practical approach is often to use a computer programmed to carry out this random selection process. Or a random sample can be obtained by using a *table of random numbers.* Each digit in such a table is determined by chance and has an equal likelihood of appearing at any single-digit space in the table. Appendix 3 at the back of the book contains a random number table.

To show the use of random numbers in simple random sampling, suppose we have a list of 200 customer accounts eligible for a consumer survey and we want a sample of 20 of them. We could obtain a simple random sample in the following manner:

1. Assign each and every customer account a number from 000 to 199. Each account should have a unique number. The first account would be labeled 000, the second 001, and so on.
2. Next, consult a table of random numbers. Table 1.1 is a brief excerpt from such a table.

TABLE 1.1 AN EXAMPLE OF A TABLE OF RANDOM NUMBERS

5124	0746	6296	9279
5109	1971	5971	1264
4379	6296	8746	5899
8194	3721	4621	3634

3. It's essential that we preplan how to select a sequence of digits from the table so that no bias enters into the selection process. In our case, we need a sequence of three digits. Let's say that our pattern of selection is the last three digits of each block of numbers and that we will work down a column.
4. Select a random number in the preplanned pattern and match the random number assigned to an account. For example, account number 124 is selected first, and then account number 109 is added to the sample. If we have a random number we cannot use, as in the case of 379, we proceed to the next random number, 194, and continue this process of selection until there's a sample of 20 accounts.

Systematic Samples. Suppose we have a list of 1,000 registered voters in a community, and we want to pick a probability sample of 50. We can use a random number table to pick 1 of the first 20 voters (1,000/50 = 20) on our list. If the table gave us the number 16, then the 16th voter on the list would be the first to be selected. We would then pick every 20th name after this random start (the 36th voter, the 56th voter, and so on) to produce a **systematic sample**.

Stratified Samples. If a population is divided into relatively homogeneous groups, or strata, and a sample is drawn from each group to produce an overall sample, this overall sample is known as a **stratified sample**. Stratified sampling is usually performed when there's a large variation within the population and the researcher has some prior knowledge of the structure of the population that can be used to establish the strata. The sample results from each stratum are weighted and calculated with the sample results of other strata to provide an overall estimate.

As an illustration, suppose our population is a university student body and we want to estimate the average annual expenditures of a college student for nonschool items. Assume we know that, because of different lifestyles, juniors and seniors spend more than freshmen and sophomores, but there are fewer students in the upper classes than in the lower classes because of some dropout factor. To account for this variation in lifestyle and group size, the population of students can easily be stratified into freshmen, sophomores, juniors, and seniors. A sample can be taken from each stratum and each result weighted to provide an overall estimate of average nonschool expenditures.

Cluster Samples. A **cluster sample** is one in which the individual units are groups, or clusters, of single items. It's always assumed that the individual items within each cluster are representative of the population. Consumer surveys of large cities often employ cluster sampling. The usual procedure is to divide a map of the city into small blocks, each block containing a cluster of households to be surveyed. A *number of clusters are selected for the sample,* and all the households in a cluster are surveyed. A distinct benefit of cluster sampling is savings in cost and time. Less energy and money are expended if an interviewer stays within a specific area rather than traveling across stretches of the city.

Sampling Errors. The goal in sampling is to select a portion of the population that displays all of the characteristics of the population. If we're to make a judgment about a population from sample results, we want those results to be as typical of the population as possible. Unfortunately, it's extremely difficult, if not nearly impossible, to have a sample that's completely representative of the population. And it would be unreasonable to expect a sample result to have *exactly* the same value as some population attribute. Thus, **sampling error** resulting from such factors as size and representativeness of the sample used is to be expected. You'll see in later chapters that researchers have learned how to objectively assess and cope with such error. Of course, errors may also be introduced by people as they code and record sample data. And, as we'll see in the next chapter, results obtained from a biased sample may be worthless.

Step 4—Classifying and Summarizing the Data

After the data are collected (from published or new sources), the next step is to organize or group the facts for study. Identifying items with like characteristics and arranging them into groups or classes, we've seen, is called *classifying.* Production data can be classified, for example, by product make, location of production plant, and production process used. Classifying is sometimes accomplished by a shortened, predetermined method of abbreviation known as *coding.* Code numbers are used to designate persons (social security number, payroll number), places (Zip code, precinct), and things (part number, catalog number).

Once the data are arranged into classes, it's then possible to reduce the amount of detail to a more usable form by *summarization.* Tables, charts, and numerical descriptive values such as measures of central tendency (or averages) and measures of dispersion (the extent of the spread or scatter of the data about the central value) are summarizing tools.

Step 5—Presenting and Analyzing the Data

Summarized information in tables, charts, and key quantitative measures facilitates problem understanding. Such information also helps identify relationships and allows an analyst to present important points to others. The analyst next interprets the results of the preceding steps, uses the descriptive measures computed as the basis for making any relevant statistical inferences, and employs any statistical aids that may help identify desirable courses of action. The validity of the options selected is, of course, determined by the analyst's skill and the quality of his or her information.

Step 6—Making the Decision

Finally, the analyst weighs the options in light of established goals to arrive at the plan or decision that represents the "best" solution to the problem. Again, the correctness of this choice depends on analytical skill and information quality.

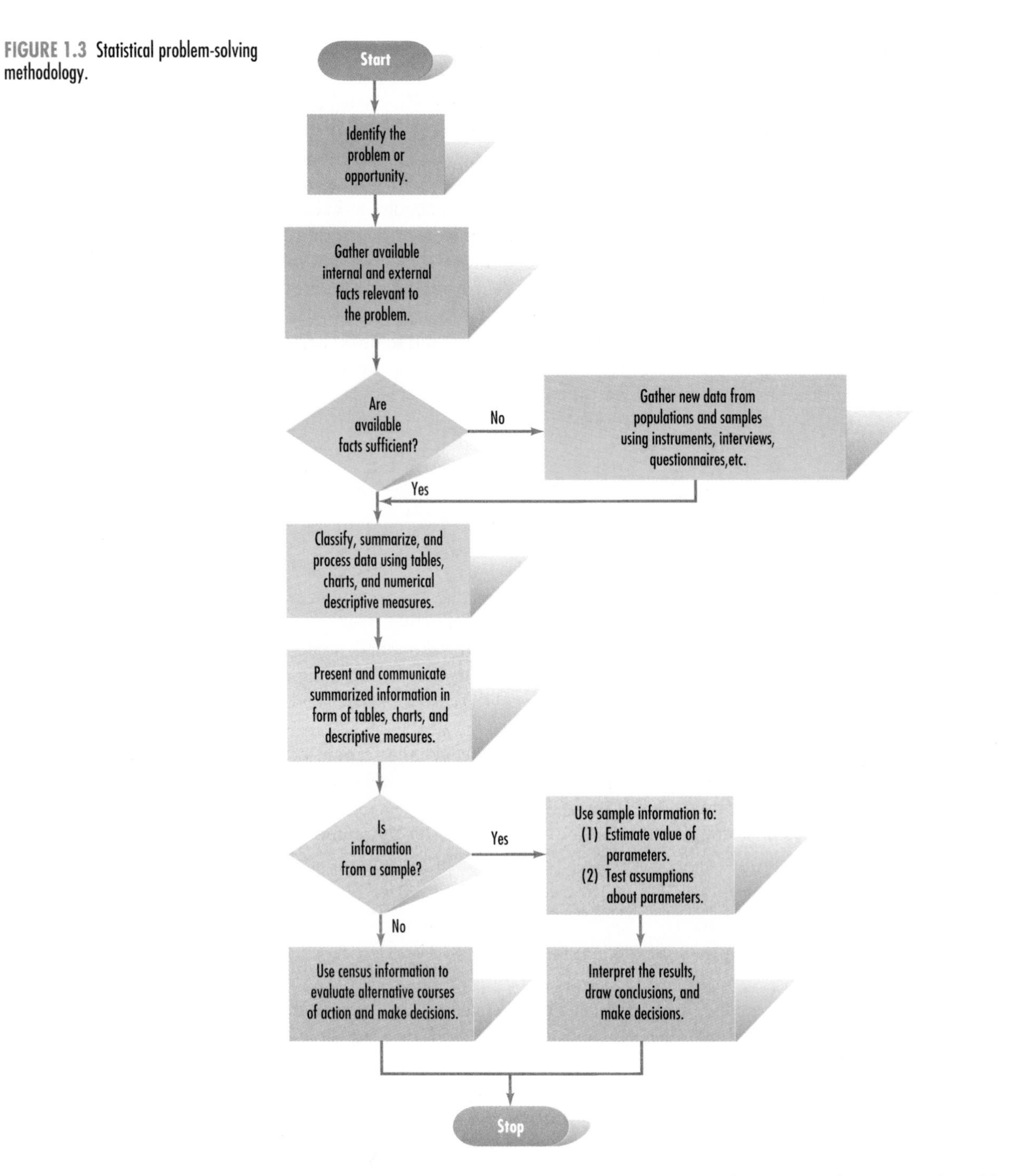

FIGURE 1.3 Statistical problem-solving methodology.

Figure 1.3 expands Figure 1.1 and summarizes the steps in the statistical problem-solving methodology.

1.5 Role of the Computer in Statistics

Computers are efficiently used when data input isn't trivial, similar tasks are performed repeatedly, and processing complexities present no practical alternative to computer use. Since many statistical problems have a large volume of input data, are repetitive, and are relatively complex, the computer is a vital tool to those who solve these problems. Procedures that could take hours, days, or weeks with a desk calculator are accurately completed in seconds or minutes with a computer.

The problems you'll encounter in this book typically use relatively small sets of input data. But even so, you'll see in a few instances that our calculations can become tedious when done by hand. It's usually desirable, though, to follow the hand calculations discussed in the text and then use similar steps to solve a few additional problems. Once you've done that, you'll understand the uses and limitations of the procedures and be able to correctly interpret the results they produce. Then you can use a computer to carry out similar future work. The meaning of those computer results probably won't be explained in any detail, but you'll then have a background to understand the output.

STATISTICS IN ACTION

Who Uses Stat Packages?

Statistical software packages are widely used now that desktop computers are everywhere. Let's look at just three examples. Bionetics Corporation biostatisticians rely on stat packages to study the effects that space flight has on the physical condition of astronauts. Demographic scientists from developing countries rely on similar software to analyze their human populations. And MGM/United Artists analysts gather data from people who watch advance screenings of new films. A stat package is then used to analyze the data and arrive at a prediction of the film's market success.

Spreadsheets and Stat Packages

Data analysis involves separating a mass of related facts into constituent parts and then studying and manipulating those parts to achieve a desired result. Two tools used for data analysis are electronic spreadsheet and statistical analysis packages.

The *spreadsheet* is a program that accepts user-supplied data values and relationships in the columns and rows of its work sheet. The intersection of a spreadsheet column and row is called a *cell.* Values in the cells can be changed to answer "what-if" questions. For example, a city manager can quickly see what may happen to monthly tax receipts if a new sales tax rate is passed by the city council. Popular spreadsheet programs include Lotus's *1-2-3* and Microsoft's *Excel.*

Statistical analysis packages are similar to spreadsheets, but stat packages are preprogrammed with many more of the specialized formulas and built-in procedures a user may need to carry out a range of statistical studies. Like spreadsheets, statistical programs can:

- Accept data from other sources.
- Copy and move data to duplicate the contents of one cell (or group of cells) into other locations or to erase the contents of one or more cells in one place and place them in another.
- Add or remove data items, columns, or rows.
- Format the way cells, rows, and columns are laid out and then save this format for future updates.
- Perform analyses on single and multiple sets of data and then print summary values and analysis results.
- Use numeric data to produce charts and graphs.

Many of the most able general-purpose stat programs such as *Minitab, SAS,* and *SPSS* were first written for large computer systems, and they've been around in various forms for many years. Each of these products is available in a version that runs on personal computers. An equally potent product—*Systat*—was designed first for personal computers and is now also available for larger systems. And many other general-purpose packages may be found. *Statgraphics,* for example, performs many statistical procedures and offers users many types of graphic output.

A recently released *Minitab* program is used throughout this book to demonstrate statistical concepts, but earlier *Minitab* versions and other software packages can easily handle the examples we'll discuss in later chapters. The latest releases are designed to be easier to learn and simpler to use. For example, the first thing a user of the *Minitab for Windows* program sees on the computer screen after the program is initiated is a *main Minitab window* (Figure 1.4*a*). This window features a *Menu bar, Data* and *Session* subwindows, and *Info* and *History icons.*

The Menu bar gives a listing of the main selection options available to the user. When one of these main options is selected, a pull-down submenu is displayed to offer

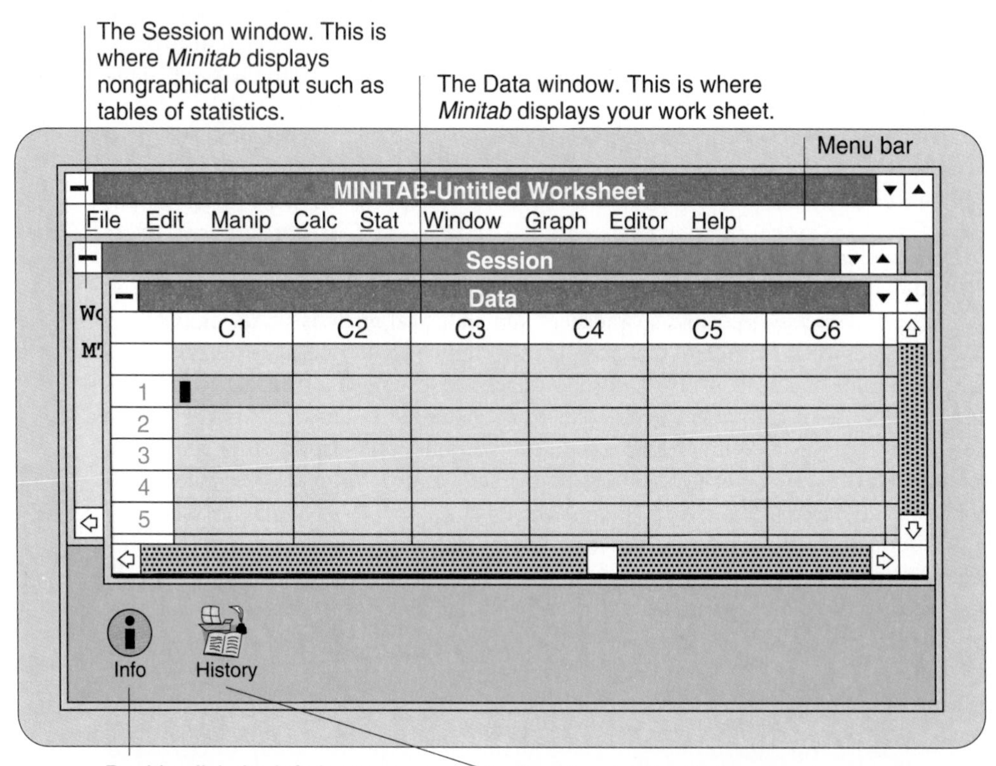
The Session window. This is where *Minitab* displays nongraphical output such as tables of statistics.
The Data window. This is where *Minitab* displays your work sheet.
Menu bar
MINITAB-Untitled Worksheet
File Edit Manip Calc Stat Window Graph Editor Help
Session
Data
C1 C2 C3 C4 C5 C6
1
2
3
4
5
Info
History
Double-click the Info icon to open the Info window, which is where *Minitab* displays a summary of your work sheet.
Double-click the History icon to open the History window, which is where *Minitab* displays the session commands that have been executed.

(*a*)

	C1	C2	C3	
	GALSOLD			
1	82.50			
2	88.50			
3	91.00			
4	93.25			
5	95.00			
6	95.00			
7	97.50			
8	99.50			

⋮

(*b*)

FIGURE 1.4 (*a*) The main window of the *Minitab for Windows* software package. (*b*) Data entry in the *Minitab for Windows* statistical package.

further choices. The Session window displays nongraphical output such as tables. The Info and History icons (see the bottom of Figure 1.4*a*) are pictures that represent options that the user can select. By using the computer's *mouse* pointing device to move an arrow on the screen to the Info icon and by then clicking a button on the mouse twice in quick succession, the user can open the Info window to get a work sheet summary. "Double-clicking" on the History icon displays commands that have been executed.

The Data window displays rows and columns, and new data are entered into the cells of this window. Once entered, the data may be saved and then called back into the Data window at a later time for use or for modification. Figure 1.4*b* shows how a user can enter a data set into the Data window. The screen *cursor* (a blinking rectangle) is positioned in the first column right below the C1 label, and a user-supplied name (GALSOLD) is selected for the data. A down arrow (↓) key is then pressed to move the cursor to the row labeled 1 in column 1 (C1), and the user enters the first data item, a value of 82.50. This procedure continues until the entire data set of 50 values has been entered into C1. (Other data sets, of course, can have multiple columns and rows.) The user can then move the cursor to File on the Menu bar, click the mouse, go down the File submenu to the "Save Worksheet" entry, and save the GALSOLD data set for further use. As you'll see in Chapter 3, we'll use this GALSOLD data set along with built-in *Minitab* formulas and procedures to produce numeric and graphic output.

Have You Hugged Your Computer Today?

Home computers are commonplace today. In 1989, 15 percent of the families in the United States had a home computer. By 1990, this figure had increased to 24 percent. By 1993, the Census Bureau estimated that 34 percent (more than one in three) families had such equipment. And the figure will likely exceed 50 percent within the next few years.

LOOKING BACK

1. The purpose of this book is to give you a grasp of statistical concepts and methods so that you can generate and then interpret statistical results. The word "statistics" refers to the body of knowledge developed to design studies, gather data, and then classify, summarize, interpret, and present these data to support needed decisions. Thus, "statistics" refers to a subject of study. The word "statistics" shouldn't be confused with the word "statistic," which is a number that describes a sample characteristic. The term "parameter" refers to a number that describes a population characteristic. You need a knowledge of both descriptive statistics and statistical inference to describe and understand numerical relationships and make better decisions.

2. The six steps in a statistical problem-solving procedure are listed and discussed in this chapter. The *first* of these steps is to identify the problem. The *second* step may be to gather available facts from internal or external sources. If the data needed by decision makers can't be gleaned from existing primary or secondary sources, the researcher must gather new data through the use of such tools as measuring instruments, personal interviews, and mail questionnaires. Although these tools can be used to gather facts from an entire population, the new data usually come from a sample taken from the population. A judgment sample may be used, but a probability sample is usually preferred. A simple random sample is one type of probability sample in which each item in the population has an equal chance of being selected. Random samples are often picked with the help of computer programs and tables of random numbers. Systematic, stratified, and cluster samples are other types of probability samples used to gather new data. Although probability sample results are subject to sampling error, researchers have learned how to objectively assess and

cope with such error. After data are collected, the *third* step is to classify and summarize the facts. Finally, the data are presented (step 4), analyzed (step 5), and a decision is made (step 6).

3. Computers are efficiently used in statistical work. Generalized prewritten software packages are commonly used to carry out statistical studies. Two software tools commonly used for data analysis are the spreadsheet and the stat package.

Review Exercises

1-8. Match each of the following terms to its correct definition:

TERMS	DEFINITIONS
1. Parameter	a. The complete collection of items under study
2. Statistical inference	b. A number that describes a sample characteristic
3. Census	c. Procedures for collecting, classifying, summarizing, and presenting data
4. Statistics	d. A number that describes a population characteristic
5. Population	e. The science of gathering and summarizing data and using results to make decisions
6. Descriptive statistics	f. A subset of the population
7. Sample	g. The process of arriving at a conclusion about a population parameter on the basis of a sample statistic
8. Statistic	h. A survey of all the elements in a population

9–12. A psychologist wants to study the behavior patterns of the 8,563 college students at State U. She decides to start by obtaining a random sample of 30 students and asking the average number of hours each member of the sample sleeps on a weekday night. For each of the following questions, identify the type of sample obtained (simple random, stratified, systematic, cluster):

9. Each student is assigned a number from 0001 to 8563. A number from 1 to 285 is randomly selected, and every 285th student on the list from that point on is then included in the sample.

10. Students are separated into academic majors, and a proportional number of students are selected from each academic major.

11. Students are listed by number (0001 to 8563), and a computer is used to generate a list of 30 numbers representing the students to be used in the sample.

12. Students are listed by their school residence locations (dormitories or apartment buildings). Three residence locations are randomly selected. Then students from each of these locations are chosen for the psychologist's sample.

13–16. The AGT Corporation has branches in three major cities with a total of 326 salespeople. The sales manager wants to obtain a random sample of 40 of his staff to determine their average gross sales per month. For each of the following, identify the type of sample obtained:

13. One of the three branches is selected, and 40 salespeople are selected from this branch.

14. Sales employees are numbered 1 to 326, and a random number table is used to produce a sample of 40.

15. Salespeople are listed alphabetically. A number from 1 to 8 is selected at random, and every 8th person from there is selected for the sample.

16. A proportional number is selected from each of the three branches.

17. There are 560 students enrolled in a statistics course at the local university. How would you use the random number table from Appendix 3 to obtain a random sample of 20 of them?

18. There are 83 members of a population. How would you use the random number table in Appendix 3 to obtain a random sample of 10 of them?

Topics for Review and Discussion

1. List three reasons why it might be more practical to obtain data values from a sample rather than from the entire population.

2. "A population parameter is a single fixed value, while sample statistics from the same population can vary." Discuss why this statement is true.

3. What's the difference between a population parameter and a sample statistic?

4. How are sample statistics used to make decisions about population parameters?

5. Outline the steps used in statistical problem solving.

6. Discuss the relationship between descriptive statistics and inferential statistics.

7. What's the difference between a spreadsheet and a statistical software program?

8. How may a statistical software program be used?

9. Discuss the reasons for sampling errors.

10. Discuss the different types of sampling described in this chapter. For each type, give a possible advantage and disadvantage.

Projects/Issues to Consider

1. Obtain an article from a recent newspaper or periodical that discusses the results of a statistical study. What is the implied population? If possible, determine how the sample was obtained.

2. Go to the library and investigate primary and secondary sources. Make a list of at least three of each type.

Computer Exercises

1. *Money* magazine recently published a study that compared the taxes that would be paid by a hypothetical family of four if they lived in each of the 50 states and the District of Columbia. It was specified in the study that the family income consisted of $72,385 in earnings, $2,782 in interest receipts, $455 in dividends, and $1,472 in capital gains. Enter and save the following information in a computer file named TAXES. Name your columns STATE and TAX. (We'll use this data set in Chapter 3.)

	State	Tax		State	Tax
1	Alabama	$5,552	27	Montana	$6,781
2	Alaska	1,632	28	Nebraska	7,728
3	Arizona	6,637	29	Nevada	3,539
4	Arkansas	7,074	30	New Hampshire	4,591
5	California	7,605	31	New Jersey	7,371
6	Colorado	7,268	32	New Mexico	5,948
7	Connecticut	8,389	33	New York	10,016
8	Delaware	5,354	34	North Carolina	7,263
9	Dist. of Col.	9,348	35	North Dakota	5,292
10	Florida	3,846	36	Ohio	7,751
11	Georgia	7,301	37	Oklahoma	6,907
12	Hawaii	8,272	38	Oregon	8,390
13	Idaho	7,634	39	Pennsylvania	6,969
14	Illinois	7,125	40	Rhode Island	8,314
15	Indiana	6,712	41	South Carolina	6,531
16	Iowa	7,006	42	South Dakota	4,284
17	Kansas	6,935	43	Tennessee	4,038
18	Kentucky	6,744	44	Texas	4,647
19	Louisiana	5,752	45	Utah	7,892
20	Maine	8,611	46	Vermont	7,962
21	Maryland	8,568	47	Virginia	7,217
22	Massachusetts	8,764	48	Washington	4,694
23	Michigan	7,493	49	West Virginia	5,981
24	Minnesota	8,311	50	Wisconsin	8,770
25	Mississippi	5,792	51	Wyoming	2,945
26	Missouri	6,047			

2. Sort the data from the file named TAXES into two new columns named STATE1 and TAX1 so that the data are listed in an ascending order (from lowest to highest) by the amount of tax paid in each state. Since you'll want the state name carried along with the appropriate tax amount, be sure to select STATE as well as TAX for your columns to be sorted.

3. There are 548 students enrolled in the management program at State University. Use a software program to generate a random sample of 25 of them.

4. There are 157 restaurants listed in a local phone book. As a public health official, your job is to inspect the kitchens of 10 of them each week. Assuming you number the restaurants 1 to 157, use a software program to select a random sample of 10 restaurants to be inspected.

Thus, the median of $13,000 represents the earnings of the middle grower in the group of 100.

Another type of average is the *mode.*

The **mode** of a group of values is the score that occurs most often. If no value occurs more than once, there's no mode. And when there's a tie between two values for the greatest count, the data set is said to be *bimodal.*

In our example of grape growers, the mode is also $13,000, but in other examples it could differ from both the mean and median. You can see from this example that the average (mean) of nearly $83,000 is misleading because it distorts the general situation. Yet it's not a lie; it has been correctly computed. The problem of the aggravating average is often traced to the word "average."

The word **average** is a broad term that applies equally to several measures of central tendency such as the arithmetic mean, the median, or the mode.

It is not uncommon for one of these averages to be used when it isn't appropriate and deliberately misleads the consumer of the information.

2.3 Disregarded Dispersions

Suppose that Karl Tell, an economist specializing in nineteenth-century German antitrust matters, is being pressured to coach the track team by the president of the small college where he teaches. Karl isn't enthused about this prospect, since it will distract him from his study of the robber barons of Dusseldorf, but the president reminds him that he doesn't yet have tenure and that his classes are not in enormous demand. Therefore, Karl does a little checking and finds that the four high jumpers can clear an average of only 4 feet and that the three pole vaulters can manage an average height of only 10 feet. Karl concludes that his first venture into athletic management is likely to result in consideral verbal abuse from both alumni and faculty colleagues. Is Karl correct? Probably, but not because of the data he has gathered. Karl has been the victim of aggravating averages (arithmetic means in this example). Had he checked further, he would have found that one of his four high jumpers consistently clears 7 feet—good enough to win every time in the competition he will face—while the other three can each only manage to stumble over 3-foot heights. Likewise, in the pole vault there is one athlete who vaults 16 feet (with a bamboo pole) and two others who can each manage to explode for only 7 feet.

The moral here is that averages alone often don't adequately describe the true picture. A measure of *dispersion* must also be considered.

Dispersion is the amount of spread or scatter that occurs in the data.

And we are simply making the further distinction here that *disregarded dispersion* exists when the scatter of the values about the central measure is such that the average tends to mislead. Of course, disregarded dispersions and aggravating averages are usually

Statistical tables and charts may summarize data, uncover relationships, and interpret and communicate numerical facts to those who can use them. (*Courtesy Hewlett Packard*)

found acting in concert to confuse and mislead. To summerize, the story is often told of the Chinese warlord who was leading his troops into battle with a rival when he came to a river. Since there were no boats and since the warlord remembered that he had read somewhere that the average depth of the water in the river was only 2 feet at that time of year, he ordered his men to wade across. After the crossing, the warlord was surprised to learn that a number of his soldiers had drowned. Although the average depth was indeed just 2 feet, in some places it was only a few inches, while in other places it was over the heads of many who became the unfortunate victims of disregarded dispersion.

2.4 The Persuasive Artist

Statistical tables and charts are prepared to summarize data, uncover relationships, and interpret and communicate numerical facts to those who can use them.

Statistical Tables

Statistical tables efficiently organize classified data into columns and rows so users can quickly find the facts needed. In Figure 2.1, for example, the data in the table are classified by department store chains and by sales per square foot of floor space.

Line Charts

A **line chart** is one in which data points on a grid are connected by a continuous line to convey information. The vertical axis in a line chart is usually measured in quantities

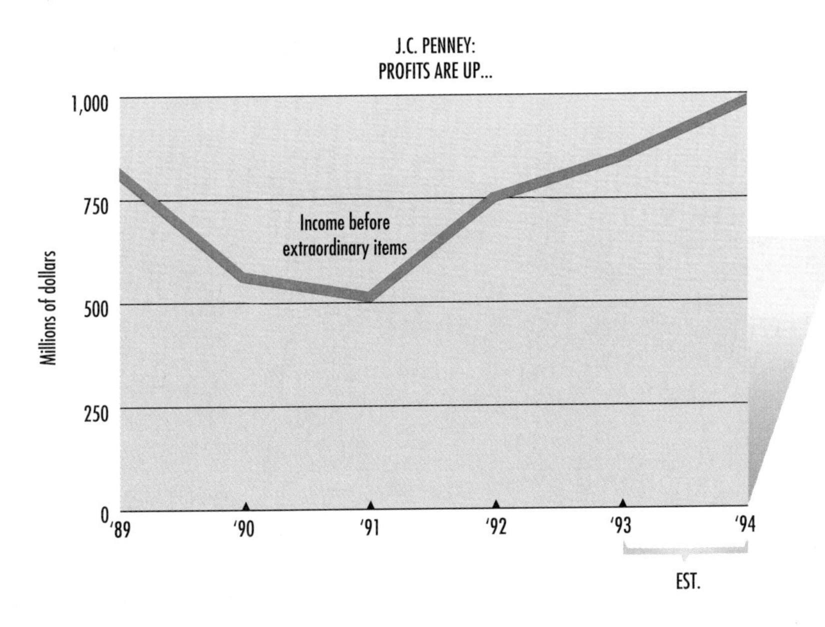

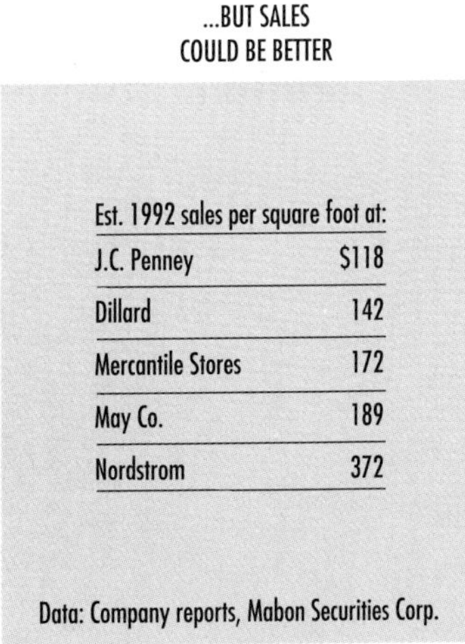

...BUT SALES COULD BE BETTER

Est. 1992 sales per square foot at:	
J.C. Penney	$118
Dillard	142
Mercantile Stores	172
May Co.	189
Nordstrom	372

Data: Company reports, Mabon Securities Corp.

FIGURE 2.1 A statistical table and a single-line chart. (*Source: Business Week*, April 5, 1993, p. 51.)

(dollars, bushels) or percentages, while the horizontal axis is often measured in units of time (and thus a line chart becomes a *time-series* chart). Line charts don't present specific data as well as tables, but they're usually able to show relationships more clearly. As you can see in Figure 2.1, both a table and a line chart are frequently used together in a presentation. The *single-line* chart in Figure 2.1 shows actual and expected company profits over a period of several years.

Of course, *multiple series* can also be depicted on line charts, as shown in Figure 2.2. The lines in these figures are all plotted against the same baseline. Note, though, that in the *component-part* (or area) line drawings in Figure 2.3, the chart is built up in layers. Thus, the debt amounts maturing at different time intervals are added to get the $2.484 trillion of privately held debt owed in March, 1993.

Bar Charts

A **bar** (or column) **chart** uses the length of horizontal bars or height of vertical columns to represent quantities or percentages. As in the case of line charts, one scale on the bar chart measures values while the other may show time or some other vari-

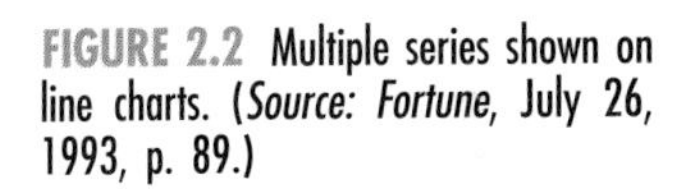

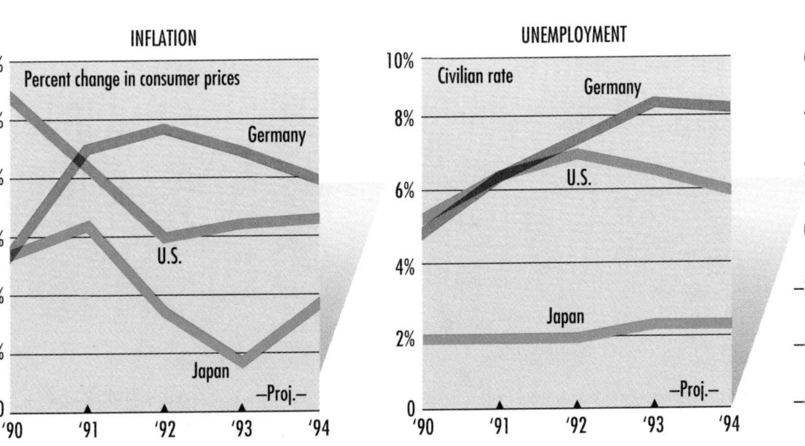

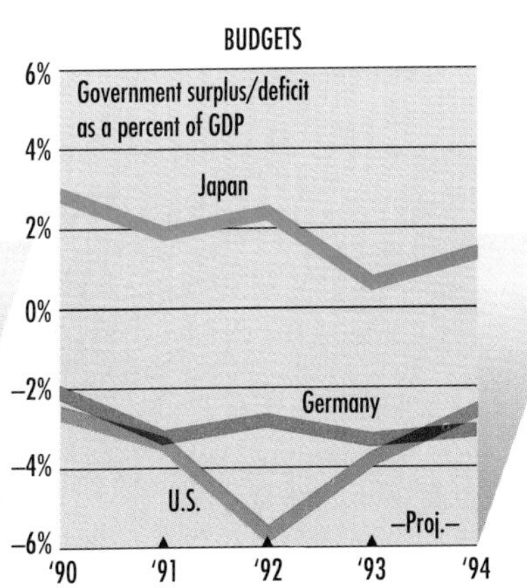

FIGURE 2.2 Multiple series shown on line charts. (*Source: Fortune*, July 26, 1993, p. 89.)

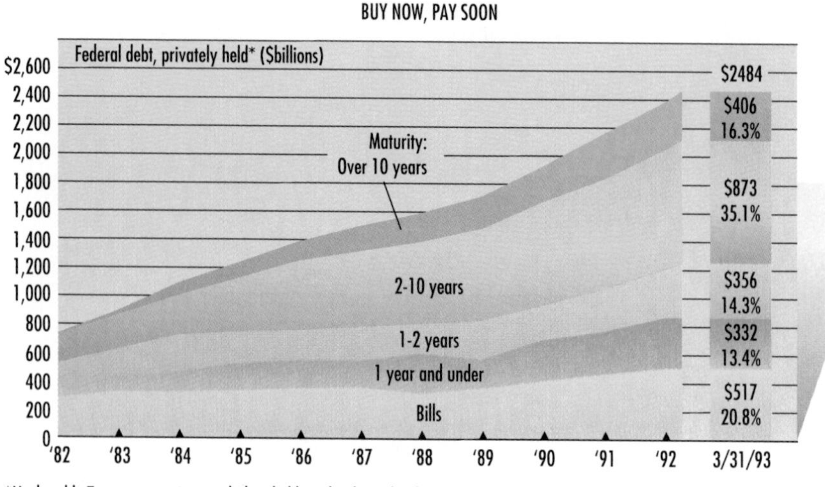

FIGURE 2.3 A component-part line drawing and a component-part bar chart. (*Source: Forbes*, June 7, 1993, p. 135.)

able. The bars typically start from a zero point and are frequently used to show multiple comparisons. A *component-part* (or *stacked*) bar chart can be used as we've seen in Figure 2.3. Or bars may be clustered together to show how identified categories of interest change over time (Figure 2.4.).

Pie Charts

Pie charts are circles divided into sectors, usually to show the component parts of a whole. Single circles can be used, or several pie charts can be drawn to compare changes in the component parts over time (Figure 2.5). A technique that's often used is to separate a segment of the drawing from the rest of the pie to emphasize an important piece of information.

Combination and Other Charts

As you've seen in Figures 2.3 and 2.5, it's possible to include a mix of the charts we've now examined in a graphics presentation. And other types of charts that are used

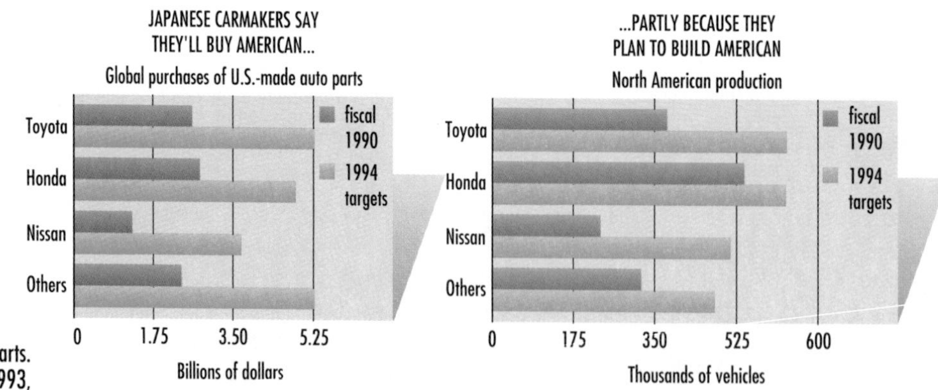

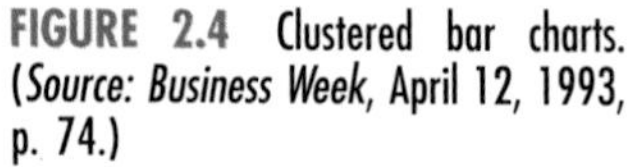
FIGURE 2.4 Clustered bar charts. (*Source: Business Week*, April 12, 1993, p. 74.)

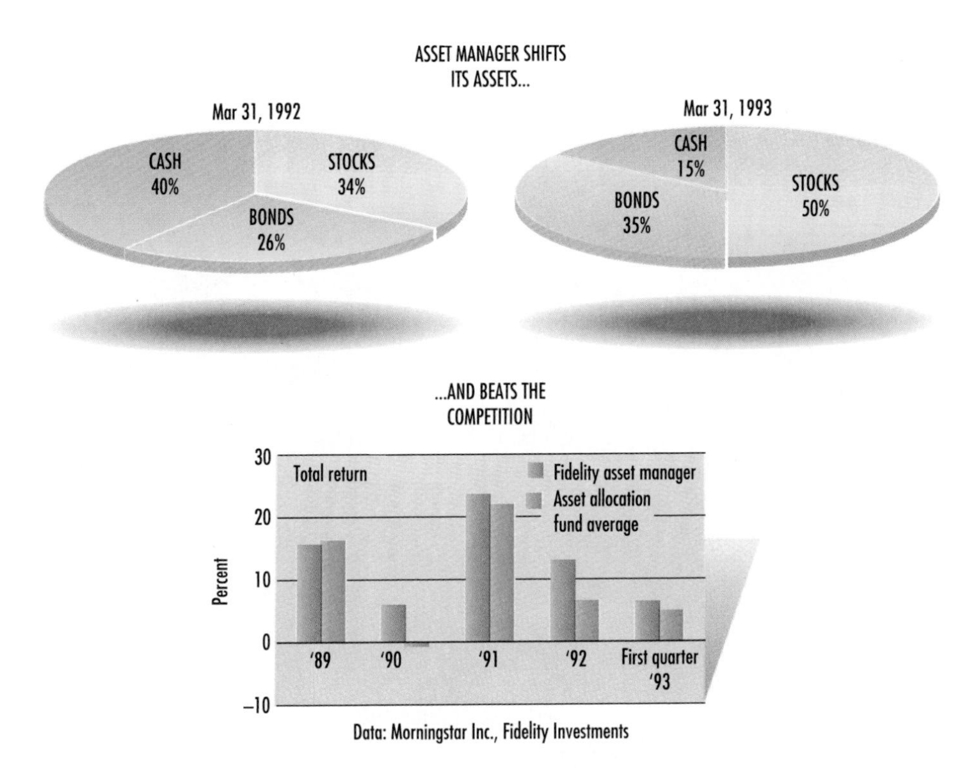

FIGURE 2.5 A combination of pie and bar charts. (*Source: Business Week*, May 3, 1993, p. 140.)

to display quantitative facts are **statistical maps** that present data on a geographical basis and **pictographs** or pictorial charts that use picture symbols to convey meaning. Pictographs must be used with caution, as we'll see in a few paragraphs.

Computer Graphics Packages

Graphics packages are software products that convert the numeric data used by computers into the visual images that people often prefer to use to communicate ideas. Statistical information is usually manipulated with graphics packages used for analysis and presentation.

An **analysis package** converts numeric data into a visual summary so that people can grasp relationships, spot patterns, and make more informed decisions. After supplying an analysis package with the data to be analyzed, the user can make a preliminary chart selection from a menu of chart formats supplied by the program. The program quickly displays the data in the selected format. The user can vary colors, add and delete lines and headings, change scales, and edit the appearance of the chart in other ways.

Decision makers use analysis packages to gain insight into the relationships, changes, and trends that are buried in their data. But they use **presentation packages** to communicate messages to an audience. A presentation package has all the features found in an analysis package. Bar, pie, line, and other charts may still be used. But a presentation package can produce multiple three-dimensional images. And some presentation packages have animation capabilities. Thus, a user can make bars grow on a screen to dramatize increases.

Statistical tables efficiently organize classified data into columns and rows so users can quickly find needed facts. (*Courtesy Hewlett Packard*)

Misusing Graphs

So much for the ways that data can be honestly presented. But the purpose of some persuasive artists is to take honest facts and create misleading impressions. How is this done? There are numerous tricks, but we'll limit our discussion here to just a few examples.

Suppose you're running for reelection to a legislative body, and during your past 2-year term appropriations have increased in your district from $8 million to $9 million. Now, as your fellow politicians know, this isn't a particularly good record, but the voters don't need to know that. In fact, you can perhaps turn this possible liability into an asset with the help of a persuasive artist. Figure 2.6*a* shows one way to present the information honestly. But since your goal is to mislead without actually lying, you prefer instead to distribute Figure 2.6*b* during your campaign. The difference between Figures 2.6*a* and *b*, of course, lies in the changing of the vertical scale in the latter figure. (The wavy line correctly indicates a break in the vertical scale, but this is often not considered by unwary consumers of this type of information.) By breaking the vertical scale and by then changing the proportion between the vertical and horizontal scales, you've given the impression that you've been doing a good job of getting appropriations. And to further impress (and mislead) your constituents, you can keep the vertical scale used in Figure 2.6*b* and then compress the horizontal (time) scale.

Having received favorable comments on your appropriations chart, you decide to employ another trick. New industry has come into your district during the past 2 years. There has been some increase in the air and water pollution as a result, but there has also been an increase in average weekly wages of unskilled workers from $160 per week

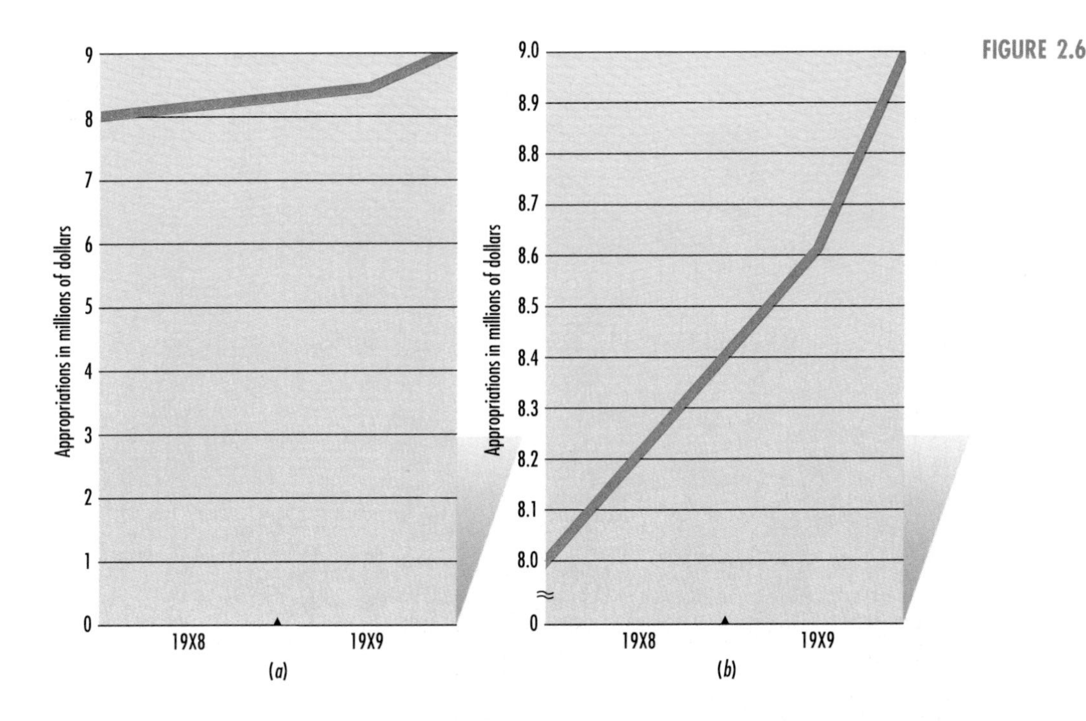

FIGURE 2.6

to $240 per week. Of course, you had little to do with bringing in the new industry, and there's also some disturbing evidence that the fact that wage increases have *followed* the new plants doesn't necessarily mean that they were *caused* by the new industry. But you see no reason to complicate matters with additional confusing facts.

How can you best communicate this wage increase information to your constituents? After trying several approaches, you decide to use the pictograph in Figure 2.7*a*. The height of the small money bag represents $160, and the height of the large bag is correctly proportioned to represent $240. What's wrong and misleading, though, is the *area* covered by each figure. The space occupied by the larger bag creates a misleading visual impression. But that was the intent, wasn't it? If you think this example is far-fetched, consider Figures 2.7*b* and *c*.

Let's now assume that in spite of your persuasive artwork the voters have seen fit to throw you out of office in favor of a write-in candidate. However, you're able to find work in your father's manufacturing company. One of the first jobs you're given is to prepare reports for stockholders and the union explaining company progress over the past year. The company has done well, and profits have amounted to 25 cents of every sales dollar. This can be accurately presented in picture form, as shown in Figure 2.8*a*. But a stronger impact can be made on stockholders if the coin is shown from the perspective of Figure 2.8*b*. Since you don't want the union members to become restless, though, you can show them the profit situation from the perspective of Figure 2.8*c*.

2.5 The *Post Hoc Ergo Propter Hoc* Trap

The Latin phrase for the logical reasoning fallacy that states that because B *follows* A, B was *caused* by A is *post hoc ergo propter hoc*, which means "after this, therefore because of this." Erroneous cause-and-effect conclusions are often drawn because of the misuse of quantitative facts. Of course, as the following examples show, some errors are easy to spot:

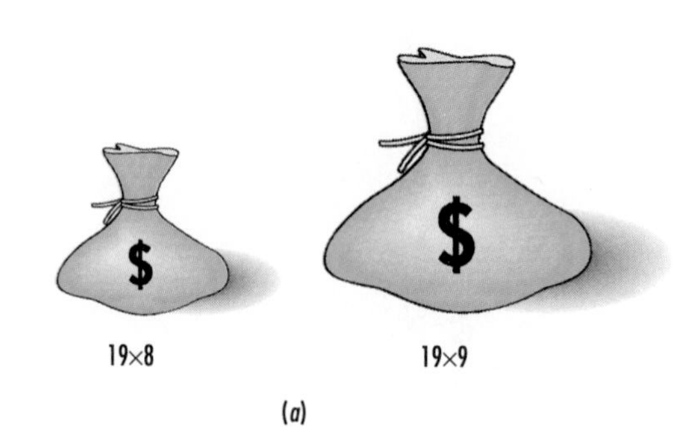

(*a*)

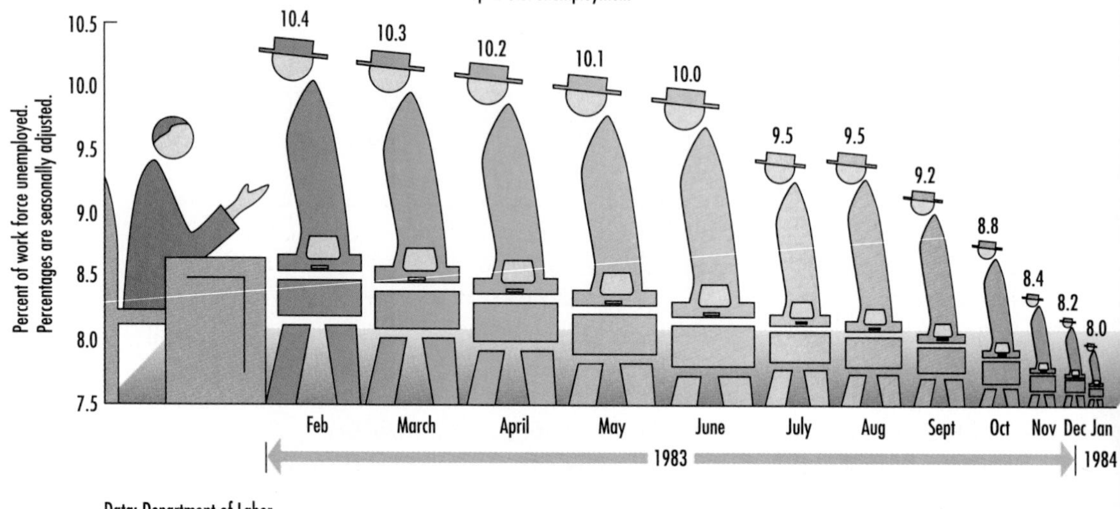

(*b*)

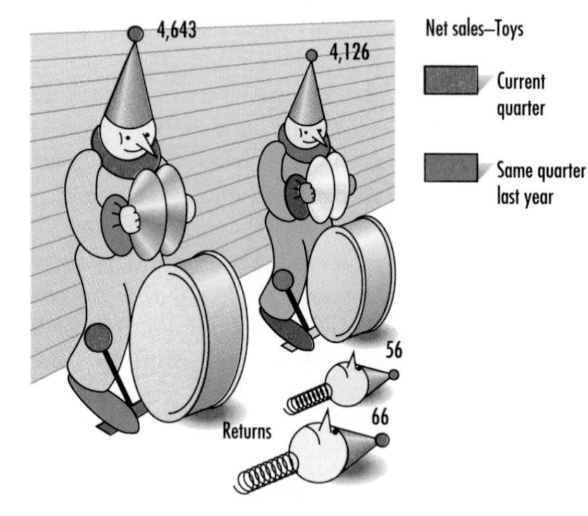

(*c*)

FIGURE 2.7 (*a*) The first of our perfidious pictographs. (*b*) Note the absence of any break in the vertical scale that compounds the area misrepresentation. (*Source: Fort Worth Star Telegram*, February 4, 1984, p. 9B. Reprinted with permission.) (*c*) And note the absence of any scale at all in this figure. (*Source: PC World*, April, 1986, p. 272. Reprinted with permission.)

FIGURE 2.8

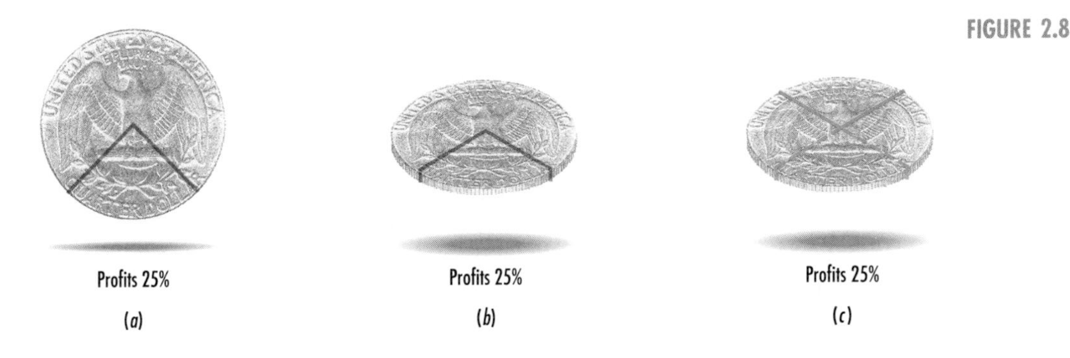

- Increased shipments of bananas into the port of Houston have been followed by increases in the national birthrate. Therefore, bananas were the cause of the increase in births.
- The average human life span has doubled in the world since the discovery of the tobacco plant. Therefore, tobacco . . . (this is just too gross to complete).

But not all examples of the *post hoc* trap are so obvious. Some, in fact, can be subtle. To illustrate, you may recall that a few pages earlier (in the discussion of bias and the *Literary Digest* poll) the following sentence appeared: "Unfortunately for the *Digest* (which ceased to exist in 1937), the ballots had been sent to persons listed in telephone directories and automobile registration records." Did you perhaps conclude from this example that since the *Digest* folded in 1937, the cause was the poor forecast about the 1936 election? Obviously, the poll didn't help the magazine's reputation, but did it cause the publication to go out of business? Isn't it likely that a number of factors combined to bring about this demise?

2.6 Antics with Semantics and Trends

What Does That Mean?

Failing to define terms that are important to a clear understanding of the message, *making improper comparisons* between unlike or unidentified things, *using an alleged statistical fact* to jump to a conclusion that ignores other possibilities or that is quite illogical, using *jargon* to cloud the message when simple words and phrases are suitable—all these antics with semantics are used to confuse and mislead. The following examples should be adequate to illustrate this unfortunate fact:

- The Chapter 1 example of different unemployment figures produced by different groups shows what a failure to define terms can do to understanding. In fact, the Bureau of Labor Statistics has published no fewer than eight versions of the unemployment rate, using different labor-force definitions. Terms such as "poverty," "population," and "living standard," to name just a few, are also subject to different definitions, and the consumer of the information should be told which is being used.
- The Federal Trade Commission (FTC) took exception to the advertised claim made by the manufacturers of Hollywood Bread that their product had fewer calories per slice than other breads. According to the FTC, the claim was misleading since the Hollywood slice was thinner than normal. Actually there was no significant difference in the number of calories when equal amounts of bread were compared.

The Numbers Show . . . What?

After World War II, Sam Gill asked people what they thought of the Metallic Metals Act. About 70 percent of those queried had strong opinions about it. Some thought it was needed legislation, some thought it had no value, and some felt that the act should be under the control of the states rather than the federal government. Years later, Sam's questions were answered by a later generation. About 64 percent of these respondents had definite opinions. There was a shift in the pattern of responses between polls, but in both cases people were giving their views about a *nonexistent* act!

Reservations About Reserves

In 1950 the world's "known" oil reserves were 75 billion barrels. Twenty years later people had sucked up 180 billion barrels, but the world's reserves were then . . . 455 billion barrels. The numbers for iron, copper, and other minerals also move around on the page. For example, reserves of iron ore were placed at 19,000 million metric tons in 1950. By 1970, however, over 9,000 millon metric tons had been mined, and reserves were then 251,000 million metric tons—a snappy increase of over 1,000 percent! "Known reserves" calculations, like figures from public opinion polls, are subject to limitations. Reserve numbers can tell us where we have been and approximately where we are, but they can't serve very well as the base for uncritical projections.

- "The Egress carburetor is up to 10 percent less polluting and up to 50 percent more efficient." Less polluting than what, a steel mill? And more efficient than what, a Boeing 747?
- "One in 10 births is illegitimate. Thus, your estimate of your fellow human is correct 10 percent of the time." Figures on many activities (including illegitimate births, rapes, marijuana smoking, etc.) are just not reliable because many cases remain unreported.
- Representative Ben Grant, in arguing for plain wording in a proposed new Texas constitution, points out how jargon can be used. If, noted Grant, a man were to give another an orange, he would simply say, "Have an orange." But if a lawyer were the donor, the gift might be accompanied with these words, "I hereby give and convey to you, all and singular, my estate and interests, right, title, claim, and advantages of and in said orange, together with its rind, juice, pulp, and pits, and all rights and advantages therein, with full power to bite, cut, suck, and otherwise eat same. . . ." Alas, the same kind of language may also be used to convey quantitative facts.

The Trend Must Go On!

Another way that a person may misuse statistical facts is to assume that because a pattern has developed in the past in a category, that pattern will certainly continue into the future. Such uncritical extrapolation, of course, is foolish. Changes in technology, population, and lifestyles all produce economic and social changes that may quickly produce an upturn or a downtrend in an existing social pattern or economic category. The invention of the automobile, for example, brought about significant growth in the petroleum and steel industries and a rapid decline in the production of buggies and buggy whips. Yet, as British editor Norman MacRae has noted, an extrapolation of the trends of the 1880s would show todays cities buried under horse manure.

One more example of the advertiser's art should be sufficent here. The California Raisin Advisory Board ran this ad in a women's magazine a few years ago: "Your husband could dance with you for 11 minutes on the energy he'd get from 49 raisins. Think what would happen if he never stopped eating them." As Stephen Campbell observed in his *Flaws and Fallacies in Statistical Thinking*, "I can think of a lot of things that might happen to a man if he never stopped eating raisins—and they are all painful."

Of course, the *judicious* projection of past patterns or trends into the future can be a very useful tool for the planner and decision maker. But the failure to apply a generous measure of common sense to extrapolations of past quantitative patterns can lead to faulty conclusions that people are seriously asked to accept.

2.7 Follow the Bouncing Base (If You Can)

In an editorial on minimum competency in English and Math published on April 7, 1978 the editors of the *Pensacola Journal* actually stated that "After all, if you give the test to four students and four flunk, that's a 50 percent failure rate."*

*Ron McCuiston, "Standard Deviations of the Square Root of Infinity," *The Matyc Journal*, Winter 1979, p. 59. Reprinted with permission of *The Matyc Journal*, Inc.

People today are often confused because they fail to follow the bouncing base—that is, *the base period used in computing percentages.* A few examples will show how failure to clarify the base may lead to misunderstanding:

➤ A worker is asked to take a 20 percent pay cut from his weekly salary of $500 during a recessionary period. Later, a 20 percent increase is given to the worker. Is he happy? The answer may depend on what has happened to the base. If the *cut* is computed using the earlier period (and the salary of $500) as the base, the reduced pay amounts to $400 ($500 × 0.80). But if the pay *increase* of 20 percent is figured on a base that has been shifted from $500 to $400, the worker winds up with a restored pay of $480 (1.20 × $400). Thus, the bouncing base has cost him $20 each week, and this isn't likely to please him.

➤ An **index number** is a measure that shows how much a composite group has changed with respect to a base period that's usually defined to have a value of 100. An important example is the Bureau of Labor Statistics' *Consumer Price Index for Urban Consumers (CPI-U).* The CPI-U is a popular series that periodically measures the cost of about 400 dissimilar things that people buy. The CPI-U uses 1982 to 1984 as its base period and assigns to the prices that prevailed in that period an index number value of 100. Later, in 1988, the annual average index value rose to 118.3, and by April, 1993, the CPI-U had reached 144.0. These numbers mean that there was an 18.3 percent increase in prices between the 1982 to 1984 period and 1988, and a 44.0 percent increase between 1982 to 1984 and April, 1993. Thus, it would take $118.30 in 1988 to buy the same quantities of goods and services that could be obtained for $100 in 1982 to 1984. So far, so good. But let's assume that a reporter now misuses these figures during an article and notes that there has been a 25.7 *percent increase* in prices between 1988 and April, 1993. It's true that the numbers 118.3 and 144.0 represent percentages, and it's also true that there's a difference of 25.7 *percentage points* between 1988 and April, 1993. But the *percentage increase* was actually 23.36 percent [(144.0 − 118.3)/118.3 = 23.36]. In this case, the reporter failed to shift the base to 1988.

➤ Percentage *increases* can easily exceed 100 percent. For example, a company whose sales have increased from $1 million in 1992 to $4 million in 1994 has had a percentage increase of 300 percent [($4 million − $1 million)/$1 million]. Of course, the sales in 1994 *relative* to the sales in 1992 were 400 percent—($4 million/ $1 million) × 100—and this *percentage relative* figure is sometimes confused with the percentage increase value. Remember, though, that percentage *decreases* exceeding 100 percent aren't possible if the origional data are positive values. For example, *Newsweek* magazine reported some years ago that the Chinese government had cut the salaries of certain officials by 300 percent. Of course, once 100 percent is gone, there isn't anything left. Embarrassed editors later admitted that the cut was 66.67 percent rather than 300 percent. And a more recent article in *Datamation* began with these words: "We certainly understood the new industry economics. Oil prices had plunged more than 100% in the past two years. . . ."

The above examples show only a few types of abuses that may be associated with the use of percentages. But they do give you an idea of the importance of following the bouncing base.

Measuring It

Economic movements are measured in countless ways. For example, one measure of U.S. economic activity is the *Forbes Index* that consists of eight equally weighted elements including new housing starts, personal income, new claims for unemployment compensation, and total retail sales. The data used for the *Forbes Index* are found in 10 series produced by the U.S. government. Another indicator of economic activity is the *Business Week Production Index* that includes 10 production measures such as the amount of steel, electric power, coal, paperboard, and lumber produced, and the number of cars and trucks made. Both indexes use 1967 as the base period.

Stay in Bed and Eat Grapes

A few years ago, Dr. James Muller, codirector of the Institute for Prevention of Cardiovascular Disease at the New England Deaconess Hospital in Boston, published a study that showed that heart attacks tend to occur in the morning. Recently, Dr. Muller was listening to a local radio talk show where the subject under discussion was another study showing that grapes could reduce the risk of heart disease. The talk show host combined the two studies and proclaimed that the best way to avoid heart attacks was to stay in bed, eating grapes, until noon.

2.8 Avoiding Spurious Accuracy and Other Pitfalls

Spurious (and Curious) Accuracy

Statistical data based on sample results are often reported in precise numbers. It's not unusual for several decimal places to be used, and the apparent precision lends an air of infallibility to the information reported. Yet the accuracy image may be false. To illustrate, W.E. Urban, a statistician for the New Hampshire Agricultural Experiment Station, wrote a letter to the editor of *Infosystems* magazine taking issue with a previously published article. "Your magazine," wrote Urban, "has provided me with an excellent example of impeccable numerical accuracy and ludicrous interpretation which I will save for my statistics classes. With a total sample size of 55, reporting percentages with two decimal places is utter nonsense." The first sample percentage quoted in the article was 31.25, but, as Urban noted, the corresponding estimate of the population was likely to have been anywhere between 12 and 62 percent! As Urban concluded: "I realize that it is painful to throw away all the nice decimals the computer has given us. . . . but who are we kidding?" The reply of the editor: "No one. You're right."

Spurious accuracy isn't limited to sample results. The *Information Please Almanac* once listed the number of Hungarian-speaking people at 13,000,000, while in the same year the *World Almanac* placed the number at 8,001,112. Thus, there was a difference of about 5 million people in the estimates. This isn't particularly surprising. But isn't it curious that the *World Almanac* figure could be so precise? Does it stand to reason that the accuracy could be so great when the figures are well up in the millions? Albert Sukoff has observed that "huge numbers are commonplace in our culture, but oddly enough the larger the number the less meaningful it seems to be. . . . Anthropologists have reported on the primitive number systems of some aboriginal tribes. The Yancos in the Brazilian Amazon stop counting at three. Since their word for 'three' is '*poettarrarorincoaroac*,' this is understandable."

Oscar Morgenstern summarizes the issue of spurious accuracy with these words:

> It is pointless to treat material in an "accurate" manner at a level exceeding that of the basic errors. The classical case is, of course, that of the story in which a man, asked about the age of a river, states that it is 3,000,021 years old. Asked how he could give such accurate information, the answer was that 21 years ago its age was given as 3 million years.

Avoiding Other Pitfalls

Harass them, harass them,
Make them relinquish the ball!—***Cheer at small but illustrious liberal arts college***

An important function of any statistics course is to help people distinguish between valid and invalid uses of quantitative tools. Thus, information found in the following chapters should help you avoid many of the pitfalls discussed in this chapter. But even after you finish this book, you'll find that it isn't always easy to recognize or cope with statistical fallacies. You must, like the team being encouraged by the cheer printed above, remain on the defense to avoid serious statistical blunders. To avoid pitfalls, you might ask yourself questions such as:

- *Who is the source of the information you are asked to accept?* Special interests have a way of using statistics to support preconceived positions. Using essentially the same raw data, labor unions might show that corporate profits are high and thus higher wage demands are reasonable, while the company management might make a case to show profit margins are low and labor productivity isn't keeping up with productivity in other industries. Also, politicians of opposing parties use the same government statistics relating to employment, taxation, national debt, welfare spending, budgets, and defense appropriations to draw surprisingly different conclusions to present to voters.
- *What evidence is offered by the source in support of the information?* Suspicious methods of data collection and/or presentation should put you on guard. And, of course, you should determine the relevancy of the supporting information to the issue being considered.
- *What information is missing?* What isn't made available by the source may be more important than what is supplied. If assumptions about trends, methods of computing percentages or making comparisons, definitions of terms, measures of central tendency and dispersion used, sizes of samples employed, and other important facts are missing, then there may be ample cause for skepticism.
- *Is the conclusion reasonable?* Have valid statistical facts or statements been used to support the jump to a conclusion that ignores other plausible possibilities? Does the conclusion seem logical and sensible?

LOOKING BACK

1. Many have uncritically accepted statistical conclusions only to discover later that they've been misled. The aim of this chapter isn't to show you how to misapply statistics so that you may better con fellow humans. Rather, the purpose has been to alert you to the possibility of misleading statistical information so that you can better distinguish between valid and invalid uses of statistical techniques.
2. Poorly worded and/or slanted questions may be used to gather data, and a "finagle factor" may be used to support preconceived opinions. You should be aware of the bias that may exist in the information you're asked to accept.
3. The word "average" is a broad term that applies to several measures of central tendency such as the arithmetic mean, the median, or the mode. It's not uncommon for one of these averages to be used where it isn't appropriate and where it's selected to deliberately mislead. But averages alone usually don't adequately describe a data set. Dispersion—the amount of spread or scatter that occurs in the data—must also be considered. Disregarded dispersions and aggravating averages are usually found acting in concert to confuse and mislead.
4. Statistical tables and charts are prepared to summarize data, uncover relationships, and interpret and communicate numerical facts to those who can use them. We've considered the legitimate uses of statistical tables and charts. But persuasive artists can easily take honest facts and create misleading impressions, as you've seen in this chapter.
5. Consumers of statistical information must be alert to the argument that because B follows A, B was caused by A. Erroneous cause-and-effect conclusions are often

drawn because of the misuse of quantitative facts. And failing to define terms, making improper or illogical comparisons, using an alleged statistical fact to jump to a conclusion, using jargon and lengthy words to cloud a message—all these antics with semantics are used to confuse and mislead. Also, assuming that because a pattern has developed in the past, that pattern will certainly continue in the future is foolish, but such assumptions aren't unusual.

6. People are often misled or confused because of a bouncing base—that is, the base period used in computing percentages. And statistical data are often reported in precise numbers, and several decimal places may be used. This apparent precision lends an air of infallibility to the information reported. But as we've seen, the accuracy may be spurious (and curious).

7. To avoid the pitfalls presented in this chapter, you might ask yourself these questions: (*a*) Who is the source of the information you're asked to accept? (*b*) What evidence is offered in support of the information? (*c*) What evidence is missing? (*d*) Is the conclusion reasonable?

Review Exercises

1. An advertisement for Guardian Fund apperared in *Money Forecast* as follows: "A $10,000 investment in Guardian Fund made in 1950 with income, dividends and capital gains reinvested would be worth over one and one-half million dollars." Analyze this statement.

2. Professor Plum and Professor Peach are offering courses in Statistics I next semester. Your friend tells you that the Statistics I class average last semester for each professor was a C. You investigate and find that Professor Plum's grades were as follows:

B C D C D B C C D
B B C D C B D D C
C B C B

and Professor Peach's grades were:

A A F A F F A F F
A A F A F A A F A
F A F F

Analyze the differences between the grades of the two professors. Which professor would you rather have?

3. *Money Forecast* listed five tax-exempt mutual funds. The minimum initial investment for each fund was as follows:

Mutual Fund	Min. Initial Investment
Templeton Tax-Free	$ 500
Strong Municipal	2,500
Fidelity Spartan Muni.	25,000
Calvert Tax-Free Res.	2,000
Evergreen Tax Exempt	2,000

Bull Marquette, your stock broker, tells you that the average minimum investment is $2,000. And your friend Anna Liszt says the average minimum investment is $6,400. How can they both be correct?

4. In a survey mailed to 2,000 readers of *Macworld*, it was found that 97 percent use Macintosh computers. Analyze the fallacy behind the following statement: "Macintosh computers are enjoying huge popularity since 97 percent of the people surveyed use them."

5. According to the Equal Employment Opportunity Commission, 10,522 people filed sexual harassment complaints in a recent year, while 6,883 filed such suits in the preceding year. Does this necessarily show that much more sexual harassment has occurred?

6. *USA Today* published a study that showed that there was one centenarian (aged 100 or

over) for every 3,961 people in Iowa. In Alaska, there was one centenarian for every 36,670 people. Is there anything fallacious about this statement: "Go to Iowa and you'll live longer"?

7. Recent studies have been made about the risk factors for heart disease. Among the findings: Walnuts may help prevent the disease, but certain types of baldness are bad. Dr. James Muller, in an article in *The New York Times*, said he was skeptical of these risk factors. He noted that the California Walnut Commission supported research demonstrating that walnuts are heart-healthy and the Upjohn Corporation (makers of a hair cream) financed the baldness study. What do you think of Dr. James's observations?

8. "According to the alumni office, the average Prestige U graduate, class of '85, makes $86,123 a year." Comment on this press release.

9. An independent laboratory test showed that Krinkle Gum toothpaste users report 36 percent fewer cavities. Discuss this advertisement.

10. "Grogain cream has been used by a quarter of a million customers to cure baldness. We have a double-your-money-back guarantee, and only 2 percent of those who used Grogain were not helped and asked for a refund." Discuss this advertisement.

11. "There are as many people with above-average income as there are people with below-average income." Discuss the type of average for which this statement is always true.

12. "Our firm's income has gone from $5 million to $10 million in just 1 year—an increase of 200 percent." Discuss this statement.

13. "Since 66 percent of all rape and murder victims were one-time friends or relatives of their assailants, you are safer at night in a public park with strangers than you are at home." Comment on this remark.

14. "A group of Texas schoolteachers took a history test and failed with an average grade of 60. Thus, Texas schoolteachers are deficient in history." Do you agree?

Topics for Review and Discussion

1. "Last year 760.67 million marijuana cigarettes were smoked in the United States—a flaunting of the law unequaled since Prohibition." Discuss this statement.

2. Why were the polls wrong in 1948 when they predicted that Dewey would defeat Truman for president? You'll have to do outside research to answer this question.

3. The word "average" is a broad term that applies to several measures of central tendency. Identify and define three such measures.

4. "Averages alone don't adeqately describe the true picture of a set of data." Why is this statement true?

5–8. What is a:

5. Statistical table?

6. Line graph or chart?

7. Bar graph or chart?

8. Circle (pie) graph?

9. Give two examples of how persuasive artists can take honest facts and create misleading impressions.

10. How may antics with semantics be used to confuse and mislead?

11. "Percentage decreases exceeding 100 percent aren't possible if the original data are positive values." Explain why this is true.

12. What questions might you ask during the evaluation of quantitative information to avoid being misled?

Projects/Issues to Consider

1. Find examples in newspapers and periodicals of misleading graphics or questionable statistical usage, and prepare a report of your findings

2. Design an advertisement that uses misleading statistics.

3. Students in Professor Larry Schuetz's

classes at Linn Benton Community College conduct Prove That Claim projects each term. A student first identifies an ad that contains a claim of superiority for a product. The student then writes a letter to a company official (names are found in library reference books) asking that person to supply information that will substantiate the claim. A copy of this letter goes to the instructor, and when a reply is received the student discusses the ad and the response in a class presentation. Follow these same steps to cary out your own Prove That Claim research project.

4. Prepare a presentation describing the rise in revenues of an organization (perhaps the one you work for). Construct two line graphs for the same information, with one showing a large increase and the other appearing to show that revenues are leveling off. Use different scales for your axes, and describe the impression that is made when these different scales are used.

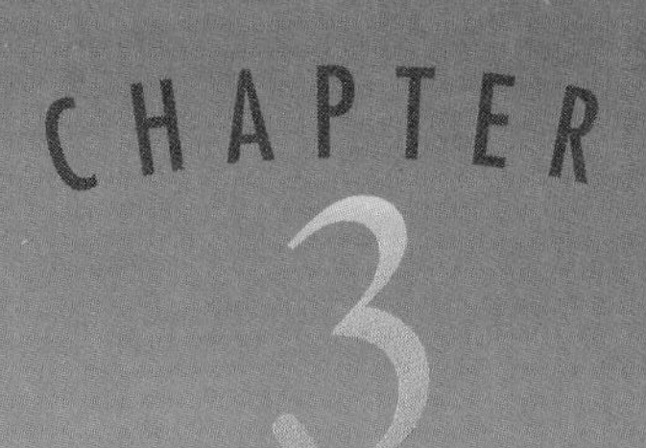

CHAPTER 3

Descriptive Statistics

All Numbers Aren't Created Equal

Some numbers are nominal measures that can be used only for identification. At a higher level are ordinal measures that place data values in a preference order. Horses finishing a race can be numbered 1, 2, 3, . . . , as they cross the finish line. But this ranking says nothing about the time separating the winner from the second (or third) horse. Similarly, we know that gold, silver, and bronze medal winners in Olympic competition finished 1, 2, and 3, but the ranking alone doesn't tell us the closeness of the competition. Calculations based on nominal or ordinal data values are meaningless. Interval and ratio measures also allow data values to be ranked in order but on a standard scale. The scale at the ratio level includes a true zero starting point, but this point is absent on an interval scale. Concepts for considering nominal and ordinal data are included in later chapters, but much of the analysis in this book involves the use of interval/ratio data.

LOOKING AHEAD

You saw in Chapter 1 that "descriptive statistics" is a term applied to the procedures of data collection, classification, summarization, and presentation. We'll now be concerned with all these aspects of statistical description. That is, in this chapter we'll (1) consider the collection and organization of raw data, (2) examine ways to classify and graphically present data in a frequency distribution format, and (3) show the procedures used to compute selected central tendency and dispersion measures.

Thus, after studying this chapter, you should be able to:

- Explain how to organize raw data into an array and how to construct and interpret a frequency distribution.
- Graphically present data in the form of a histogram, frequency polygon, stem-and-leaf display, dotplot, and boxplot.
- Present an overview of the types of measures that summarize and describe the basic properties of data sets.
- Compute such central tendency measures as the arithmetic mean, median, and mode.
- Compute such dispersion measures as the range, mean absolute deviation, and standard deviation.
- After showing computations of the quantitative measures noted above, the chapter concludes with a discussion of a measure used to summarize qualitative data.

3.1 Introduction to Data Collection

We've seen in Chapter 1 that data are gathered early in the process of statistical problem solving. Data may be classified into either attribute or numerical types (Figure 3.1). Examples of *attribute* (also called "qualitative" or "categorical") data are gender (two categories, male and female), political affiliation (Democrat, Republican, and other categories), and citizenship (United States, Canada, Mexico, Nigeria, etc.). Attribute data are often assigned code numbers before being entered into computer databases (perhaps 1 = female, 2 = male). These numbers thus become a *nominal measure*, but they should be used only for identification purposes. It's foolish, for example, to add the female-male code numbers and compute an average gender value!

Those with problems to solve generally focus on *numerical* data that represent counts or measurements. Of course, these numerical facts aren't all identical since there's little reason to study such a situation. Rather, *variables* are the focus of the analyst's attention.

A **variable** is a characteristic of interest—one that can be expressed as a number—that's possessed by each item under study. The value of this characteristic is likely to change or vary from one item in the data set to the next.

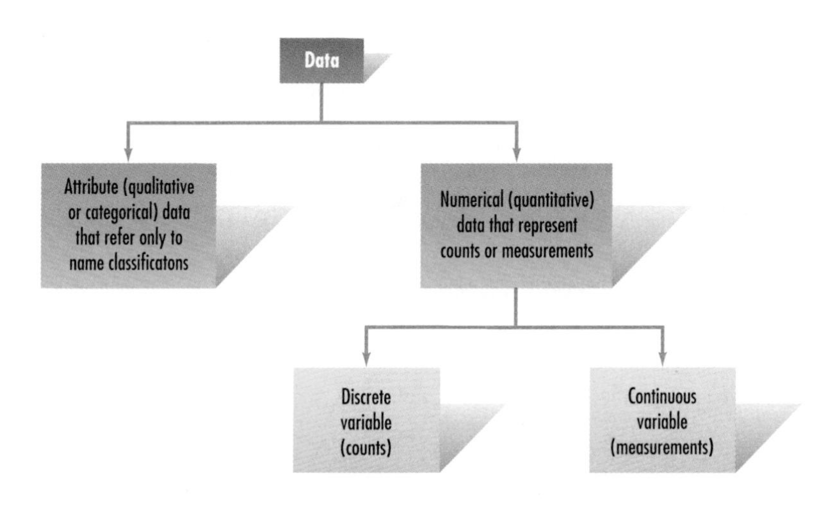

FIGURE 3.1 Basic types of data.

A value for a variable of interest is typically obtained either by counting or by using some measuring device. The number of mailboxes in branch post offices, for example, is found by *counting*. Similarly, counting is used to find the number of automobiles in a student parking lot. Thus, a **discrete variable** is generally one that has a countable or finite number of distinct values.

A **continuous variable** is one that can assume any one of the countless number of values along a line interval. The value of a continuous variable is typically found with an instrument that *measures* the variable to some predetermined degree of accuracy. Two such instruments are the automobile odometer that measures distance traveled and the bathroom scale that shows weights of individuals. The measured quantities produced by such instruments are continuously variable, but they are only approximate values. The pointer on a bathroom scale, for example, might read 145 pounds. But if the pointer is lengthened and sharpened and if the scale is calibrated more precisely, the reading might be 144.5 pounds. Further refinements or better instruments might give readings of 144.42 pounds, 144.4234 pounds, and so on.

Self-Testing Review 3.1

The Lemon Marketing Corporation has asked you for information about the car you drive. For each question, identify each of the types of data requested as either attribute data or numeric data. When numeric data are requested, identify the variable as discrete or continuous.

1. What is the weight of your car?
2. In what city was your car made?
3. How many people may be seated in your car?
4. What's the distance traveled from your home to your school?
5. What's the color of your car?
6. How many cars are in your household?
7. What's the length of your car?
8. What's the normal operating temperature (in degrees Fahrenheit) of your car's engine?

9. What gas mileage (miles per gallon) do you get in city driving?
10. Who made your car?
11. How many cylinders are there in your car's engine?
12. How many miles have you put on your car's current set of tires?

3.2 Data Organization and Frequency Distributions

Assuming that the data have been collected, let's now see how these facts may be organized.

The Raw Data

A listing of the units produced by each worker would likely be of little value to a production manager who's trying to determine overall worker productivity. To be meaningful, such unorganized raw data should be arranged in some systematic order.

An example of raw sales data is presented in Table 3.1. Since we'll be computing various measures to summarize and describe these sales facts later in this chapter, let's make sure we understand this data set. The Slimline Beverage Company makes and sells a line of dietetic soft drink products. These products are sold in bottles and cans. In addition, soft drink syrups are sold to restaurants, theaters, and other outlets that mix small amounts of the syrup with carbonated water and sell the result in paper cups. The sales manager wants to see how a new Fizzy Cola syrup is selling, and so the raw

A computer makes it easier for this doctor to collect, classify, and summarize statistical data for radiation therapy planning purposes. (*Will and Deni McIntyre/Photo Researchers*)

TABLE 3.1 RAW DATA (GALLONS OF FIZZY COLA SYRUP SOLD BY 50 EMPLOYEES OF SLIMLINE BEVERAGE COMPANY IN 1 MONTH)

Employee	Gallons Sold	Employee	Gallons Sold
P.P.	95.00	R.N.	148.00
S.M.	100.75	S.G.	125.25
P.T.	126.00	A.D.	88.50
P.U.	114.00	R.O.	133.25
M.S.	134.25	E.Y.	95.00
F.K.	116.75	Y.O.	104.50
L.Z.	97.50	O.U.	135.00
F.E.	102.25	U.S.	108.25
A.N.	110.00	L.T.	122.50
R.J.	125.00	E.A.	107.25
O.O.	144.00	A.T.	137.00
U.Y.	112.00	R.I.	114.00
T.T.	82.50	N.S.	124.50
G.H.	135.50	I.T.	118.00
R.I.	115.25	N.I.	119.00
O.S.	128.75	G.C.	117.25
U.S.	113.25	A.S.	93.25
P.O.	132.00	N.C.	115.00
O.R.	105.00	Y.A.	116.50
F.T.	118.25	T.N.	99.50
W.O.	121.75	H.B.	106.00
O.F.	109.25	I.E.	103.75
R.T.	136.00	N.F.	115.25
K.H.	124.00	G.U.	128.50
E.I.	91.00	X.N.	105.00

sales data on gallons of syrup sold were gathered as shown in Table 3.1. (The continuous variable—gallons sold—in this example has been measured to the nearest ¼ gallon.) This is the GALSOLD data set that was entered into the *Minitab* statistical package in Chapter 1 to produce some interesting output, but in its present form this unorganized mass of numbers probably isn't of much value to the manager. And what if there had been 500 values rather than just 50?

The Data Array

Perhaps the simplest device for systematically organizing raw data is the **array**—an arrangement of data items in either an *ascending* (from lowest to highest value) or *descending* (from highest to lowest value) order. An array of the Fizzy Cola sales data given in Table 3.1 is presented in Table 3.2. This array, of course, is in an ascending order. Statistical packages such as *Minitab* have built-in SORT commands and are easily able to sort an unordered raw data set into an array.

There are several *advantages in arraying raw data:*

- We see in Table 3.2 that the sales vary from 82.50 to 148.00 gallons. The difference between the highest and lowest values—called the **range**—is 65.50 gallons.

TABLE 3.2 DATA ARRAY (GALLONS OF FIZZY COLA SYRUP SOLD BY 50 EMPLOYEES OF SLIMLINE BEVERAGE COMPANY IN 1 MONTH)

Gallons Sold	Gallons Sold
82.50	115.25
88.50	116.50
91.00	116.75
93.25	117.25
95.00	118.00
95.00	118.25
97.50	119.00
99.50	121.75
100.75	122.50
102.25	124.00
103.75	124.50
104.50	125.00
105.00	125.25
105.00	126.00
106.00	128.50
107.25	128.75
108.25	132.00
109.25	133.25
110.00	134.25
112.00	135.00
113.25	135.50
114.00	136.00
114.00	137.00
115.00	144.00
115.25	148.00

Source: Table 3.1.

- The lower one-half of the values are distributed between 82.50 and 115.25 gallons, and the upper 50 percent of the values vary between 115.25 and 148.00 gallons.
- An array can show the presence of large concentrations of items at particular values. In Table 3.2, no single value appears more than twice, but in other arrays there may be a pronounced concentration.

In spite of these advantages, though, the array is still a rather awkward data organization tool, especially when the number of data items is large. Thus, there's often a need to arrange the data into a more compact form for analysis and communication purposes.

Frequency Distributions

The purpose of a *frequency distribution* is to organize the data items into a compact form without obscuring essential facts. This purpose is achieved by grouping the arrayed data into a relatively small number of classes. Thus:

> A **frequency distribution** (or **frequency table**) groups data items into classes and then records the number of items that appear in each class.

In Table 3.3, for example, we've grouped the gallons sold into seven classes and then indicated the number of employees whose sales have turned up in each of the seven classes. (The term "frequency distribution" comes from this frequency of occurrence of values in the various classes.)

TABLE 3.3 FREQUENCY DISTRIBUTION (GALLONS OF FIZZY COLA SYRUP SOLD BY 50 EMPLOYEES OF SLIMLINE BEVERAGE COMPANY IN 1 MONTH)

Gallons Sold	Number of Employees (Frequencies)
80 and less than 90	2
90 and less than 100	6
100 and less than 110	10
110 and less than 120	14
120 and less than 130	9
130 and less than 140	7
140 and less than 150	2
	50

Source: Table 3.2.

You'll notice in Table 3.3 that the data are now arranged in a compact form. A quick glance at the frequency distribution shows, for example, that the sales of about two-thirds of the employees ranged from 100 to 130 gallons (the sales of 33 of the 50 employees are distributed in the middle three classes). In short, Table 3.3 gives us a reasonably good view of the overall sales *pattern* of Fizzy Cola syrup. But the compression of the data has resulted in some loss of detailed information. We no longer know, for example, exactly how many gallons each employee sold. And we don't know from Table 3.3 that the values have a range or spread of exactly 65.50 gallons. All we know about these matters is that there are two employees in the first class, for example, whose sales were somewhere between 80 and less than 90 gallons and that the range of values is going to be somewhere between 50 and 70 gallons. On balance, though, the advantage of gaining new insight into the data patterns that may exist through the use of a frequency distribution can often outweigh this inevitable loss of detail.

To construct a frequency distribution, it's necessary to determine (1) the number of classes that will be used to group the data, (2) the width of these classes, and (3) the number of observations—or the **frequency**—in each class. In the next section we'll look at the first two interrelated considerations. The last step is a routine transfer of information from an array to a distribution, and so we'll not consider it here.

Classification Considerations

It's usually desirable to consider the following criteria when creating a frequency distribution:

1. In formal presentations, the number of classes used to group the data generally varies from a minimum of 5 to a maximum of 18. The actual number of classes used depends on such factors as the number of observations being grouped, the purpose of the distribution, and the arbitrary preferences of the analyst. One could group the data in Table 3.3 into many classes, with each class having a small width. Such a distribution can be useful for preliminary analysis, as we'll see in a few pages. But a distribution with many small classes may contain too much detail to be used in a formal data presentation. And at the other extreme, a grouping of the data in Table 3.3 into only three classes with intervals of 22 gallons each would result in the loss of important detail.
2. Classes must be selected to *conform to two rules:* (*a*) All data items from the smallest to the largest must be included, and (*b*) each item must be assigned to one *and only one* class. Possible gaps and/or overlaps between successive classes that could cause this second rule to be violated must be avoided.
3. Whenever possible, the *width* of each class—that is, the **class interval**—should be equal. (It's also often desirable to use class intervals that are multiples of numbers such as 5, 10, 100, 1,000, and so on.) Although unequal class intervals may be needed in frequency distributions where large gaps exist in the data, such intervals may cause difficulties. For example, if frequencies in a distribution with unequal intervals are compared, the observed variations may merely be related to interval sizes rather than to some underlying pattern. Other difficulties of using unequal intervals can arise during the preparation of graphs. Our Table 3.3 has arbitrarily been prepared with seven classes of equal size. How was the interval width of 10 gallons determined? You ask very perceptive questions. The following simple formula first estimates a preliminary interval, and this interval is then *rounded up* to a convenient value:

$$\text{Width} = \frac{\text{range}}{\text{number of classes}} \quad \text{(and then round up)}$$

Of course, as we've seen in Table 3.2, the Fizzy Cola sales data range from a low of 82.50 gallons to a high of 148.00 gallons. Thus:

$$\text{Width} = \frac{(148.00 - 82.50)}{7} \quad \text{(and then round up)}$$

$= 9.36,$ and this figure is then rounded up to the class interval size of 10 gallons used in Table 3.3.

4. Whenever possible an *open-ended class interval*—one with an unspecified upper or lower class limit—should be avoided. Table 3.4 has such an interval, and so it's an example of an *open-ended distribution.* An open-ended class may be needed when a few values are extremely large or small compared with the remainder of the more concentrated observations or when confidential information might be revealed by stating an upper limit. For example, placing an upper limit on the data in Table 3.4 might tend to reveal the income of an easily identifiable family in a small community. But open-ended classes should be used sparingly because of graphing problems and because (as we'll soon see) it's impossible to compute such important descriptive measures as the arithmetic mean and the standard deviation from an open-ended distribution.

TABLE 3.4 OPEN-ENDED DISTRIBUTION (TOTAL INCOME REPORTED BY SELECTED FAMILIES)

Total Income	Number of Families
Under $10,000	6
$10,000 and under $20,000	14
20,000 and under 30,000	18
30,000 and under 40,000	10
40,000 and under 50,000	5
50,000 and under 60,000	4
60,000 and over	3
	60

5. When there's a concentration of raw data around certain values, it's desirable to construct the distribution so that these points of concentration fall at the **class midpoint** or middle of a class interval. (The reason for this will become apparent later when we compute the arithmetic mean for data found in a frequency table.) In Table 3.3, the midpoint of the class "110 and less than 120" is 115 gallons, the lower limit of that class is 110 gallons, and the upper limit is 119.999 . . . gallons. Of course, another analyst could gather additional raw sales data for Fizzy Cola syrup and could then round the sales to the *nearest* gallon. This analyst might then set up a distribution similar to the one in Table 3.3 with class intervals of 80 to 89, 90 to 99, 100 to 109, 110 to 119, and so on. In this case, the *stated* limits are only 9 gallons apart, but the size of these class intervals is still 10 gallons. Why? Because the class "110 to 119" has a real lower limit or *lower boundary* of 109.5 and a real upper limit

or *upper boundary* of 119.5. A **class boundary** is thus a number that doesn't appear in the stated class limits but is rather a value that falls midway between the upper limit of one class and the lower limit of the next larger one. With the class "110 to 119," the class interval still has a width of 10 gallons, but the class midpoint in this case is 114.5 gallons.

Crime and the Numbers

The President's Commission on Organized Crime recently suggested that organized crime in the United States had receipts of \$50 billion and net profits of \$30 billion. If these guesses (how could precise data be acquired?) are true, organized crime profits equal about 30 percent of the after-tax profits of all domestic U.S. corporations. *Fortune* magazine notes that by reducing competition and diverting money away from legitimate investments, organized crime costs the country 40,000 jobs and each of us over \$75 a year in disposable income.

Self-Testing Review 3.2

1–5. Bill Alott, a management consultant with Global Technologies, Inc., needs to prepare his Global Office Technologies Consulting Hourly Analysis (GOTCHA) report. He must first determine the number of billable hours that his staff has "clocked" for the previous week. The billable hours listed below represent the time (rounded to the nearest hour) each of his staff has worked on a project for the week of March 15. We'll assume that the 50 employees represent a sample of all Global consultants.

Employee ID Number	Billable Hours	Employee ID Number	Billable Hours
670	38	250	49
561	25	410	57
828	38	571	54
580	40	505	32
153	44	265	42
127	43	504	47
484	42	399	46
798	40	742	55
519	64	730	35
422	46	188	47
433	36	607	50
770	49	873	47
711	38	570	46
576	44	964	47
216	30	442	43
453	42	420	56
779	41	898	44
706	51	705	37
363	50	443	38
721	37	953	32
969	51	962	41
955	48	187	43
849	48	884	58
929	37	169	38
951	44	167	43

1. Arrange the above data in an ascending array.

2. What is the range of values?

3. Organize the data items according to billable hours into a frequency distribution with eight classes beginning with "25 and less than 30," then "30 and less than 35," . . . and finally, "60 and less than 65."

4. Is it possible to use 6 or 10 classes rather than 8 classes in the above frequency distribution?

5. What would have been reasonable class interval widths if you had prepared frequency distributions using 6 or 10 classes rather than 8?

6–8. The following listing (from the *Business Week 1000*) shows analysts' earnings-per-share estimates (in dollars) for 59 top U.S. firms:

Firm	Earnings/Share	Firm	Earnings/Share
IBM	11.18	Pac Tel	2.91
Exxon	4.18	Texaco	5.56
Philip Morris	4.72	Walt Disney	6.35
GE	5.22	Southwest Bell	3.89
Merck	5.35	Ford	−0.54
Wal-Mart	1.41	Schlumberger	3.05
Bristol-Myers	4.00	NYNEX	6.30
A.T.&T.	2.66	US West	3.22
Coca-Cola	2.39	Eastman Kodak	4.09
P & G	5.10	Dow Chemical	3.93
Johnson & Johnson	4.40	Anheuser-Busch	3.29
Amoco	3.95	American Express	2.69
Chevron	5.71	McDonald's	2.43
Mobil	4.93	Hewlett-Packard	3.40
PepsiCo	1.63	Microsoft	3.43
Du Pont	3.13	Schering-Plough	2.96
Bellsouth	3.70	Sears, Roebuck	3.39
GM	−1.36	Fed. Natl. Mortge	5.01
Eli Lilly	4.70	Marion Merrell	2.01
GTE	2.21	Warner-Lambert	4.20
Atlantic Richfield	11.04	Pacific Gas & Elec	2.10
Waste Mgmt	1.78	Intel	3.54
Abbott Labs	2.57	Kellogg	4.68
3-M Corporation	6.20	Digital Equipment	4.15
American Inter	7.29	H. J. Heinz	2.37
Bell Atlantic	3.73	Emerson Electric	2.89
American Home	4.24	American Brands	4.14
Pfizer	5.45	Baxter Internatl	1.98
Ameritech	5.00	The Limited	1.29
Boeing	4.90		

6. Arrange the above data in a descending array.

7. What is the range of values?

8. Organize the data items according to earnings per share into a frequency distribution with seven classes, beginning with "$–2.00 < $0.00," "$0.00 < $2.00," "$2.00 < $4.00," . . . "$10.00 < $12.00."

9–13. Brock and Parse Lee, owners of the Lee Health Foods Company, are studying the size of the orders placed by a sample of customers in an outlying county. In the past week, the following 30 orders have been received:

$42.50	$45.00	$47.75	$52.10	$29.00	$31.25
21.50	56.30	55.60	49.80	35.55	42.30
43.50	34.60	65.50	45.10	40.25	58.00
30.30	44.80	36.50	55.00	59.20	36.60
38.50	41.10	46.00	39.95	25.35	49.50

9. Arrange the above data in an ascending array.
10. What is the range of values?
11. Organize the data items according to order size into a frequency distribution having the five classes "$20 and under $30," "$30 and under $40," . . . and "$60 and under $70."
12. Would it be possible to use six or seven classes rather than five classes in the above frequency distribution?
13. What would have been a reasonable class interval if you had prepared a frequency distribution using eight classes rather than five?

14–16. *Business Week* recently published the price of a share of common stock issued by 59 top U.S. firms. These firms (and their share prices to the nearest dollar) are as follows:

Firm	Share Price	Firm	Share Price
IBM	131	Pac Tel	42
Exxon	55	Texaco	65
Philip Morris	67	Walt Disney	127
GE	67	Southwest Bell	54
Merck	106	Ford	34
Wal-Mart	36	Schlumberger	64
Bristol-Myers	77	NYNEX	76
A.T.&T.	33	US West	38
Coca-Cola	52	Eastman Kodak	46
P & G	87	Dow Chemical	53
Johnson & Johnson	90	Anheuser-Busch	47
Amoco	53	American Express	27
Chevron	75	McDonald's	34
Mobil	65	Hewlett-Packard	49
PepsiCo	33	Microsoft	103
Du Pont	38	Schering-Plough	51
Bellsouth	52	Sears, Roebuck	32
GM	40	Federal Nat. Mo.	46
Eli Lilly	83	Marion Merrell	39
GTE	32	Warner-Lambert	78
Atlantic Richfield	130	Pacific Gas & Elec	25
Waste Mgmt	43	Intel	52
Abbott Labs	47	Kellogg	83
3-M Corporation	91	Digital Equipment	82
American Inter	94	H. J. Heinz	37
Bell Atlantic	48	Emerson Electric	43
American Home	57	American Brands	46
Pfizer	108	Baxter Internatl	33
Ameritech	65	The Limited	25
Boeing	50		

14. Arrange the data in an ascending array.

15. What is the range of values?

16. Organize the data items according to share price into a frequency distribution table using eight classes. Use "$25 < $40" for the first class.

17–19. *U.S. News & World Report* recently published the 25 top-ranked graduate schools of business and gave the average General Management Aptitude Test (GMAT) scores of their students. The data are:

School	Avg. GMAT Score	School	Avg. GMAT Score
Stanford	675	UCLA	640
Harvard	644	Carnegie-Mellon	620
Penn	644	Yale	657
Northwestern	642	North Carolina	620
M.I.T.	650	New York University	609
Chicago	635	Indiana	610
Michigan	630	Texas	631
Columbia	635	USC	606
Duke	630	Rochester	608
Dartmouth	643	Purdue	601
Virginia	617	Pittsburgh	597
Cornell	640	Vanderbilt	602
UC Berkeley	635		

17. Arrange the data in an ascending array.

18. What is the range of values?

19. Organize the data items according to GMAT scores into a frequency distribution table using nine classes. Use "595 < 605 (GMAT points)" for the first class.

20–22. Cerebral vascular accident (CVA)—otherwise known as "stroke"—is an interruption of the flow of blood to the brain that's caused either by blockage or by rupture of an artery. CVA may be accompanied by partial or complete paralysis of the arms and legs. Many who suffer from CVA must undergo occupational therapy to rehabilitate paralyzed limbs. An occupational therapist working at a nursing home in Cincinnati recorded the number of weeks each of 30 patients, selected from a random sample of CVA patients, underwent occupational therapy. The data are:

8 21 6 9 4 15 10 9 7 9 6 17 8 9 9

2 8 8 3 10 16 13 5 3 2 1 9 4 13 7

20. Arrange the data in an ascending array.

21. What is the range of values?

22. Organize the data items (number of weeks of therapy) into a frequency distribution table using five classes. Use "1 < 5 weeks" for the first class.

23–25. In a recent issue, *Money* magazine created a hypothetical family that earned

$72,385 annually and also had additional income from interest ($2,782), dividends ($455), and capital gains ($1,472). The state and local taxes that would have to be paid on this income in each state and the District of Columbia were then shown as follows:

State	Taxes	State	Taxes
Alabama	5,552	Montana	6,781
Alaska	1,632	Nebraska	7,728
Arizona	6,637	Nevada	3,539
Arkansas	7,074	New Hampshire	4,591
California	7,605	New Jersey	7,371
Colorado	7,268	New Mexico	5,948
Connecticut	8,389	New York	10,016
Delaware	5,354	North Carolina	7,263
Dist. of Columbia	9,348	North Dakota	5,292
Florida	3,846	Ohio	7,751
Georgia	7,301	Oklahoma	6,907
Hawaii	8,272	Oregon	8,390
Idaho	7,634	Pennsylvania	6,969
Illinois	7,125	Rhode Island	8,314
Indiana	6,712	South Carolina	6,531
Iowa	7,006	South Dakota	4,284
Kansas	6,935	Tennessee	4,038
Kentucky	6,744	Texas	4,647
Louisiana	5,752	Utah	7,892
Maine	8,611	Vermont	7,962
Maryland	8,568	Virginia	7,217
Massachusetts	8,764	Washington	4,694
Michigan	7,493	West Virginia	5,981
Minnesota	8,311	Wisconsin	8,770
Mississippi	5,792	Wyoming	2,945
Missouri	6,047		

23. Arrange the data in an ascending array.

24. What is the range of values?

25. Organize the data items (taxes to be paid) into a frequency distribution table with eight classes. Use "$1,600 < $2,800" for the first class.

3.3 Graphic Presentations of Frequency Distributions

Once data are grouped into a more compact form, the frequency distribution can be used for analysis, interpretation, and communication purposes. It's often possible to prepare a graphic presentation of a frequency distribution to achieve one or more of these aims. Such a presentation of the data found in a frequency table is more likely to get the attention of the casual observer, and it may show trends or relationships that might be overlooked in a table. So how can graphic presentations of frequency distributions be prepared, you eagerly ask? Well let's see . . .

The Histogram

One popular graphic tool is the *histogram.*

A **histogram** is a bar graph that can portray the data found in a frequency distribution.

Figure 3.2 is a histogram of the Fizzy Cola syrup sales data found in Table 3.3. As you can see, this histogram simply consists of a set of vertical bars. Values of the variable being studied—in this case gallons of syrup sold—are measured on an arithmetic scale on the horizontal axis. The bars in Figure 3.2 are of equal width and correspond to the equal class intervals in Table 3.3; the height of each bar in Figure 3.2 corresponds to the frequency of the class it represents. Thus, the area of a bar above each class interval is proportional to the frequencies represented in that class.*

Most statistical software packages are programmed to produce histograms. A user merely enters a data set and then issues an appropriate command to tell the program what it should do. You saw in Chapter 1 how the gallons of Fizzy Cola syrup sold by the 50 Slimline employees was entered into the *Minitab* package and stored under the GALSOLD name. Now we can use a HISTOGRAM command with *Minitab* to produce graphic results.

Figure 3.3*a* shows the *Minitab default chart*—one that's automatically produced for a data set by the software without any formatting instructions from the user. You'll notice that the "bars" of the *Minitab* graph run horizontally rather than vertically. You can also see that the program has created eight rather than seven classes and has used class midpoints of 80, 90, . . . , rather than 85, 95, ... To produce the same results as those shown in Figure 3.2—that is, to produce Figure 3.3*b*—the *Minitab* user can modify the default graph with some simple instructions that we need not consider here.

The Frequency Polygon

Another popular graphic tool is the *frequency polygon.*

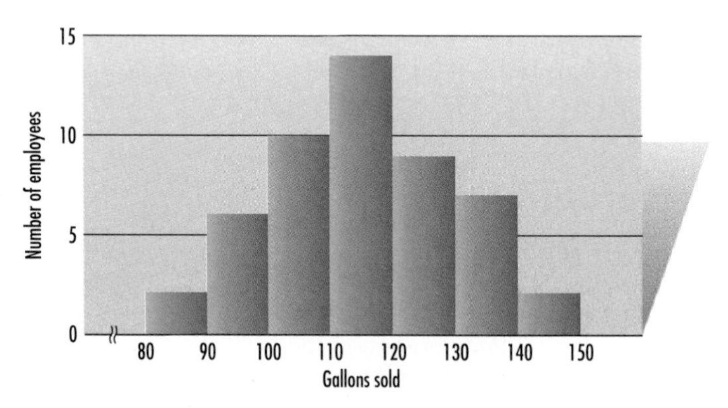

FIGURE 3.2 Histogram of frequency distribution of gallons of Fizzy Cola syrup sold by 50 employees of Slimline Beverage Company in 1 month.

*If unequal class intervals were used in a frequency distribution, the *areas of the bars above the various class intervals would still have to be proportional to the frequencies represented in the classes*—e.g., if the third interval is twice as wide as each of the first two, the frequency of the third interval must be divided by 2 to get the appropriate height for the bar.

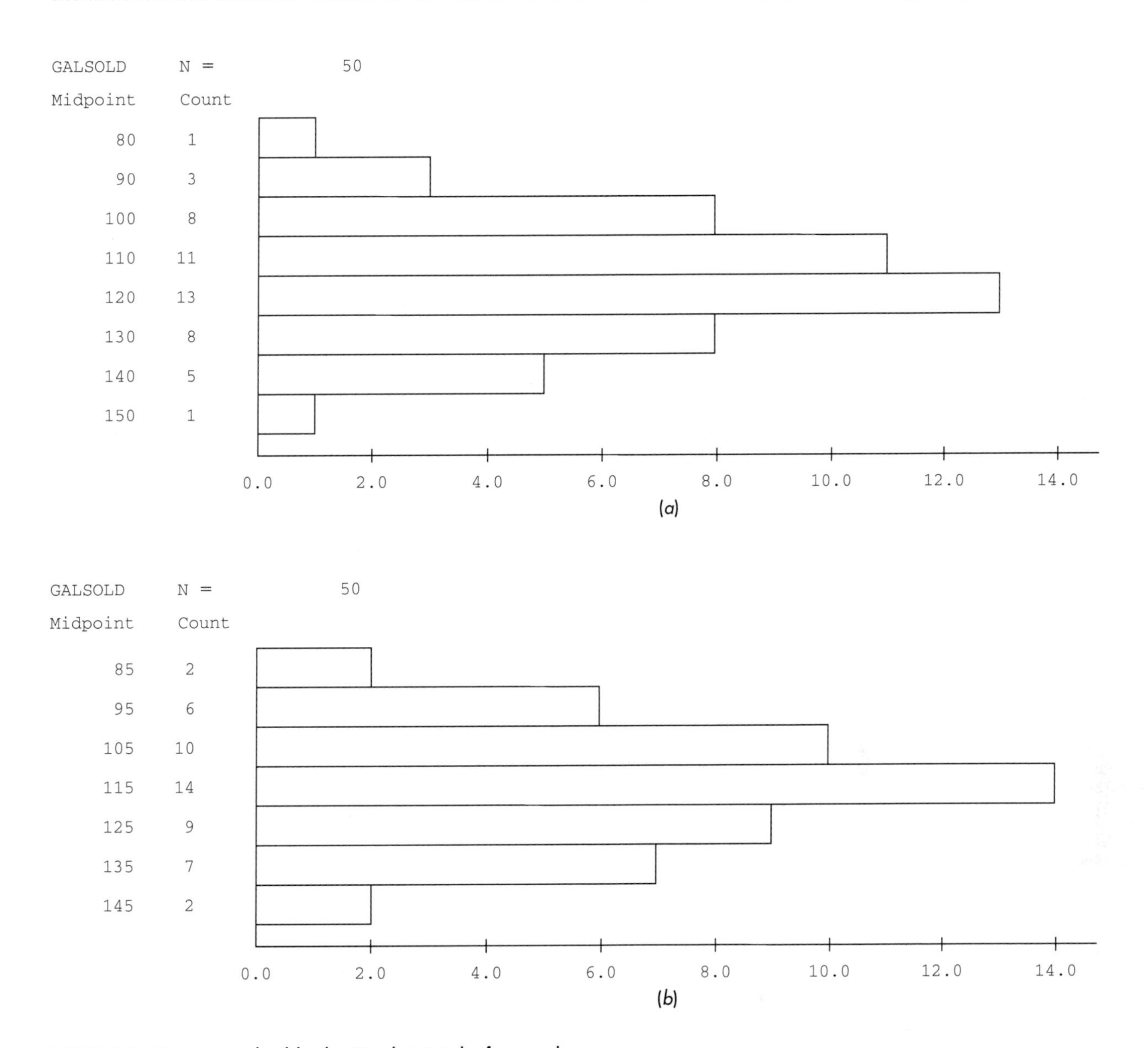

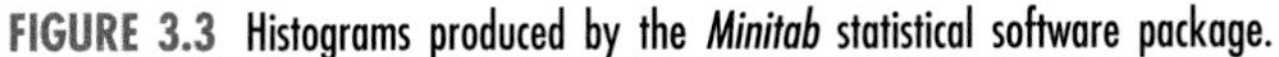
FIGURE 3.3 Histograms produced by the *Minitab* statistical software package.

A **frequency polygon** is a line chart that depicts the data found in a frequency distribution. It is thus a graphic presentation tool that may be used as an alternative to the histogram.

Figure 3.4 is a frequency polygon using the same data and plotted on the same scales as the histogram in Figure 3.2. (In fact, Figure 3.2 has been lightly reproduced as background in Figure 3.4.) As you can see, points are placed at the midpoints of each class interval. The height of each plotted point in Figure 3.4, of course, represents the frequency of the particular class. These points are then connected by a series of straight lines. It's customary to close the polygon at both ends by (1) placing points on the baseline half a class interval to the left of the first class and half a class interval to the

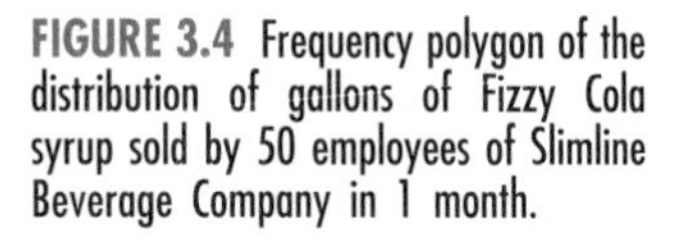
FIGURE 3.4 Frequency polygon of the distribution of gallons of Fizzy Cola syrup sold by 50 employees of Slimline Beverage Company in 1 month.

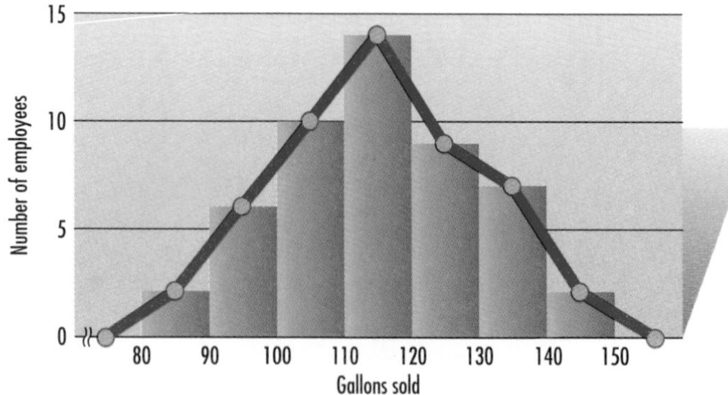

right of the last class, then (2) drawing lines from the points representing the frequencies in the first and last classes to these baseline points (see Figure 3.4).

If the class intervals in a distribution are continuously reduced in size and if the number of items in the distribution is continually increased, the frequency polygon will resemble a smooth curve more and more closely. Thus, if the frequency polygon in Figure 3.4 represented only a small *sample* of all the available data on Fizzy Cola syrup sales made by hundreds of employees and if the frequency distribution—that is, the *population* frequency distribution—that could be prepared to account for all these data were made up of very narrow class intervals, the resulting population distribution curve might resemble the one shown in Figure 3.5. This bell-shaped curve, known as a **normal curve**, describes the distribution of many kinds of variables in the physical sciences, social sciences, medicine, agriculture, business, and engineering. The normal curve is very important in statistics and will be reintroduced in many later chapters.

TABLE 3.5 CUMULATIVE FREQUENCY DISTRIBUTION (GALLONS OF FIZZY COLA SYRUP SOLD BY 50 EMPLOYEES OF SLIMLINE BEVERAGE COMPANY IN 1 MONTH)

Gallons Sold	Number of Employees
Less than 80	0
Less than 90	2
Less than 100	8
Less than 110	18
Less than 120	32
Less than 130	41
Less than 140	48
Less than 150	50

A Cumulative Frequency Graph

It's sometimes useful to find the number of data items that fall above or below a certain value rather than within a given interval. In such cases, a regular frequency distribution may be converted to a cumulative frequency distribution—one that adds the number of frequencies as shown in Table 3.5. As you can see, we've merely arranged the data from Table 3.3 in a different form. The eight employees who sold less than 100 gallons, for example, are the two who sold less than 90 plus the six in the class of "90 and less than 100" gallons. An **ogive** (pronounced "oh-jive") is a graphic presentation of a cumulative frequency distribution. The ogive for Table 3.5 is shown in Figure 3.6, and each point represents the number of employees having sales of less

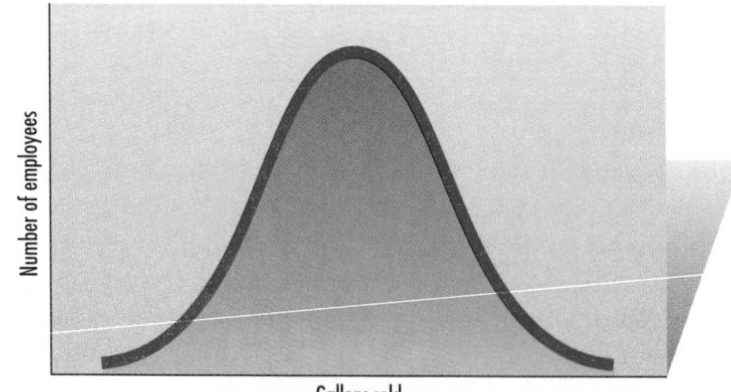

FIGURE 3.5 Generalized normal population distribution.

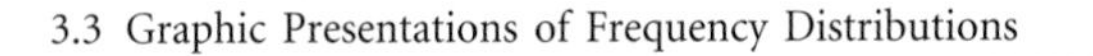

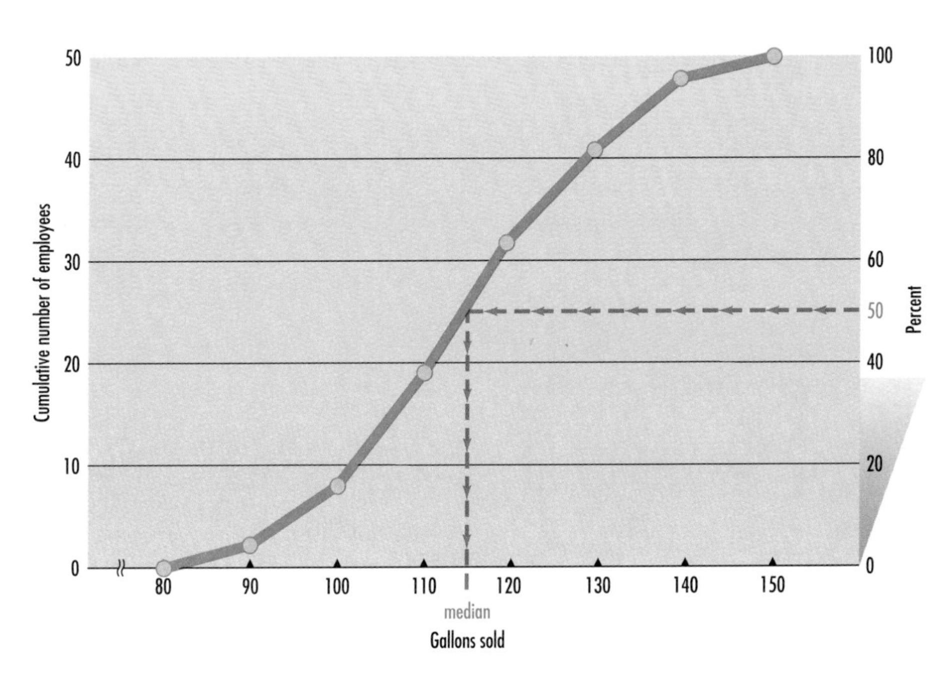

FIGURE 3.6 Ogive for the distribution of gallons of Fizzy Cola syrup sold by 50 employees of Slimline Beverage Company in 1 month.

than the gallons indicated on the horizontal scale. Since each employee represents 2 percent of the total ($1/50 \cdot 100$), we can double the scale on the left vertical axis to get the percentage scale to the right of the ogive. Then, it's possible to graphically approximate or estimate the median. If, as shown in Figure 3.6, we draw a line from the 50 percent point on the percentage scale over to where it intersects with the ogive line and if we then draw a perpendicular line from this intersection to the horizontal scale, we're able to read the median value of about 115 gallons, the approximate amount of syrup sold by the middle employee in the arrayed group of 50.

Some Exploratory Data-Analysis Presentations

The term **exploratory data analysis (EDA)** refers to several techniques that analysts can use to get a feel for the data being studied. These EDA techniques can be used before (or in place of) more traditional analysis approaches. *Stem-and-leaf displays, dotplots,* and *boxplots* are examples of EDA tools in common use. We'll look at the first two tools here and discuss boxplots later in the chapter.

Stem-and-Leaf Display. The **stem-and-leaf display** uses the actual data items in a data set to create a plot that looks like a histogram. The data array of the Fizzy Cola sales figures found in Table 3.2 is used in Table 3.6 to produce a stem-and-leaf display. A *stem* is a number to the left of the vertical line in Table 3.6, and it represents the leading digit(s) of all data items that are listed on the same row. A *leaf* is a single number to the right of the vertical line that represents the trailing digit of a value. Thus, in Table 3.6 there are 50 leaf values representing the 50 items in our data set. The first stem value in the top row of Table 3.6 is 8, and the first leaf value is 2. These numbers are combined to give a sales figure of 82 gallons. (The lowest sales value in Table 3.2 is 82.50 gallons, but fractional units are dropped in Table 3.6.) The second leaf figure in the top row is used with its stem to produce a sales figure of 88 gallons. In the second row, the stem of 9 is combined with each of the leaves in that row to get six sales figures ranging from 91 to 99 gallons. The third-row stem of 10, when combined with leaves of

TABLE 3.6 A STEM-AND-LEAF DISPLAY OF THE SLIMLINE BEVERAGE COMPANY SALES FIGURES FOUND IN TABLE 3.2

Stem	Leaf Values	Number of Data Items
8	28	2
9	135579	6
10	0234556789	10
11	02344555667889	14
12	124455688	9
13	2345567	7
14	48	2

0, 2, 3, . . . , produces data items of 100, 102, 103, . . . , and so it goes throughout the display. The appearance of our stem-and-leaf plot would have looked just the same if we had used the raw sales data from Table 3.1 rather than the arrayed values from Table 3.2. But our leaf values wouldn't have been in sequence.

Stem-and-leaf plots are simple to prepare, but it's even easier to turn the task over to a computer. Table 3.7*a* shows *Minitab*'s *default* output when it executes its STEM-AND-LEAF command. You'll notice that *Minitab* lists each stem on two lines. Leaf values of 0 through 4 are printed on the first of the two lines, and leaves 5 through 9 appear on the second. The first column (to the left of the stems) in the *Minitab* output supplies additional information. This column begins at the top by accumulating the number of data values that have been accounted for in each line. There's 1 item in row 1, a total of two items in rows 1 and 2, four values through row 3, and 23 items are accounted for in the top seven rows. The parentheses in the next row show that we've arrived at the line that holds the median value, and the number in the parentheses gives a count of the leaves on the line. Since there are 50 items in the data set and since we've accounted for 23 of them in the first seven rows, then the median is one of the early values in row 8 (about 115 in this case). After the median is reached, the figures in column 1 then show how many items remain on that line and on the lines below it in the data set. A simple subcommand in Table 3.7*b* allows *Minitab* to duplicate Table 3.6.

Dotplot. A **dotplot** is a preliminary data-analysis tool that groups the study data into many small classes or intervals and then shows each data item as a dot on a chart. Dotplots are usually generated by statistical software programs, and they often help analysts compare two or more data sets.

You're certainly aware by now of the sales of Fizzy Cola syrup made by 50 Slimline Beverage employees (the data arrayed in Table 3.2). Let's assume, though, that the Slimline sales manager wants to compare the cola syrup sales with the syrup sales of Plum Natural—a carbonated diet drink with a fruit juice component. The sales of Plum Natural syrup by 50 Slimline employees is arrayed in Table 3.8. After entering the sales data for the two soft drink products into a *Minitab* program, the sales manager can use a DOTPLOT command to initiate a built-in charting operation. Dotplots for the Fizzy Cola data (labeled GALSOLD1) and the Plum Natural data (labeled GALSOLD2) are shown in Figure 3.7. The manager can now see at a glance that cola sales are concentrated just below the 120-gallon mark, while the heaviest sales of Plum Natural syrup fall below 80 gallons. It's also obvious that the "average" sale of the

TABLE 3.7 STEM-AND-LEAF DISPLAYS PRODUCED BY *MINITAB*

```
MTB > STEM-AND-LEAF 'GALSOLD'

Stem-and-leaf of GALSOLD   N = 50
Leaf Unit = 1.0

    1    8 2
    2    8 8
    4    9 13
    8    9 5579
   12   10 0234
   18   10 556789
   23   11 02344
  (9)   11 555667889
   18   12 1244
   14   12 55688
    9   13 234
    6   13 5567
    2   14 4
    1   14 8
```

(*a*)

```
MTB > STEM-AND-LEAF 'GALSOLD';
SUBC> INCREMENT 10.

Stem-and-leaf of GALSOLD   N = 50
Leaf Unit = 1.0

    2    8 28
    8    9 135579
   18   10 0234556789
 (14)   11 02344555667889
   18   12 124455688
    9   13 2345567
    2   14 48
```

(*b*)

TABLE 3.8 ARRAY OF GALLONS OF PLUM NATURAL SYRUP (SOLD BY 50 EMPLOYEES OF SLIMLINE BEVERAGE COMPANY IN 1 MONTH)

Gallons Sold			
58.25	75.00	94.50	111.25
63.50	77.75	96.00	114.00
69.75	78.75	97.00	114.75
69.75	80.00	99.50	116.25
70.00	82.50	100.25	119.50
71.25	84.00	102.50	121.00
71.25	84.50	102.50	125.75
72.50	85.75	102.50	128.00
72.50	87.85	104.00	130.75
72.50	90.75	104.50	130.75
73.75	91.50	106.75	135.00
73.75	91.50	108.00	
74.00	91.50	110.50	

```
MTB > DOTPLOT 'GALSOLD1' 'GALSOLD2';
SUBC> SAME.

                                   .
                                   .
                              .   ...   .    .
                         . . .... ... ....  ..
                  .  . ................... ....  . .
-------+---------+---------+---------+---------+---------GALSOLD1
             .
             .
            ...         .    .
            ... .  .   .   ...    .       .
      .  .  ........................ ..  ...  .
-------+---------+---------+---------+---------+---------GALSOLD2
      60        80       100       120       140       160
```

FIGURE 3.7

cola syrup (as measured by both the arithmetic mean and the median) exceeds the "average" sales result obtained with the Plum Natural product.

Self-Testing Review 3.3

1–5. The following exercises use the data and your results from problems 1 and 3 in Self-Testing Review (STR) 3.2.

1. Prepare a histogram of the frequency distribution of billable hours for the GOTCHA report that you created in problem 3.
2. Draw a frequency polygon of the frequency distribution of billable hours for the GOTCHA report that you created in problem 3.
3. Prepare an ogive of the frequency distribution of billable hours for the GOTCHA report that you created in problem 3.
4. Prepare a stem-and-leaf display using the arrayed billable hours data developed in problem 1.
5. Prepare a dotplot using the arrayed billable hours data developed in problem 1.

6–8. The following exercises use the data and your results from problems 6 and 8 in STR 3.2.

6. Prepare a histogram for the frequency distribution of the earnings-per-share data that you developed in problem 8.
7. Draw a frequency polygon for the frequency distribution of the earnings-per-share data that you developed in problem 8.
8. Prepare a stem-and-leaf display using the arrayed earnings-per-share data developed in problem 6.

9–11. The following exercises use the data and your results from problems 9 and 11 in STR 3.2.

9. Draw a histogram of the frequency distribution prepared for the Lee Health Foods Company data in problem 11.

10. Draw a frequency polygon of the frequency distribution prepared for the Lee Health Foods Company data in problem 11.

11. Prepare a stem-and-leaf display using the arrayed data produced in problem 9.

12–14. The following exercises use the data and your results from problems 14 and 16 in STR 3.2.

12. Draw a histogram for the frequency distribution of share price data that you developed in problem 16.

13. Draw a frequency polygon for the frequency distribution of share price data that you developed in problem 16.

14. Construct a stem-and-leaf display using the arrayed share price data that you developed in problem 14.

15–17. The following exercises use the data and your results from problems 17 and 19, STR 3.2.

15. Draw a histogram for the frequency distribution of GMAT scores that you developed in problem 19.

16. Draw a frequency polygon for the frequency distribution of GMAT scores that you developed in problem 19.

17. Construct a stem-and-leaf display using the arrayed GMAT scores that you developed in problem 17.

18–20. The following exercises use the data and your results from problems 20 and 22 in STR 3.2.

18. Draw a histogram for the frequency distribution of number of weeks of therapy for victims of CVA that you developed in problem 22.

19. Draw a frequency polygon for the frequency distribution of number of weeks of therapy for victims of CVA that you developed in problem 22.

20. Construct a stem-and-leaf display using the arrayed weeks of therapy data that you developed in problem 20. Use two lines per stem.

21–23. The following exercises use the data and your results from problems 23 and 25 in STR 3.2.

21. Draw a histogram for the frequency distribution of taxes to be paid that you developed in problem 25.

22. Draw a frequency polygon for the frequency distribution of taxes to be paid that you developed in problem 25.

23. Construct a stem-and-leaf display using the arrayed taxes-to-be-paid data that you developed in problem 23. Use a leaf unit of 100 (ignore the ten's and one's places).

You're Not Average, I'm Not Average

The word "average" conjures up images that are so, . . . , so commonplace, so uneventful. And those images certainly don't fit us. Still, we're fascinated with averages because we like to measure ourselves against these benchmarks. A few examples: The average height of American men is 5 feet, 9.5 inches, and of women is 5 feet, 3.6 inches. The average American college grad recognizes about 30,000 words—about twice as many as the average American high school grad. Each year the average American spends 7 days sick in bed. The average American nudist is a married 35-year-old person. And here's a sobering thought: About half the people are *below* average. Of course, that's not you or me, that's them!

3.4 Computing Measures of Central Tendency

A Preview

You know that the word "average" applies to several measures of central tendency. The purpose of these averages is to summarize in a single value the typical size, middle property, or central location of a set of values. The most familiar average is, of course, the *arithmetic mean,* which is simply the sum of the values of a group of items divided by the number of such items. But you also saw in Chapter 2 that the *median* and *mode* are other measures of central tendency that are commonly used. Figure 3.8 shows some possibilities that could exist in different data sets. Suppose in Figure 3.8*a* that we have the monthly sales distributions of two Slimline products—Outrageous Orange (OO) and Rowdy Root Beer (RR). Although the spread of the sales data in each distribution looks the same, it's obvious that the average sales of root beer are greater than the average sales of the orange beverage; that is, the root beer sales are concentrated around a higher value than the orange sales.

The data sets in Figure 3.8*a* have **symmetrical distributions**: If you draw a perpendicular line from the peaks of these curves to the baselines, you'll divide the area of the curves into two *equal* parts. As you can see in Figure 3.8*b*, however, curves may be skewed rather than symmetrical. A **skewed distribution** occurs when a few values are much larger or smaller than the typical values found in the data set. For example, distribution P in Figure 3.8*b* might be the curve resulting from Professor Nastie's first statistics test. Most of the test scores are concentrated around the lower values, but a few curve breakers made extremely high grades. When the extreme values tail off to the right (as in distribution P), the curve represents a *positively skewed distribution.* Distribution N in Figure 3.8*b*, on the other hand, might be a curve of the test scores obtained by Professor Sweet's statistics students. As you can see, most of the students made high scores (although a few unfortunates had extremely low grades). When extreme values tail off to the left (as in distribution N), the curve shows a *negatively skewed distribution.*

Data often have a tendency to congregate about some central value, and this central value may then be used as a summary measure to describe the general data pattern. If the collected facts are to be processed by noncomputer methods and are limited in number, an analyst may prefer to work directly with the *ungrouped data*—facts that haven't been put in a distribution format—rather than group them together in a frequency table. Likewise, if a computer is used, very large lists of ungrouped data can be easily processed in a few seconds or minutes without any need for a frequency

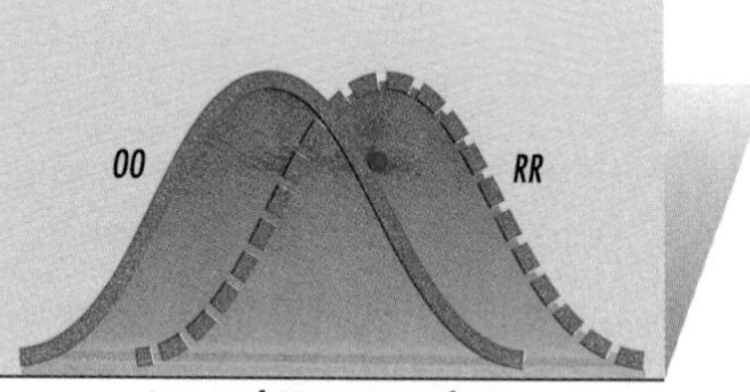

Average of $RR >$ average of OO; dispersion the same

(*a*)

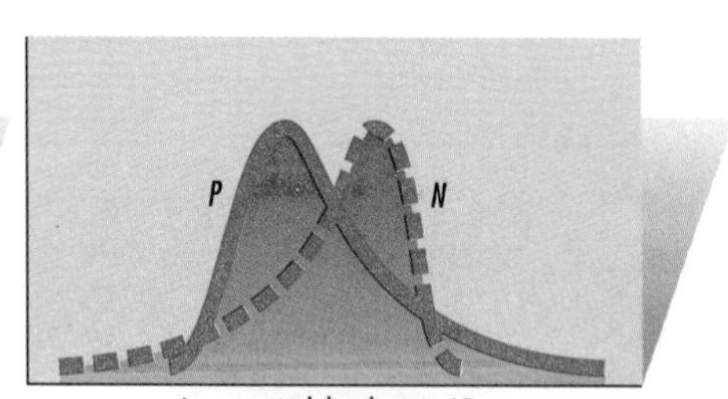

Asymmetrical distributions (P is positively skewed; N is negatively skewed.)

(*b*)

FIGURE 3.8 (*a*) Average of $RR >$ average of OO; dispersion the same. (*b*) Asymmetrical distributions (P is positively skewed; N is negatively skewed).

distribution. But if we're using secondary data that have been compressed into a frequency table to make them more easily understood, we should be able to compute the desired measures from the data in that format.

In the following sections we'll first compute measures of central tendency using *ungrouped data*, and we'll look at methods of computing the same measures when the *data are grouped* into a frequency table. Then, we'll look at how measures of dispersion may be determined for ungrouped and grouped data. Our computations may show characteristics of a *population* (parameters), or they may produce results taken from a *sample* (statistics). Statisticians maintain the distinction between parameters and statistics through the use of different symbols. Greek letters are generally used to denote parameters, while lowercase italic letters denote sample statistics. Table 3.9 shows some common symbols. As you can see, a population mean or percentage is designated by μ (mu) or π (pi), while a sample mean or percentage is denoted by $\bar{x}$ or p. Similarly, a commonly used measure of dispersion—the standard deviation—is identified by σ (lowercase sigma) if it's computed from a population and by s if sample data are used.

What's This, . . . , A Numerical Misstep? And by the U.S. Government?

A study prepared for the Joint Economic Committee of Congress showed that the nation's wealth owned by the richest one-half of 1 percent of the population had increased from 25 to 35 percent between 1963 and 1983. The resulting political outcry and indignation was soon cut short, however, by the discovery that a coding error in the study had changed one household's assets from \$2 million to \$200 million. Since this household figured prominently in the sample taken for the study, the coding error distorted everything. Fixing the error showed that there had been little significant change in the distribution of wealth during the period.

TABLE 3.9 DISTINCTIONS BETWEEN A POPULATION AND A SAMPLE

Area of Distinction	Population	Sample
Definition	Defined as a total of the items under consideration by the researcher.	Defined as a portion of the population selected for study.
Characteristics	Characteristics of a population are parameters.	Characteristics of a sample are statistics.
Symbols	Greek letters or capitals: μ = population mean σ = population standard deviation N = population size π = population percentage	Lowercase italic letters: $\bar{x}$ = sample mean s = sample standard deviation n = sample size p = sample percentage

Measuring Central Tendency for Ungrouped Data

The Arithmetic Mean. When people use the word "average," they're usually referring to the arithmetic mean. And when you have totaled the test grades you've made in a subject during a school term and divided by the number of tests taken, you've computed the *arithmetic mean.* The arithmetic mean is the most commonly used average.

Let's review the computation of the mean by considering the statistics grades made by Peter Parker* during one agonizing semester (the grades have been *arrayed* in a descending order):

*A name selected in memory of another loser, Sir Peter Parker, the British naval commander during the Revolutionary Battle of Sullivan's Island outside Charleston, South Carolina. During this battle, while giving orders aboard *HMS Bristol,* Sir Peter had the "unspeakable mortification" to have a cannon ball carry away the seat of his pants. (According to an old ballad, it "propelled him along on his bumpus.")

75
75
61
50
40
25
10
5
1
342 total of all grades

It's customary to let the letter x represent the values of a variable (such as Peter's grades). Thus, the formulas to compute the *mean* are:

$$\mu = \frac{\Sigma x}{N} \quad \text{(for a population)} \tag{3.1}$$

$$\bar{x} = \frac{\Sigma x}{n} \quad \text{(for a sample)} \tag{3.2}$$

where
μ = arithmetic mean of a population
$\bar{x}$ = arithmetic mean of a sample
Σ (capital sigma) = "the sum of"
N = number of x items in the population
n = number of x items in the sample

Since, in the case of Peter's grades, Σx is 342, Peter's mean semester grade is 38 ($342/N$ or $342/9 = 38$). (These scores represent the population of all grades made by Peter in the course, but if a sample of test scores made by all students in a class had yielded the same data, the sample mean would be exactly the same.)

The Median. You've seen that the *median* is a measure of central tendency that occupies the *middle position* in an *array* of values. That is, half the data items fall below the median, and half are above that value. Note that the word "array" has been emphasized; it's necessary to put the data into an ascending or descending order before selecting the median value. In the example of Peter's grades, the middle value in the array, and thus the median grade, is 40:

75
75
61
50

40 median

25
10
5
1

The median *position*—but not the median value—is found by using the formula $(n + 1)/2$, where n in this example is 9. Thus, $(9 + 1)/2$ is 5, which is the median position in Peter's grade array. Although the median in this example differs by a small amount from the mean, the ultimate result of using either the mean or the median as the semester average grade is the same—Peter doesn't have a clue about the general subject of statistics and has failed the course. As we saw in the example of the grape growers' income in the last chapter, however, one or a few extremely high (or extremely low) values in a data set can cause a substantial difference between the mean and the median.

What if Peter's instructor had dropped Peter's lowest grade before computing the median?

75
75
61

50
40

$$\frac{50 + 40}{2} = 45$$

25
10
5

In that event, the middle position in Peter's grade array would be the mean of the two middle scores 40 and 50—i.e., the median value would then be 45 (a change that doesn't do a thing for Peter's final grade).

The Mode. The *mode*, by definition, is the *most commonly occurring value* in a series. Thus, in the example of Peter's grades, the mode is 75—an average that appeals to Peter but not to his professor. Although not of much use in our grade example, the mode may be an important measure to a clothing manufacturer who must decide how many dresses of each size to make. Obviously, the manufacturer wants to produce more dresses in the most commonly purchased size than in the other sizes.

Other Measures. In addition to the mean, median, and mode, there are other specialized measures of central tendency that are occasionally used. The **weighted arithmetic mean,** for example, is a modification of the measure we have been computing that assigns weights or indications of relative importance to the values to be averaged. Thus, if you get grades of 83 and 87 on hourly statistics exams and a grade of 95 on the final, if the hourly exams each carry a weight of 25 percent of your semester grade and if the final counts for 50 percent of your grade, your weighted mean or semester average is found as follows:

$$\text{Weighted mean} = \frac{\Sigma\,(x \cdot w)}{\Sigma\, w} \tag{3.3}$$

$$= \frac{83(25) + 87(25) + 95(50)}{100} = 90 \qquad \text{(congratulations!)}$$

Measuring Central Tendency for Grouped Data

Computers can easily and accurately process huge lists of ungrouped data if such data are available. But if you're using grouped data supplied by others (such as government employees) and the raw data are unavailable, you have no choice but to compute values from a frequency table. Remember, though, that such values aren't exact but are *only approximations.* Certain assumptions (outlined later) are required in computations, and the validity of these assumptions in any given problem determines the accuracy of the results.

Let's now use the Slimline Company data found in Table 3.3 to demonstrate how to compute an estimated mean for grouped data.

The Arithmetic Mean. Computing the arithmetic mean for a frequency distribution is similar to computing the mean for ungrouped data. But since the compression of data in a frequency table results in the loss of the actual values of the observations in each class in the frequency column, it's necessary to make an assumption about these values. The assumption (or estimate) is that *every observation in a class has a value equal to the class midpoint.* Thus, in Table 3.10, it's assumed that the two employees (f) in the first class each sold 85 gallons (m) of Fizzy Cola syrup, giving a total of 170 gallons (fm) sold. Of course, we have the advantage of knowing from Table 3.2 that neither employee sold 85 gallons but their actual total sales of 171 gallons is only 1 gallon over our estimate. And although our assumption in this first class has caused us to *underestimate* the true figure, it's quite possible that a similar error occurring in another class may cause us to *overestimate* that amount. Therefore, throughout a properly constructed distribution, the effect may be that most of these errors will cancel out. For example, we've slightly overstated the sales of the seven employees in the sixth class because it's assumed that they sold a total of 945 gallons (see the fm column in Table 3.10). In fact, their sales in Table 3.2 amounted to 943 gallons.

The computation of the mean of 115.2 gallons is shown in Table 3.10. (We'll assume

TABLE 3.10 COMPUTATION OF ARITHMETIC MEAN (GALLONS OF FIZZY COLA SYRUP SOLD BY 50 EMPLOYEES OF SLIMLINE BEVERAGE COMPANY IN 1 MONTH)

Gallons Sold	Number of Employees (f)	Class Midpoints (m)	fm
80 and less than 90	2	85	170
90 and less than 100	6	95	570
100 and less than 110	10	105	1,050
110 and less than 120	14	115	1,610
120 and less than 130	9	125	1,125
130 and less than 140	7	135	945
140 and less than 150	2	145	290
	$n = \Sigma f = 50$		5,760

$$\bar{x} = \frac{\Sigma fm}{n} = \frac{5{,}760}{50} = \underline{\underline{115.2}} \text{ gallons sold}$$

that the sales of the 50 employees represents a sample of the sales of all Slimline representatives.) The formulas for computing the mean for grouped data are:

$$\mu = \frac{\Sigma fm}{N} \quad \text{(for a population)} \tag{3.4}$$

$$\bar{x} = \frac{\Sigma fm}{n} \quad \text{(for a sample)} \tag{3.5}$$

where f = frequency or number of observations in a class
m = midpoint of a class and the assumed value of every observation in the class
N = total number of frequencies or observations in the population distribution
n = total number of frequencies in the sample distribution

How does our estimate of 115.2 gallons compare to the true mean sales of the 50 Slimline employees? If you add all the sales data arrayed in Table 3.2, you'll find that the total is 5,770 gallons. Thus, our estimate of 115.2 is close to the true sample mean of 115.4 gallons (5,770 gallons/50 = 115.4 gallons).

In the discussion of classification considerations a few pages earlier, you saw that (1) open-ended classes should be avoided if possible and (2) points of data concentration should fall at the midpoint of a class interval. Perhaps the reasons for these comments may now be clarified. *First*, the uses of open-ended distributions are limited because it's impossible to compute the arithmetic mean from such distributions. Why is this true? Because, as you can see in Table 3.4, we cannot make any assumption about the income of each of the three families in the "$60,000 and over" class. Since there's no upper limit in this class, there's no midpoint value that we can assign to represent the total income for each of the three families. And *second*, if the raw data values are concentrated at the lower or upper limits of several classes rather than at the class midpoints, the assumption that we've made to compute the approximate value of the mean is incorrect and can lead to distorted results. For example, if the raw data are concentrated around the lower limits of several classes, the computed mean can overstate the true mean by a significant amount.

Getting a Feel for the Median and Mode. Since the actual values of a data set are lost when a distribution is constructed, it's only possible to approximate the median value from grouped data. Figure 3.9 helps us understand what we are looking for.

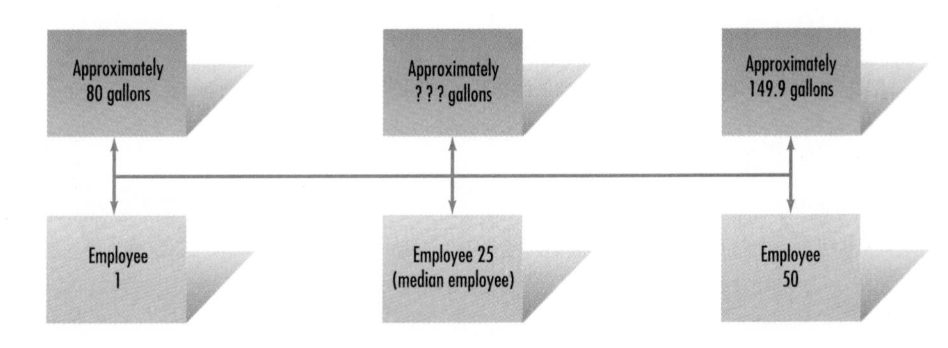

FIGURE 3.9 What are we looking for?

We might assume from the distribution data in Table 3.10 that the employee whose sales quantity was lowest (call him employee 1) sold approximately 80 gallons, and we might guess that the highest-selling employee (call her employee 50) sold approximately 149.9 gallons. What we are looking for, however, is the approximate quantity sold by the middle (let's use the 25th) employee. (The middle position in our data set is between workers 25 and 26, but we're dealing with an *estimate* here so we can settle for the sales of the 25th employee.)

A first step is to locate the *median class*—the class that contains the 25th worker in the group of 50. As you can see in Table 3.10, the 18 employees in the first three classes sold less than 110 gallons, and the 32 employees in the first four classes sold less than 120 gallons. Thus, the 25th employee must be one of the 14 in the fourth or median class. Which of the 14 is the median employee? If 18 are accounted for in the first three classes and if we're looking for number 25, that employee must be the *seventh* one in the group of 14 in the fourth class (that is, $25 - 14 = 7$). It just happens that the median employee is found 7/14 or one-half of the way through the median class. This is just a coincidence. The median observation could have been anywhere in the median class.

But if we assume (as we must) that the sales of the 14 employees in the median class are *evenly distributed throughout the class* and if our median worker happens to be the middle one in the class, then his or her sales should be halfway through the class interval. In short, the median value should be about 115 gallons sold by the 25th employee. (You can verify in Table 3.2 that the true median is 115.25; therefore, our approximation of 115.0 gallons is close.)

When actual data values are unknown, the class in a distribution with the largest number of frequencies is often referred to as the *modal class*, and the mode may be arbitrarily defined to be the midpoint of that class. If two (or more) classes share the

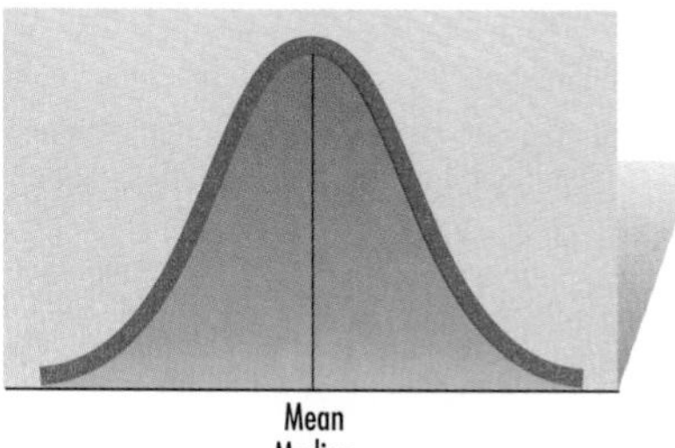

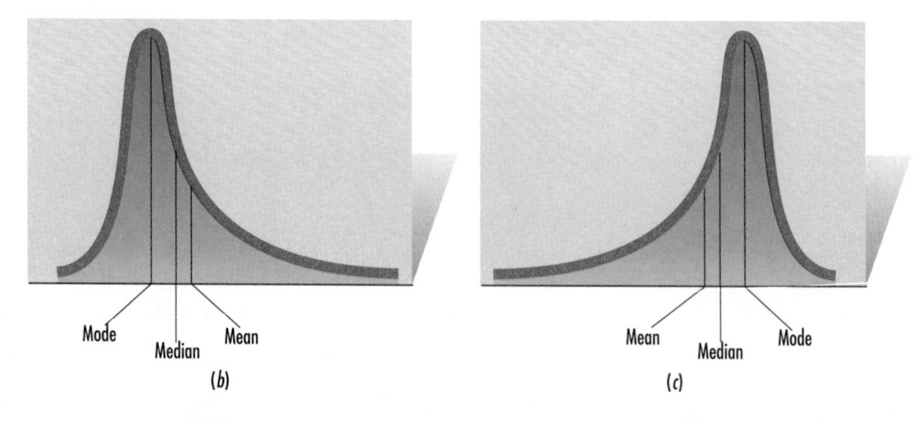

FIGURE 3.10 Typical locations of the principal measures of central tendency. (*a*) Symmetrical distribution. (*b*) Positively skewed distribution. (*c*) Negatively skewed distribution.

Stouffer's	490
Budget Gourmet	290
Healthy Choice	280
Lean Cuisine	290
Weight Watchers	230
Le Menu	250

38. Compute the mean number of calories.

39. Compute the median number of calories.

40. Determine the mode if there is one.

41–43. There was an article in *The Lancet* that described a study of 12 heart patients. The ages of the patients were:

63 59 31 54 66 51 61 66 37 66 53 50

41. Compute the mean age.

42. Compute the median age.

43. Determine the mode if there is one.

44–45. *WordPerfect The Magazine* recently published the amount of hard drive space (in megabytes) needed by seven popular electronic dictionaries. The data are:

American Heritage	3.8	*Funk & Wagnall's*	5.9
Instant Definitions	2.3	*Multilex*	12.4
Random House Webster's	8.7	*Reference Lib. CD-ROM*	.19
The Writer's Toolkit	7.2		

44. Calculate the mean number of megabytes.

45. Calculate the median number of megabytes.

46–48. *Bowling Magazine* has published the standings for a number of bowling teams. The team averages were as follows:

Team	Average	Team	Average
Duds	525	Zebkuhler	553
So What?	535	Jerry's Kids	572
Vintage Rock & Roll	559	Skid Row Hook	546
Up in Smoke	535	Tri-Right	543
Two D's and a Sub	538	Bo's Bach	474
A Beautiful Thing	520	Woulda Coulda Did	593
3UDSU	520	VHS	565

46. Calculate the mean team score.

47. Calculate the median team score.

48. Determine the mode if there is one.

49–51. As part of an experiment conducted in a Biology II lab, the air volume exhaled after each student had inhaled fully was measured for the 13 students in the lab. The air volume figures are:

1.7 3.2 0.5 1.2 1.5 2.6 1.7 0.5 0.5 1.7 0.4 4.4 2.5

49. Calculate the mean air volume exhaled.

50. Calculate the median air volume exhaled.

51. Determine the mode if there is one.

52–54. *Climbing* magazine gave the weights (in ounces) of 15 popular climbing shoes. The data are:

20 14 20 13 15 20 22 16 16 18 16 14 12 16 18

52. Calculate the mean weight.

53. Calculate the median weight.

54. What is the modal weight?

55–56. There are nine branches of The Gap in a local area. Each day the manager takes a "midday read" of store sales in the area. One day the "midday read" for the nine stores is:

$1,256 2,726 1,224 2,588 3,294 1,893 2,537 3,177 2,460

55. Calculate the mean midday read.

56. Calculate the median midday read.

57–58. *Consumer Reports* has presented the overall miles per gallon of various sporty cars. These cars and their miles per gallon (mpg) ratings are:

Mazda MX-6	25	Hyundai Scoupe	32
Dodge Stealth	20	VW Corrado	25
Toyota Celica	27	Isuzu Impulse	29
Mazda Miata	30	Geo Storm	29
Toyota MR2	28	Mercury Capri	28
Mitsubishi Eclipse	29		

57. Calculate the mean miles per gallon.

58. Calculate the median miles per gallon.

59–60. *Fortune* magazine has identified a list of 10 top-rated publishing and printing companies. A numerical rating was given to each company based on such characteristics as quality of management, quality of products or services, innovativeness, and so on. The companies and their ratings are:

Company	Score	Company	Score
Berkshire Hathaway	7.92	R. R. Donnelley	6.94
Reader's Digest	7.14	New York Times	6.67
Gannett	7.12	Tribune	6.67
Dow Jones	7.04	Times Mirror	6.66
Knight-Ridder	6.95	McGraw-Hill	5.95

59. Calculate the mean rating.

60. Calculate the median rating.

61–62. Many American drug producers have operations in Puerto Rico. *The New York Times* has published a list of 10 of these producers and has given the tax savings (in millions of dollars) that each has realized from their operations in Puerto Rico. The data are:

Company	Tax Savings	Company	Tax Savings
Johnson & Johnson	158.0	American Home Products	105.6
Abbott Laboratories	87.0	Eli Lilly	88.3
Bristol-Myers Squibb	219.4	SmithKline Beecham	67.5
Warner-Lambert	22.7	Merck	163.5
Schering-Plough	65.0	Pfizer	124.7

61. Calculate the mean savings (in millions of dollars).

62. Calculate the median savings.

3.5 Computing Measures of Dispersion

A Preview

What if two data sets have the same average? Does this mean that there's no difference in these sets? Perhaps, but then again, perhaps not. In Figure 3.11, distributions X and Y would have the same average, but they're certainly not identical. The difference lies in the amount of spread, scatter, or dispersion of the values in each distribution as measured along the horizontal axis. Obviously, the dispersion in distribution X is greater than the spread of the values in distribution Y.

A **measure of dispersion** is one that gauges the variability that exists in a data set. Such measures are usually expressed in the units of the original observations—such as gallons sold, dollars earned, or miles driven. And these measures may be computed for ungrouped data and for data grouped into frequency distributions.

FIGURE 3.11

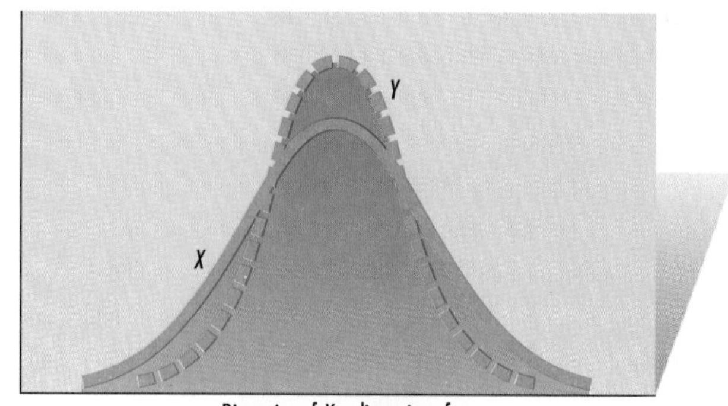

Dispersion of $X >$ dispersion of Y; average the same

There are at least two reasons for measuring dispersion. The *first* reason is to form a judgment about how well the average value depicts the data. For example, if a large amount of scatter exists among the items in a data set, the average size used to summarize those values may not be representative of the data being studied. This isn't a new thought; we saw in Chapter 2 the dangers of disregarded dispersions.

A *second* reason for measuring dispersion is to learn the extent of the scatter so that steps may be taken to control the existing variation. For example, a tire maker tries to produce a product that has a long average mileage life. But the manufacturer also wants to build tires of a uniform high quality so there isn't a wide spread in tire mileage results to alienate customers. (You're pleased with the 40,000 miles you got from a set

Statistical data for decision making are gathered as industrial processes are carried out. (*Shambroom/Photo Researchers*)

of these tires, but I'm really steamed with the 12,000 miles I got with my set of the same tires.) By measuring the existing variation, the manufacturer may see a need to improve the uniformity of the product through better inspection and other quality control procedures.

Measuring Dispersion for Ungrouped Data

Three measures of dispersion are often computed from ungrouped data. These measures are the *range*, the *mean absolute deviation*, and the *standard deviation*.

The Range. The *range* is the simplest measure of dispersion, and we've seen that it's merely the difference between the highest and lowest values in an array. The range is used to report the movement of stock prices over a time period, and weather reports typically state the high and low temperature readings for a 24-hour period.

The Mean Absolute Deviation. Let's assume that after his recent academic ordeal our friend Peter Parker decides to use his mechanical skills to rebuild and then sell a pickup truck (preferably a 4 × 4). This activity will improve Peter's morale, and he expects that it will help restore his badly bent bank balance. Unfortunately, Peter doesn't yet own a truck, and he can't afford newer models or older "collectible" vehicles. What his budget *can* manage is a clunker in the 15- to 20-year age bracket. A fellow student who works part-time at a vehicle auction house tells Peter that her company is planning sales at two locations. At Peter's urging, his friend reports that the mean age of all pickups to be auctioned at sale A is 19, while the mean age of trucks to be sold at sale B is 31. Peter quickly decides to attend sale A. The actual ages of the trucks to be sold at each auction are as follows:

Ages of Trucks at Sale A	Ages of Trucks at Sale B
2	18
2	19
2	19
4	19
5	19
7	19
10	20
11	20
11	45
34	45
35	46
35	47
50	48
58	50
266 total age	434 total age

$$\bar{x} = \frac{266}{14} = 19 \qquad \bar{x} = \frac{434}{14} = 31$$

When Peter arrives at the sale A site, he learns to his dismay that all the trucks are likely to be beyond his financial reach. If Peter had looked beyond the mean age of the sample of trucks to be sold at each sale, it's likely that he would have made a different decision. What Peter apparently wanted was a mean age of 19 and very little spread or scatter of the individual ages about the mean. In short, Peter would have preferred a small measure of dispersion to go along with the mean of 19. (Of course, Peter made a grade of 5 on the test covering measures of dispersion.) One measure of dispersion that would have alerted Peter to the spread of ages is the **mean absolute deviation (MAD)**—an average of the absolute deviations of the individual items about their mean. Before tackling this measure, though, let's look at an important property of the mean.

The difference between each data item and the mean of all the data items in a set is called a *deviation*. An important property of the arithmetic mean is that the algebraic sum of all the deviations is always zero. That is, $\Sigma (x - \bar{x}) = 0$ when we have sample data, and $\Sigma (x - \mu) = 0$ when we have population values. To illustrate this property, suppose that we have the following observations: 2, 3, 4, 7, and 9. The mean of these items is 25/5 or 5. Thus,

x	$(x - \bar{x})$	$(x - \bar{x})^2$
2	−3	9
3	−2	4
4	−1	1
7	+2	4
9	+4	16
25	0	34

As you can see, $\Sigma (x - \bar{x})$ *must* equal zero (this is really a definition of the mean). And if we square and add the deviations—that is, if we complete the last column to produce a $\Sigma (x - \bar{x})^2$ result—we eliminate negative values and get a total of 34.

Now let's calculate the mean absolute deviation. To do this, it's necessary to (1) compute the mean of the items being studied; (2) determine the *absolute deviation*, which is the numeric difference of each item from the mean *without regard to the algebraic sign*; and (3) compute the mean of these absolute deviations. The appropriate formula is:

$$\text{MAD} = \frac{\Sigma |x - \bar{x}|}{n} \quad \text{(for a sample)} \tag{3.6}$$

where MAD = mean absolute deviation
x = values of the sample observations
$\bar{x}$ = mean of the sample observations
$||$ = algebraic signs of the deviations are to be ignored (we consider only absolute values)
n = total number of sample observations

Table 3.11 shows how to compute the MAD for the ages of pickup trucks to be auctioned at sale B. You can verify that if the signs of the deviations about the mean aren't ignored, the sum of these deviations is zero. Thus, it's impossible to compute the MAD unless absolute values are used. The MAD in our example is 13.57 years. (Would

TABLE 3.11 COMPUTATION OF THE MEAN ABSOLUTE DEVIATION (PETER PARKER'S PICK LACKED PERSPICACITY)

Ages of Trucks at Sale B (1)	Arithmetic Mean Age (2)	$(x - \bar{x})$ (1) − (2)	$\lvert x - \bar{x}\rvert$ |(1) − (2)|
18	31	−13	13
19	31	−12	12
19	31	−12	12
19	31	−12	12
19	31	−12	12
19	31	−12	12
20	31	−11	11
20	31	−11	11
45	31	14	14
45	31	14	14
46	31	15	15
47	31	16	16
48	31	17	17
50	31	19	19
434		0	190

$$\bar{x} = \frac{434}{14} = 31$$

$$\text{MAD} = \frac{\Sigma\,|x - \bar{x}|}{n} = \frac{190}{14} = 13.57 \text{ years}$$

the mean deviation have been larger or smaller if we had computed it for those trucks being sold at auction A?)

Unlike the range, the MAD takes every observation into account and shows the average scatter of the data items about the mean; however, it's still relatively simple to understand and compute.

The Standard Deviation. The standard deviation is also used with the mean, and it is generally the most important and useful measure of dispersion. In a precise sense, the population **standard deviation** is the square root of the average of the squared deviations of the individual data items about their mean. What a tongue twister! In easier-to-understand terms, though, the standard deviation is a measure of how far away items in a data set are from their mean. As we'll see later, a majority of the values in the data set will likely fall no more than 1 standard deviation away from their mean, and only a few will lie more than 3 standard deviations from the mean.

Like the calculation of the MAD, the computation of the standard deviation is based on, and is representative of, the deviations of the individual data items about the mean of those values. And another similarity with the mean absolute deviation is that, as the actual observations become more widely scattered about their mean, the standard deviation becomes larger and larger. Of course, if all the items in a series are identical in value—that is, if there is no spread or scatter of values about the mean—the standard deviation is zero. We disregarded algebraic signs to calculate the MAD, but to compute the standard deviation, we square all deviations to eliminate negative values.

Before we can compute a standard deviation, though, we must determine if our data set represents a population or a sample. We must know this fact so that the correct formula can be used. To calculate the standard deviation for a *population*, we use:

$$\sigma = \sqrt{\frac{\Sigma (x - \mu)^2}{N}} \tag{3.7}$$

where σ = population standard deviation
x = values of the observations
μ = mean of the population
N = number of observations in the population

And when a *sample* data set is used, the standard deviation is found with:

$$s = \sqrt{\frac{\Sigma(x - \bar{x})^2}{n - 1}} \tag{3.8}$$

where s = sample standard deviation
x = values of the observations
$\bar{x}$ = mean of the sample
n = number of observations in the sample

As you can see, formulas 3.7 and 3.8 are similar but a denominator of $n - 1$ (rather than n) is used to compute the sample measure. The sample standard deviation is often used to estimate the value of an unknown population standard deviation, and, as you'll see in Chapter 7, the use of $n - 1$ produces better estimates.

Table 3.12 shows the calculation of the standard deviation for the ages of trucks to be auctioned at sale B. These trucks represent a sample of those Peter could consider, so we'll use formula 3.8. The steps in the computation are:

1. The arithmetic mean of the data is computed. (We've seen that it is 31.)
2. The mean is subtracted from each individual age in column 1. (See column 3.)
3. The deviations of the individual ages about the mean (column 3) are squared, thus eliminating negative values (see column 4). The squared deviations are then totaled. This total, $\Sigma (x - \bar{x})^2$, a mathematical property of the mean, is always a minimum value. (So—given our data—if any value other than 31 is used in step 2, the total in column 4 will be larger than 2,654.)
4. Divide the total in column 4 (it's 2,654) by $(n - 1)$—in this case $(14 - 1)$ or 13. The value obtained here (204.16) is the square of the standard deviation and is called the **variance**. The symbols for sample and population variances are s^2 and σ^2, respectively. Take another look at these steps needed to compute the variance, for it's an important statistical measure in its own right. In Chapters 7 and 8 we'll see how to estimate and test hypotheses about the population variance, and then in later pages we'll compute the variances of several samples as a part of an *analysis of variance* procedure that's used to see if the arithmetic means of several populations are likely to be equal. And note that although the variance measures the amount of

TABLE 3.12 COMPUTATION OF THE SAMPLE STANDARD DEVIATION; UNGROUPED DATA (PETER PARKER'S PICK LACKED PERSPICACITY)

Ages of Trucks at Sale B (x) (1)	Arithmetic Mean Age ($\bar{x}$) (2)	$(x - \bar{x})$ (1) – (2) (3)	$(x - \bar{x})^2$ $[(1) - (2)]^2$ (4)
18	31	−13	169
19	31	−12	144
19	31	−12	144
19	31	−12	144
19	31	−12	144
19	31	−12	144
20	31	−11	121
20	31	−11	121
45	31	14	196
45	31	14	196
46	31	15	225
47	31	16	256
48	31	17	289
50	31	19	361
434		0	2,654

$$\bar{x} = \frac{434}{14} = 31$$

$$s = \sqrt{\frac{\Sigma (x - \bar{x})^2}{n - 1}} = \sqrt{\frac{2,654}{14 - 1}} = \sqrt{204.16} = 14.29 \text{ years}$$

variability that exists about the mean of a data set, it's not expressed in the units of the original data. That is, the variance in our example is 204.16, but this value represents the variability of *squared* ages. Thus, to obtain a measure of dispersion expressed in terms of the original values, the following final step is needed.

5. *The standard deviation is computed by taking the square root of the variance.* As you can see in Table 3.12, the standard deviation for our example is 14.29 years of age. The standard deviation is always larger than the average deviation for the same data set because the squaring of deviations puts more emphasis on extreme values. Would the standard deviation figure have been larger or smaller if we had computed it for the trucks to be sold at auction A?

It's not too difficult to use formula 3.8 with a small sample data set. But when many more items are included, the procedure becomes tedious. In that case, you might prefer to use a "shortcut" variation of formula 3.8 (the symbols haven't changed). Although it can be shown algebraically that the following formula, 3.9, is equivalent to formula 3.8, we'll spare you the proof:

$$s = \sqrt{\frac{n(\Sigma x^2) - (\Sigma x)^2}{n(n - 1)}} \tag{3.9}$$

Note: (Σx^2) means to square each of the values of x and then compute the sum, and ($\Sigma x)^2$ means to first add the values of x and then square the sum.

TABLE 3.13 AN ALTERNATIVE COMPUTATION OF THE SAMPLE STANDARD DEVIATION; UNGROUPED DATA

Ages of Trucks at Sale B (x) (1)	$(x)^2$ (2)
18	324
19	361
19	361
19	361
19	361
19	361
20	400
20	400
45	2,025
45	2,025
46	2,116
47	2,209
48	2,304
50	2,500
434	16,108

$$s = \sqrt{\frac{n(\Sigma x^2) - (\Sigma x)^2}{n(n-1)}} = \sqrt{\frac{14(16{,}108) - (434)^2}{14(14-1)}}$$

$$= \sqrt{\frac{225{,}512 - 188{,}356}{182}} = \sqrt{204.16} = 14.29 \text{ years}$$

```
MTB > DESCRIBE 'AGES'

                N      MEAN    MEDIAN    TRMEAN     STDEV    SEMEAN
AGES           14     31.00     20.00     30.50     14.29      3.82

              MIN       MAX        Q1        Q3
AGES        18.00     50.00     19.00     46.25
```

FIGURE 3.12 Output produced by the *Minitab* statistical package when supplied with Peter Parker's sale B data set. *Minitab* is programmed to produce a "trimmed" or modified mean (TRMEAN) that averages the middle 90 percent of the values in a data set. The meanings of Q_1 and Q_3 are discussed later in this chapter. An explanation of SEMEAN must wait until Chapter 6.

Table 3.13 shows that this alternative approach yields the same results achieved in Table 3.12. Of course, if you have a statistical software package, you can simply enter the data set from column 1 of Table 3.13 into the program, issue an appropriate command, and await the results. In Figure 3.12 we can see how *Minitab* has processed our data.

The Standard Deviation for Grouped Data

As you read this section, keep in mind the two points about grouped data computations raised earlier. *First,* the ability of computers to easily process lengthy lists of ungrouped data has eliminated the computational advantages of using frequency distributions. And *second,* the standard deviation obtained from a frequency distribution can only be an approximate value.

When data grouped in a frequency distribution must be processed, however, the primary measure of dispersion is the standard deviation, which is used along with the mean for descriptive purposes. In demonstrating the procedures for computing approximations of the standard deviation, we'll once again use the Slimline Beverage Company data found in Table 3.3. We'll also assume that we have a sample data set—that is, our data items represent a sample of all those of interest to the sales manager.

Computing the standard deviation from a frequency distribution is similar to calculating the measure from ungrouped data. The formulas used to approximate the standard deviation are:

$$\sigma = \sqrt{\frac{\Sigma f(m - \mu)^2}{N}} \quad \text{(for a population)} \qquad (3.10)$$

$$s = \sqrt{\frac{\Sigma f(m - \bar{x})^2}{n - 1}} \quad \text{(for a sample)} \qquad (3.11)$$

where f = frequency or number of observations in a class
m = midpoint of a class and the assumed value of every observation in the class
N = total number of observations in the population distribution
n = total number of observations in the sample distribution

As you can see in Table 3.14, the standard deviation for the Slimline Company data is 14.78 gallons. The class intervals and the (f), (m), and (fm) columns of Table 3.14 duplicate the same columns in Table 3.10, page 66, that were used to compute the mean. But the last three columns in Table 3.14 are new. The mean of the distribution (115.2 gallons) is subtracted from each of the class midpoints in the $(m - \bar{x})$ column to get a deviation amount. Each of these deviations is squared in the $(m - \bar{x})^2$ column. And each of these squared deviations is multiplied by the frequencies in each class in the last $f(m - \bar{x})^2$ column. The total of this last column (10,698.00) is then divided by $n - 1$ or $50 - 1$, the variance of 218.33 is obtained, and the square root of this value—the standard deviation of 14.78 gallons—finally emerges.

You've undoubtedly noticed that using formula 3.11 to compute the standard deviation requires several columns and a number of tedious calculations. You can reduce the workload by using the following shortcut variation of formula 3.11:

$$s = \sqrt{\frac{n[\Sigma f(m)^2] - [\Sigma fm]^2}{n(n - 1)}} \qquad (3.12)$$

where $\Sigma f(m)^2$ = sum of fm times m for each class, and *not* Σf times $(\Sigma m)^2$

TABLE 3.14 COMPUTATION OF STANDARD DEVIATION (GALLONS OF FIZZY COLA SYRUP SOLD BY 50 EMPLOYEES OF SLIMLINE BEVERAGE COMPANY IN 1 MONTH)

Gallons Sold	Number of Employees (f)	Class Midpoints (m)	(fm)	Deviation ($m - \bar{x}$)	$(m - \bar{x})^2$	$f(m - \bar{x})^2$
80 and less than 90	2	85	170	−30.2	912.04	1,824.08
90 and less than 100	6	95	570	−20.2	408.04	2,448.24
100 and less than 110	10	105	1,050	−10.2	104.04	1,040.40
110 and less than 120	14	115	1,610	−0.2	0.04	0.56
120 and less than 130	9	125	1,125	9.8	96.04	864.36
130 and less than 140	7	135	945	19.8	392.04	2,744.28
140 and less than 150	2	145	290	29.8	888.04	1,776.08
	50		5,760			10,698.00

$$\bar{x} = \frac{\Sigma fm}{n} = \frac{5{,}760}{50} = 115.2 \text{ gallons}$$

$$s = \sqrt{\frac{\Sigma f(m - \bar{x})^2}{n - 1}} = \sqrt{\frac{10{,}698.00}{50 - 1}} = \sqrt{218.33} = 14.78 \text{ gallons}$$

Table 3.15 shows the use of this shortcut method. The results of Tables 3.14 and 3.15 must agree, and they do, as you can verify. Only one additional column beyond those used to calculate the mean is needed in Table 3.15. The figures in this column—labeled $f(m)^2$—are found by multiplying the (m) and (fm) values in each table row. For example, in the first row 85 is multiplied by 170 to get the $f(m)^2$ result of 14,450. (Squaring the m value of 85 and then multiplying by the f value of 2 will, of course, produce the same product.) Once the $f(m)^2$ column is completed and totaled, all figures needed for formula 3.12 are available.

Interpreting the Standard Deviation

We know that dispersion is the amount of spread or scatter that occurs in a data set. If, for example, the values in the set are clustered tightly about their mean, the measured dispersion—in this case the standard deviation—is small. But if we have other data sets where the values become more and more scattered about their means, the standard deviations for those sets become larger and larger. To summarize, then, if a standard deviation is small, the items in the data set are bunched about their mean, and if the standard deviation is large, the data items are widely dispersed about their mean. To drive home this generalization in a more tangible way, let's first consider *Chebyshev's theorem.*

Russian mathematician P. P. Chebyshev has been dead for a century now, but his theorem still lives on.

TABLE 3.15 COMPUTATION OF STANDARD DEVIATION USING FORMULA 3.12 (GALLONS OF FIZZY COLA SYRUP SOLD BY 50 EMPLOYEES OF SLIMLINE BEVERAGE COMPANY IN 1 MONTH)

Gallons Sold	Number of Employees (f)	Class Midpoints (m)x	(fm) =	$f(m)^2$
80 and less than 90	2	85	170	14,450
90 and less than 100	6	95	570	54,150
100 and less than 110	10	105	1,050	110,250
110 and less than 120	14	115	1,610	185,150
120 and less than 130	9	125	1,125	140,625
130 and less than 140	7	135	945	127,575
140 and less than 150	2	145	290	42,050
	50		5,760	674,250

$$\bar{x} = \frac{\Sigma fm}{n} = \frac{5{,}760}{50} = 115.2 \text{ gallons}$$

$$s = \sqrt{\frac{n[\Sigma f(m)^2] - [\Sigma fm]^2}{n(n-1)}} = \sqrt{\frac{50[674{,}250] - [5{,}760]^2}{50(49)}}$$

$$= \sqrt{\frac{33{,}712{,}500 - 33{,}177{,}600}{2{,}450}} = \sqrt{218.33} = 14.78 \text{ gallons}$$

Chebyshev's Theorem

Chebyshev's theorem states that the proportion or percentage of *any* data set that lies within k standard deviations of the mean (where k is any positive number greater than 1) is *at least* $1 - (1/k^2)$.

Thus, if we substitute 2 for k in the theorem, we get $1 - (1/k^2) = 1 - (1/2^2) = 1 - (1/4) = 3/4 = 75$ percent. And this result means that *at least* 75 percent of the items in *any* data set (no matter how skewed it is) must lie within two standard deviations of the mean. And *at least* 88.9 percent $[1 - (1/3^2)$ or $8/9]$ of the items in *any* data set must fall within 3 standard deviations of the mean.

Chebyshev's theorem shows us how the standard deviation is related to the scatter of data items. But it tells us only the minimum percentage of items that must fall between given intervals in any data set. We've seen earlier (and in Figure 3.5), though, that many data sets have values that are found to be distributed or scattered about their means in reasonably symmetrical ways.

For such bell-shaped distributions, the *empirical rule* applies and is of greater significance that Chebyshev's theorem:

The Empirical Rule

The **empirical rule** for distributions that are generally bell shaped is that:

- About 68 percent of all data items lie within 1 standard deviation of the mean:

  $(\mu \pm 1\sigma$ or $\bar{x} \pm 1s)$

- About 95 percent of all data items lie within 2 standard deviations of the mean:

  $(\mu \pm 2\sigma$ or $\bar{x} \pm 2s)$

- About 99.7 percent of all data items lie within 3 standard deviations of the mean:

  $(\mu \pm 3\sigma$ or $\bar{x} \pm 3s)$

Let's look at an example of an application of this empirical rule. Suppose that many people are given a new type of IQ test, and the resulting raw scores are organized into a frequency distribution. A frequency polygon is prepared from the distribution and is found to be symmetrical in shape. The arithmetic mean of this mound-shaped distribution is 100 points, and the standard deviation is 10 points. In this situation, the mean IQ score is directly under the peak of the curve, and the following relationships exist: (1) About 68 percent of the test scores fall within *1* standard deviation of the mean—that is, about 68 percent of the people have test scores between 90 and 110 points; (2) about 95 percent of the test scores fall within *2* standard deviations of the mean—that is, about 95 percent of those taking the test have scores between 80 and 120 points; and (3) virtually all (99.7 percent) of the test scores fall within *3* standard deviations of the mean (scores between 70 and 130). Figure 3.13 shows these relationships.

The relationships that exist between the mean and the standard deviation in a bell-shaped distribution may also be used for analysis purposes with distributions that are

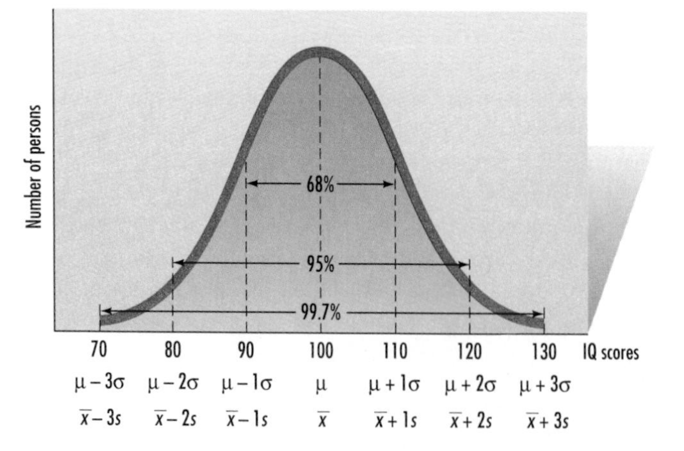

FIGURE 3.13

only approximately symmetrical. Let's return to our Slimline Fizzy Cola example and interpret the meaning of the standard deviation of 14.78 gallons, since that distribution is approximately normal. We can conclude that about the middle two-thirds of the 50 employees sold syrup quantities between $\bar{x} \pm 1s$, that is, between 115.20 gallons ± 14.78 gallons (or from 100.42 to 129.98 gallons). Furthermore, about 95 percent of the employees sold syrup quantities between $\bar{x} \pm 2s$, or between 85.64 and 114.76 gallons. You can verify from the data array in Table 3.2, page 45, that 66 percent of the employees sold between 100.42 and 129.98 gallons and that 96 percent of them sold between 85.64 and 144.76 gallons. And all the 50 employees in our sample sold syrup quantities between $\bar{x} \pm 3s$.

Measures of Position

Like the range, the **quartile deviation (QD)** is a measure that describes the existing dispersion in terms of the *distance* between selected observation points. With the range, the observation points are simply the highest and lowest values. In determining the quartile deviation, though, statisticians find an **interquartile range (IQR)** that includes approximately the *middle 50 percent* of the values. Thus, our observation points, as shown in Figure 3.14, are at the *first* (Q_1) and *third* (Q_3) quartile positions.

The interquartile range is simply the distance or difference between Q_3 and Q_1. The *first quartile* (Q_1) position is the point that separates the *lower* 25 percent of the values from the upper 75 percent. And the *third quartile* (Q_3) position is the point that separates the *upper* 25 percent of the values from the lower 75 percent. Thus, the lower and upper 25 percent of the values aren't considered in the computation of the quartile deviation. Note in Figure 3.14 that the first quartile value is the same as the 25th percentile, the third quartile value is the same as the 75th percentile, and the *second quartile* (Q_2) is just another name for the median or 50th percentile. (Also, the median is sometimes called the "fifth decile" . . . it's enough to make you cry.)

The idea of dividing a data set into four essentially equal parts seems simple enough. But different books may follow slightly different approaches, and the results may also be slightly different. In Figure 3.12 we saw the values of Q_1 and Q_3 that were produced by the *Minitab* statistical program from Peter Parker's "Trucks at Sale B" data set. These 14 trucks are once again arrayed by age from 18 to 50 years in Table 3.16. *Minitab* then locates Q_1 at position $(n + 1)/4$, and Q_3 is positioned at $3(n + 1)/4$.

A Case of Confusing Comparisons

The mean annual earnings of a sample of firefighters is \$30,000, and the data show a standard deviation of \$800. The mean annual income of a sample of master plumbers is \$56,000, and their standard deviation is \$1,400. A comparison of these standard deviations seems to suggest that firefighters' earnings are more uniform while greater income variability exists in the plumbing ranks. Following a dispute among firefighters and plumbers at a local watering hole, a statistics student decided to use the *coefficient of variation* (CV) to measure the relative dispersion that actually exists among the groups.

For the firefighters;

$$CV = \frac{s}{\bar{x}}(100) =$$

$$\frac{\$800}{\$30{,}000}(100) = 2.67 \text{ percent}$$

And for the plumbers:

$$CV = \frac{s}{\bar{x}}(100) =$$

$$\frac{\$1{,}400}{\$56{,}000}(100) = 2.50 \text{ percent}$$

Thus the annual earnings of the plumbers are actually slightly *more* uniform (there's less relative dispersion) than are the earnings of the firefighters.

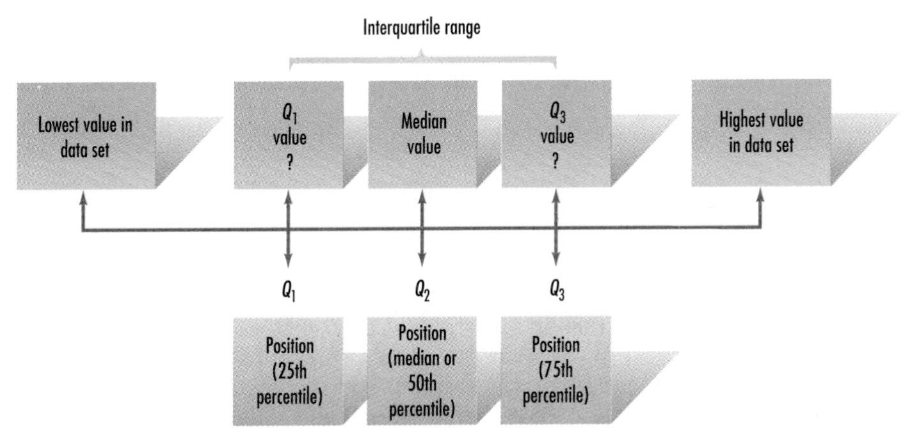

FIGURE 3.14

TABLE 3.16 DETERMINING THE QUARTILE DEVIATION

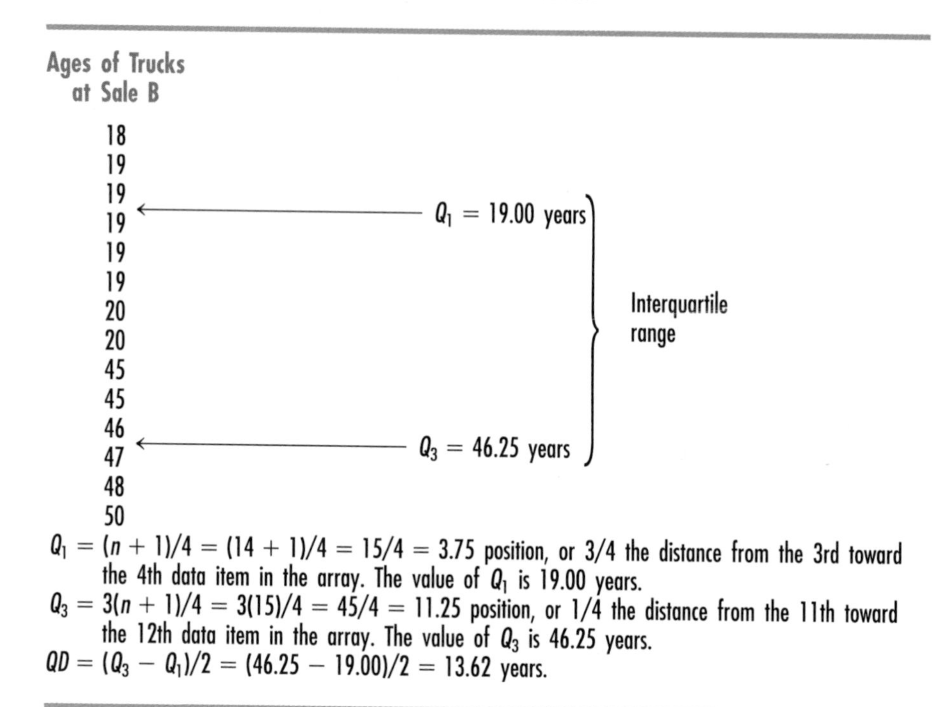

Ages of Trucks at Sale B

18
19
19
19 ← Q_1 = 19.00 years
19
19
20
20
45
45
46
47 ← Q_3 = 46.25 years
48
50

$Q_1 = (n + 1)/4 = (14 + 1)/4 = 15/4 = 3.75$ position, or 3/4 the distance from the 3rd toward the 4th data item in the array. The value of Q_1 is 19.00 years.
$Q_3 = 3(n + 1)/4 = 3(15)/4 = 45/4 = 11.25$ position, or 1/4 the distance from the 11th toward the 12th data item in the array. The value of Q_3 is 46.25 years.
$QD = (Q_3 - Q_1)/2 = (46.25 - 19.00)/2 = 13.62$ years.

Thus, the location of Q_1 is shown in Table 3.16 to be the 3.75 position in the array—that's three-quarters of the way from the 3rd value toward the 4th value. Since both the 3rd and 4th values equal 19 years, the value of Q_1 is also 19 years. The location of Q_3 is shown in Table 3.16 to be the 11.25 position in the array. Since the 11th value is 46 years, and the 12th value is 47, the value of Q_3 is interpolated to be .25 of the distance between 46 and 47. So Q_3 is 46.25, as you can verify in the *Minitab* printout in Figure 3.12.

We now know that about the middle 50 percent of the trucks at sale B range in age from 19 to 46.25 years, and the interquartile range ($Q_3 - Q_1$) of 27.25 years shows that the middle 50 percent of the trucks varied with an age spread of 27.25 years.

The quartile deviation is simply *one-half the interquartile range.* That is:

$$\text{QD} = \frac{(Q_3 - Q_1)}{2} = \frac{(46.25 - 19.00)}{2} = 13.62 \text{ years}$$

The smaller the QD, the greater the concentration of the middle half of the observations in the data set. If the data in a set can be used to create a symmetrical and bell-shaped distribution, then 50 percent of the values are found in the range of the median ± 1 QD because the values of Q_1 and Q_3 are equal distances from the median. This relationship can also be used for analysis purposes with distributions that are approximately normal. Thus, in our Slimline example discussed earlier, we can conclude that approximately the middle 50 percent of the employees sold syrup quantities between median ± 1 QD. You can verify from the data array in Table 3.2, page 45, that the position of Q_1 is $(50 + 1)/4$ or 12.75, and 0.75 of the distance between the 12th value (104.50 gallons) and the 13th value (105.00 gallons) is 104.87 gallons. Likewise, the position of Q_3 is $3(50 + 1)/4$ or 38.25, and 0.25 of the distance between the 38th value (125.25 gallons) and the 39th value (126.00 gallons) is 125.44 gallons. Thus, the QD for this data set is (125.44 − 104.87)/2 or 10.28, and the median is 115.25. Can you now verify from Table 3.2 that about the middle 50 percent of the sales figures are found between the median (115.25) ± the quartile deviation (10.28), or between 104.97 and 125.53 gallons?

Box-and-Whiskers Display. You saw earlier that stem-and-leaf displays and dotplots are exploratory data-analysis tools that analysts use to get a feel for the data being studied. Another graphic technique used in exploratory analysis is the box-and-whiskers display. A **box-and-whiskers display** (also called a **boxplot**) shows the middle half of the values in a data set—the values that lie in the interquartile range—as a *box* and then draws lines, or *whiskers*, extending to the left and right from the box to indicate the remaining 50 percent of the data items. Figure 3.15 shows the boxplot generated by the *Minitab* software package when supplied with our Slimline Company data set.

As you can see in Figure 3.15, the box is a rectangle. The line inside the box shows the median location. Each end of the box is called a **hinge**. For our purposes, the left or **lower hinge** of a box is basically the same as Q_1, and the right or **upper hinge** is essentially the same as Q_3. (Different approaches are used in texts and software packages to define and calculate hinges and quartiles. The results are only slightly different, so we'll not go into these details.) To understand hinge terminology a little better, try this experiment: (1) Put a column of evenly spaced and arrayed numbers on a sheet of paper, with the lowest value placed at the top edge and the largest value placed at the bottom edge of the sheet. (2) Fold the paper in the middle so it's half as long. (3) Now fold the paper in the middle again, crease it, and then unfold it to its original shape. You'll find that the two outer creases—or hinges—should be about where the Q_1 and Q_3 data values are located, and the middle crease should locate the median. In other words, in this experiment the median divides the entire data set into a lower and upper half, the lower hinge is the median of the lower half, and the upper hinge is the median of the upper half.

The boxplot thus gives analysts a quick pictorial representation of what's sometimes called a **five-number summary** of the data set. As we've seen, these numbers are the median, the two hinges, and the smallest and largest values. Boxplots can also tell at a

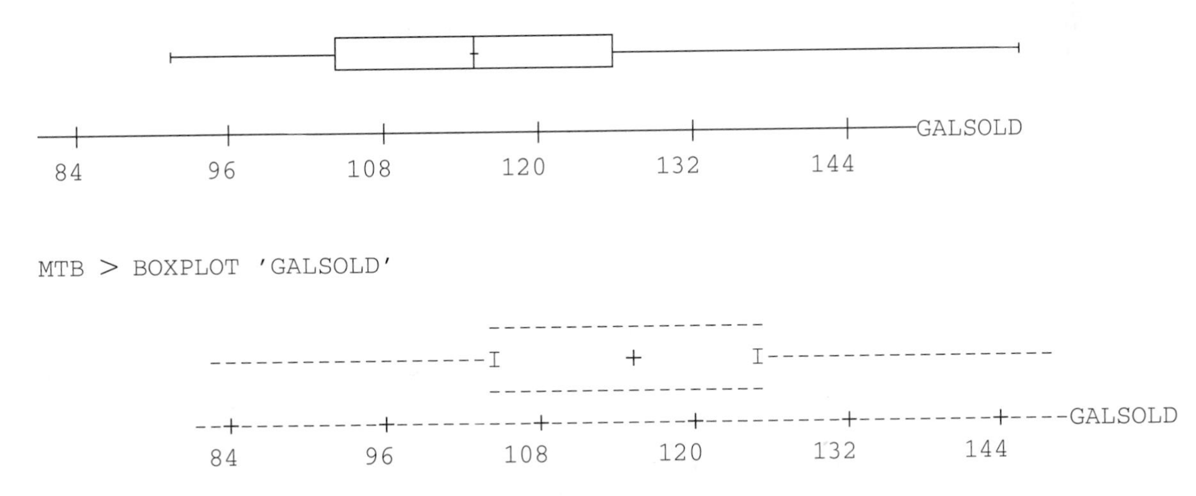

FIGURE 3.15 Two versions of the *Minitab* box-and-whiskers display of the Slimline Company data set.

How Do You Measure a Skew?

Many measurements in the social sciences tend to produce skewed distributions. For example, if the data from many thousands of marriages are analyzed, a distribution of the ages of bridegrooms will peak when the grooms are in their 20s. This distribution will then tail off to the right (or in a positive direction) as the ages of bridegrooms increase. The separation of mean and median values in skewed situations may be used to calculate a *coefficient of skewness (Sk)*. This measure gives the direction (negative or positive) of the skew, and also indicates its degree with the formula:

$$Sk = \frac{3(\bar{x} - Md)}{s}$$

If the mean and median are the same (no skew present), the formula gives a value of zero because $(\bar{x} - Md) = 0$. As the mean and median become separated, the value of *Sk* moves from zero in a positive or negative direction (it will seldom exceed ± 1.00). The skew in our Slimline Beverage Company example is:

$$Sk = \frac{3(115.4 - 115.25)}{14.91} = +.030,$$

a small degree of positive skewness.

glance if a data set is reasonably symmetrical, or if it's skewed. How? A median mark that's about in the center of the box, as it is in the plot in Figure 3.15, tells the analyst that the data set is reasonably symmetrical. But a median mark that's to the right side of the box indicates negative skewness, while a mark that's placed on the left side of the box suggests positive skewness. Also, skewness is indicated if one whisker line is appreciably longer than the other. For example, a longer right whisker suggests positive skewness. Data items that are extremely large or extremely small compared to the rest of the data set are called *outliers*, and these outliers are often identified by special symbols (for example an asterisk) on the boxplot.

Summary of Comparative Characteristics

There's no rule that will always identify the proper dispersion measure to use. In picking a measure, the characteristics of each must be considered, and the type of data available must be evaluated. A summary of comparative characteristics for each measure discussed in the preceding pages is given below.

The Range. Some of the characteristics of the range are:

1. *It's the easiest measure to compute.* Since its calculation involves only one subtraction, it is also the easiest measure to understand.
2. *It emphasizes only the extreme values.* Because the more typical items are completely ignored, the range may give a very distorted picture of the true dispersion pattern.

The Mean Absolute Deviation. Some of the characteristics of the MAD are:

1. *It gives equal weight to the deviation of every observation.* Thus, it's more sensitive than measures such as the range or quartile deviation that are based on only two values.
2. *It's an easy measure to compute.* It is also not difficult to understand.
3. *It's not influenced as much by extreme values as the standard deviation.* The squaring of deviations in the calculation of the standard deviation places more emphasis on the extreme values.
4. *Its use is limited in further calculations.* Because the algebraic signs are ignored, the average deviation is not as well suited as the standard deviation for further computations.
5. *It can't be computed from an open-ended distribution.*

The Standard Deviation. Included among the characteristics of the standard deviation are:

1. *It's the most frequently encountered measure of dispersion.* Because of the mathematica properties it possesses, it's more suitable than any other measure of dispersion for further analysis involving statistical inference procedures. We shall use the standard deviation extensively in later chapters.
2. *It's a computed measure whose value is affected by the value of every observation in a series.* A change in the value of any observation will change the standard deviation value.

3. *Its value may be distorted too much by a relatively few extreme values.* Like the mean, the standard deviation can lose its representative quality in badly skewed data sets.
4. *It can't be computed from an open-ended distribution in the absence of additional information.* As formulas 3.7 and 3.8 show, if the mean cannot be computed, neither can the standard deviation.

The Quartile Deviation. Some of the characteristics of the quartile deviation are:

1. *It's similar to the range in that it's based on only two values.* As we've seen, these two values identify the range of the middle 50 percent of the values.
2. *It's frequently used in badly skewed data sets.* The quartile deviation will not be affected by the size of the values of extreme items, and so it may be preferable to the average or standard deviation when a data set is badly skewed.
3. *It may be computed in an open-ended distribution.* Since the upper and lower 25 percent of the values are not considered in the computation of the quartile deviation, an open-ended distribution presents no problem.

Self-Testing Review 3.5

1–5. Let's suppose that Peter Parker repeats the statistics course, and this time his grades are:

50 85 45 60 1 85 50 85 55

1. What is the population mean of this set of grades?
2. What is the range of Peter's grades?
3. What is the value of the mean absolute deviation?
4. What is the value of the population variance?
5. What is the value of the population standard deviation?

6–12. Let's again consider Bill Alott's GOTCHA report that shows the number of billable hours that his staff has clocked (these are but a sample of the hours clocked by Global consultants). For your convenience, we've reproduced below the frequency distribution and grouped data sample mean calculations that you prepared in Self-Testing Review (STR) 3.2 and 3.4:

Class	Midpoint (m)	Frequency (f)	fm
25 < 30	27.5	1	27.5
30 < 35	32.5	3	97.5
35 < 40	37.5	10	375.0
40 < 45	42.5	15	637.5
45 < 50	47.5	11	522.5
50 < 55	52.5	5	262.5
55 < 60	57.5	4	230.0
60 < 65	62.5	1	62.5
		50	2,215

$$\bar{x} = \frac{\Sigma fm}{n} = \frac{2{,}215}{50} = 44.3$$

6. Use the grouped data formula to estimate the value of the sample variance.
7. Estimate the value of the sample standard deviation.
8. From Chebyshev's theorem, we know that *at least* 75 percent of all billable hours in Bill's GOTCHA report will fall between what values?
9. From Chebyshev's theorem, we know that *at least* 89 percent of all billable hours in the report will fall between what values?
10. The histogram and frequency polygon for Bill's data that you prepared in STR 3.3, problems 1 and 2, show that the data are approximately symmetrical and bell shaped. Thus, using the empirical rule, we can say that about 68 percent of the values will fall between what values?
11. About 95 percent of the billable hours will fall between what values?
12. About 99.7 percent of the billable hours will fall between what values?

13–14. In STR 3.4, problem 24, you used a frequency distribution to compute the sample mean for the orders placed by customers of the Lee Health Foods Company.

13. Now, compute the sample standard deviation for the Lee Company distribution.
14. Use the empirical rule and the values of the mean and standard deviation to analyze this distribution.

15–16. In STR 3.4, problem 26, you used a frequency distribution to compute the sample mean of share prices of common stock.

15. Now, compute the sample standard deviation for this distribution.
16. Use Chebyshev's theorem and the values of the mean and standard deviation to analyze the distribution.

17–18. In STR 3.4, problem 28, you used a frequency distribution to compute the mean GMAT test score posted by students at the 25 top-ranked graduate business schools.

17. Now, compute the standard deviation GMAT value for this population of top-ranked schools.
18. Use the empirical rule and the values of the population mean and standard deviation to analyze this data set.

19–22. For the data on 17 stock "buys" recommended by *Money* magazine that were presented in STR 3.4, problems 34 and 35:

19. Calculate the range.
20. Compute the mean absolute deviation.
21. Compute the sample variance.
22. Compute the sample standard deviation.

23–26. For the data on starting salaries for graduates from five top graduate departments of finance that were presented in STR 3.4, problems 36 and 37:

23. Calculate the range.
24. Calculate the mean absolute deviation.

Company	Rating Score	Company	Rating Score
3M Corporation	8.12	Eastman Kodak	6.19
Xerox	6.67	EG&G	6.18
Bausch & Lomb	6.64	Honeywell	5.96
Becton Dickinson	6.46	Tektronix	5.64
Baxter Internatl.	6.40	Polaroid	5.60

60. Compute the mean rating score (assume that this is a population of interest).

61. Compute the median rating score.

62. Compute the range.

63. Determine the mean absolute deviation.

64. Compute the population standard deviation.

65–72. The following scores were made by Professor Shirley A. Meany's accounting students on a test:

68 52 49 56 69 74 41 59 79
81 42 57 60 88 87 47 65 55
68 65 50 78 61 90 85 65 66
72 63 95

65. Arrange the grades in an ascending array.

66. What is the range of the values?

67. Organize the grade data into a frequency distribution having classes "40 < 50," "50 < 60," . . . , "90 < 100."

68. Use the grouped data formula to estimate the population mean.

69. Use the grouped data formula to estimate the population standard deviation.

70. Compute the interquartile range.

71. Compute the quartile deviation.

72. Construct a stem-and-leaf display of the grades.

73–77. *Nation's Business* has published results of a study made to determine an employer's health care costs for various plans. The following data set gives the average cost per employee (in dollars) for a health maintenance organization (HMO) plan in 12 major cities:

City	Cost ($)	City	Cost ($)
Atlanta	3,259	Minneapolis/St. Paul	2,673
Chicago	3,133	New York Metro.	3,254
Cleveland	3,465	Philadelphia	2,882
Dallas/Ft. Worth	2,963	Richmond	2,448
Houston	3,295	San Francisco	2,939
Los Angeles	3,025	Seattle	2,624

73. Compute the sample mean cost.

74. Compute the median cost.

75. Compute the interquartile range and the QD.

76. Determine the range.

77. Compute the sample standard deviation.

78–81. Many American drug producers have operations in Puerto Rico. *The New York Times* recently published a list of 10 such companies and gave the number of employees that each had in Puerto Rico. The data are:

Company	Employees	Company	Employees
Johnson & Johnson	2,829	American Home Prod.	1,301
Abbott Laboratories	2,359	Eli Lilly	1,267
Bristol-Myers Squibb	1,784	Smithkline Beecham	873
Warner-Lambert	1,569	Merck	655
Schering-Plough	1,517	Pfizer	750

78. Compute the sample mean number of employees.

79. Compute the median number of employees.

80. Determine the range.

81. Compute the sample standard deviation.

82–86. Fast food is a fact of life, but such food is often saturated with fat. Healthy selections were made from the breakfast menus at four popular fast-food restaurants, and the number of calories in each selection was determined. The data are:

Restaurant	Calories
Burger King	615
Jack-In-The-Box	387
McDonald's	567
Wendy's	440

82. Compute the sample mean number of calories.

83. Compute the median number of calories.

84. Compute the mean absolute deviation.

85. Determine the range.

86. Compute the sample standard deviation.

87–91. The average temperatures for 10 randomly selected cities in the eastern United States on June 17 are listed below:

City	Temperature (°F)	City	Temperature (°F)
Atlanta	89	Hartford	84
Baltimore	82	Miami Beach	88
Boston	76	New York	80
Burlington, Vt.	81	Orlando	91
Columbia, S.C.	87	Philadelphia	82

87. Compute the sample mean temperature.

88. Compute the median temperature.

89. Determine the mode.

90. Determine the range.

91. Compute the sample standard deviation.

92–96. Brock Lee, president of Lee Health Foods Company, prepared the following distribution of the miles traveled last year (rounded to the nearest whole mile) by a sample of 96 route trucks:

Miles Traveled	Number of Trucks
5,000 < 7,000	5
7,000 < 9,000	10
9,000 < 11,000	12
11,000 < 13,000	20
13,000 < 15,000	20
15,000 < 17,000	14
17,000 < 19,000	11
19,000 < 21,000	4

92. Construct a histogram and a frequency polygon of this mileage distribution.

93. Construct an ogive, and graphically locate the value of the median.

94. Estimate the mean number of miles traveled.

95. Find the median class.

96. Use the grouped data formula to estimate the sample standard deviation.

97–99. The following are the figures for absences in Dr. Flower's environmental science class:

Days Absent	No. of Students
1 < 4	9
4 < 7	6
7 < 10	13
10 < 13	9
13 < 16	1

97. Estimate the mean number of days absent with a grouped data formula.

98. What is the median class for the number of days absent?

99. Estimate the population standard deviation for the number of days absent.

100. Greg's test scores for two semesters of calculus are listed below (the percentage of each semester's grade represented by each score is also given):

Calculus I	Calculus II	Percent of Grade
72	87	15
68	68	15
74	71	15
68	69	15
90	88	40

Compute the weighted arithmetic mean for each semester.

101. A study in *The Journal of Abnormal Psychology* reported on the results obtained with a family assessment device that was given to functional and dysfunctional families. On the Family Environment Scale for Cohesion, the *functional* families had a mean score of 7.17 scale points and a standard deviation of 1.49. According to Chebyshev's theorem, at least 75 percent of all functional families should score between what two values?

102. According to Chebyshev's theorem, at least 89 percent of all functional families should score between what two values?

103. When the Family Environment Scale for Cohesion was measured for a sample of *dysfunctional families*, there was a mean of 5.57 scale points and a standard deviation of 2.49. According to Chebyshev's theorem, at least 75 percent of all dysfunctional families should score between what two values?

104. According to Chebyshev's theorem, at least 89 percent of all dysfunctional families should score between what two values?

105. To prepare a government report, a university must determine the percentage of men and women faculty members in its several colleges. The faculty data are as follows:

Faculty	College A	College B	College C	College D
Men	148	64	12	102
Women	32	42	26	48

What is the percentage of women faculty members in each college?

106–107. The following scores were made by students on a recent statistics test:

Test Scores	Number of Students
40 and under 50	10
50 and under 60	5
60 and under 70	7
70 and under 80	3
80 and under 90	16
90 and under 100	12

106. Compute the mean test score.

107. Compute the population standard deviation.

108–113. The miles traveled by a sample of six students attending an evening class at a local college are as follows:

Student	Miles Traveled
A	1
B	4
C	9
D	8
E	5
F	6

108. What is the mean distance traveled?

109. What is the median distance traveled?

110. What is the modal distance traveled?

111. What is the mean absolute deviation for this data set?

112. What is the sample variance?

113. What is the sample standard deviation value?

114–115. The number of traffic tickets issued for a sample period by five Crossville police officers is as follows:

Officer Name	No. of Tickets
R. Oldman	16
A. Trapper	9
L. Perez	10
J. Ketchum	8
F. Wheeler	5

114. What is the sample mean number of tickets written?

115. What is the median number of tickets written?

116–120. After six holes, a sample of 25 golfers at the Duffers International Tournament had the following scores:

71	68	85	96	12	92	37	41	54	25
66	15	73	23	14	55	65	43	88	92
19	22	51	62	84					

116. Array the scores in an ascending array.

117. What is the range of the scores?

118. What is the sample mean for the above scores?

119. What is the median for the above scores?

120. What is the sample standard deviation value for the scores?

121. A study is conducted to find the mean income of a population of salespersons. The study concludes that for the population of 100 salespersons the mean income is $33,000. It's discovered later that the income of the last person in the group was incorrectly reported to be $20,000 when it should have been $50,000. What's the true mean income of the population?

122–125. The following scores were made by a sample of Professor Shirley A. Meany's accounting students who dared to take an optional makeup test:

72 65 43 50 68 62

122. What is the range for this data set?

123. What is the mean absolute deviation?

124. What is the variance?

125. What is the sample standard deviation?

126–130. Entrance exam scores for a sample of applicants to Jeopardy University are tabulated below:

990 1,403 1,059 1,213 763 1,352
898 999 1,181 1,264 269 428 582
381 1,141 760 455 345

126. Compute the range of these scores.

127. Compute the sample mean score.

128. Compute the mean absolute deviation.

129. What is the value of the sample variance?

130. What is the value of the sample standard deviation?

131. In Chemistry I there are two tests that are each worth 20 percent of the term grade, a midterm worth 25 percent of the grade, lab assignments that account for 5 percent of the grade, and a final exam that's worth 30 percent of the grade. Ella Meant scored 82 and 75 on the two tests, 87 on the midterm, 75 on the lab work, and 85 on the final. What is her weighted mean grade for the term?

Topics for Review and Discussion

1. Discuss the differences between discrete data and continuous data. Give three examples of each type of data.

2. What are the advantages of putting raw data into an array? What information can be readily found from an array?

3. Give an advantage of organizing data into a frequency distribution. What is a disadvantage?

4. Discuss some basic criteria to use when organizing a frequency distribution.

5. What difficulty may be encountered when using open-ended classes in distributions?

6. What is a histogram? How does it differ from a bar graph?

7. What is a frequency polygon? What are the advantages of using a frequency polygon?

8. Discuss the differences and similarities of a stem-and-leaf display and a histogram.

9. Discuss several situations where a stem-and-leaf display should not be used.

10. What is a cumulative frequency distribution? How does an ogive give a picture of a frequency distribution?

11. Discuss the three measures of central tendency described in this chapter. What are the characteristics of each?

12. What basic assumption is needed to approximate the arithmetic mean from grouped data?

13. Discuss the relationship between the mean, median, and mode in a symmetric, mound-shaped distribution. Discuss the relationship among the three measures of central tendency when the distribution is positively skewed. What about the relationship when it is negatively skewed?

14. Discuss the differences in qualitative and quantitative data. What types of statistics can only be found with quantitative data? How can we describe quantitative data?

15. Why is it necessary to measure dispersion to describe a data set?

16. Why is it necessary to eliminate the algebraic signs of the deviations when computing measures of dispersion? Discuss the different ways of doing this when computing the mean absolute deviation and the standard deviation.

17. Discuss the relationship between the variance and the standard deviation. Why is the standard deviation a more commonly used measure?

18. Identify the main characteristics of the range, mean absolute deviation, standard deviation, and quartile deviation.

19. Discuss the key elements of a box-and-whisker plot. Where is the interquartile range in such a plot?

Projects/Issues to Consider

Go to your school library and obtain a set of raw data (not statistics) that are of interest to you. You might want to use a current periodical or a journal from your field of study. Write a report describing the data, and include the following in your report:

1. Identify your source. Include the date of publication.

2. What is the population for your data? How was the sample obtained?

3. Arrange the data in array form.

4. Find the range, Q_1, the median, and Q_3.

5. Find the interquartile range and the quartile deviation.

6. Does your data have a mode?

7. Construct a box-and-whisker plot.

8. Construct a stem-and-leaf display of your data. You may find it easier to round off your data for a better display.

9. Construct a frequency distribution using a convenient number of classes.

10. Construct a histogram and a frequency polygon.

11. Construct an ogive.

12. Compute an approximation of the mean and standard deviation using grouped data formulas.

13. From your histogram and your box-and-whisker plot, discuss the shape of your distribution. Is it symmetric, skewed left, or skewed right?

Computer Exercises

1–5. After you've retrieved the computer file named TAXES that you created for the Chapter 1 computer exercise, use your software package to produce:

1. A default histogram (one the package produces automatically), and then construct a histogram with 7 bars.

2. A stem-and-leaf display for this data set.

3. A dotplot for this data set.

4. A boxplot for this data set.

5. The basic statistical measures (mean, median, standard deviation, minimum value, maximum value, Q_1, Q_3, . . .) for this data set.

6–10. Locate the billable hours for the sample of 50 workers that Bill Alott used to prepare his GOTCHA report in STR 3.2, problems 1 to 5. Now, enter these 50 data items into your software package to produce:

6. An array of the data.

7. A histogram for this data set.

8. A stem-and-leaf display for this data set.

9. A boxplot for this data set.

10. The basic statistical measures (mean, median, standard deviation, minimum value, maximum value, Q_1, Q_3, . . .) for this data set.

(You may want to save this data set since it will be used in computer exercises sections in later chapters.)

11–15. Locate the sales order data for the sample of 30 customers of the Lee Health Foods Company in STR 3.2, problems 9 to 13. Now, enter these 30 data items into your software package to produce:

11. An array of the data.

12. A histogram for this data set.

13. A stem-and-leaf display for this data set.

14. A boxplot for this data set.

15. The basic statistical measures (mean, median, standard deviation, minimum value, maximum value, Q_1, Q_3, . . .) for this data set.

16–20. Enter the *Consumer Reports* bathroom scale price data used for end-of-chapter exercises 1 to 10 into your software package and then produce:

16. An array of the data.

17. A histogram for this data set.

18. A stem-and-leaf display for this data set.

19. A boxplot for this data set.

20. The basic statistical measures (mean, median, standard deviation, minimum value, maximum value, Q_1, Q_3, . . .) for this data set.

21–25. A listing of a year's tuition at the 25 top-ranked graduate schools of business is given for end-of-chapter exercises 11 to 20. Use these tuition figures and your software package to produce:

21. An array of the data.

22. A histogram for this data set.

23. A stem-and-leaf display for this data set.

24. A boxplot for this data set

25. The basic statistical measures (mean, median, standard deviation, minimum value, maximum value, Q_1, Q_3, . . .) for this data set.

26–30. A tally of the daily sales figures for The Children's Place was provided for end-of-chapter exercise 21 to 30. Use these sales figures and your software package to produce:

26. An array of the data.

27. A histogram for this data set.

28. A stem-and-leaf display for this data set.

29. A boxplot for this data set.

30. The basic statistical measures (mean, median, standard deviation, minimum value, maximum value, Q_1, Q_3, . . .) for this data set.

Answers to Odd-Numbered Self-Testing Review Questions

Section 3.1

1. Numeric, continuous *3.* Numeric, discrete *5.* Attribute *7.* Numeric, continuous *9.* Numeric, continuous *11.* Numeric, discrete

Section 3.2

1. 25, 30, 32, 32, 35, 36, 37, 37, 37, 38, 38, 38, 38, 38, 40, 40, 41, 41, 42, 42, 42, 43, 43, 43, 43, 44, 44, 44, 44, 46, 46, 46, 47, 47, 47, 47, 48, 48, 49, 49, 50, 50, 51, 51, 54, 55, 56, 57, 58, 64

3.

Billable Hours	Number of Employees (Frequencies)
25 < 30	1
30 < 35	3
35 < 40	10
40 < 45	15
45 < 50	11
50 < 55	5
55 < 60	4
60 < 65	1
	50

5. With 6 classes, you can use a width of 7, which is a round-up of 39/6. Classes could be 25 and less than 32, that is, 25 < 32, 32 < 39, . . . , 61 < 68. With 10 classes, you can use a width of 4, which is a round-up of 39/10. Classes could be 25 < 29, 29 < 33, . . . , and so on.

7. The range = \$11.18 − (−\$1.36) = \$12.54

9. 21.50 25.35 29.00 30.30 31.25 34.60 35.55 36.50 36.60 38.50 39.95 40.25 41.10 42.30 42.50 43.50 44.80 45.00 45.10 46.00 47.75 49.50 49.80 52.10 55.00 55.60 56.30 58.00 59.20 65.50

11.

Size of Orders (\$)	Number of Orders (Frequencies)
20 < 30	3
30 < 40	8
40 < 50	12
50 < 60	6
60 < 70	1
	30

13. Use a class width of 6, the round-up of 44/8.

15. The range = \$131 − \$25 = \$106

17. 597 601 602 606 608 609 610 617 620 620 630 630 631 635 635 635 640 640 642 643 644 644 650 657 675

19.

Average GMAT Score	Number of Schools (Frequencies)
595 < 605	3
605 < 615	4
615 < 625	3
625 < 635	3
635 < 645	9
645 < 655	1
655 < 665	1
665 < 675	0
675 < 685	1
	25

21. The range = 21 − 1 = 20 weeks.

23. \$1,632 2,945 3,539 3,846 4,038 4,284 4,591 4,647 4,694 5,292 5,354 5,552 5,752 5,792 5,948 5,981 6,047 6,531 6,637 6,712 6,744 6,781 6,907 6,935 6,969 7,006 7,074 7,125 7,217 7,263 7,268 7,301 7,371 7,493 7,605 7,634 7,728 7,751 7,892 7,962 8,272 8,311 8,314 8,389 8,390 8,568 8,611 8,764 8,770 9,348 10,016

25.

Taxes (in \$)	Number of States (Frequencies)
1,600 < 2,800	1
2,800 < 4,000	3
4,000 < 5,200	5
5,200 < 6,400	8
6,400 < 7,600	17
7,600 < 8,800	15
8,800 < 10,000	1
10,000 < 11,200	1
	51*

*Includes the District of Columbia.

Section 3.3

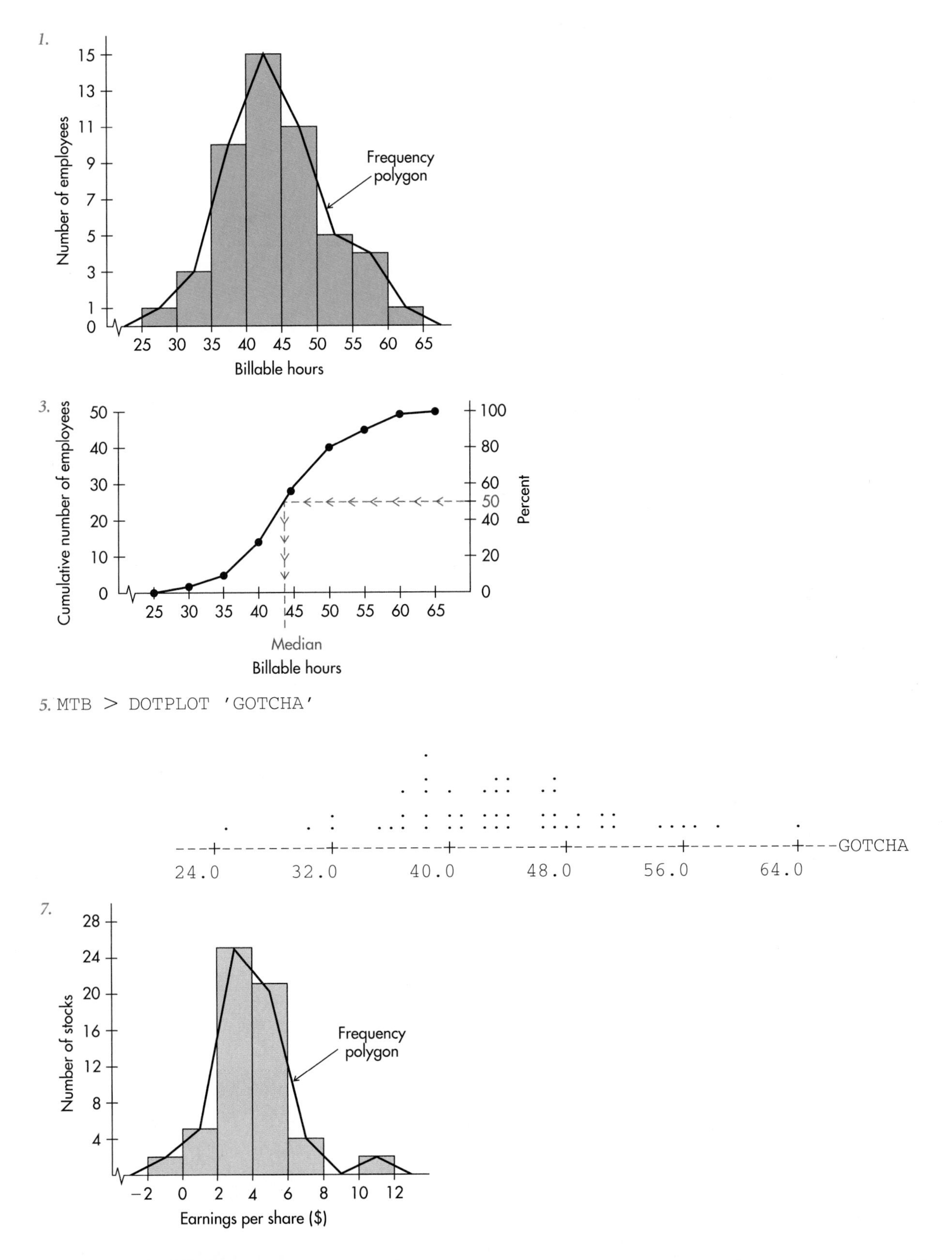
1.
Number of employees
15
13
11
9
7
5
3
1
0
Frequency polygon
25 30 35 40 45 50 55 60 65
Billable hours
3.
Cumulative number of employees
50
40
30
20
10
0
100
80
60
50
40
20
0
Percent
25 30 35 40 45 50 55 60 65
Median
Billable hours
5. MTB > DOTPLOT 'GOTCHA'
GOTCHA
24.0 32.0 40.0 48.0 56.0 64.0
7.
Number of stocks
28
24
20
16
12
8
4
Frequency polygon
−2 0 2 4 6 8 10 12
Earnings per share ($)

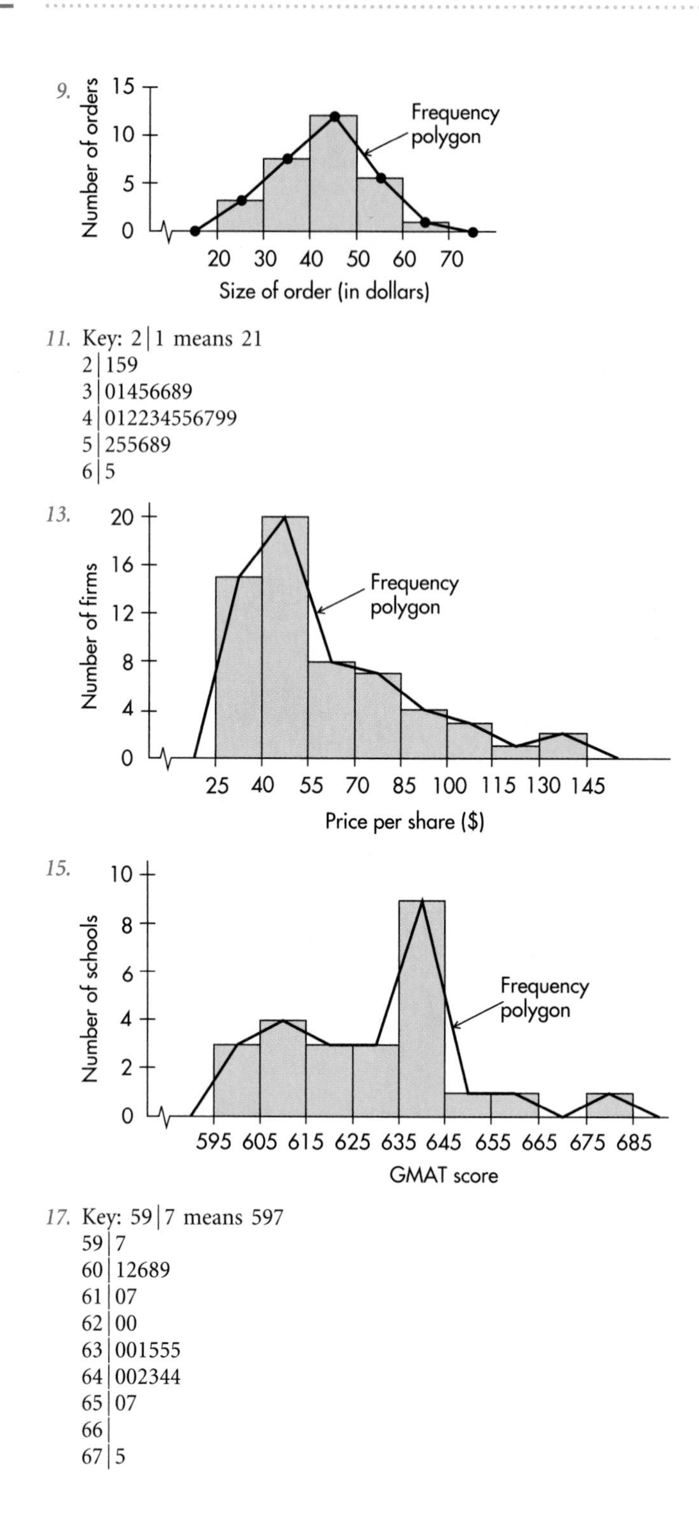

11. Key: 2|1 means 21

2|159
3|01456689
4|012234556799
5|255689
6|5

17. Key: 59|7 means 597

59|7
60|12689
61|07
62|00
63|001555
64|002344
65|07
66|
67|5

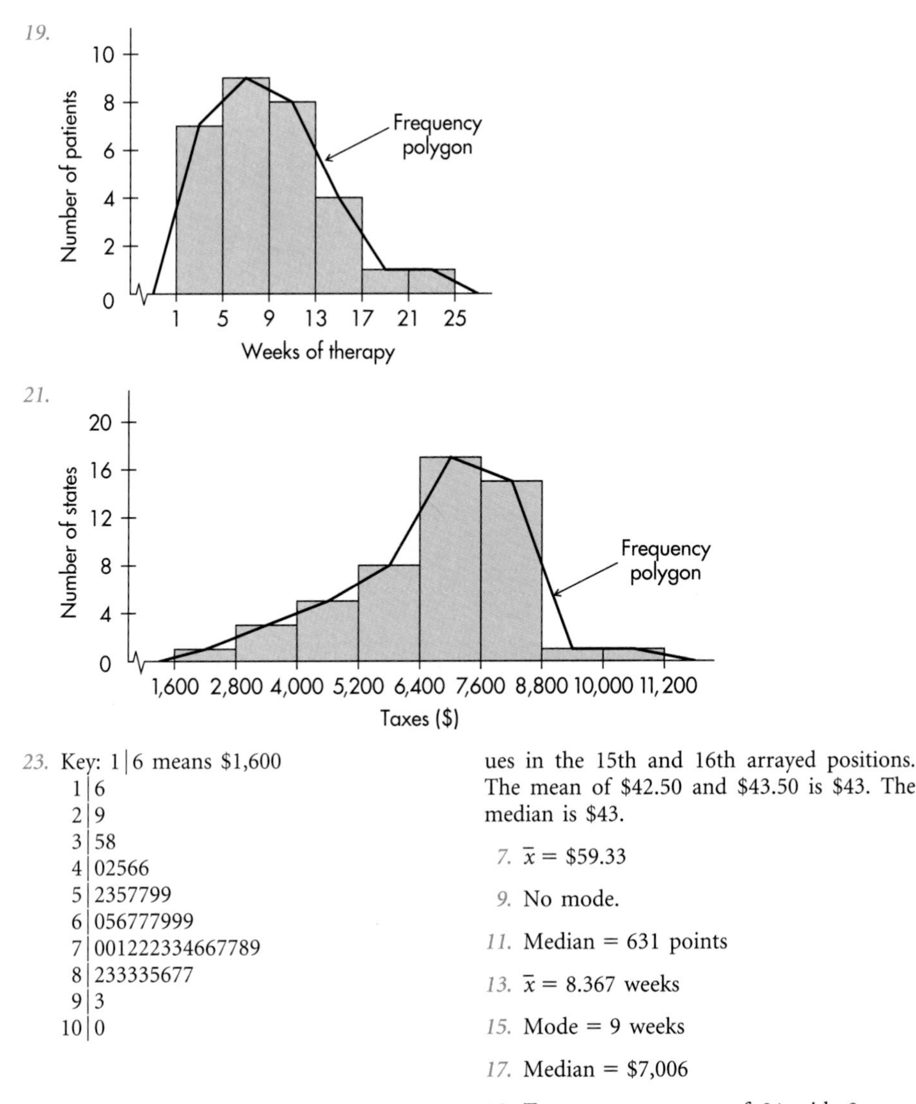

23. Key: 1 | 6 means $1,600

Stem	Leaves
1	6
2	9
3	58
4	02566
5	2357799
6	056777999
7	001222334667789
8	233335677
9	3
10	0

Section 3.4

1. $\Sigma x = 2{,}193$ and $n = 50$, so $\bar{x} = 2{,}193/50 = 43.86$ hours.

3. The mode is 38 (value occurs five times).

5. The median value is the mean of the values in the 15th and 16th arrayed positions. The mean of $42.50 and $43.50 is $43. The median is $43.

7. $\bar{x} = \$59.33$

9. No mode.

11. Median = 631 points

13. $\bar{x} = 8.367$ weeks

15. Mode = 9 weeks

17. Median = $7,006

19. To earn an average of 84 with 9 tests, Peter's classmate must have a total of 84×9 or 756 points. The given values add to 662 points. Since an additional 94 points ($756 - 662 = 94$) is needed, the instructor deleted a grade of 94.

21. Median = 34 assemblies

23.

Hours	Number of Employees (Frequencies)	Midpoints (*m*)	*fm*
25 < 30	1	27.5	27.5
30 < 35	3	32.5	97.5
35 < 40	10	37.5	375.0
40 < 45	15	42.5	637.5
45 < 50	11	47.5	522.5
50 < 55	5	52.5	262.5
55 < 60	4	57.5	230.0
60 < 65	1	62.5	62.5
	50		2,215.0

We can estimate that $\bar{x} = 2{,}215/50 = 44.3$ hours. The true mean using actual values is 43.86 hours. The difference is -0.44 hours.

25.

Size of Orders ($)	Number of Orders (Frequencies)	Midpoint (*m*)	*fm*
20 < 30	3	25	75
30 < 40	8	35	280
40 < 50	12	45	540
50 < 60	6	55	330
60 < 70	1	65	65
	30		1,290

We can estimate that $\bar{x} = \$1{,}290/30 = \$43$. The real sample mean using individual values is $43.28. The difference is $0.28.

27.

Price Per Share ($)	Number of Firms (Frequencies)	Midpoint (*m*)	*fm*
25 < 40	15	32.50	487.50
40 < 55	19	47.50	902.50
55 < 70	8	62.50	500.00
70 < 85	7	77.50	542.50
85 < 100	4	92.50	370.00
100 < 115	3	107.50	322.50
115 < 130	1	122.50	122.50
130 < 145	2	137.50	275.00
	59		3,522.50

The estimated $\bar{x} = \$3{,}522.50/59 = \$59.70$. The true mean is $59.33, so there's a difference of $-\$0.37$.

29.

Average GMAT Score	Number of Schools (Frequencies)	Midpoint (m)	fm
595 < 605	3	600	1,800
605 < 615	4	610	2,440
615 < 625	3	620	1,860
625 < 635	3	630	1,890
635 < 645	9	640	5,760
645 < 655	1	650	650
655 < 665	1	660	660
665 < 675	0	670	0
675 < 685	1	680	680
	25		15,740

The estimated $\mu = 15{,}740/25 = 629.6$ score points. The true population mean is 628.84. The difference is -0.76 point.

31.

Weeks of Therapy	Number of Patients (Frequencies)	Midpoint (m)	fm
1 < 5	7	3	21
5 < 9	9	7	63
9 < 13	8	11	88
13 < 17	4	15	60
17 < 21	1	19	19
21 < 25	1	23	23
	30		274

The estimated $\bar{x} = 274/30 = 9.133$ weeks. The true mean is 8.367 weeks, so the difference is -0.766.

33.

Taxes (in $)	Number of States (Frequencies)	Midpoint (m)	fm
1,600 < 2,800	1	2,200	2,200
2,800 < 4,000	3	3,400	10,200
4,000 < 5,200	5	4,600	23,000
5,200 < 6,400	8	5,800	46,400
6,400 < 7,600	17	7,000	119,000
7,600 < 8,800	15	8,200	123,000
8,800 < 10,000	1	9,400	9,400
10,000 < 11,200	1	10,600	10,600
	51*		343,800

*Includes the District of Columbia.

The estimated μ = \$343,800/51 = \$6,741.18. The true mean is \$6,698. The difference is −\$43.18.

35. \$38.75

37. \$57,000

39. 285 calories

41. 54.75 years

43. 66 years

45. 5.9 megabytes

47. 540.5 pins

49. 1.723

51. Two modes: 0.5 and 1.7

53. 16 ounces

55. \$2,351

57. 27.455 miles per gallon

59. 6.906 rating points

61. \$110.2 million in savings

Section 3.5

1. $\mu = (50 + 85 + 45 + 60 + 1 + 85 + 50 + 85 + 55)/9 = 516/9 = 57.33$ points

3. MAD = (7.33 + 27.67 + 12.33 + 2.67 + 56.33 + 27.67 + 7.33 + 27.67 + 2.33)/9 = 171.33/9 = 19.0367

5. $\sigma = \sqrt{638} = 25.26$

7.

Hours	No. Employees (Frequencies)	Midpoints (m)	fm	$m - \bar{x}$	$(m - \bar{x})^2$	$f(m - \bar{x})^2$
25 < 30	1	27.5	27.5	−16.8	282.2	282.2
30 < 35	3	32.5	97.5	−11.8	139.2	417.6
35 < 40	10	37.5	375.0	−6.8	46.2	462.0
40 < 45	15	42.5	637.5	−1.8	3.2	48.0
45 < 50	11	47.5	522.5	3.2	10.2	112.2
50 < 55	5	52.5	262.5	8.2	67.2	336.0
55 < 60	4	57.5	230.0	13.2	174.2	696.8
60 < 65	1	62.5	62.5	18.2	331.2	331.2
	50		2,215.0			2,686.0

s = 7.40 hours

9. 44.3 ± 3(7.40) or between 22.1 and 66.5 hours.

11. Since data are symmetrical and mound shaped, 95 percent will fall within 2 standard deviations or between 29.5 and 59.1 hours.

13.

Size of Orders (\$)	No. of Orders (Frequencies)	Midpoint (m)	fm	$m - \bar{x}$	$(m - \bar{x})^2$	$f(m - \bar{x})^2$
20 < 30	3	25	75	−18	324	972
30 < 40	8	35	280	−8	64	512
40 < 50	12	45	540	2	4	48
50 < 60	6	55	330	12	144	864
60 < 70	1	65	65	22	484	484
	30		1,290			2,880

Note: You'll recall that $\bar{x}$ = 1,290/30 = \$43.

s^2 = 2,880/29 = 99.3103, so s = \$9.97.

15.

Price per Share ($)	Number of Firms (Frequencies)	Midpoint (m)	fm	$m - \bar{x}$	$(m - \bar{x})^2$	$f(m - \bar{x})^2$
25 < 40	15	32.50	487.50	−27.20	739.84	11,097.60
40 < 55	19	47.50	902.50	−12.20	148.84	2,827.96
55 < 70	8	62.50	500.00	2.80	7.84	62.72
70 < 85	7	77.50	542.50	17.80	316.84	2,217.88
85 < 100	4	92.50	370.00	32.80	1,075.84	4,303.36
100 < 115	3	107.50	322.50	47.80	2,284.84	6,854.52
115 < 130	1	122.50	122.50	62.80	3,943.84	3,943.84
130 < 145	2	137.50	275.00	77.80	6,052.84	12,105.68
	59		3,522.50			43,415.56

Note: You'll recall that $\bar{x}$ = \$3,522.50/59 = \$59.70.

$s^2 = 43,415.56/58 = 748.5441$, so $s = \$27.36$

17.

Avg. GMAT Score	No. of Schools (Frequencies)	Midpoint (m)	fm	$m - \mu$	$(m - \mu)^2$	$f(m - \mu)^2$
595 < 605	3	600	1,800	−29.60	876.16	2,628.48
605 < 615	4	610	2,440	−19.60	384.16	1,536.64
615 < 625	3	620	1,860	−9.60	92.16	276.48
625 < 635	3	630	1,890	0.40	0.16	0.48
635 < 645	9	640	5,760	10.40	108.16	973.44
645 < 655	1	650	650	20.40	416.16	416.16
655 < 665	1	660	660	30.40	924.16	924.16
665 < 675	0	670	0	40.40	1,632.16	0.00
675 < 685	1	680	680	50.40	2,540.16	2,540.16
	25		15,740			9,296.00

Note: You'll recall that μ = 15,740/25 = 629.6 points.

$\sigma^2 = 9,296/25 = 371.84$, so $\sigma = 19.28$ points

19. Range = \$66.75 − \$23 = \$43.75

21. 278.067

23. Range = \$66,000 − \$53,000 = \$13,000

25. \$5,244

27. Range = 490 − 230 = 260 calories

29. 93.8 calories

31. 8.75

33. Range = 12.4 − 0.19 = 12.21 megabytes

35. 4.12 megabytes

37. 746.0612

39. Range = 4.4 − 0.4 = 4

41. 1.2

43. 8.5238

45. Range = \$2,070

47. \$748

49. 10.2720

51. Range = 1.97 rating points

53. 0.498 rating point

55. \$54.61 million

57. 47.06 and 81.74

59. 52.84 and 75.96

61. 49 and 61°

63. 52 and 58°

65. 46 and 64°

67. 48.25 hours

69. 38.375 hours to 48.625 hours

71. \$77

73. \$33 to \$71

75. 609.5

77. IQR = 33, and QD = 16.5

79.

MTB > BOXPLOT 'GMAT'

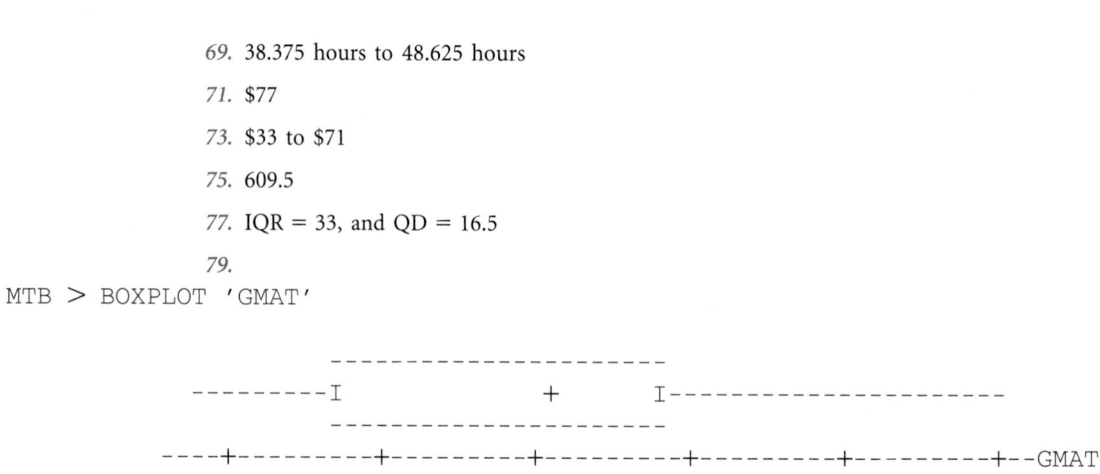

81. 10 weeks

83.

MTB > BOXPLOT 'WEEKS'

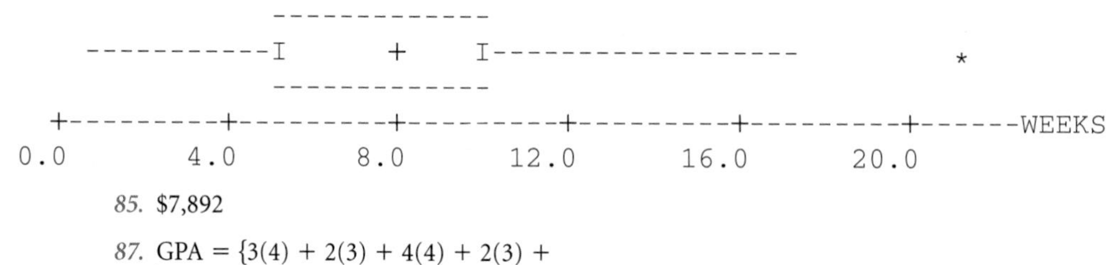

85. \$7,892

87. GPA = $\{3(4) + 2(3) + 4(4) + 2(3) + 1(2)\}/(4 + 3 + 4 + 3 + 2) = 42/16 = 2.625$

Section 3.6

1. The percent of defective microchips in batch 1 = 13.14 percent, in batch 2 = 9.50 percent, in batch 3 = 7.62 percent, in batch 4 = 11.95 percent, and in batch 5 = 13.00 percent.

CHAPTER 4

Probability Concepts

LOOKING AHEAD

In this chapter you'll learn the basic probability concepts that form the foundation for statistical inference. There are many statistical applications for which it is necessary to use probability concepts to make decisions under conditions of uncertainty. For example, probability studies have been made in the health fields that link second-hand smoking to asthma in young children. In the business world, other such studies have sought to identify marketing strategies to attract those in the 18 to 25 age group. And psychologists have used probability studies to try to learn if attention deficit disorder is due to heredity or to environmental factors. In this chapter we'll look at (1) basic probability definitions and the methods used to assign probabilities for simple events, (2) methods of computing probabilities for compound events, and (3) the concepts of random variables, probability distributions, and expected value.

Thus, after studying this chapter, you should be able to:

- ➤ Define probability terms and explain how simple probabilities may be assigned.
- ➤ Explain the concepts of conditional, independent, and mutually exclusive events.
- ➤ Correctly use multiplication and addition rules to perform probability computations.
- ➤ Differentiate between discrete and continuous random variables.
- ➤ Discuss the concept of a probability distribution.
- ➤ Compute the expected value of a discrete random variable.

4.1 Some Basic Considerations

Deep thinkers have often been fascinated by probability concepts. Consider the following quotes:

It is truth very certain that, when it is not in our power to determine what is true, we ought to follow what is most probable.—***René Descartes***

But to us, probability is the very guide of life.—***Bishop Joseph Butler (1736)***

The principal means for ascertaining truth—induction and analogy—are based on probabilities; so that the entire system of human knowledge is connected with the theory of probability.—***Pierre Simon, Marquis de Laplace (1819)***

These thinkers have certainly elevated the importance of probability to a high level. Perhaps when you finish this and the later chapters in the book, you'll agree with them. Before we can discuss the subject, though, we should define a few basic terms.

> A **probability experiment** is any action for which an outcome, response, or measurement is obtained.

Let's suppose you toss a coin and note the side that lands face up. Then you roll a die (one-half of a pair of dice) and count the number of dots on the top. Finally, you choose an employee of a corporation at random and determine which type of health plan the person subscribes to. In each of these instances, you have performed a probability experiment!

That wasn't so bad, so let's look at another term.

A **sample space** is the set of all possible simple outcomes, responses, or measurements of an experiment.

For example, if you toss a coin, the possible outcomes are heads or tails (we don't consider that the coin might land on edge). When you roll a die, all the possible outcomes are 1, 2, 3, 4, 5, and 6. And when you choose an employee from Titan Corporation to determine his or her type of health plan—the firm's choices are a health maintenance organization (HMO), Blue Cross/Blue Shield insurance, or no plan—the sample space is HMO, Blue Cross/Blue Shield, and none.

Now, let's consider two other related terms.

An **event** is a subset of the sample space.

and

A **simple event** is one that can't be broken down any further.

If you toss a die, you could be interested in the event that you roll a 3. Specific events are symbolized with capital letters in this chapter. Thus, we might let T represent the simple event of having the die land on a 3. Or we could simply note that T = {3}. Later in this chapter we'll investigate events that consist of more than one simple outcome. If we roll a die, for example, we might want to analyze the event A that the roll is *less than* a 5. In this case we could say A = {1, 2, 3, 4}. The sample space and event A in this example can be presented with a graphic aid called a *Venn diagram* in this way:

A 1 3 4 2 5 6

As you can see, the sample space = {1, 2, 3, 4, 5, 6}, and event A = {1, 2, 3, 4}.

Assigning Simple Probabilities

The **probability** of a specified event, say, E, may be defined as the relative likelihood of that event occurring. We denote this as *P*(E), which is read as "*P* of E." Let's look now at three methods—*a priori*, relative frequency, and intuition—that can be used to assign probabilities.

Screaming Headlines

The headlines scream that three Americans are killed by terrorists in a foreign land, while the back pages of the newspaper tell of the daily automobile accidents that claim lives. In this context, being killed by a terrorist may seem to be a major risk of traveling abroad, while automobile accidents are the routine results of local travel. But available statistics show that there's only about 1 chance in 700,000 that you'll be killed by a terrorist if you travel abroad, while your chances of being killed near your home in a car wreck are 1 in 5,300. Also, there's 1 chance in 96,000 that if you ride a bicycle you'll be killed in a bicycle crash.

What Are the Chances?

For centuries schizophrenics were believed to be possessed by devils or even angels. Research now indicates that the tendency to develop schizophrenia is hereditary. While the average child has a 1 percent chance of being stricken, the child of a schizophrenic parent faces 10 times those odds. If both parents are affected, the likelihood of a child's being stricken jumps to about 40 percent. But children of parents with schizophrenia raised by adoptive parents who do not have the disorder have a somewhat reduced factor. If one identical twin has the condition, there is only a 50 percent chance the other will. Although genetics play a major role, environmental factors may help trigger the disorder.

A Priori Probability. In some experiments, such as the tossing of a coin or the rolling of a die, the probability of various events can be determined before the fact or ***a priori*** (prior to any statistical experiments). Intuitively we know that when a coin is tossed, the two outcomes are equally likely, so it makes sense to assign a probability of 1/2 or .5 to each of the two outcomes. This doesn't mean that if you flipped a coin twice, you would get one head and one tail. It does mean, though, that if you flipped the coin a large number of times, the proportion of heads would approach .5.

You might try to flip a quarter 10 times and record the number of heads. The *Minitab* program was used to simulate this experiment, and it produced the sample results shown in Figure 4.1 (0 = head, and 1 = tail). For this experiment, we got 6 heads in 10 "tosses" for a proportion of .6. Note that the first 4 tosses were heads. But as the number of trials increased, the proportion of heads got closer to .50. We then had *Minitab* simulate tossing the quarter 500 times. The results are given in Figure 4.2. This time we got 245 heads in 500 trials for a relative proportion of .49. Continuing along, we simulated the tossing of the quarter 1,000 times. For this experiment, we got 505 heads for a relative proportion of .5050. Figure 4.3 tallies the results of our three experiments. To summarize, then, when we say $P(\text{H}) = .5$, we mean that if we performed the experiment a very large number of times (some statisticians believe this should be at least 10,000 times), the proportion of heads would be very close to .50.

Similarly, it makes sense to assign a probability of 1/6 to each of the 6 equally likely simple outcomes from the roll of a fair (not loaded) die. Let's denote the probability of a certain event E as $P(\text{E})$. If we know in advance that the sample space consists of equally probable outcomes, we can assign the value to $P(\text{E})$ as follows:

$$P(\text{E}) = \frac{\text{number of simple outcomes favorable to the event}}{\text{total number of simple outcomes}} = \frac{f}{n} \tag{4.1}$$

Thus, if A = {the die lands on a number less than 5}, then $P(\text{A}) = 4/6$ since there are 4 ways of being successful (1 or 2 or 3 or 4) out of the 6 possible outcomes.

Relative Frequency (Empirical) Probability. Suppose we randomly choose an employee from Titan Corporation to find the probability that this person subscribes to an HMO health plan. In this instance, we must have some empirical data to determine the answer. We do some research and find that Titan has 386 employees. Of these, 184 subscribe to an HMO, 127 have Blue Cross/Blue Shield coverage, and 75 are not insured. We can now use the following method to assign probabilities to each simple event:

$$P(\text{E}) = \frac{\text{number of favorable outcomes}}{\text{total number of trials}} = \frac{f}{n} \tag{4.2}$$

FIGURE 4.1

```
MTB > Print 'FLIP 10'.

FLIP 10
   0    0    0    0    1    1    1    0    0    1
```

```
MTB > Print 'FLIP 500'.

FLIP 500
   0   0   1   1   0   0   0   0   0   0   1   0   1   0   1
   0   0   0   1   1   0   1   1   0   0   1   0   1   1   0
   1   0   1   1   1   1   1   0   0   0   0   0   0   1   1
   0   0   1   0   1   0   0   1   1   0   0   1   1   0   0
   1   0   0   0   0   1   1   1   1   1   0   1   0   0   1
   0   0   0   0   1   0   1   1   0   1   1   1   0   0   0
   1   0   1   0   0   1   1   0   1   1   0   0   0   1   1
   1   1   0   0   1   1   0   1   0   0   0   1   1   0   1
   1   1   0   0   1   0   1   1   0   0   1   1   0   1   1
   1   1   0   1   1   0   0   1   0   0   0   1   0   0   1
   0   0   1   1   1   0   0   0   1   0   1   0   0   1   0
   0   1   1   0   1   0   1   0   0   1   1   0   0   1   0
   1   1   1   0   1   0   1   1   1   0   1   0   1   0   0
   0   1   1   0   0   0   0   0   0   0   1   0   1   0   1
   0   0   1   0   1   0   0   1   1   1   0   1   1   0   1
   1   1   1   1   1   0   1   1   1   1   0   0   1   0   0
   1   1   0   0   0   1   0   1   1   0   0   1   0   0   0
   1   0   1   1   0   1   1   0   0   1   1   0   0   1   0
   1   1   1   0   1   1   0   0   0   1   1   1   1   1   1
   0   0   1   0   0   0   0   0   0   1   1   0   0   1   1
   0   1   0   0   0   0   1   0   1   1   1   0   0   0   1
   1   1   0   1   1   1   0   1   0   1   0   0   1   1   0
   1   1   1   0   1   0   0   1   1   1   0   0   1   1   1
   1   0   1   0   1   1   0   0   0   0   0   1   0   0   0
   1   1   0   1   1   1   0   0   0   0   1   1   0   0   1
   0   0   1   1   0   0   0   1   1   0   1   1   0   1   0
   0   0   1   0   0   0   0   0   1   0   1   1   0   0   0
   1   0   0   1   0   1   0   1   1   1   1   1   0   0   1
   0   0   1   1   1   1   1   1   1   0   1   1   0   1   0
   0   1   1   1   1   0   0   0   1   0   1   0   0   0   0
   1   1   0   1   0   1   0   0   1   1   0   0   1   1   1
   0   1   1   1   1   1   0   0   1   0   1   1   0   1   0
   1   0   1   1   1   1   1   1   1   1   0   1   1   1   0
   1   1   0   1   1
```

FIGURE 4.2

```
MTB > Tally 'FLIP 10'-'FLIP1000';
SUBC>   Counts;
SUBC>   Percents.

 FLIP 10  COUNT PERCENT   FLIP 500  COUNT PERCENT  FLIP1000  COUNT PERCENT
       0      6   60.00          0    245   49.00         0    505   50.50
       1      4   40.00          1    255   51.00         1    495   49.50
      N=     10                 N=    500                N=   1000
```

FIGURE 4.3

A Hit and a Little Miss?

"Economic forecaster Philip Braverman prides himself on the fact that he has accurately predicted the sex of all seven of his grandchildren. The odds against such success (as long as the odds against guessing seven consecutive coin tosses) are more than 100 to 1. 'My method,' he says, 'is simple. In each case my kids have picked a name for a boy and a girl. I choose the name I like better—and go with that preference. God smiles and gives me what I preferred.'"

—Gene Epstein, "Growth Recession," *Barron's*, May 4, 1992, p. 12.

For the event H = {the employee subscribes to an HMO}, we assign $P(\text{H}) = 184/386$ or a probability of .4767. Likewise, if B = {the employee subscribes to Blue Cross/Blue Shield}, we assign $P(\text{B}) = 127/386$, which is a probability value of .3290. And if N = {the employee has no health insurance}, then $P(\text{N}) = 75/386$ or .1943. Each of the three values just computed is an example of a **relative frequency** (or **empirical**) **probability**—one that's determined by observation and/or experimentation.

Intuition. A third method used to assign probabilities is intuition. A salesperson might make an educated guess and say, "There's a 40 percent probability that I'll make a sale to my next customer." This last method isn't scientifically based, but rather, it is based on a subjective hunch. Or a plant manager may believe that there's a .60 probability that the union will call a strike next week. This probability is the manager's subjective estimate of the likelihood of a strike and is not an *a priori* or empirical value. The accuracy of the strike estimate, of course, depends on the manager's experience and skill.

Regardless of how probability values are assigned, though, the following two properties must apply:

1. Probability is measured on a scale from 0 to 1. That is, a probability of interest—say, $P(\text{E})$—is always $0 \leq P(\text{E}) \leq 1$. If an event is *impossible*, we assign 0 as its probability. An event that is *certain to occur* has a probability of 1. All other probabilities—those of interest to us—are found between these extremes.
2. The probabilities of all the simple events within a sample space must add up to 1. That is, $\Sigma\, P(\text{E}) = 1$.

Let's look at some examples of these conditions:

- When we roll a fair die, we assign $P(1) = P(2) = P(3) = P(4) = P(5) = P(6) = 1/6$. Thus, the probabilities of all possible events are $1/6 + 1/6 + 1/6 + 1/6 + 1/6 + 1/6 = 6/6$ or 1.
- If we had reason to believe the die was not fair, we could assign the following probabilities: $P(1) = .2$, $P(2) = .1$, $P(3) = .3$, $P(4) = .2$, $P(5) = .1$, $P(6) = .1$. Each of the assigned probabilities is between 0 and 1, and the sum of the probabilities is still 1.
- The probabilities of $P(1) = .3$, $P(2) = .1$, $P(3) = .1$, $P(4) = .2$, $P(5) = .1$, and $P(6) = .3$ could *not* be assigned to the outcomes for the roll of a loaded die since the sum of the probabilities is not 1.
- The probabilities of $P(1) = .4$, $P(2) = .1$, $P(3) = .3$, $P(4) = -.2$, $P(5) = .1$, and $P(6) = .3$ could *not* be assigned to the outcomes for the roll of a loaded die since each probability must be between 0 and 1, and $P(4) = -.2$ isn't valid.
- A few paragraphs earlier we assigned probabilities of 184/386, 127/386, and 75/386 to the events that Titan employees subscribe to an HMO, to Blue Cross/Blue Shield coverage, or to no health insurance. Note that each probability is between 0 and 1 and the sum of the probabilities is 386/386 or 1.

Self-Testing Review 4.1

1. A card is drawn at random from a standard deck of 52 cards. What is the probability that it is a club? (Figure 4.4 is included here for you non–card

A Standard Deck of Cards

52 cards
4 suits (♥, ♦, ♣, ♠)
13 cards in each suit
(A, 2, 3, 4, 5, 6, 7, 8, 9, 10, J, Q, K)

Hearts	Diamonds	Clubs	Spades
A♥ - Ace of hearts	A♦ - Ace of diamonds	A♣ - Ace of clubs	A♠ - Ace of spades
2♥ - Two of hearts	2♦ - Two of diamonds	2♣ - Two of clubs	2♠ - Two of spades
3♥ - Three of hearts	3♦ - Three of diamonds	3♣ - Three of clubs	3♠ - Three of spades
4♥ - Four of hearts	4♦ - Four of diamonds	4♣ - Four of clubs	4♠ - Four of spades
5♥ - Five of hearts	5♦ - Five of diamonds	5♣ - Five of clubs	5♠ - Five of spades
6♥ - Six of hearts	6♦ - Six of diamonds	6♣ - Six of clubs	6♠ - Six of spades
7♥ - Seven of hearts	7♦ - Seven of diamonds	7♣ - Seven of clubs	7♠ - Seven of spades
8♥ - Eight of hearts	8♦ - Eight of diamonds	8♣ - Eight of clubs	8♠ - Eight of spades
9♥ - Nine of hearts	9♦ - Nine of diamonds	9♣ - Nine of clubs	9♠ - Nine of spades
10♥ - Ten of hearts	10♦ - Ten of diamonds	10♣ - Ten of clubs	10♠ - Ten of spades
J♥ - Jack of hearts	J♦ - Jack of diamonds	J♣ - Jack of clubs	J♠ - Jack of spades
Q♥ - Queen of hearts	Q♦ - Queen of diamonds	Q♣ - Queen of clubs	Q♠ - Queen of spades
K♥ - King of hearts	K♦ - King of diamonds	K♣ - King of clubs	K♠ - King of spades

FIGURE 4.4

players. There are four suits—clubs, diamonds, hearts, and spades—in the deck, and there are 13 cards in each suit.)

2. If a card is drawn at random from a standard deck of 52 cards, what is the probability it is a face card? (There are 3 face cards—jack, queen, and king—in each of the four suits.)

3. In the general population there are four types of blood: A, B, AB, and O. The relative proportions of people having each type are .42, .10, .03, and .45, respectively. What is the probability that a person selected at random will have type O blood? (For the purpose of blood donation, type O is called the "universal donor" since it can safely be transfused to people with any blood type.)

4. A person is selected at random. What is the probability he or she has type AB blood (refer to the probabilities in problem 3)?

5. In the annual reader survey made by *Home-Office Computing*, it was found that 61 percent of the magazine's readers were male. A reader of this magazine is selected at random. What is the probability that the reader is male?

6. The National Center for Educational Statistics of the U. S. Department of Education predicts that in 199*x* there will be a demand for 169 thousand new teachers. Out of this total there will be a need for 88 thousand elementary teachers and 81 thousand secondary teachers. What's the probability that a new teaching job in 199*x* will be for a secondary teacher?

7. The Federal Highway Administration reports that there were a total of 9,898 thousand passenger cars sold in a recent year. Of these, 7,073 thousand were domestic models and 2,895 thousand were imports. If a car purchased during the year is selected at random, find the probability it is a domestic model.

8. The Federal Highway Administration reports that there were a total of 4,941 thousand trucks sold in a recent year. Of these, 4,403 thousand were domestic models and 538 thousand were imports. If a truck purchased during the year is selected at random, find the probability it is a domestic model.

9. The Bucks County District Attorney reported 6,358 cases involving drugs in a recent year. From these cases, there were 791 arrests. Find the probability that in that year a drug case resulted in an arrest.

10. In the annual reader survey of *Home-Office Computing*, it was found that 68 percent of the respondents run a home-based business, 6 percent run a business that is not home based, 9 percent plan to start a business, and 17 percent work at home part of the week. A reader is selected at random. What is the probability that he or she runs a business that is home based?

11. On the television game show *The Price Is Right*, there is a segment called "Showcase Showdown." The contestant is asked to spin a wheel which has 20 equally spaced sectors marked in amounts 5, 10, 15, . . . , 100. If a pointer on the wheel lands on the sector marked 100, the contestant wins $1,000. What's the probability that the contestant spins the wheel and wins $1,000?

12. On the "Showcase Showdown" wheel, two sectors are green. If a contestant spins the wheel, what's the probability the pointer will land on a green sector?

13. In a statistics class, 47 students say that they are right-handed and 5 say they are left-handed. If a student from this class is picked at random, what is the probability the student is left-handed?

14. The *Statistical Abstract of the United States* reported that in a recent year there were a total of 2,347 thousand arrests made for serious crimes. Of these, there were 18 thousand arrests for murder and nonnegligent manslaughter, 31 thousand for forcible rape, 134 thousand for robbery, 355 thousand for aggravated assault, 357 thousand for burglary, 1,254 thousand for larceny, 183 thousand for motor vehicle theft, and 15 thousand for arson. If a person arrested for a serious crime during that year is selected at random, what is the probability that he or she was arrested for motor vehicle theft?

15–19. A loaded die is tossed. Determine which of the following assignments could be made for the probabilities of each simple outcome. If an assignment cannot be made, tell why.

15. $P(1) = .3$ $P(2) = .2$ $P(3) = .3$ $P(4) = .1$ $P(5) = .2$ $P(6) = .1$

16. $P(1) = .1$ $P(2) = 1.2$ $P(3) = .2$ $P(4) = .1$ $P(5) = .2$ $P(6) = .1$

17. $P(1) = .1$ $P(2) = .1$ $P(3) = .1$ $P(4) = .1$ $P(5) = .3$ $P(6) = .3$

18. $P(1) = .2$ $P(2) = .3$ $P(3) = .2$ $P(4) = -.1$ $P(5) = .2$ $P(6) = .2$

19. $P(1) = .3$ $P(2) = .3$ $P(3) = .3$ $P(4) = .1$ $P(5) = 0$ $P(6) = 0$

20. *The Chronicle of Higher Education* estimated that 2,470 thousand students graduated from high school in a recent year. From this group, it's estimated that 2,210 thousand graduate from a public high school and 255 thousand graduated from a private school. If a high school graduate of that year is selected at random, what is the probability that he or she graduated from a private high school?

21. *The Chronicle of Higher Education* estimated that 14,366 thousand students were enrolled in college in a recent year. Of these, about 6,531 thousand were men and 7,835 thousand were women. If a college student enrolled that year is selected at random, what's the probability that the student is a male?

4.2 Probabilities for Compound Events

We've just looked at ways to determine the probabilities of *simple events*. But we're often interested in the relative chance that some *combination of events* may occur. Such a combination is called a *compound event.*

A **compound event** is one that combines two or more simple events.

There are three categories of compound events of interest to us. In the following pages, we'll determine the probability that in a single experimental trial:

1. Event A occurs, given that (or *on the condition that*) event B has happened. The notation for this conditional outcome is *P*(A, given B) or *P*(A|B).
2. Both events A and B happen—the notation here is *P*(A and B).
3. Either event A occurs, or event B occurs, or they both occur. The notation for this outcome is *P*(A or B).

But What Was His Average in August?

At the beginning of a recent baseball season, New York Yankee center fielder Roberto Kelly had a batting average of .335. Against right-handed pitchers, Kelly was hitting .336, and against left-handers he was hitting .333. However, with teammates in scoring position, Kelly had 16 hits in 45 attempts for an average of .356.

Conditional Probability Concepts

When additional facts are available for an experiment, probability values for an event can be reassessed in the light of new information. For example, suppose you are playing a game (such as "Trivial Pursuit") involving the use of a die. Your opponent rolls the die but does not let you see the outcome. You will win the game if your opponent rolls a 3, so you are interested in the probability of this occurrence. In the previous section we saw that it's reasonable to assign *P*(die lands on a 3) = *P*(T) = 1/6. However, suppose you are given additional information about the outcome of that roll, namely, that the roll has landed on an odd number. If we designate the event that the die lands on an odd number as D, then D = {1, 3, 5}. We can now assess the relative chance that the roll will be a 3 in light of this new information.

Since our sample space has been reduced to three equally probable outcomes (1, 3, or 5), we can now say that the relative chance that the roll was a 3, *given* that the roll was an odd number, is 1/3. This is thus an example of a *conditional probability.*

A **conditional probability** is the probability that one event will occur given that another has happened.

In our example, we have the conditional probability of T given D, which we can denote as *P*(T, given D) or *P*(T|D). Note that for this example, the conditional probability is not the same as the simple probability. That is, $P(\text{T}|\text{D}) \neq P(\text{T})$.

Let's look at another situation involving conditional probability. In problem 3 in STR 4.1, it's shown that the probability of randomly selecting a person with type O blood from the general population is .45. Now suppose when selecting an individual that you have additional information that the person selected is a female. What is the probability that the person has type O blood, given that the person is a female? That is, what is *P*(type O|female)? Intuitively, we can see that *P*(type O|female) = .45. The proportion of females with type O blood is .45. In other words, the additional information (that the person is female) didn't change the value of the probability of selecting a person who has type O blood. For this situation the conditional probability is the same as the simple probability—$P(\text{type O}|\text{female}) = P(\text{type O})$.

Now let's pay Titan Corporation another visit to get additional employee information. This time we're interested in the type of medical insurance employees subscribe to as well as whether or not they have dependent children. The information gathered can be placed in the following table. This table is called a **contingency table** because it shows all the classifications of the variables being studied—that is, it accounts for all contingencies in a particular situation.

	HMO(H)	BC/BS(B)	NONE(N)	TOTAL
Dependent Children (D)	145	85	23	253
No Dependent Children	39	42	52	133
Totals	184	127	75	386

We've seen that the probability that a Titan employee subscribes to an HMO [$P(\text{H})$] is 184/386 or .4767. Now we'll investigate the conditional probability that an employee subscribes to an HMO given that he or she has dependent children (D). To do this, we only consider the 253 employees who have dependent children.

	HMO(H)	BC/BS(B)	NONE(N)	TOTAL
Dependent Children (D)	145	85	23	253

Within this revised sample space consisting only of employees with dependent children, 145 are HMO subscribers, so we can say P(employee subscribes to HMO, given that employee has dependent children) or $P(\text{H}|\text{D}) = 145/253 = .5731$.

Joint Probability—Multiplication Rule for Computing P(A and B)

When a probability experiment is performed, it's sometimes necessary to find the probability that *both* events A *and* B occur. To do so, we use the following formula:

$$P(\text{A and B}) = P(\text{A}) \times P(\text{B}|\text{A}) \qquad (4.3)$$

Thus, we must *multiply* the probability of the first event times the conditional probability of the second event, given that the first has occurred.

Some examples should help clarify the use of this **multiplication rule**:

➤ We have 10 pieces of candy in a dish. We know that 5 pieces are red, 3 are green, and 2 are yellow. If we choose 2 pieces at random without looking (since we are busy concentrating on learning about probabilities), what's the probability that both are green? Think of this as two actions—(1) choosing the first piece of candy

Architects can use statistical methods to analyze stresses and evaluate designs. (*Courtesy Hewlett Packard*)

The following examples show the use of the *addition rule* when events aren't mutually exclusive:

- If we draw a card from a standard 52-card deck, what's the probability that the card is a 7 or it is red? We know that *P*(card is a 7) is 4/52. And because half the cards in the deck are red (the others are black), *P*(card is red) is 26/52. Since the events are *not* mutually exclusive, we must also determine *P*(the card is both a 7 and it is red). A few non–card players may not know that there are two red 7s (the 7 of hearts and the 7 of diamonds), so *P* (the card is a 7 and it is red) is 2/52. We then use the addition formula for non–mutually exclusive events to find the probability that at least one of the specified events occurs. That is, $P(7 \text{ or red}) = P(7) + P(\text{red}) - P(7 \text{ and red}) = 4/52 + 26/52 - 2/52 = 28/52$. Note that the two red 7s have been counted twice—once as a 7 and once as a red card. The subtraction of 2/52 occurs so that the same cards are not double-counted.

- What's the probability that an employee chosen at random from Titan either subscribes to an HMO or has dependent children? The following contingency table will help give us the answer:

	HMO(H)	BC/BS(B)	NONE(N)	TOTAL
Dependent Children (D)	145	85	23	253
No Dependent Children	39	42	52	133
Total	184	127	75	386

As you can see, $P(\text{H}) = 184/386$, $P(\text{D}) = 253/386$, and $P(\text{H and D}) = 145/386$.

$$P(\text{H or D}) = P(\text{H}) + P(\text{D}) - P(\text{H and D})$$
$$= \frac{184}{386} + \frac{253}{386} - \frac{145}{386} = \frac{292}{386} \text{ or } .7567$$

The Complement of an Event

We've seen earlier that the probabilities of all the simple events in a sample space must equal 1. That is, $\Sigma P(\text{E}) = 1$. So when we toss a fair coin 1 time, there's a 1/2 chance we'll get a head, a 1/2 chance we'll get a tail, and the probabilities of the mutually exclusive outcome resulting from a single coin flip (1/2 + 1/2) must equal 1. Let's assume now that we toss the coin 4 times and want to find the probability that we get *at least* 1 head. A direct and somewhat tedious way to solve this problem is discussed in the next chapter, but with a little logic we can tackle it here by first computing the probability that we get *all tails* and no heads in the 4 tosses. Since the tosses are independent events, $P(\text{T}) = (1/2)\,(1/2)\,(1/2)\,(1/2) = 1/16$. Thus, if the probability that we get 4 tails in 4 flips is 1/16, then the probability that we can expect at least 1 head at some time in the 4 tosses must be $1 - 1/16$ or 15/16.

We've just used the *complement* of an event to solve a problem.

> The **complement** of event E—denoted $\overline{\text{E}}$—consists of all possible outcomes from the sample space that are *not* in event E.

Thus, the complement of event E is the event that E *doesn't* happen. If we roll a die and let E = {the roll of the die is less than 5}, then $\overline{\text{E}}$ = {the roll of the die is 5 or more}. And if we choose an employee at random from Titan Corporation and let H = {the employee subscribes to an HMO plan}, then $\overline{\text{H}}$ = {the employee doesn't subscribe to the HMO plan}.

Events E and $\overline{\text{E}}$ must be mutually exclusive, so $P(\text{E or } \overline{\text{E}}) = P(\text{E}) + P(\overline{\text{E}})$. Since it's certain that event E will either occur or it will not occur, $P(\text{E or } \overline{\text{E}}) = 1$. Thus, $P(\text{E}) + P(\overline{\text{E}}) = 1$, and solving this equation for $\overline{\text{E}}$ gives us a formula for computing the probability of the complement of an event:

$$P(\overline{\text{E}}) = 1 - P(\text{E}) \tag{4.7}$$

Perhaps some examples will help about now:

- We know the probability that A (the roll of a die is less than 5) is 4/6. So the probability that the roll is *not* less than a 5 is $1 - 4/6$ or 2/6. And the event $\overline{\text{A}}$ (the roll is a 5 or a 6) has a probability of 2/6.

➤ Let's let $P(\text{H})$ represent the probability that we randomly select a Titan employee who subscribes to an HMO. Thus, $P(\text{H})$ is 184/386. And if we let $P(\overline{\text{H}})$ represent the complement in this example—that is, the employee doesn't belong to the HMO plan—then $P(\overline{\text{H}}) = 1 - P(\text{H}) = 1 - 184/386 = 386/386 - 184/386 = 202/386$. This is the same result we obtained earlier when we computed the probability that the employee had Blue Cross/Blue Shield insurance or did not have any health coverage. As you can see, it's often easier to use the formula for complementary events to compute various probabilities.

➤ A study published in *Physical Therapy* reported on the relationship that existed between a subject's age and the normal active range of motion of hip and knee joints. The sample consisted of 1,683 people. Of the subjects in the group, 821 were male and 862 were female, 433 were 25 to 39 years of age, 727 were 40 to 59 years of age, and 523 were 60 to 74 years of age. A person is picked from this study at random (we'll assume that gender and age are independent variables). Find the probability that the subject is:

a) Age 25 to 39.
b) A female.
c) A female aged 25 to 39.
d) Not a female aged 25 to 39.
e) Age 25 to 39 or is a female.

The answers are:

a) There were 433 people aged 25 to 39 out of the sample of 1,683. Thus, 433/1,683 =.2573.
b) There were 862 females, so 862/1,683 = .5122.
c) To compute this joint probability (no pun intended), we'll use the multiplication rule for independent events. Thus, the probability is (.2573)(.5122) = .1318.
d) Using the rule for complements, the probability is 1 − .1318 = .8682.
e) We use the addition rule for non–mutually exclusive events in this example. Thus, P (age 25 to 39 or female) = .2573 + .5122 − .1318 = .6377. (We have to remember to subtract here so that the females in the given age category are not counted twice.)

Self-Testing Review 4.2

1–4. Three jars are used in the Pennsylvania Lottery Lucky Number drawing. Each jar contains 10 Ping-Pong balls, and each ball is marked with a different digit 0, 1, 2, . . ., 9. The balls are mixed, and 1 is randomly selected from each jar.

1. What is the probability that the ball selected from the first jar is a 7?
2. What is the probability that the ball selected from the second jar is a 7, given the ball from the first was a 7?
3. What's the probability the ball selected from the third jar is a 7, given that each of the balls in the first two jars was a 7?
4. On January 27, 1993, the lucky number for the Pennsylvania lottery was 777. What is the probability that this three-digit number will be drawn on any given day?
5. On the same day, the number in the New York Lottery was also 777. What's the probability that the number 777 will be selected in *both* states on the same day?

6. A person is selected from the general population. What's the probability that he or she has either type A or type AB blood? (The simple probabilities of having type A, B, AB, or O blood are .42, .10, .03, and .45, respectively.)

7. What's the probability that a person selected at random does not have type O blood?

8–11. A bowl contains 6 brown, 3 red, 4 yellow, and 2 blue pieces of candy. If you choose 2 at random (without replacing the first) what's the probability that:

8. The first is red and the second is yellow?

9. The first is yellow and the second is red?

10. One is red and the other is yellow (in either order)?

11. Both are brown?

12–14. A single die is thrown. What is the probability that it is:

12. At least a 4?

13. An odd number?

14. A 4 or an odd number?

15–18. A card is drawn from a standard deck of 52 cards. What's the probability that it is:

15. Red?

16. A face card (a jack, queen, or king)?

17. A red face card?

18. Red or it's a face card?

19–20. A survey conducted by *TV Guide* found that men control the TV remote control unit in 41 percent of the homes, women control the unit in 19 percent of the homes, and control is shared in 27 percent of the homes. (In the remaining households, the fight for control was still on or there was no remote unit.) If a household is selected at random, find the probability that the remote unit is:

19. Shared.

20. Controlled by men or is shared.

21. The probability of a person's being left-handed is .12. If two unrelated people are selected at random, what's the probability they are both left-handed?

22. In the 1992 U.S. presidential election all three major candidates were left-handed. What's the probability of selecting three unrelated people who are all left-handed?

23–24. There are 77 students in the M.B.A. program at Systems Tech. In this class, 39 students are taking statistics, 45 are taking finance, and 21 are taking both statistics and finance. If a student is chosen at random, find the probability that he or she is:

23. Taking either statistics or finance.

24. Is not taking statistics.

25–30. A box of parts contains 8 good items and 2 defective items. If 2 are selected at random *with replacement*—that is, the first item *is* replaced before the second is drawn—find the probability that:

25. Both items are good.

26. Both items are defective.

27. The first is good and the second is defective.

28. The first is defective and the second is good.

29. One is defective and the other isn't.

30. At least 1 is defective.

31–36. A box of parts contains 8 good items and 2 defective ones. If 2 are selected at random *without replacement*—that is, the first item *isn't* replaced before the second is drawn—find the probability that:

31. Both items are good.

32. Both items are defective.

33. The first is good and the second is defective.

34. The first is defective and the second is good.

35. One is defective and the other isn't.

36. At least 1 is defective.

37–38. The following table shows the opening songs performed by The Grateful Dead for the 20 concerts played during the first half of 1992:

	Frequency of Use
"Bucket"	3
"Jack Straw"	3
"Touch of Grey"	3
"Cold Rain"	2
"Good Times Roll"	2
"Help on the Way"	2
"Bertha"	1
"Box of Rain"	1
"Greatest Story"	1
"Half Step"	1
"Stranger"	1

37. If you recorded one of these concerts at random, what's the probability that either "Touch of Grey" or "Bertha" was the opening song?

38. What's the probability that "Good Times Roll" wasn't the opening song?

39–40. A study in *The Journal of Clinical Psychology* analyzed the sociodemographic characteristics of male Vietnam veterans. There were 319 veterans who suffered from posttraumatic stress disorder (PTSD). Their marital status is reported as follows:

Marital Status	Frequency
Married	200
Living as married	44
Separated	27
Divorced	17
Widowed	0
Never married	31

What is the probability that a male Vietnam veteran with PTSD is:

39. Separated or divorced?

40. Not married?

41–42. The study in *The Journal of Clinical Psychology* described in problems 39 to 40 also analyzed the marital status of 871 male Vietnam veterans who did not suffer from PTSD. The findings were:

Marital Status	Frequency
Married	678
Living as married	45
Separated	9
Divorced	96
Widowed	1
Never married	42

What's the probability that a male Vietnam veteran who doesn't suffer from PTSD is:

41. Separated or divorced?

42. Not married?

43–50. The following contingency table describes the educational background of male Vietnam veterans with and without PTSD. (Information is from *The Journal of Clinical Psychology.*)

Educational Background	No PTSD	PTSD	Total
Less than high school	45	35	80
High school grad	299	110	409
Some college	363	154	517
College grad	71	9	80
Graduate school	93	11	104
Total	871	319	1,190

What is the probability that a male Vietnam vet has:

43. PTSD?

44. Less than a high school education?

45. PTSD and less than a high school education?

46. PTSD or less than a high school education?

47. PTSD, given that he has less than a high school education?

48. Less than a high school education, given that he has PTSD?

Now, show that the events of having PTSD and having less than a high school education are not:

49. Mutually exclusive.

50. Independent.

51–59. A study of public school dropouts in Southeastern Pennsylvania counties that was conducted by the Pennsylvania Department of Education produced the following dropout figures:

County	Male	Female	Total
Bucks	379	247	626
Chester	283	151	434
Delaware	399	263	662
Montgomery	322	222	544
Philadelphia	5,477	4,259	9,736
Total	6,860	5,142	12,002

A dropout is chosen at random from this area. Find the probability that the dropout is:

51. From Bucks County.

52. A male.

53. From Delaware or Montgomery county.

54. Not from Philadelphia County.

55. A male from Bucks County.

56. Either from Montgomery County or is a male.

57. A female, given that she is from Philadelphia County.

Now, are the events of being female and being from Philadelphia County:

58. Mutually exclusive? Show why or why not.

59. Independent? Show why or why not.

Just Wait a Little Longer . . .

"Statistical projections are invoked so thoughtlessly that it wouldn't be surprising to see someday that the projected waiting period for an abortion is a year. . . . Somehow too many Americans escape education in mathematics with only the haziest feel for numbers and probability and for the ways in which these notions are essential to understanding a complex world."
—*John Allen Paulos*

4.3 Random Variables, Probability Distributions, and Expected Value

You saw in Chapter 3 that a *variable* is a characteristic of interest that's possessed by each item under study. Of course, the value of this characteristic will likely vary from one item in the data set to the next. You also saw in Chapter 3 that we can group observed data items into a *frequency distribution* and then graphically portray the results using a *histogram* or *frequency polygon.* And in this chapter you've learned some basic probability concepts. We can now build on what you've learned to introduce you to random variables and probability distributions.

Random Variables

Many probability experiments have outcomes that are *numerical* observations, counts, or measurements.

> A **random variable** (identified by a symbol such as x) has a single numerical value for each outcome of a probability experiment.

Thus, x can assume any of the numbers associated with the possible outcomes of the experiment, and the particular value that x assumes in any single trial of an experiment is a chance or random outcome.

You saw in Chapter 3 that a discrete variable has a countable or finite number of values. Similarly, a **discrete random variable** is one in which all possible values can be counted or listed. And like a continuous variable that can assume any one of the countless number of values along a line interval, a **continuous random variable** has an infinite number of values that can fall, without interruption, along an unbroken interval. Continuous random variables are typically created when measurements are involved. But since measurements can be produced to give any desired number of decimal points, it's impossible to list all the possible values.

Let's look at some probability experiment examples now to help clarify the terms we've just considered (this text has already included many corny examples, so a few more shouldn't bother you):

- We count the number of kernels on an ear of corn. The variable is discrete since the values of 1, 2, 3, . . . , can be listed.
- We measure the length of an ear of corn. The variable is continuous since possible values include 5.325 inches, 4.9873 inches, and so on. All such values couldn't possibly be listed.
- We time the period needed to harvest the corn. We are measuring time, so the variable is continuous.
- We determine the number of trucks needed to ship the corn. The variable is discrete. We can count the number of trucks, and we certainly can't send a fractional part of a truck to the grain buyer.

Probability Distributions

If we list all the possible outcomes of a probability experiment and further list all the probability values associated with these outcomes, then we have created a *probability distribution.*

> A **probability distribution** gives the probability for each of the values of a random variable.

Let's create a probability distribution for a discrete random variable by considering all the possible outcomes that you could get in a single roll of your favorite pair of dice. Your little cousin has just used fingernail polish to paint one die red (!), but that won't ruin our example. The spots facing up in a single roll of the dice must total 2, 3, 4, . . . , 11, or 12. What's the probability that in a single roll your result is 2? Since there's a 1/6 chance that the white die will produce a 1, and a 1/6 chance that the red die will show a 1, and since these are independent results, the probability that both yield a 1 to produce a total of 2 is (1/6) (1/6) or 1/36. Now what's the probability that in your next roll the result is 3? You can get this result either by rolling a 1 on the white die and a 2 on the red die or by getting a 2 on the white die and a 1 on the red die. The probability of a 1 on the white die and a 2 on the red die is computed as follows:

$$P(\text{1 on white and 2 on red}) = P(\text{1 on white}) \times P(\text{2 on red}) = \left(\frac{1}{6}\right)\left(\frac{1}{6}\right) = \frac{1}{36}$$

The probability of 2 on white and 1 on red is:

$$P\ (\text{2 on white and 1 on red}) = P(\text{2 on white}) \times P(\text{1 on red}) = \left(\frac{1}{6}\right)\left(\frac{1}{6}\right) = \frac{1}{36}$$

Since we want to know the probability that one or the other of these mutually exclusive events will occur, we now use the *addition rule* to determine the probability of throwing a 3 with the pair of dice:

$$\begin{aligned} P(3) &= P\ (\text{1 on white and 2 on red}) + P\ (\text{2 on white and 1 on red}) \\ &= \frac{1}{36} + \frac{1}{36} = \frac{2}{36} \end{aligned}$$

The probabilities of throwing a 4, 5, 6, 7, 8, 9, 10, 11, or 12 may be computed in the same way. The complete probability distribution is shown in Figure 4.6 and in the following table:

FIGURE 4.6

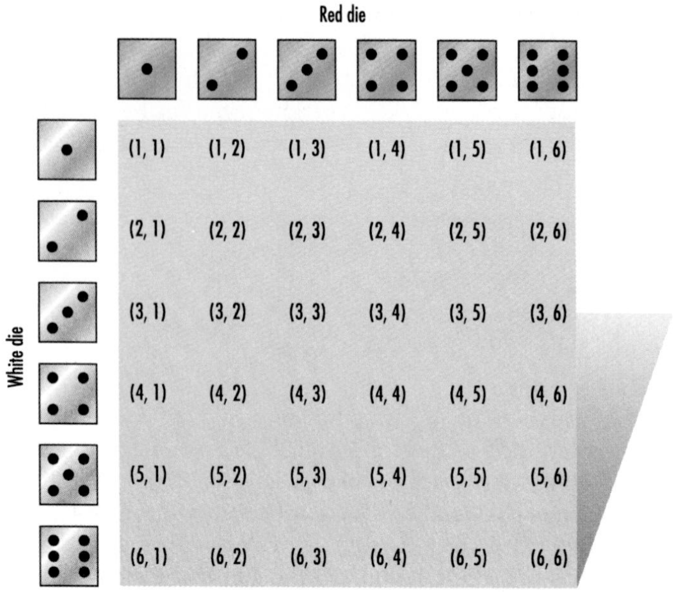

Red die

White die

	1	2	3	4	5	6
1	(1, 1)	(1, 2)	(1, 3)	(1, 4)	(1, 5)	(1, 6)
2	(2, 1)	(2, 2)	(2, 3)	(2, 4)	(2, 5)	(2, 6)
3	(3, 1)	(3, 2)	(3, 3)	(3, 4)	(3, 5)	(3, 6)
4	(4, 1)	(4, 2)	(4, 3)	(4, 4)	(4, 5)	(4, 6)
5	(5, 1)	(5, 2)	(5, 3)	(5, 4)	(5, 5)	(5, 6)
6	(6, 1)	(6, 2)	(6, 3)	(6, 4)	(6, 5)	(6, 6)

Sum	2	3	4	5	6	7	8	9	10	11	12	Total
Number of ways	1	2	3	4	5	6	5	4	3	2	1	36

PROBABILITY DISTRIBUTION OF RESULTS OF THROWING A PAIR OF DICE

Result	Probability
2	1/36
3	2/36
4	3/36
5	4/36
6	5/36
7	6/36
8	5/36
9	4/36
10	3/36
11	2/36
12	1/36
Total	36/36 = 1

There are two rules that apply to this or any other probability distribution:

1. The values in a probability distribution are numbers that are on the interval from 0 to 1. That is, the possible distribution values—let's label them $P(X)$—are always $0 \leq P(X) \leq 1$. In our example, the values range from 1/36 (.0278) to 6/36 (.167).
2. All of the possible values in a probability distribution must add up to 1. That is, $\Sigma P(X) = 1$. Since the distribution in our example is a complete listing of *all* outcomes of the experiment, the probability that one or the other of these outcomes will occur is a certainty.

These rules should seem familiar to you since they are quite similar to the probability properties discussed earlier in the chapter.

Relative Frequency Distributions. You saw earlier in this chapter that probabilities could be assigned based on the number of times each event of interest occurred in a total number of trials. If we didn't intuitively know that 1/6 was the probability of rolling a given face on a die, we could roll the die 6,000 times and observe the results. If the die is fair or true, the resulting frequency distribution would show that each face appeared *about* (not exactly) 1,000 times, as in column 2 in the following table:

Die Face	Face Appearances	Proportion of Times Face Appears
1	985	.1642
2	1,005	.1675
3	1,020	.1700
4	979	.1632
5	992	.1653
6	1,019	.1698
Total	6000	1.0000

If we then divided each of these face totals by 6,000, we would get the proportion of times the surface faces up, and we would have a **relative frequency distribution** (column 3 in the table). As we'll see in later chapters, several important probability distributions are used to approximate the relative frequency distributions of the populations being studied. For now, let's look at the following example of a relative frequency distribution.

The Chronicle of Higher Education Almanac has reported attitudes and characteristics of college freshmen. When asked for the number of colleges they applied to *other than the one they were attending*, the student responses, together with their relative frequencies, were as follows:

Number of Other Colleges Applied to (x)	Proportion of Times Response Was Given [$P(x)$]
0	.376
1	.147
2	.158
3	.137
4	.078
5	.047
6	.056
Total	1.000

Let's consider these questions relating to this example:

- What's the probability that a student applied to 4 colleges other than the one he or she was attending? From the table, we see that $P(x = 4)$ is .078.
- What is the probability a student applied to *at least* 4 colleges other than the one she or he was attending? In this context, "at least 4" means 4 *or* 5 *or* 6. Applying to 4 other colleges is mutually exclusive to applying to 5 other schools, which, in turn, is mutually exclusive to applying to 6 others. So, $P(\text{at least } 4) = P(4) + P(5) + P(6) = .078 + .047 + .056 = .181$.
- What's the probability that a student applied to *at least* one other college? We could add the probabilities of $x = 1, 2, 3, 4, 5$, and 6 to get the answer. A more efficient way, however, is to use the concept of complementary events. The complement to "applying to at least one other college" is "applying to no other college." From the table we see that $P(x = 0) = .376$, and $1 - .376 = .624$.

A Probability Distribution Graph. We've seen how probability distributions can be presented in tables. But discrete probability distributions are frequently shown graphically through the use of histograms (bar graphs) and frequency polygons (line graphs). We'll demonstrate the use of a histogram here, but other graphs depicting continuous probability distributions are also shown in the next chapter.

The probability distribution table giving the results of throwing a pair of dice was shown a few paragraphs earlier. The histogram for this same discrete probability distribution is shown in Figure 4.7. The possible numbers that could result when the pair of dice are rolled are centered in each bar on the horizontal axis. (These possible results are 2, 3, 4, . . . , 11, 12.) And the probability values associated with each possible outcome are plotted on the vertical scale. Thus, the height of each bar corresponds to the probability of getting the associated numerical result when the dice are thrown.

Expected Value

Another way to describe a probability distribution is to determine the theoretical average (mean) value that we could expect if we conducted an infinite number of trials. This measure of central tendency for a random variable is called the **expected value.**

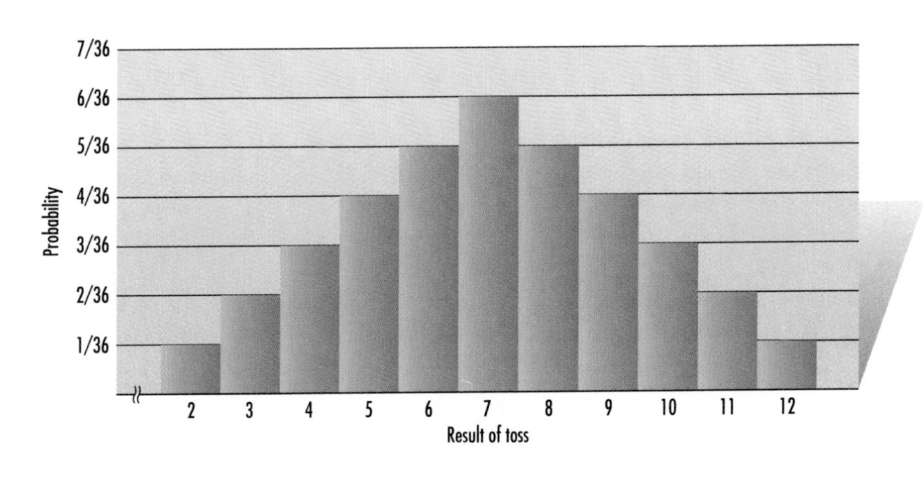

FIGURE 4.7 A histogram of the probability distribution of the results of throwing a pair of dice.

And the formula used to compute the expected value for a discrete random variable is:

$$E = \Sigma[x \cdot P(x)] \tag{4.8}$$

where E = the expected value
x = the value of each possible outcome
$P(x)$ = the probability of that possible outcome.

That is,

E = [(value of possible outcome 1) (probability of outcome 1) + (value of possible outcome 2) (probability of outcome 2) + ⋯ + (value of possible outcome n) (probability of outcome n)]

A little earlier we were given the following information that was published in *The Chronicle of Higher Education Almanac*:

Number of Other Colleges Applied to (x)	Proportion of Times Response Was Given [$P(x)$]
0	.376
1	.147
2	.158
3	.137
4	.078
5	.047
6	.056
Total	1.000

Now, let's calculate the expected value—that is, the mean number of college applications freshmen sent out to schools other than the ones they are attending:

$$E = \Sigma\,[x \cdot P(x)]$$
$$E = 0(.376) + 1(.147) + 2(.158) + 3(.137) + 4(.078) + 5(.047) + 6(.056) = 1.757$$

Of course, this doesn't mean that a particular student sent out 1.757 applications in addition to the one sent to the school being attended. (How would he or she submit three-fourths of an application?) What it does mean is that if the responses of a large number of students are analyzed, the average would be about 1.757 applications submitted per student.

Let's now consider a few other examples that show how a knowledge of expected value can help you make better choices about how to risk your money:

- Suppose you're asked to buy a \$25 raffle ticket for a car valued at \$9,500 and you're told that "only" 500 tickets will be sold. Is this a good way to spend your money? There are two possible outcomes. You'll either win the car and gain \$9,475 (the \$9,500 car less the \$25 that you don't get back), or you'll lose the raffle and your \$25. Thus, your expected value is:

$$E = \Sigma\,[x \cdot P(x)]$$
$$= \left[\left(\$9{,}475\right)\left(\frac{1}{500}\right) + (-\$25)\left(\frac{499}{500}\right)\right] = [(\$18.95) + (-\$24.95)] = -\$6.00$$

Since the expected value is negative, on the average you'll lose by buying the raffle ticket. You'll effectively contribute \$6 for each ticket you buy. In fact, even if you buy all 500 tickets to assure a win, you'll still lose big time.

- Now what if Peter Parker of Chapter 3 fame offers to pay you a dollar for each spot that appears on the surface of a die in a single toss. The price charged by Peter for each roll of the die is \$3. In Peter's game there are six possible outcomes, and six different payout amounts, as shown in the following table:

Die Surface (and Dollar Value) (x)	Probability of Occurrence of Surface P(x)	x · P(x)
1	1/6	.17
2	1/6	.33
3	1/6	.50
4	1/6	.67
5	1/6	.83
6	1/6	1.00
Total	1.00	3.50

$$E = \Sigma\,[x \cdot P(x)]$$
$$= \$3.50, \qquad \text{the expected average win}$$

If you played Peter's game *long enough*, you could expect an average win of \$3.50 on each roll of the die. It would be impossible, of course, to win \$3.50 on a single roll, but you ought to be able to eventually clean Peter out. If, for example, you play

1,000 games , your expected value will be (1,000) ($3.50) or $3,500, and your cost to play is (1,000) ($3.00) or $3,000. Peter again is victimized by statistical confusion.

Self-Testing Review 4.3

1–2. During their concerts held in the first half of 1992, The Grateful Dead played 115 different songs. A probability distribution for the number of times each of the 115 songs was played is given below:

Number of Times Song Was Played (x)	$P(x)$
1	.174
2	.339
3	.235
4	.191
5	.017
6	.009
7	.009
8	.000
9	.017
10	.009
Total	1.000

1. Construct a probability histogram for x (the number of times a song was played).

2. Compute the expected value for x.

3–6. A recent study in *The American Journal of Psychiatry* dealt with the clinical characteristics of autistic children. Twenty-one children were evaluated and given a developmental score from 1 to 5, where 1 represented the absence of signs (normal development) and 5 represented the maximal severity of signs (severe retardation). The probability distribution for this study is as follows:

Developmental Score (x)	$P(x)$
1	.19
2	.05
3	.14
4	.48
5	.14
Total	1.00

3. What's the probability that an autistic child has a developmental score of 3?

4. What's the probability that an autistic child has a developmental score of at least 3?

5. Construct a probability histogram for x (the developmental score).

6. Find the expected value of x.

7–10. A committee was formed to study the problem of parking spaces (or lack thereof) at Studywell College. Research was done to study enrollment patterns. It was found that 8 percent of the students were enrolled in classes 1 day a week, 19 percent had classes 2 days a week, 35 percent attended classes on 3 days, 29 percent were in class 4 days a week, and 9 percent had classes on 5 days.

7. Construct a probability distribution table for x (the number of days a student is enrolled in classes at Studywell).

8. Construct a probability histogram for x.

9. What's the probability a student is enrolled at least 3 days a week?

10. Calculate the expected value for x.

11–13. Grades for the Advanced Placement Examination are reported on a 5-point scale ranging from 1 to 5. Grade distributions for the 77,557 students who recently took the Calculus AB exam were reported as follows: 14,578 students earned a grade of 5, 15,976 scored a 4, 20,291 made a 3, 13,509 scored a 2, and 13,203 made a 1.

11. Calculate the probability that a student taking this test would score at least a 3. (This is the score that most colleges and universities accept for credit.)

12. What's the probability that a student taking this test scored less than a 4?

13. Compute the expected value for this test.

14. An investor has a .60 probability of making a \$20,000 profit and a .40 probability of suffering a \$25,000 loss. What is the expected value? Should she make the investment based on the expected value?

15–18. An automobile dealership in Denver has compiled the following sales data over the past year:

Number of Cars Sold per Day	Relative Frequency
0	.20
1	.30
2	.30
3	.15
4	.05
Total	1.00

53. Planning to attend a community college?

54. From Bucks County?

55. Going to a 4-year college, given that he or she is from Philadelphia County?

56. Going to attend a community college and is from Bucks County?

57. From Chester or Montgomery counties?

58. Going to attend a community college or is from Delaware County?

59. Not going to attend a 4-year college?

60. From Montgomery County, given that he or she is going to attend a 4-year college?

61–63. A survey is conducted to determine eye and hair color in a population. The results are shown in the following contingency table:

Hair Color	Blue Eyes	Brown Eyes	Total
Blond	10	30	40
Black	40	100	140
Red	5	25	30
Total	55	155	210

What's the probability that a person selected at random from this survey has:

61. Red hair?

62. Blue eyes and black hair?

63. Blue eyes or blond hair?

64–70. For each of the following, is the random variable discrete or continuous?

64. The number of credit hours you are currently taking.

65. The length of time you slept last night.

66. The distance you live from school.

67. The number of students in your statistics class.

68. The amount of gasoline you bought last week.

69. The number of tires you own.

70. The thickness of your statistics book.

71–72. The following statement appeared in a questionnaire: "A career in international business requires people skills." Respondents were asked to respond to this statement by using a 5-point scale (where 5 = strongly agree and 1 = strongly disagree). The responses of females were tabulated as follows:

x	$P(x)$
1	.00
2	.11
3	.11
4	.61
5	.17

71. Construct a probability histogram for the random variable x.

72. Compute the expected value for x.

73–74. When males were asked to respond to the same statement given in problems 71 to 72, the results were as follows:

x	$P(x)$
1	.02
2	.08
3	.17
4	.43
5	.30

73. Construct a probability histogram for the random variable x.

74. Compute the expected value for x.

75–77. In the game "Plinko" on *The Price Is Right*, a contestant drops a chip which randomly falls into a slot that is marked with a dollar designation. There are nine slots, two of which are marked $0, two are marked $100, two are marked $500, two are marked $1,000, and one is labeled $5,000.

75. If a chip is dropped, what's the probability that it will land in a $1,000 slot?

76. What's the probability a chip will land in a slot valued at less than $1,000?

77. What is the expected value for this game?

78. The marketing department of Gogetum, Inc., has determined that for the next fiscal year there is a probability of .40 that the company will lose $10,000, a probability of .50 that it will gain $20,000, and a probability of .10 that it will gain $30,000. What is the expected value in this situation?

Topics For Review and Discussion

1. (*a*) What is probability? (*b*) Discuss three applications of probability in everyday life.

2. (*a*) Discuss the three methods of assigning probability values. (*b*) Explain why a probability value can never be less than 0 or more than 1.

3. (*a*) When do you use a multiplication rule to determine probabilities? (*b*) When do you use an addition rule?

4. (*a*) Discuss the concept of independence of variables. (*b*) Give an example of two variables that are independent. (*c*) Give an example of two variables that are not independent.

5. (*a*) Discuss the concept of mutually exclusive and non–mutually exclusive variables. (*b*) Give two examples for each of these types of variables.

6. Discuss the relationship between conditional probability and independent events.

7. Give an example of a probability distribution of the variable x, where $x =$ the number that turns up when you roll a die.

8. (*a*) Discuss the procedure used to find the expected value of a discrete probability distribution. (*b*) Interpret the probable outcome of a game that has a negative expected value.

Projects/Issues to Consider

1. Locate an article in a periodical that discusses a study involving the concepts of probability. Prepare a brief summary of the article for a class presentation.

2. Research the contributions of a mathematician who pioneered in the development of probability concepts. Some possible names are: Thomas Bayes, Jacques Bernoulli, Francis Galton, William Gossett, Karl Friedrich Gauss, Christian Huygens, Pierre-Simon de Laplace, Abraham De Moivre, Blaise Pascal, Karl Pearson, and Simeon Denis Poisson.

3. Roll a die 60 times and have a partner keep a record of which face lands up. Did you get what you expected? Do you think your die is fair? If you did not get 10 rolls on each of the six numbers, explain why you still might think the die is fair. Now switch and have your partner roll the die 60 times while you keep the record of the outcomes. Combine the results. Write a paragraph discussing your findings.

Computer Exercises

1–3. Review the simulation exercises shown in Figures 4.1, 4.2, and 4.3, and then use your software to simulate the tossing of the coin 50 times.

1. How many heads did you get? How many tails?

2. Use your software to simulate the

tossing of the coin 200 times. How many heads did you get? How many tails?

3. What conclusions can you draw from your experiments?

4–6. Use your software to simulate the rolling of a fair die 180 times.

4. About how many times would you expect each of the six sides to appear face up in such an experiment?

5. How many times did each of the six sides actually appear in your simulation?

6. What conclusions can you draw from this experiment?

Answers to Odd-Numbered Self-Testing Review Questions

Section 4.1

1. There are 13 ways to be "successful" with 52 cards. The probability is 13/52 or 1/4 or .25.

3. .45

5. .61

7. 7,073/9,898 = .7146

9. 791/6,358 = .1244

11. 1/20 = .05

13. There are 47 + 5 = 52 students and 5 are left-handed, so the probability is 5/52 = .0962.

15. No, since the sum of the probabilities is not 1.

17. Yes, the sum of the probabilities is 1, and each probability is between 0 and 1.

19. Yes

21. 6,531/14,366 = .4546

Section 4.2

1. 1/10

3. 1/10

5. The events are independent, so we multiply to determine the joint probability: (1/1,000)(1/1,000) = 1/1,000,000 = .000001. It's incredible, but it actually happened!

7. Use the rule for complementary events: 1 − .45 = .55

9. (4/15)(3/14) = 2/35 = .0571

11. This translates to the first is brown and the second is brown. We have dependent events, so we multiply the probability that the first is brown times the conditional probability that the second is brown, given that the first is brown: (6/15)(5/14) = 1/7 = .1429.

13. We determine the probability that the die lands on a 1 or a 3 or a 5. These events are mutually exclusive, so we add 1/6 + 1/6 + 1/6 = 3/6 = .5.

15. 26/52 = 1/2 = .5

17. You multiply the probability that it's a face card by the conditional probability that it's a red card, given that it is a face card. Six of the 12 face cards are red, so (12/52)(6/12) = 6/52 or .1154.

19. .27

21. (.12)(.12) = .0144

23. This is an *or* question and the events aren't mutually exclusive, so after adding the individual probabilities, we must subtract the joint probability to make sure that there has been no double counting. Thus, 39/77 + 45/77 − 21/77 = 63/77 = .8182.

25. This question is equivalent to "the first is good *and* the second is good." Since the first is replaced before the second is drawn, the events are independent. Thus, (8/10)(8/10) = 64/100 = .64.

27. (.8)(.2) = .16

29. Since the order isn't specified, we must consider that the first is good and the second is defective, or the first is defective and the second is good. Adding the results from problems 27 and 28, we get .16 + .16 = .32.

31. The probabilities on the second draw now depend on the probabilities of the first draw. Thus, (8/10)(7/9) = 56/90 = .62.

33. (8/10)(2/9) = 16/90 = .1778

35. Consider the two different orders and add 16/90 + 16/90 = 32/90 = .3556.

37. Since the events are mutually exclusive, add 3/20 + 1/20 = 4/20 = .2.

39. 27/319 + 17/319 = 44/319 = .1379

41. 9/871 + 96/871 = 105/871 = .1206

43. 319/1,190 = .2681

45. 35/1,190 = .0294

47. 35/80 = .4375

49. Events are not mutually exclusive because having less than a high school education does not exclude the same individual from suffering from PTSD. In fact, there were 35 male Vietnam vets who had less than a high school education and who also suffered from PTSD.

51. 626/12,002 = .0522

53. 662/12,002 + 544/12,002 = .1005

55. 379/12,002 = .0316

57. 4,259/9,736 = .4374

59. P(female) = 5,142/12,002 = .4284. P(female, given from Philadelphia) = .4374. Although .4284 ≠ .4372, the values are so close that we can say that the events may be independent.

Section 4.3

1.

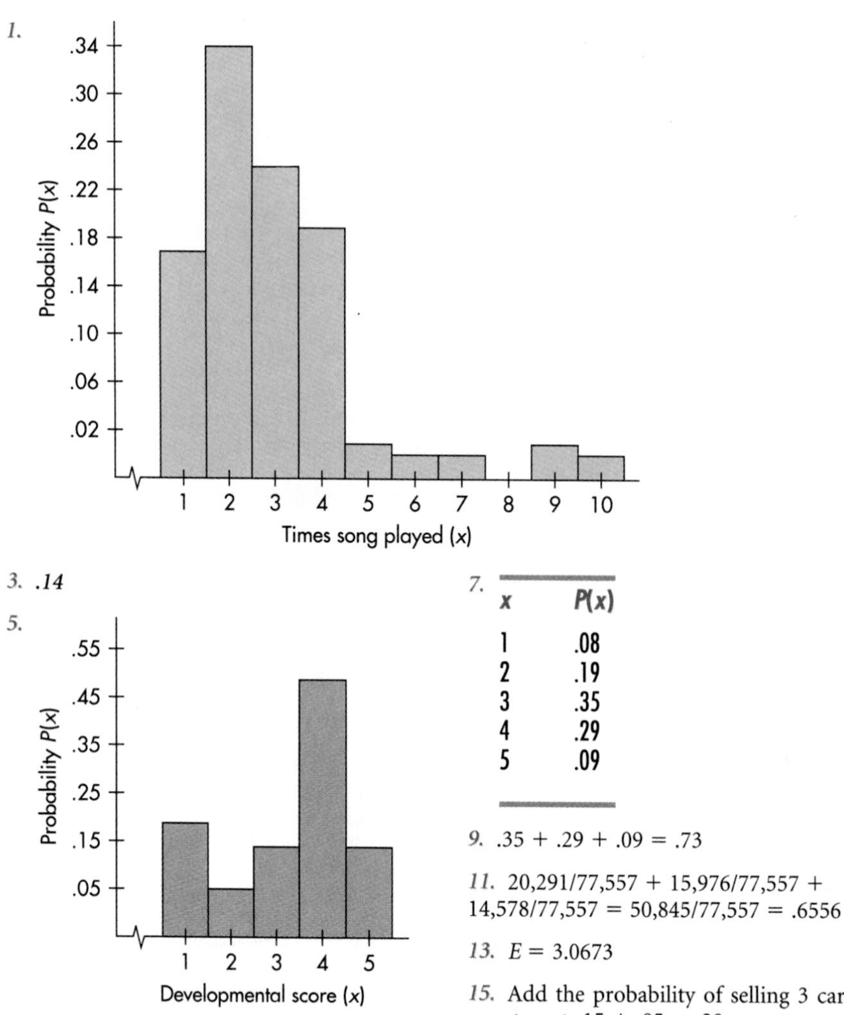

3. .14

5.

7.

x	P(x)
1	.08
2	.19
3	.35
4	.29
5	.09

9. .35 + .29 + .09 = .73

11. 20,291/77,557 + 15,976/77,557 + 14,578/77,557 = 50,845/77,557 = .6556

13. $E = 3.0673$

15. Add the probability of selling 3 cars or 4 cars to get .15 + .05 = .20.

17. $E = 0(.20) + 1(.30) + 2(.30) + 3(.15) + 4(.05) = 1.55$

Probability Distributions

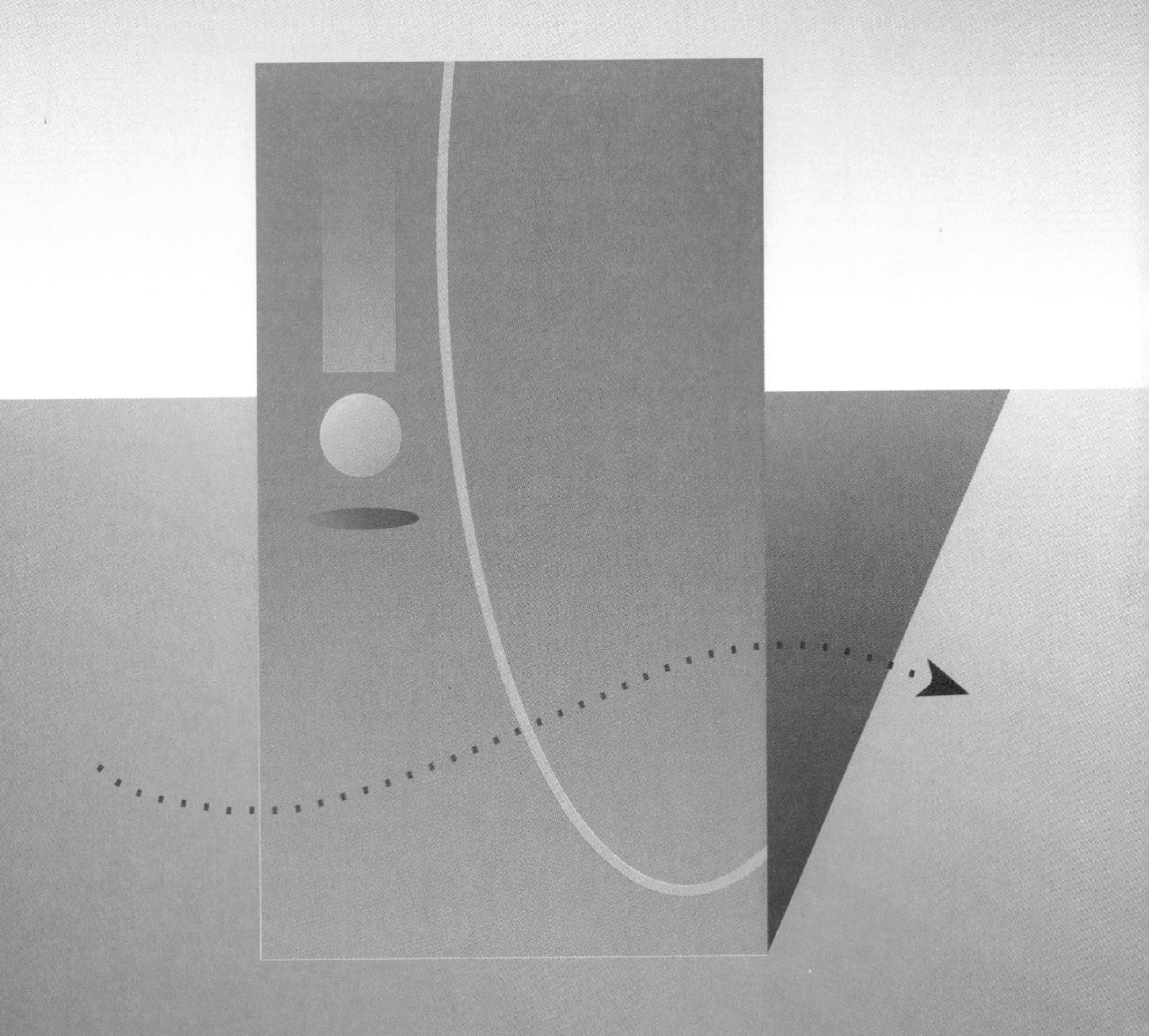

Quotable Quotations

"Lest men suspect your tale untrue, keep probability in view."—*John Gay*

"A reasonable probability is the only certainty."—*E. W. Howe*

"A thousand probabilities do not make one fact."—*Italian proverb*

"The laws of probability, so true in general, so fallacious in particular."—*Edward Gibbon*

"It is probable that many things will happen contrary to probability."—*Anonymous*

Source: From Hardeo Sahai, "Some Quotable Quotations Usable in a Probability and Statistics Class," *School Science and Mathematics,* October 1979, pp. 486–492. Reprinted with permission of *School Science and Mathematics.*

LOOKING AHEAD

You learned basic concepts of probability and probability distributions in Chapter 4. Now, we'll consider three probability distributions that are commonly used in statistics. These distributions are the *binomial, Poisson,* and *normal.* The binomial and Poisson are examples of discrete probability distributions, and the normal distribution is continuous. You'll see in the chapters that follow that an understanding of probability distributions—particularly the normal distribution—is essential to the study of sampling and statistical inference.

Thus, after studying this chapter, you should be able to:

- Explain what a binomial distribution is, identify binomial experiments, and compute binomial probabilities.
- Compute the expected value and standard deviation of a binomial experiment.
- Explain what a Poisson distribution is, identify Poisson experiments, and compute Poisson probabilities.
- Explain the properties of a normal distribution, and use the mean and standard deviation of a normally distributed random variable to translate random variable values into standard scores.
- Compute probabilities for a normal distribution and explain their relationship to areas under the standard normal probability curve.
- Determine standard z scores from specified probability requirements.

5.1 Binomial Experiments

You learned in Chapter 4 that a probability experiment is any action for which an outcome, response, or measurement is obtained. A binomial experiment is simply a particular type of probability experiment. In a **binomial experiment**:

1. The same action (trial) is repeated a fixed number of times.
2. Each trial is independent of the others.
3. For each trial, there are just two outcomes of interest. One outcome may be designated "success," and the other "failure."
4. The probability of success remains constant for each trial.

The random variable r represents a count of the number of *successes* in n trials in the experiment. Since we can list the possible values of r, it is a discrete random variable.

Knowing how eager you are to participate in a binomial experiment, we've prepared an example that requires you to take the following QuickQuiz:

QuickQuiz

Answer each of the following. There are four choices for each question, one of which is correct.

1. What is the 11th digit after the decimal point for the irrational number e?
   a) 2 *b)* 7 *c)* 4 *d)* 5
2. What was the Dow Jones Industrial Average on February 27, 1993?
   a) 3265 *b)* 3174 *c)* 3285 *d)* 3327
3. How many students from Sri Lanka studied at U.S. universities in 1990 to 1991?
   a) 2,320 *b)* 2,350 *c)* 2,360 *d)* 2,240
4. How many kidney transplant operations were performed in 1991?
   a) 2,946 *b)* 8,972 *c)* 9,943 *d)* 7,341
5. The *American Heritage Dictionary* (second college edition) has how many words?
   a) 60,000 *b)* 80,000 *c)* 75,000 *d)* 83,000

THE CHRONICLE OF HIGHER EDUCATION MISCHA RICHTER AND HARALD BAKKEN

"On the other hand, if you're not interested in good and evil, this one would give you a good understanding of statistical probability."

Assuming you're unable to answer with certainty any of these questions (and who could!), you can think of each question as a painful trial where guessing the correct answer is a success and guessing a wrong answer is a failure. Thus, this nightmare of a test serves our purpose because it illustrates a binomial experiment.

For such a binomial experiment, we'll let n equal the number of trials. The probability of success is the same for each trial, and we'll designate p as the probability of success in a *single* trial. Next, we'll let q equal the probability of failure in a *single* trial. Finally, since success and failure are complementary events, $p + q$ must equal 1 (or 100 percent).

For our QuickQuiz example, each question constitutes a trial, so n is 5. And since each question has 4 equally likely choices, one of which is correct, p is 1/4 and q is 3/4. The random variable r, which counts the number of correct answers (successes), can equal 0, 1, 2, 3, 4, or 5.

Now let's look at another binomial experiment example from a more useful (although still unpleasant) situation. Advertisements claim that 80 percent of the hypertension patients who respond to a certain drug have their condition controlled by taking 5 milligrams of the product per day. Ten such patients are given the drug to see if their hypertension has been controlled. This is a binomial experiment because each patient constitutes a trial, p is .80, q is .20, and $r = 0, 1, 2, 3, 4, 5, 6, 7, 8, 9$, or 10.

But not all experiments with repeated trials are binomial. Let's suppose that a bowl contains 5 green, 3 red, and 7 blue chips. Two chips are picked at random, one at a time and without replacement, to see if they are both blue. This isn't a binomial problem because the "trials" (picking a chip) aren't independent. The probability of picking a blue chip for the second trial depends on whether or not the first chip is blue.

Self-Testing Review 5.1

Identify the binomial experiments in the following examples. If the experiment is binomial, describe a trial, find n, p, and q, and list the values of r. If it isn't binomial, explain why.

1. An advertisement for *Vantin* claims a 77 percent end-of-treatment clinical success rate for flu sufferers. *Vantin* is given to 15 flu patients who are later checked to see if the treatment was a success.
2. According to *USA Today*, 32 percent of new health club members joined in the winter, 24 percent joined in the spring, 21 percent enrolled in the summer, and 23 percent joined in the fall. Seven people who joined health clubs last year are selected at random and asked for the season of the year they joined.
3. According to a study recently published in *The Journal of Abnormal Psychology*, 76 percent of seventh graders have consumed alcoholic beverages in their lifetime. Five seventh graders are randomly selected across the country and asked if they have ever had an alcoholic drink.
4. A recent study showed that 83 percent of the patients receiving heart transplants after 1988 were still alive. The files of six heart recipients were selected at random to see if each of the patients was still alive.
5. In a study of frequent fliers (those who made at least three domestic trips or one foreign trip per year), it was found that 67 percent had an annual income of over $35,000. Twelve frequent fliers are selected at random, and their income level is determined.

5.2 Determining Binomial Probabilities

Let's return to our terrible QuickQuiz. Suppose we're interested in finding the probability of correctly guessing exactly 3 questions on this test. This would give us a 60 percent correct score and would thus produce a barely passing grade at most schools. [Incidentally, the correct answers are: (1) *d*, (2) *a*, (3) *b*, (4) *c*, and (5) *b*. How many did you guess correctly?] We'll designate the outcome of answering the first 3 questions correctly (*S*) and the last 2 incorrectly (*F*) as *SSSFF*. Since each trial (answering a question) is independent, we can then compute *P*(*SSSFF*) using the multiplication rule for independent events that you learned in Chapter 4. Thus, $P(SSSFF) = P(S) \cdot P(S) \cdot P(S) \cdot P(F) \cdot P(F) = 1/4 \cdot 1/4 \cdot 1/4 \cdot 3/4 \cdot 3/4 = .00879$.

This .00879 value is the probability of getting the first 3 correct and the last 2 incorrect. Of course, as you know, there are other ways of getting exactly 3 correct out of the 5 questions (two such ways are *SFSFS* and *FFSSS*). Each of these ways is called a "combination of 5 items taken 3 at a time" (or choosing 3 out of 5). We can list all the combinations of three successes out of five tries and see how many there are, but let's first discuss a formula that does just that.

Combinations

A **combination** is a selection of *r* items from a set of *n* distinct objects *without regard to the order* in which the *r* items are picked. The symbol ${}_nC_r$ is used to designate the number of ways to choose *r* items from a group of *n* objects. The formula to find ${}_nC_r$ is:

$$ {}_nC_r = \frac{n!}{r!(n-r)!} \qquad (5.1)$$

where *n*! (or *n factorial*) represents the product of all integers from *n* down to 1
r! (or *r factorial*) represents the product of all integers from *r* down to 1

If *n* is 6, then $n! = 6 \cdot 5 \cdot 4 \cdot 3 \cdot 2 \cdot 1 = 720$. And if *r* is 4, then $r! = 4 \cdot 3 \cdot 2 \cdot 1 = 24$. For convenience, we define 0! to have a value of 1. Many calculators have a ! key.

Let's use our combination formula now to solve some example problems.

Example 5.1 How many ways can we choose 3 items out of 5—that is, what is the number of combinations of 5 things taken 3 at a time? You'll recognize that this is the number of ways we could guess the correct answer to 3 of the 5 QuickQuiz questions.

The answer is:

$$ {}_nC_r = \frac{n!}{r!(n-r)!}$$

$$ {}_5C_3 = \frac{5!}{3!(5-3)!} = \frac{5!}{3!2!} = \frac{5 \cdot 4 \cdot 3 \cdot 2 \cdot 1}{(3 \cdot 2 \cdot 1)(2 \cdot 1)} = 10$$

The 10 combinations for the QuickQuiz example are: *SSSFF*, *SSFSF*, *SFSFS*, *SFFSS*, *SFSSF*, *FFSSS*, *FSFSS*, *FSSFS*, *SSFFS*, and *FSSSF*. As you can imagine, it's usually impractical to try to list all combinations.

Example 5.2. How many ways can we choose 2 items out of 7? That is, what's the number of combinations of 7 items taken 2 at a time?

The answer is:

$$_nC_r = \frac{r!}{r!(n-r)!}$$

$$_7C_2 = \frac{7!}{2!(7-2)!} = \frac{7 \cdot 6 \cdot 5 \cdot 4 \cdot 3 \cdot 2 \cdot 1}{(2 \cdot 1)(5 \cdot 4 \cdot 3 \cdot 2 \cdot 1)} = 21$$

Calculating Binomial Probabilities with a Formula

Let's return now to the problem of determining the probability of correctly guessing *exactly* 3 of the 5 questions on the QuickQuiz. We've seen that the probability of guessing the first 3 correctly and then missing the last 2 is .00879. We found this by multiplying 3 factors of 1/4 (the probability of success) times 2 factors of 3/4 (the probability of failure). Now, since each of the 10 combinations of 3 successes and 2 failures has exactly the same probability of occurring, the probability of getting *any* 3 correct and 2 incorrect, no matter what the order, can be found by multiplying 10 times .00897 to get the answer of .0897.

In general, for a binomial experiment with n trials, where p is the probability of success and q is the probability of failure in a *single* trial, the probability of *exactly r successes* in the n trials is given by this formula:

$$P(r) = {_nC_r} p^r q^{n-r} \qquad (5.2)$$

To develop a probability distribution for r, the number of correctly guessed questions in the QuickQuiz, we can use formula 5.2 to compute $P(r)$ when r is 0, 1, 2, 3, 4, and 5 as follows:

$$P(r=0) = {_5C_0}\left(\frac{1}{4}\right)^0\left(\frac{3}{4}\right)^5 = \frac{5!}{0!(5-0)!}\left(\frac{1}{4}\right)^0\left(\frac{3}{4}\right)^5 = 1 \cdot 1 \cdot (.2373) = .2373$$

$$P(r=1) = {_5C_1}\left(\frac{1}{4}\right)^1\left(\frac{3}{4}\right)^4 = \frac{5!}{1!(5-1)!}\left(\frac{1}{4}\right)^1\left(\frac{3}{4}\right)^4 = 5 \cdot (.25)(.3164) = .3955$$

$$P(r=2) = {_5C_2}\left(\frac{1}{4}\right)^2\left(\frac{3}{4}\right)^3 = \frac{5!}{2!(5-2)!}\left(\frac{1}{4}\right)^2\left(\frac{3}{4}\right)^3 = 10 \cdot (.0625)(.4219) = .2637$$

$$P(r=3) = {_5C_3}\left(\frac{1}{4}\right)^3\left(\frac{3}{4}\right)^2 = \frac{5!}{3!(5-3)!}\left(\frac{1}{4}\right)^3\left(\frac{3}{4}\right)^2 = 10 \cdot (.0156)(.5625) = .0879$$

$$P(r=4) = {_5C_4}\left(\frac{1}{4}\right)^4\left(\frac{3}{4}\right)^1 = \frac{5!}{4!(5-4)!}\left(\frac{1}{4}\right)^4\left(\frac{3}{4}\right)^1 = 5 \cdot (.0039)(.75) = .0146$$

$$P(r=5) = {_5C_5}\left(\frac{1}{4}\right)^5\left(\frac{3}{4}\right)^0 = \frac{5!}{5!(5-5)!}\left(\frac{1}{4}\right)^5\left(\frac{3}{4}\right)^0 = 1 \cdot (.0009)(1) = .0010$$

Thus, the probability distribution for r is:

r	$P(r)$
0	.2373
1	.3955
2	.2637
3	.0879
4	.0146
5	.0010
	1.0000

And this probability distribution can be described in a histogram as shown in Figure 5.1.

Using a Table to Determine Binomial Probabilities

We've now laboriously calculated the probabilities $[P(r)]$ of correctly guessing 0, 1, 2, 3, 4, and 5 of our QuickQuiz answers. Given the tedious nature of these calculations, it's not surprising that tables have been prepared to produce binomial values for a selected number of trials using predetermined probability levels. Such a table is presented in Appendix 1 in the back of this book. To get the same results from the table that we've just computed using formula 5.2, you first read down the left-hand column of the table to get to $n = 5$. Then scan across to the values under the $p = .25$ column. The entire probability distribution that we've just calculated is listed, and you can select any value you need. That's a lot easier than using the formula, isn't it?

The probabilities for p listed in Appendix 1 only go to .50. But we can translate questions for greater values of p if we use the properties of a binomial experiment. Suppose you need to compute the probability of getting 6 successes in an experiment in which n is 9 and p is .70. To locate the correct entry, consider the equivalent probability of getting 3 failures in 9 trials in which the probability of failure is .30. You can check the table and verify that this answer is .2668.

Now let's consider some other binomial probability examples.

Example 5.3. What's the probability of correctly guessing *at least* 3 of our QuickQuiz answers? You can use formula 5.2 or the table in Appendix 1 to find the probabil-

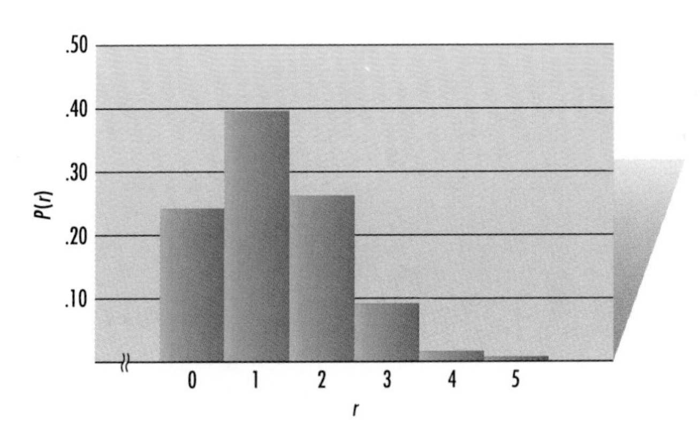

FIGURE 5.1 Our QuickQuiz probability distribution.

A Puff a Day Keeps the Doctor in Pay

To gauge the association between passive smoking (breathing the smoke produced by others) and adult lung cancer, the Environmental Protection Agency (EPA) recently compiled the results of 30 studies made in eight countries around the world. Each study compared lung cancer rates for two classes of nonsmoking women—those living with smokers and those living with nonsmokers. The EPA researchers now estimate that Americans who live or work among smokers experience a 20 to 30 percent increase in lung cancer risk. Other findings: Each year, environmental tobacco smoke causes 3,000 lung cancer deaths, helps to produce 150,000 to 300,000 respiratory infections in babies that result in 7,500 to 15,000 hospitalizations, triggers 8,000 to 26,000 new cases of asthma in previously unaffected children, and intensifies the symptoms in 400,000 to 1 million asthmatic children.

ity that $r \geq 3$—that is, $P(r = 3 \text{ or } 4 \text{ or } 5)$. You find this probability by adding the corresponding probabilities. Thus, $P(r \geq 3) = .0879 + .0146 + .0010 = .1035$. The next time you think about taking a multiple-choice test without studying, remember that there is roughly a 10 percent chance of correctly guessing at least 3 questions out of 5 when 4 options are offered for each question.

Example 5.4. What's the probability of correctly guessing *from 2 to 4* questions in our QuickQuiz? You can again use formula 5.2 or Appendix 1 to see that $P(2 \leq r \leq 4) = P(2) + P(3) + P(4) = .2367 + .0879 + .0146 = .3662$.

The Expected Value, Variance, and Standard Deviation of a Binomial Distribution

You'll remember from Chapter 4 that the *expected value* is the theoretical mean value for a random variable. And like any other probability distribution, each binomial random variable has an expected value. Since a binomial distribution is discrete, we can use the formula discussed in Chapter 4. Thus, to obtain the expected value for the number of questions we could correctly guess, we compute the value as follows:

$$E = \Sigma\, [x \cdot P(x)] = 0(.2373) + 1(.3955) + 2(.2637) + 3(.0879) + 4(.0146) + 5(.0010)$$
$$= 1.25$$

If many of your classmates take our QuickQuiz, the average number of questions answered correctly in the long run will be 1.25 per student. Thus, if 100 students take the test, we would expect about 125 correct answers, and if 1,000 take the test, we would expect to find about 1,250 correct responses.

Happily, though, there's a quicker way to find the expected value when you're working with binomial probabilities. The formula that makes this possible is:

$$E = np \tag{5.3}$$

To find the expected value of a binomial distribution, simply multiply the number of trials by the probability of success in a single trial. Thus, in our QuickQuiz example, $E = np = 5(.25) = 1.25$ questions.

You'll recall from Chapter 3 that the variance is a measure of dispersion and is the square of the standard deviation. The variance of a binomial distribution is found with this formula:

$$\sigma^2 = npq \tag{5.4}$$

So the variance in our QuickQuiz example is $\sigma^2 = (5)(.25)(.75) = .9375$. And the standard deviation of a binomial experiment is found with this formula:

$$\sigma = \sqrt{npq} \tag{5.5}$$

So the standard deviation in our QuickQuiz example is $\sigma = \sqrt{npq} = \sqrt{.9375} = .9682$ question.

Now let's consider one more example.

Example 5.5. According to the *Chronicle of Higher Education Almanac,* 37.5 percent of the college students in the United States use a personal computer. If 20 students are selected at random, what's the expected number who use a computer? What's the variance and standard deviation for this distribution? The expected value is $E = np = 20(.375) = 7.5$ students. The variance is $\sigma^2 = npq = (20)(.375)(.625) = 4.6875$. And the standard deviation is $\sigma = \sqrt{npq} = \sqrt{4.6875} = 2.165$ students.

Self-Testing Review 5.2

1–4. An advertisement for *Cold-Ex* claims an 80 percent end-of-treatment clinical success rate for the treatment of flu. Twelve flu patients are given *Cold-Ex* and are later checked to see if the treatment was successful. Let's assume that the claim is correct. Find the probability that:

1. Exactly 8 have been cured.

2. At least 10 have been cured.

3. Fewer than 5 have been cured.

4. Between 6 and 9 (including 6 and 9) have been cured.

5–7. According to a recent study published in *The Journal of Abnormal Psychology,* 1 percent of seventh graders use marijuana weekly. Ten seventh graders are randomly selected across the country and are asked if they use marijuana weekly. Assuming that they answer truthfully (?), find the probability that:

5. Exactly 1 uses marijuana weekly.

6. Fewer than 4 use marijuana weekly.

7. At least 5 use marijuana weekly.

8–10. A recent study found that of the patients receiving heart transplants after 1988, 83 percent were still alive. The files of 6 heart recipients are selected at random, and a follow-up study is made to see if each of the patients is alive. Find the probability that:

8. Exactly 4 are alive.

9. At least 5 are alive.

10. Four or fewer are alive.

11–13. A study was made of frequent fliers (those who made at least three domestic trips or one foreign trip per year). It was found that 67 percent had an income of over $35,000 a year. Twelve frequent fliers are selected at random and their income is recorded. Find the probability that:

11. Exactly 10 had an income of over $35,000 a year.

12. At least 10 had an income of over $35,000.

13. Nine or fewer had an income of over $35,000.

14–16. A fair coin is flipped 8 times. What's the probability of getting:

14. Exactly 4 heads?

15. Exactly 5 heads?

16. Four or 5 heads?

17–19. If 60 percent of a population are Democrats, what's the probability that a sample of 6 from this population will contain:

17. Exactly 4 Democrats?

18. Exactly 5 Democrats?

19. At least 4 Democrats?

20–22. A *J. D. Power* report shows that 30 percent of the new cars sold today are in the $13,500 to $17,400 price range. If a dealership sells 10 cars, what is the probability that:

20. Exactly 7 of them are in this price range?

21. At least 7 of them are in this price range?

22. No more than 2 of them are in this price range?

23–25. *The New York Times* recently reported that 45 percent of the households in Manhattan District 5 participated in recycling. If 20 households are selected at random from this district, find the probability that:

23. At least 15 participate in recycling.

24. Between 8 and 10 (including 8 and 10) participate in recycling.

25. Fewer than 5 participate in recycling.

26–28. In a poll conducted by the Olsten Corporation, a temporary personnel firm, 46 percent of the employers replied that their employees were less willing to give up free time today than they were 5 years ago. If 14 employees from these firms are selected at random, find the probability that:

26. Exactly 10 will be less willing to give up free time.

27. Exactly 11 will be less willing to give up free time.

28. Ten or 11 would be more willing to give up free time.

29–31. The Wyatt Company recently conducted a survey on the changing benefits packages offered by midsize and large companies. It was found that 45 percent of the companies offered family leave to care for a sick spouse. If 17 mid- to large-size companies are selected at random, find the probability that:

29. Exactly 10 will offer this type of leave.

30. At least 10 will offer this type of leave.

31. Fewer than 6 will offer this type of leave.

32–34. A multiple-choice test has 20 questions, and each question has 5 choices, 1 of which is correct. You don't have a clue about the subject matter and guess your way through the test. What's the probability you guess:

32. At least 12 questions correctly?

33. Fewer than 5 questions correctly?

34. At least 10 questions correctly?

35–36. An advertisement for *Cold-Ex* claims an 80 percent end-of-treatment clinical success rate for the treatment of flu. Twelve patients are given *Cold-Ex* and then later checked to see if the treatment was successful. Let's assume that the claim is correct:

35. What's the expected number of patients who are successfully treated?

36. What is the standard deviation of this binomial distribution?

37–38. In a Gallup poll, 53 percent of the eligible women responding said they would not date a co-worker. If 350 employed women are selected at random and asked if they would date a co-worker, what is the:

37. Expected number who would date a co-worker?

38. Standard deviation of this binomial distribution?

39–40. According to a recent study published in *The Journal of Abnormal Psychology*, 1 percent of seventh graders use marijuana weekly. Fifty seventh graders are randomly selected across the country and asked if they use marijuana weekly. Assuming that they answer truthfully (?), what is the:

39. Expected number of seventh graders who use marijuana weekly?

40. Standard deviation of the binomial distribution?

41–42. A recent study shows that of the patients receiving heart transplants after 1988, 83 percent are still alive. The files of 6 heart recipients are selected at random. A follow-up study is done to see if each of the patients is alive. What is the:

41. Expected number of patients still alive?

42. Standard deviation of this binomial distribution?

43–44. If 6.4 percent of the households in Manhattan District 10 participate in recycling, and if 1,000 households are selected at random from this district, find the:

43. Expected number of households that recycle.

44. Standard deviation of this binomial distribution.

5.3 The Poisson Distribution

We've just seen that if we know the probability of success in a single trial, the binomial probability distribution can tell us the probability of getting a specified number of successes for a given number of repeated trials. But other situations lend themselves to different probability distributions.

Suppose, for example, that we're interested in the number of specified occurrences that take place *within a unit of time or space* rather than during a given number of trials. In that case, it's appropriate to use the **Poisson distribution** (named after Simeon Denis Poisson, the French mathematician who developed it). The Poisson distribution can be used to calculate the probability that there will be a specified number of automobile accident claims coming to an insurance company during a *period of time.* Or it

Electrical engineers use statistical data to anticipate the fluctuations in the demand for electrical power. (*Comstock*)

may be used to calculate the probability that there will be a specific number of flaws found on the *surface space* of a sheet-metal panel used in the production of a space satellite.

Other situations with a time-unit reference that call for the use of the Poisson distribution in probability computations include demand for a product, demand for a service, number of accidents, and number of arrivals at tollbooths, supermarket stands, and airports. And another space-related application is the number of toxicants found in the volume of air emitted from a manufacturing plant. In these and other situations, it's appropriate to use a Poisson distribution when the following conditions are met:

- An experiment consists of counting the number of times a certain event occurs during a given unit of time or space.
- The probability that an event occurs is the same for each unit of time or space.
- The number of events that occur in one unit of time or space is independent of the number that occur in other such units.

Like the binomial, the Poisson distribution is discrete since the values of the random variable can be listed.

When the Poisson distribution is appropriate, the probability of observing exactly x number of occurrences per unit of measure (hour, minute, cubic centimeter, page) can be found using this formula:

$$P(x) = \frac{\mu^x e^{-\mu}}{x!} \tag{5.6}$$

where $P(x)$ = the probability of exactly x number of occurrences
μ = the mean number of occurrences per unit of time or space
e = a constant value of 2.71828 . . .

Note that many calculators have a key for computing e^x.

To see how this formula is used, let's look at an example.

Example 5.6. An average of 3 cars arrive at a highway tollgate every minute. If this rate is approximated by a Poisson process, what's the probability that exactly 5 cars will arrive in a 1-minute period? The answer is:

$$P(x) = \frac{\mu^x e^{-\mu}}{x!}$$

$$P(x = 5) = \frac{(3^5)(e^{-3})}{5!}$$ (If your calculator permits, find e^{-3} by keying -3 and then e^x.)

$$= \frac{(243)(.0498)}{120} = \frac{12.10}{120} = .1008$$

Thus, the probability that 5 cars arrive in 1 minute is .1008. Other results may be computed to show the probability of arrival of 0, 1, 2, 3, 4, 6, . . . , cars at the tollgate.

Appendix 10 at the back of the book has a table of Poisson probabilities for specified values of μ and x, and it gives us an easier way to solve this problem. The columns in this table represent selected values of μ, so our first step is to locate the column with a μ value of 3.0. Having located that column, we next identify the row with an x value of 5 that corresponds to the arrival of 5 cars in our example. The answer to our problem, then, is the value of .1008 that's found at the intersection of the identified column and row. The other entries in the μ column give the probability values for the arrival of 0, 1, 2, 3, . . . , cars at the tollgate. Since the total of all the entries under a specific μ column is approximately 1.00, these entries collectively represent a probability distribution.

One other interesting feature of the Poisson distribution is that the variance is equal to the mean. This, of course, implies that for this distribution $\sigma = \sqrt{\mu}$.

Self-Testing Review 5.3

1–2. A company reports that their computer is "down" an average of 1.2 times during an 8-hour shift. What's the probability that the computer will:

1. Be down 3 times during an 8-hour shift?

2. Not be down during an 8-hour shift ($x = 0$)?

3–5. Getrich Bank records show that an average of 20 people arrive at a teller's counter during an hour. What's the probability that:

3. Exactly 30 will arrive during a 1-hour period?

4. Ten or fewer will arrive during a 1-hour period?

5. More than 10 will arrive during a 1-hour period?

6–7. The Colorall Paint factory uses agent A in the paint manufacturing process. There's an average of 3 particles of agent A in a cubic foot of the air emitted during the production process. What's the probability that there will be:

6. Five particles of agent A in a cubic foot of air emitted from the factory?

7. No agent A in a cubic foot of air emission?

8–9. If a keyboard operator averages 2 errors per page of newsprint and if these errors follow a Poisson process, what's the probability that:

8. Exactly 4 errors will be found on a given page?

9. At least 2 errors will be found on a given page?

10–11. Crossville police records show that there has been an average of 4 accidents per week on Crossville's new freeway. If these accidents follow a Poisson process, what's the probability that the police must respond to:

10. Exactly 6 accidents in a week?

11. Fewer than 2 accidents in a week?

12–13. Bigrig Trailer Corporation uses large panels of sheet metal in the manufacture of tandem trailers. If there is an average of 3 blemishes per panel and if the blemishes follow a Poisson process, what is the probability that there will be:

12. No blemishes on a given panel?

13. Exactly 2 blemishes on a given panel?

14–16. The student health center at Oklahoma State University in Stillwater treats an average of 10 cases of severe alcohol poisoning a semester. Assuming a Poisson distribution, find the probability that the health center treats:

14. Twelve cases of severe alcohol poisoning a semester.

15. Fewer than 5 cases of severe alcohol poisoning in a semester.

16. More than 15 cases of severe alcohol poisoning a semester.

Let's Take in a Museum

Market research is used by the Metropolitan, Modern, and Guggenheim museums in New York City to track the economic impact these facilities have on the city. Research indicated that during a 3-month Seurat exhibit period, 74 percent of the museum visitors were out-of-towners who brought an estimated $313 million into the city. After seeing the exhibit, 61 percent of these out-of-towners dined in a restaurant, 57 percent went shopping, 33 percent caught a Broadway show, and 10 percent visited the Statue of Liberty. Similar research led the Philadelphia Museum to begin advertising its new late hours on a local rock-and-roll radio station.

5.4 The Normal Distribution

Both the binomial and Poisson distributions are **discrete probability distributions**—that is, they both have a countable or finite number of possible values. But what if the outcomes for a probability experiment consist of an uncountably infinite number of values—such as the measure of a seventh grader's height (where we could obtain values like 57.3 inches or 57.34 inches or 57.347 inches or 57.34719 inches)? In that case we need to use a **continuous probability distribution**—one that allows us to measure our variable to whatever degree of precision is required.

By far the most common and most important continuous probability distribution is the **normal** (or gaussian) **distribution**. An expression of its importance—and its shape—was presented by W. J. Youden of the National Bureau of Standards as shown in Figure 5.2*a*. And as indicated in Figure 5.2*a*, the normal curve is *symmetrical* (that is, the mean = median = mode) and, because of its appearance, it is sometimes called a "bell-shaped" curve. It's actually not a single curve but a family of curves, and each curve extends infinitely in either direction from the peak. Figure 5.2*b* shows three normal curves with the same mean but with different standard deviation values; Figure 5.2*c* shows three normal curves with the same standard deviation, but with different mean values.

A normal distribution curve is defined by the following formula which is based on its mean (μ) and its standard deviation (σ):

$$f(x) = \frac{1}{\sigma\sqrt{2\pi}} e^{-\frac{(x-\mu)^2}{2\sigma}}$$

But we don't need to concern ourselves with this formula. And we don't need to use the advanced mathematical techniques that would show you that the total area under any normal distribution curve, as in all other probability distribution curves, is equal to 1.

Normal Distribution Probabilities

You saw in Chapter 4 (and in Figure 4.7) that a histogram could be used to graphically present a discrete probability distribution. And you may recall that the probability that the variable takes on a *specified value* is equal to the area found in the corresponding bar of the probability histogram (assuming the width of the bar is 1 unit). But for continuous distributions, we don't find the probability that the variable assumes a specified value (after all, the probability of finding a seventh grader with a height of 57.34719 inches is virtually zero). Rather, we investigate the probability that the variable assumes any value within a given *interval* of values. We might, for instance, find the probability that a seventh grader's height is between 57.2 and 57.8 inches.

FIGURE 5.2 Normal distributions.

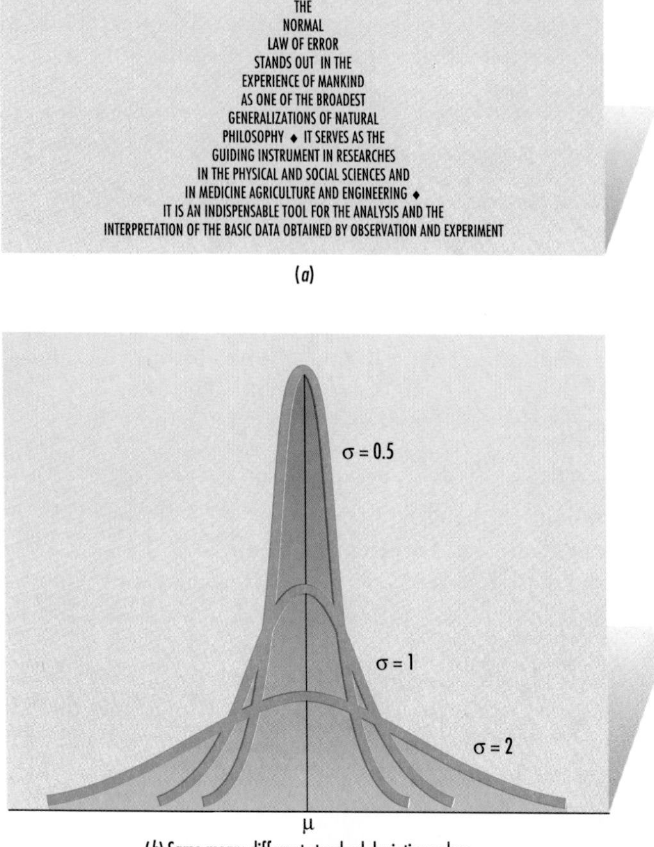
THE
NORMAL
LAW OF ERROR
STANDS OUT IN THE
EXPERIENCE OF MANKIND
AS ONE OF THE BROADEST
GENERALIZATIONS OF NATURAL
PHILOSOPHY ♦ IT SERVES AS THE
GUIDING INSTRUMENT IN RESEARCHES
IN THE PHYSICAL AND SOCIAL SCIENCES AND
IN MEDICINE AGRICULTURE AND ENGINEERING ♦
IT IS AN INDISPENSABLE TOOL FOR THE ANALYSIS AND THE
INTERPRETATION OF THE BASIC DATA OBTAINED BY OBSERVATION AND EXPERIMENT
σ = 0.5
σ = 1
σ = 2
μ

(*a*)

(*b*) Same mean; different standard deviation values

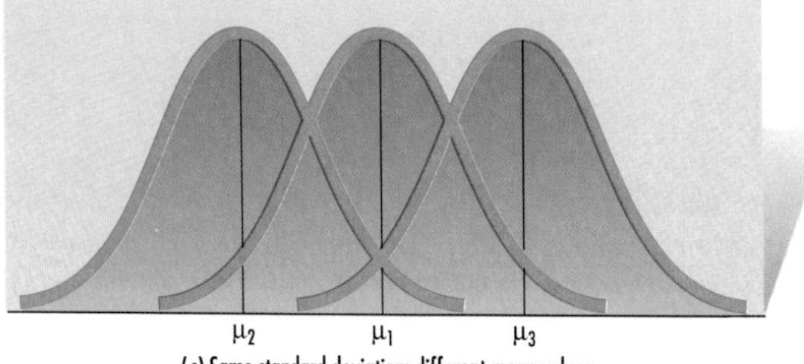
μ2
μ1
μ3

(*c*) Same standard deviation; different mean values

Probabilities for continuous distributions are represented by areas under the curve. That is, the *probability* that the variable will have a value *between a and b is the area under the curve between* two vertical lines erected at points *a and b.* For example, if the breaking strength of a material is normally distributed with a mean of 110 pounds and a standard deviation of 10 pounds, the probability that a piece of this material has a breaking strength between 110 and 120 pounds is the area under the curve covered by this interval, as shown in Figure 5.3.

Although there are an infinite number of different normal distribution curves (one each for any given pair of values of its mean and standard deviation), mathematicians have simplified things for us by calculating areas under a special normal distribution

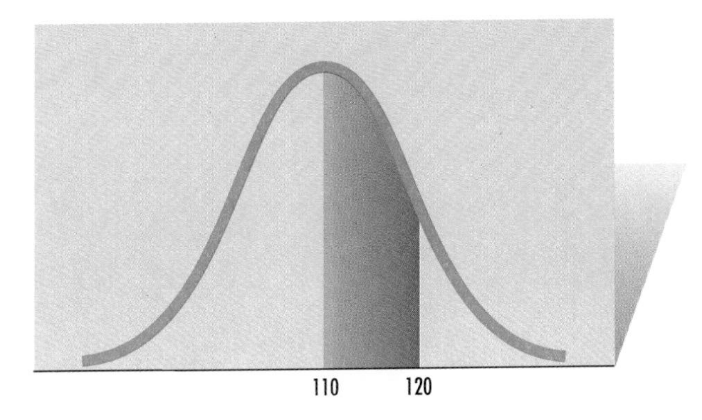

FIGURE 5.3 Probability of breaking strength between 110 and 120.

curve that has a mean (μ) of 0 and a standard deviation (σ) of 1. This specific curve is known as the *standard normal distribution.* It's of particular use because its values represent standard deviation units. The random variable for the standard normal curve is represented by the symbol z. Thus, a **z value** of $+ 2.00$ indicates 2 standard deviation units above its mean, and a z value of -1.40 represents 1.40 standard deviation units below its mean. A z value of 0 corresponds to the mean.

These points are illustrated in Figure 5.4, where it is shown that an interval of a given number of standard deviations from the mean covers the same area in any normal curve. Thus, the interval from 50 to 70 for a normal curve with a mean of 50 and a standard deviation of 20 covers the same area as the interval from 170 to 200 in a normal curve with a mean of 170 and a standard deviation of 30. Both of these intervals cover a distance of 1 standard deviation from the mean—that is, both intervals extend from the mean (0) out to a z value of $+1.00$ on our standard normal distribution scale.

Calculating Probabilities for the Standard Normal Distribution

To compute the probability that a z value falls within a given interval of the standard normal curve, we determine the area under the curve within the specified interval. A *table of area values* for the standard normal curve is found in Appendix 2 at the back of this book. The *first column* in the table gives values of z to the nearest tenth. The *remaining columns* give the second decimal place. The area values in the body of Appendix 2 represent the probability that a z value chosen at random will fall between the standardized mean of 0 and a specified z value which we'll call z_0. This probability can be stated in symbol form as $P(0 < z < z_0)$.

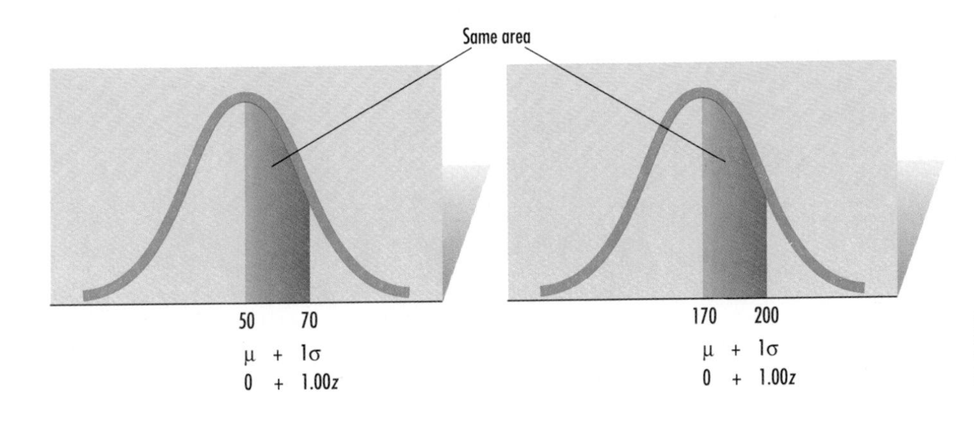

FIGURE 5.4 Both intervals extend from the mean ($z = 0$) to 1 standard deviation above the mean ($z = 1.00$).

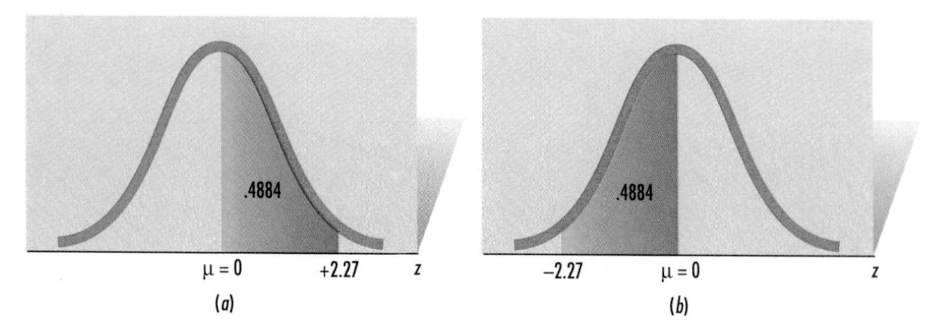

FIGURE 5.5 The probability that a z value selected at random will fall between 0 and 2.27 is .4884.

Let's use the table in Appendix 2 to find the area under the normal curve between the mean and a z value of 2.27. Go down the first column in the table until you get to 2.2, and then move across to the column labeled .07. The area given by the table is .4884. Thus, the probability that a z value selected at random will fall between 0 and 2.27 is .4884. Or in abbreviated form, $P(0 < z < 2.27) = .4884$ (see Figure 5.5*a*). It's important to note here that since the normal curve is symmetric, area values for *negative* z scores are identical to their corresponding positive scores. Thus, the area under the curve between the mean of 0 and a z value of -2.27 is also .4884 (see Figure 5.5*b*).

Now let's consider some additional examples of the use of the table of area values.

Example 5.7. What is the area under the normal curve between vertical lines drawn at $z = -1.73$ and $z = +2.45$? To answer this question, we must look up two separate areas. We first find the area between the mean of zero and a z value of -1.73 (remember that this area is the same as the area between the mean and a z value of $+1.73$). This area or probability is .4582. Next, we find that the area between the mean of zero and a z value of 2.45 is .4929. The required area obtained by adding .4582 and .4929 is .9511. Thus, $P(-1.73 < z < 2.45) = .9511$ (see Figure 5.6).

Example 5.8. What's the area under the normal curve between a z value of -1.54 and a z value of $-.76$? The area between the mean and a z value of -1.54 is .4382, and the area between the mean and a z value of $-.76$ is .2764. Since the given areas overlap, we subtract .2764 from .4382 to get .1618, which is the required area. Thus, $P(-1.54 < z < -.76) = .1618$ (see Figure 5.7).

Example 5.9. What's the area under the curve to the left of a z value of -1.96? From the table you can see that the area between the mean and a z value of -1.96 is .4750. Since the curve is symmetric and the total area is 1, we know the total area to the

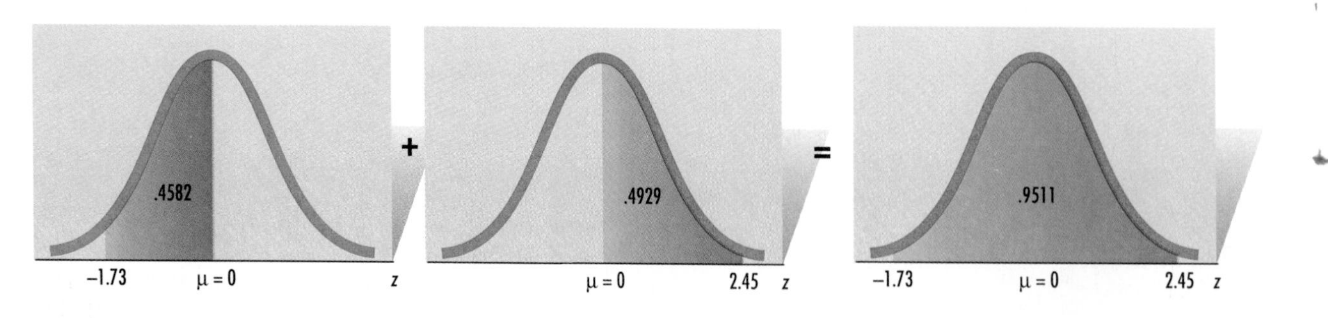

FIGURE 5.6 The area under the normal curve between vertical lines drawn at $z = -1.73$ and $z = +2.45$ is .9511.

left of the mean is .5. Thus, to find the required area, we subtract .4750 from .5000 to get .0250. So $P(z < -1.96) = .0250$ (see Figure 5.8).

Example 5.10. What's the area under the curve to the left of a z value of 1.42? We know that the area to the left of the mean is .5. And from Appendix 2 we find that the area between the mean and a z value of 1.42 is .4222. So the required area = .5000 + .4222 = .9222—that is, $P(z < 1.42) = .9222$ (see Figure 5.9).

You may have noticed that we've followed these steps to find the probability of getting a value within a particular interval:

1. We've identified the z value for each limit of the interval.
2. From the table of areas in Appendix 2 we've found the *area* for each z value.
3. If both limits of the interval are on *opposite sides* of the mean, we *add* the areas found in step 2; if the limits are on the *same side* of the mean, we *subtract* the smaller area from the larger one.

Sketching a picture of the needed area is helpful in such problems.

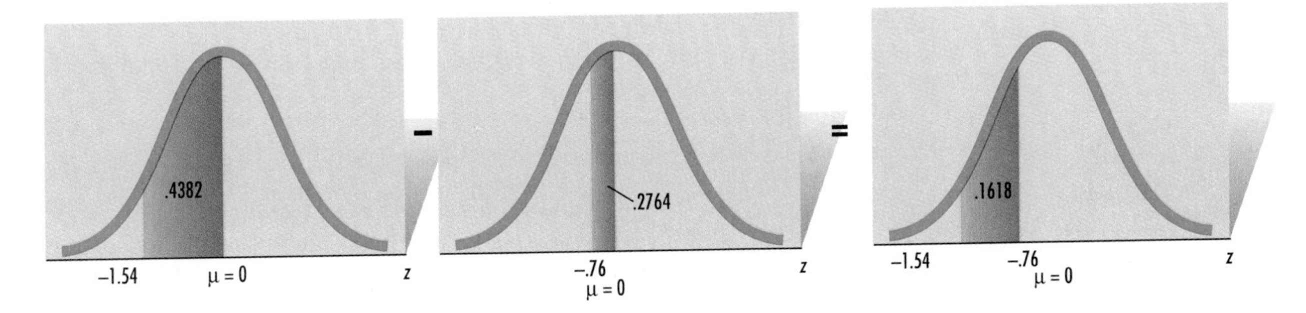

FIGURE 5.7 The area under the normal curve between a z value of −1.54 and a z value of −.76 is .1618.

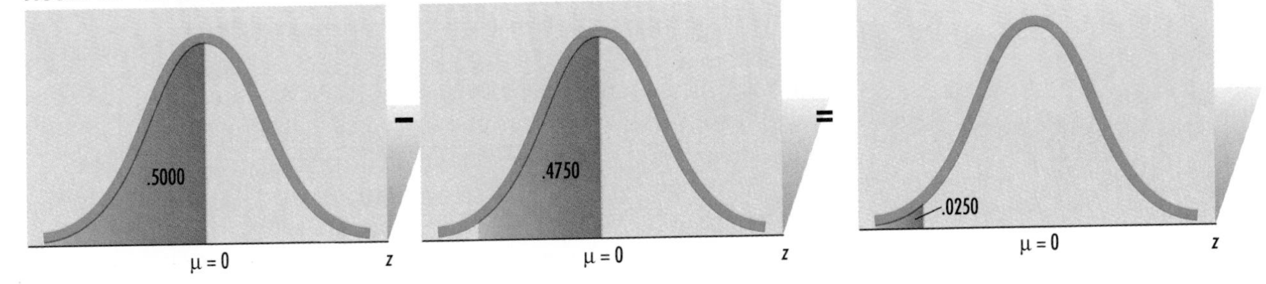

FIGURE 5.8 The area under the normal curve to the left of a z value of −1.96 is .0250.

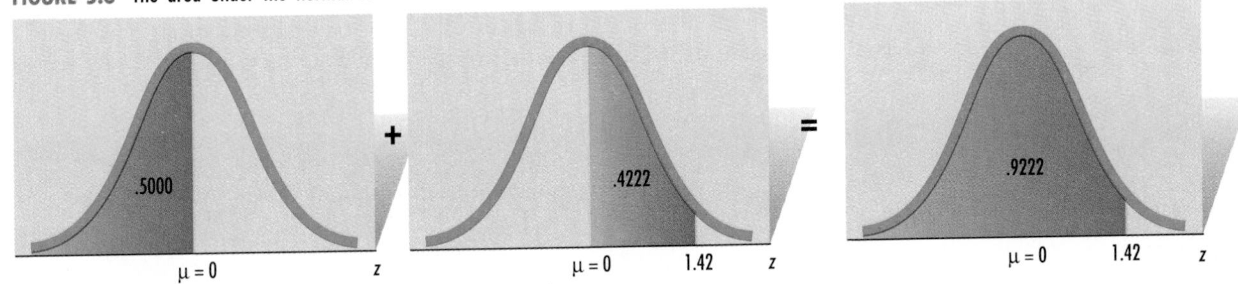

FIGURE 5.9 The area under the normal curve to the left of a z value of 1.42 is .9222.

Will the Next President Come from the NBA?

The New York Times has reported that prior to the 1992 election, the taller candidate was elected President of the United States in 18 of the preceding 22 elections. Candidates for the 1992 election were George Bush (6 feet, 2 inches), Ross Perot (5 feet, 6 inches), and Bill Clinton (6 feet, 2.5 inches). History will record that this pattern continues.

Computing Probabilities for Any Normally Distributed Variable

Few normally distributed real-world data sets have a mean of 0 and a standard deviation of 1! As you can imagine, it would be impossible to include a separate table for each combination of mean and standard deviation that one might expect to find in such data sets. Fortunately, though, we have a way to compute areas (and thus probabilities) for normal curves that aren't standard. Since the z value (the random variable for the standard normal curve) corresponds to the number of standard deviations a score is from its mean, we can see how many standard deviations *any* normally distributed variable (x) is from its mean by using the following formula:

$$z = \frac{x - \mu}{\sigma} \tag{5.7}$$

Thus, if we have a value of x for any normally distributed random variable and if we know the mean and standard deviation of its distribution, we can find the z score or *standard score*. Once we know the standard score, we can use Appendix 2 to find the required area(s) as we did in the previous section.

Now let's look at how to compute z scores and probabilities for any normally distributed random variable.

Example 5.11. If x is a normally distributed variable with a mean of 24 and a standard deviation of 3, what's the z score that corresponds to an x value of 19? Using formula 5.7 we get:

$$z = \frac{x - \mu}{\sigma} = \frac{19 - 24}{3} = -1.67$$

The score of 19 is 1.67 standard deviations *below* the mean of 24.

Example 5.12. If x is a normally distributed variable with a mean of 150 and a standard deviation of 24, what's the z score corresponding to an x value of 182? Well:

$$z = \frac{x - \mu}{\sigma} = \frac{182 - 150}{24} = 1.33$$

The score of 182 is 1.33 standard deviations *above* the mean of 150.

Example 5.13. If x is a normally distributed variable with a mean of 100 and a standard deviation of 15, translate the following interval into an interval of z scores:

$$70 < x < 130$$

We find the z score that corresponds to each endpoint of the interval in this way:

$$z = \frac{70 - 100}{15} = -2.00 \quad \text{and} \quad z = \frac{130 - 100}{15} = 2.00$$

FIGURE 5.10

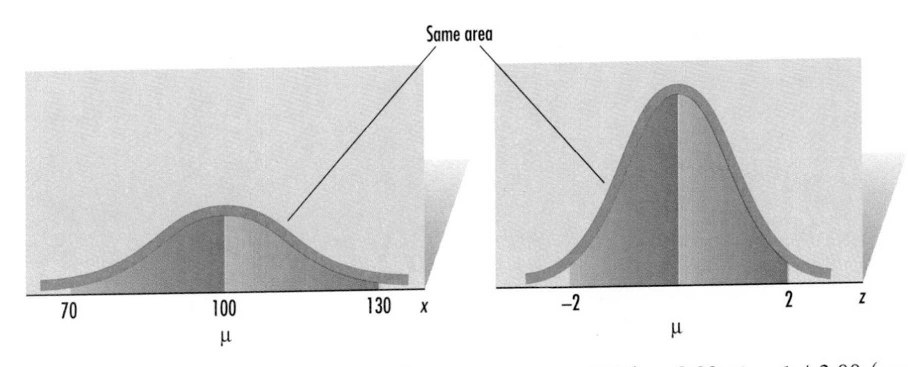

Thus, the z-score interval corresponding to $70 < x < 130$ is $-2.00 < z < +2.00$ (see Figure 5.10).

Example 5.14 For a certain IQ test, results are normally distributed with a mean of 100 points and a standard deviation of 15 points. What's the probability that a person chosen at random will have an IQ score between 70 and 130? What we're looking for is $P(70 < x < 130)$. We don't have a table of areas for a normal distribution with a mean of 100 and a standard deviation of 15. But in Example 5.13 we found the z-score interval that corresponded to this interval. So $P(70 < x < 130) = P(-2.00 < z < +2.00)$. We look up the required area figures in Appendix 2 for z values of $\pm$ 2.00 and find that it is .4772. Thus, the total area is .4772 + .4772 or .9544. About 95 percent of the population has an IQ score on this test between 70 and 130. It's important to realize that no matter what the normally distributed variable represents, about 95 percent of the population will fall within $\pm$2.00 standard deviations from the mean (see Figure 5.11).

Example 5.15. According to *The American Journal of Nutrition*, the mean midarm muscle circumference (MAMC) for males is a normally distributed random variable with a mean of 273 millimeters and a standard deviation of 29.18 millimeters. The MAMC for Clark Kant has been measured at 341 millimeters. What's the probability that a male in the population will have a MAMC measure *greater than* Clark's 341 millimeters? We obviously don't have a table that lists areas under the normal curve with $\mu = 273$ and $\sigma = 29.18$. So to find the number of standard deviations Clark's measure is from the mean, we use the z-score formula:

$$z = \frac{x - \mu}{\sigma} = \frac{341 - 273}{29.18} = 2.33$$

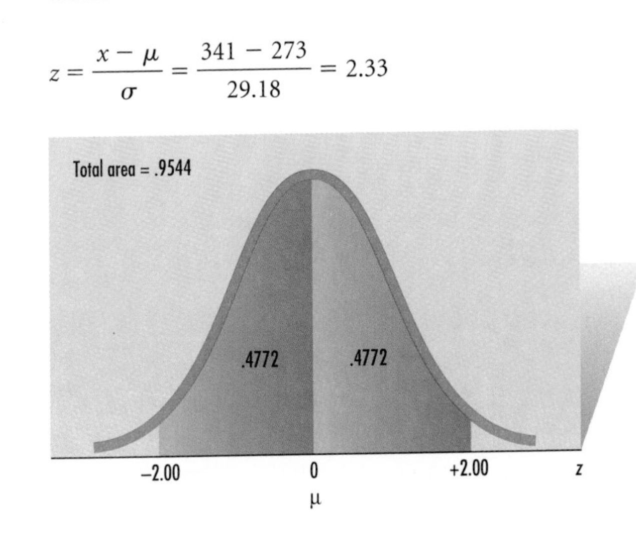

FIGURE 5.11

FIGURE 5.12

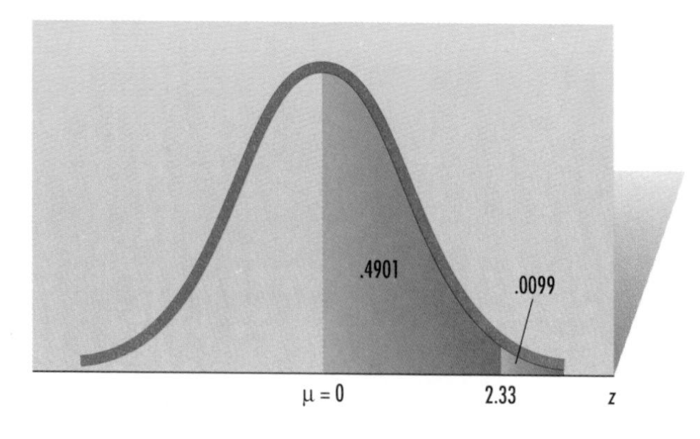

Once we know that Clark's measure is 2.33 standard deviations above the mean, we can use Appendix 2 to find the area value that corresponds to 2.33. This value is .4901. And the area above (or greater than) .4901 is .0099 (.5000 − .4901). Thus, the probability of a male's MAMC being greater than Clark's is .0099. That is, less than 1 percent of adult males have an MAMC measure greater than Clark's. In symbols: $P(x > 341) = P(z > 2.33) = .0099$ (see Figure 5.12).

Finding *z* Scores from Given Probabilities

We've now seen how to find the area under the standard normal curve when we're given the *z*-score boundaries. Now let's reverse the problem and find the *z*-score boundaries when a specific area (probability) is given. You'll recall that the areas (which correspond to the probability that a standard normal score will fall between the mean and the specified *z* score) can be found in the body of the table in Appendix 2.

Example 5.16. An area of .4370 lies under the standard normal curve between the mean and a given positive *z* score (see Figure 5.13). What is the value of that *z* score? The area values in the body of Appendix 2 are in ascending order, so we find .4370. The *z* score that corresponds to it is 1.53, and this is the answer.

Example 5.17. An area of .4808 lies under the normal curve between the mean and a given negative *z* score (see Figure 5.14). What's the value of that *z* score? We look for .4808 in the body of Appendix 2 and see that the required *z* score is −2.07.

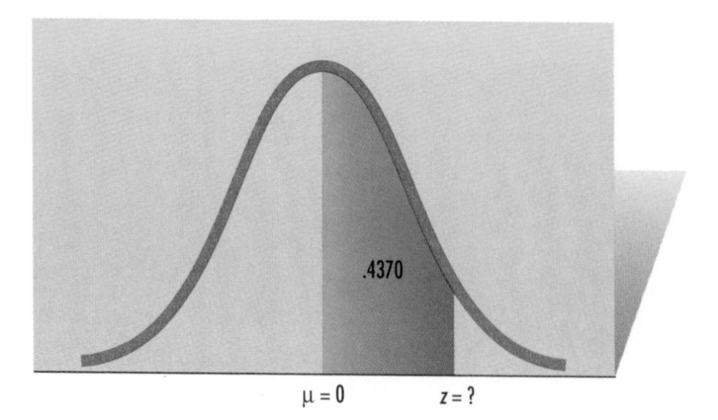

FIGURE 5.13

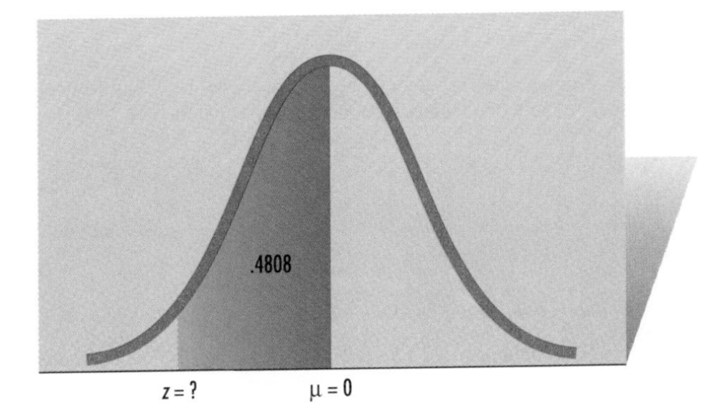

FIGURE 5.14

Example 5.18. Ninety percent of the normal curve lies to the left of a particular z score (see Figure 5.15). What's the value of that score? The required z score must be to the right of the mean (since 50 percent is to the left of the mean). That leaves .4000 of the curve between the mean and the required z score. So now we must find the z value that corresponds to an area of .4000 in the table (.9000 − .5000). There's no entry in Appendix 2 for exactly .4000, but the closest entry to it, an area of .3997, corresponds to a z value of 1.28.

Finding Cutoff Scores for Normally Distributed Variables

In many practical situations involving a normally distributed variable, it's necessary to find a cutoff value. That is, it's necessary to locate a value that separates the acceptable members of the population from those that aren't acceptable. Let's look at two examples involving cutoff values.

Example 5.19. Scores for a particular civil service exam are normally distributed with a mean of 137 points and a standard deviation of 17.2 points. Applicants for civil service jobs must take this test, and the top 10 percent can be offered jobs. What is the cutoff score that separates the highest 10 percent of the test scores from the others? We must first find the standard z score that separates the lower 90 percent of the curve from the upper 10 percent (see Figure 5.16). To locate that value in Appendix 2, it's necessary to look up an area of .4000 in the table. (If we want an area of 10 percent to be to the right of our z score, then 40 percent must be between the mean and the z score since the total area in the right half of the curve is 50 percent.) The z score that

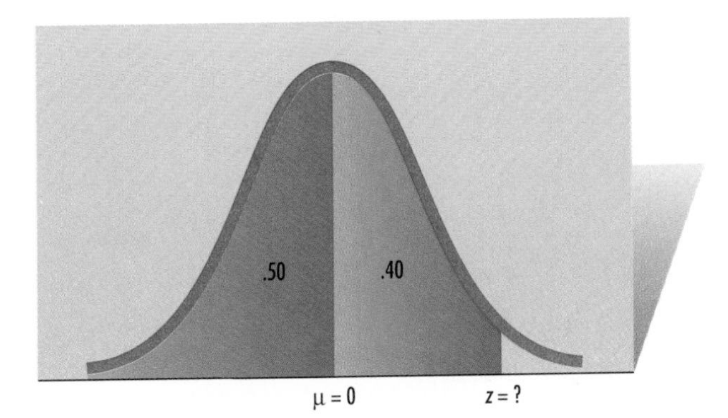

FIGURE 5.15

FIGURE 5.16

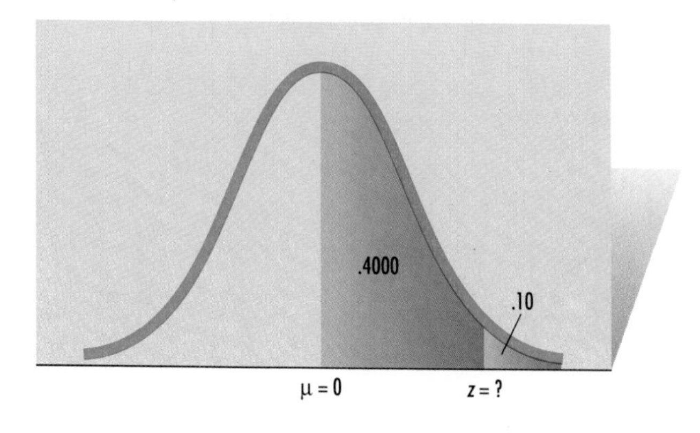

comes closest to our requirements is 1.28. We now use the known values of $\mu = 137$, $\sigma = 17.2$, and $z = 1.28$ to determine x. We know that:

$$z = \frac{x - \mu}{\sigma}$$

so we can multiply both sides of this equation by σ and then add μ to both sides of the equation to get this result:

$$x = \mu + z \cdot \sigma \tag{5.8}$$

Thus, $x = 137 + (1.28)(17.2) = 159$, and anyone who scores above 159 on this civil service test is in the top 10 percent of the job applicants.

Example 5.20. The time it takes members of the track team at Fasttrack University to run a 1-mile course is normally distributed with a mean of 5.6 minutes and a standard deviation of .76 minute. The coach has decided that the 5 percent of the team who can run the course in the least time will be sent to participate in a national track meet. What is the cutoff score that will decide which members of the team will qualify (see Figure 5.17)? If the area to the left of the required z score is .05, then the area between that z score and the mean is .4500. Looking up an area of .4500 in Appendix 2, we see it is halfway between the area entries of .4495 and .4505. The required z score is

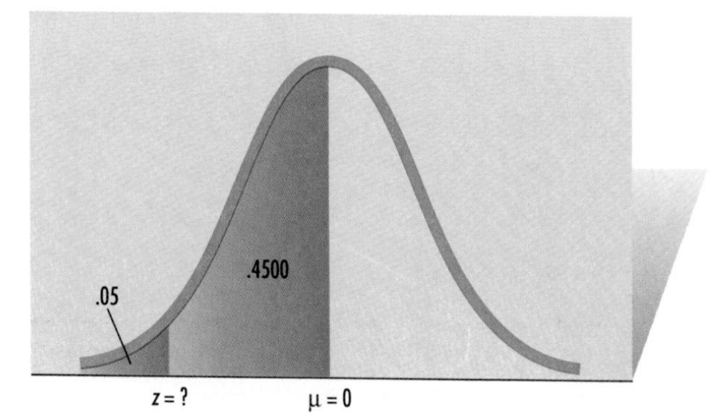

FIGURE 5.17

of binomial probabilities in Appendix 1. The expected value (E) of a binomial distribution can be computed using $E = np$, and the standard deviation of such a distribution can be found using the formula $\sigma = \sqrt{npq}$.

2. The Poisson distribution is another useful discrete probability distribution. It's used to find the probability of a specific number of occurrences that take place per unit of time or space rather than during a given number of trials. The formula:

$$P(x) = \frac{\mu^x e^{-\mu}}{x!}$$

or the table in Appendix 10 can be used to calculate the probability of x occurrences within a given time or space unit.

3. The most important continuous probability distribution is the normal distribution. A normal distribution curve is symmetrical and bell shaped. The total area under the curve is 1. The probability that a normally distributed random variable will take on a value between a and b is the area under the normal curve between two vertical lines erected at points a and b. The area covered by an interval under the normal curve can be found by using a table of areas such as the one found in Appendix 2. This table shows the area between the mean of a normal curve and a given number of standard deviations from that mean. The z score (standard score) of a variable is used to represent the number of standard deviations. A z score is calculated by the formula:

$$z = \frac{x - \mu}{\sigma}$$

Review Exercises

1–3. A binomial experiment is conducted with $n = 13$ and $p = .35$. Find the probability that the number of successes is:

1. Exactly 5.

2. At least 5.

3. Between 8 and 10, including 8 and 10.

4–6. A binomial experiment is conducted with $n = 6$ and $p = .40$. Find the probability that:

4. The number of successes is at least 3.

5. There are no successes.

6. There is at least 1 success.

7–9. It's found in an experiment that there's an average of 3.9 occurrences during a specified time interval. Assume a Poisson distribution applies and find the probability that there are:

7. Exactly 4 occurrences during the interval.

8. Exactly 3 occurrences during the interval.

9. Three or 4 occurrences during the interval.

10–12. It's found in an experiment that there's an average of 6.7 occurrences within a given space. Assume a Poisson distribution applies and find the probability that there:

10. Are no occurrences within the space.

11. Is at least 1 occurrence within the space.

12. Are more than 16 occurrences within the space.

13–18. If z is the standard normal variable, find the following probabilities:

13. $P(0 \leq z \leq 1.53)$

14. $P(-2.45 \leq z \leq 1.91)$

15. $P(z \leq 1.96)$

16. $P(-3.05 \leq z \leq -2.73)$

17. $P(z \geq 2.58)$

18. $P(z \leq -1.64)$

19–21. Physical Therapy recently reported that it has been estimated that 80 percent of all lower back pain cases are caused by weak trunk muscles. If 15 patients with lower back pain are examined, what's the probability that:

19. Exactly 10 have pain caused by weak trunk muscles?

20. More than 8 have pain caused by weak trunk muscles?

21. Fewer than 6 have pain caused by weak trunk muscles?

22–24. The U.S. National Center for Health Statistics reports that 60 percent of those with artificial hip and knee joints are female. If 20 people with artificial joints are selected at random, find the probability that:

22. Fewer than 10 are female.

23. All are female.

24. Between 12 and 15 (including 12 and 15) are female.

25–27. The response time for advanced life support ambulance service in Hilltown is normally distributed with a mean of 5.109 minutes and a standard deviation of 1.539 minutes. Find the probability that the time it takes to dispatch an ambulance delivering advanced life support is:

25. Less than 2 minutes.

26. More than 10 minutes.

27. Between 2.5 and 6.5 minutes.

28–30. The I.C.T. Telemarketing Group reports that the average employee makes 5 sales in an hour. Assuming a Poisson distribution applies, find the probability that an employee makes:

28. Three sales in an hour.

29. Four sales in an hour.

30. Three or 4 sales in an hour.

31–33. The New York Times recently reported that 200 years after the beheading of Louis XVI, 80 percent of French citizens said that the monarchy was a thing of the past. If 7 French citizens are questioned, find the probability that:

31. More than 4 think the monarchy is a thing of the past.

32. Fewer than 5 say the monarchy is a thing of the past.

33. Between 3 and 6 say the monarchy is a thing of the past.

34–36. In a recently published consumer research study investigating how restaurant customers responded to being touched by those serving them, it was reported that males who were touched by their server gave the restaurant a mean rating of 2.69 with a standard deviation of .83. Assuming the rating variable is normally distributed, find the probability that a male who was touched by a server gave the restaurant a rating of:

34. Between 2 and 4.

35. Less than 3.

36. More than 4.

37–39. In the study described in problems 34 to 36, it was found that males who were not touched by their server gave the restaurant mean and standard deviation ratings of 2.14 and .63, respectively. Find the probability that a male who wasn't touched by a server gave the restaurant a rating of:

37. Between 2 and 4.

38. Less than 3.

39. More than 4.

40–42. The American Academy of Orthopedic Surgeons reports that 15 percent of the operations performed are on the hip. If 17 orthopedic patients are selected at random, find the probability that:

40. More than 9 had hip operations.

41. Between 5 and 8, including 5 and 8, had hip operations.

42. Fewer than 3 had hip operations.

43–45. In a Gallup poll, 60 percent of the eligible men questioned said they would date a co-worker. If 10 employed men are questioned, find the probability that:

43. More than 7 would say they would date a co-worker.

44. Fewer than 4 would say they would date a co-worker.

45. At least 9 would say they would date a co-worker.

46–47. An ad for NORVASC claims: "Only 1.5% of patients in placebo-controlled experiments discontinued NORVASC due to adverse effects." If $n = 1{,}730$, find:

46. The expected number of patients who discontinued the drug due to adverse effects.

47. The standard deviation for the number of patients who discontinued the drug due to side effects.

48–49. *The American Journal of Psychiatry* published a study of World War II Pacific Theater combat veterans and POW survivors. For these veterans, 43 percent had alcohol abuse and/or dependence at some time during their life. If 28 of these veterans are randomly selected, find:

48. The expected number who reported alcohol abuse and/or dependence at some time.

49. The standard deviation for this distribution.

50–52. A study of teenagers and their drug use was recently published in *The Journal of Abnormal Psychology*. Forty percent of the teenagers in the study were not living with both natural parents. If a sample of 9 of these teenagers is selected, find the probability that:

50. More than 5 are not living with both natural parents.

51. At least 1 is living with both natural parents.

52. Fewer than 3 are not living with both natural parents.

53–55. The Kawasaki Zephyr is reported to have a mean fuel capacity of 4.5 gallons and a standard deviation of .3 gallon. If a Kawasaki Zephyr is picked at random, find the probability that it has a fuel capacity of:

53. Less than 4 gallons.

54. Between 4 and 6 gallons.

55. More than 5 gallons.

56–58. The fire department in Crossville can put out a fire in 1 hour, and the average number of alarms per hour is 2.4. Assuming a Poisson distribution applies, find the probability that:

56. No alarms are received for 1 hour.

57. At least 1 alarm is received in 1 hour.

58. Two alarms are received in 1 hour.

59–61. Asked to estimate their starting salaries after graduation, 70 percent of working students at a school said they expected to earn from $20,000 to $30,000 annually. If 12 working students from this school are selected at random, find the probability that:

59. At least 5 expect their starting salary to be between $20,000 and $30,000.

60. Fewer than 4 expect their starting salary to be between $20,000 and $30,000.

61. At least 2 expect their starting salary to be between $20,000 and $30,000.

62–64. A recent *Glamour Magazine* poll found that 35 percent of women ages 18 to 24 had not visited a gynecologist in the past year. If 15 women in this age group are selected at random, what's the probability that:

62. Fewer than 5 had not seen a gynecologist?

63. More than 10 had not seen a gynecologist?

64. Between 10 and 15, including 10 and 15, had not seen a gynecologist?

65–67. The Associated Press recently reported that small business owners will use more machines and work longer days rather than hire more people if demand for their products increases. Only 5 percent of the small businesses surveyed expect to hire more workers as the economy picks up. If 6 small businesses are selected at random, what is the probability that:

65. At least 1 of them will hire more workers?

66. None of them will hire new workers?

67. Two or more will hire new workers?

68–70. The Census Bureau reported that 55 percent of all 3- to 5-year-old children attended preschool programs at least a portion of the day. If 18 children aged 3 to 5 are chosen at random, what's the probability that:

68. At least 10 attend such a program?

69. Fewer than 6 attend such a program?

70. Between 9 and 14, including 9 and 14, attend such a program?

71–73. A recently published study in an education journal compared opinions of preservice teachers in their sophomore year, preservice teachers in their senior year, and in-service teachers. The average amount of time the sophomores thought should be spent in testing per week was 10.43 hours with a standard deviation of 3.72 hours. Assume these times are normally distributed, and find the probability that a preservice sophomore thinks that:

71. Between 5 and 9 hours should be spent in testing.

72. More than 10 hours should be spent in testing.

73. Fewer than 6 hours should be spent in testing.

74–76. In the study described in problems 71 to 73, the mean time that in-service teachers thought should be spent in testing per week was 4.37 hours with a standard deviation of 1.55 hours. What's the probability that an in-service teacher thinks that:

74. Between 5 and 9 hours should be spent in testing?

75. More than 9 hours should be spent in testing?

76. Fewer than 6 hours should be spent in testing?

77–79. *The American Journal of Public Health* recently reported on a survey that was designed to assess the extent to which soft plastic bread wrappers could be reused. Lead was detected in the printing on these wrappers. The mean amount of lead for these bags was 26 milligrams with a standard deviation of 6 milligrams. A soft plastic bread wrapper is selected at random. Find the probability (assuming a normal distribution) that the wrapper has:

77. More than 30 milligrams of lead.

78. Less than 15 milligrams of lead.

79. Between 15 and 30 milligrams of lead.

80–82. The Smart Potato Chip Company is conducting a marketing research study at the Allfood SuperMart. Customers are asked to taste two Smart chips and two chips made by Hisss, their leading competitor. If 65 percent of those participating in the taste test preferred Smart to Hisss, and if 10 consumers are later selected at random, what is the probability that:

80. They will all prefer Smart chips?

81. At least 5 will prefer Smart chips?

82. Between 5 and 10 will prefer Smart chips?

83–85. A study of the effect of predator introduction to artificial ponds was published in *Ecology*. Experimenters found that the average density of zooplankton in the pond was 4.60 individuals per centiliter. Assuming a Poisson distribution applies, what is the probability that a centiliter of fluid from the artificial pond in the study had:

83. Five individuals?

84. No individuals?

85. Four individuals?

86–88. A nationwide study of academic dishonesty among junior and senior college students was recently published in *Psychology Today*. The survey found that 70 percent of men and women confessed to cheating during high school. If a sample of 20 college students is surveyed, what's the probability that:

86. More than 12 confess to high school cheating?

87. Fewer than 10 confess to high school cheating?

88. Between 10 and 15 (including 10 and 15) confess to high school cheating?

89–91. A study conducted by the Higher Education Research Institute at UCLA shows that 2 out of every 5 freshmen participated in an organized demonstration last year. If 10 college freshmen were selected at random, what's the probability that:

89. More than 5 participated in a demonstration?

90. Fewer than 2 participated in a demonstration?

91. At least 1 participated in a demonstration?

92–94. The Higher Education Research

Institute at UCLA found that 30 percent of freshmen based their college choice on tuition costs. If 8 freshmen are selected at random, find the probability that:

92. Exactly 4 selected their college based on tuition.

93. Fewer than 5 selected their college based on tuition.

94. None selected their college based on tuition.

95–97. The New England Journal of Medicine recently published a study about patients who were treated by angioplasty after a heart attack. The number of days between the heart attack and successful angioplasty was normally distributed with a mean of 12 days and a standard deviation of 2 days. Find the probability that a patient's time between a heart attack and successful angioplasty treatment is:

95. More than 7 days.

96. Fewer than 14 days.

97. Between 7 and 14 days.

98–100. A *U.S. News & World Report* article showed that the average GMAT score of students entering Stanford University's graduate school of business in a recent year was normally distributed with a mean of 675 points and a standard deviation of 75 points. If a student entering Stanford's business school that year is randomly selected, find the probability that his or her GMAT score is:

98. Over 600.

99. Over 700.

100. Between 600 and 700.

101–103. A *U.S. News & World Report* article showed that the acceptance rate at the University of Chicago's graduate school of business was 30 percent in a recent year. If 10 applicants from that year are selected at random, find the probability that:

101. More than 5 were selected.

102. Fewer than 8 were selected.

103. Between 5 and 8 (including 5 and 8) were selected.

104–106. In a study of depressed patients, it was found that the age at onset of depression for those in dysfunctional families was normally distributed with a mean and standard deviation of 30.6 years and 9.1 years, respectively. A depressed person from a dysfunctional family is selected at random. Find the probability that the patient was:

104. More than 21 years of age at the onset of depression.

105. Less than 45 years of age at the onset of depression.

106. Between 21 and 45 years of age at the onset of depression.

107–109. In the study described in problems 104 to 106, it was found that the length of stay of depressed patients in a hospital was normally distributed with a mean of 23.3 days and a standard deviation of 7.7 days. A patient from a dysfunctional family, suffering from depression, is selected at random. What's the probability that his or her hospital stay was:

107. More than 30 days?

108. Fewer than 10 days?

109. Between 10 and 30 days?

110–112. Advertising Age has reported that the mean salary for creative directors working for ad agencies in a recent year was $85,000. Assuming these salaries are normally distributed with a standard deviation of $5,000, find the probability that if a creative director is selected at random in that year, his or her salary will be:

110. More than $75,000.

111. Less than $100,000.

112. Between $75,000 and $100,000.

113–115. The length of time sixth graders watch TV each day was found to be normally distributed with a mean and standard deviation of 118.3 minutes and 43.1 minutes, respectively. If a sixth grader is selected at random, find the probability that he or she watches:

113. More than 4 hours of TV.

114. Less than 1 hour.

115. Between 3 and 4 hours.

116–117. The Wyatt Company conducted a survey to learn the types of employee benefits offered by mid- and large-size companies. It was found that 39 percent of the companies offered guaranteed job reinstatement after maternity leave. If 250 such companies are selected at random, find the:

116. Expected number of companies that guarantee job reinstatement after maternity leave.

117. Standard deviation of this distribution.

118–119. Health, U.S., gives the most recent

analyses of data on health habits from the National Health Interview Survey. It was reported that 26 percent of the population 25 years and older smoked in a recent year. If 530 adults are selected at random, find the:

118. Expected number of those who smoked in that year.

119. Standard deviation of this distribution.

120–125. Find the z score such that:

120. 30 percent of the curve lies to the left of z.

121. 10 percent of the curve lies to the right of z.

122. 90 percent of the curve lies between $-z$ and $+z$.

123. 20 percent of the curve lies to the right of z.

124. 5 percent of the curve lies to the left of z.

125. 95 percent of the curve lies between $-z$ and $+z$.

126. The scores of an entrance exam are normally distributed with a mean and standard deviation of 550 and 95, respectively. Find the entrance score that separates the top 10 percent from the lower 90 percent.

127. If IQ scores on a particular test are normally distributed with a mean and standard deviation of 100 and 15, respectively, find the IQ score above which the top 5 percent of the population falls.

Topics for Review and Discussion

1. What is a binomial probability distribution? Discuss the conditions and requirements for a binomial experiment.

2. Why is the formula for combinations incorporated into the formula for computing binomial probabilities?

3. How may a Poisson distribution be used? What are the conditions and requirements for the use of such a distribution?

4. "The binomial and Poisson distributions are examples of discrete probability distributions." Discuss this sentence.

5. "Probabilities for continuous probability distributions are represented by areas under the curve." Discuss this sentence.

6. Discuss the procedure for computing normal curve probabilities.

7. Discuss the procedure for determining a z score (standard score) when a specified area under the standard normal curve is given. Use several examples in your explanation.

Projects/Issues to Consider

Conduct a library search to find examples of binomial, Poisson, and normally distributed variables. Write a brief report outlining your findings.

Computer Exercises

1. Use your computer software package to simulate the results of 100 students taking the QuickQuiz presented in this chapter. (If your software permits, you may use a RANDOM command and a BINOMIAL subcommand with $n = 5$ and $p = .25$.) Discuss your results.

2. The average number of customers who go to a teller's station at the Crossville National Bank in an hour is 12. Assuming a Poisson distribution, use your software package to simulate the number of customers who go to a teller's station each hour during a 40-hour time period. Discuss your results.

3. The midarm muscle circumference (MAMC) for males is normally distributed with a mean of 273 millimeters and a standard deviation of 29.18 millimeters. Simulate the MAMC measure for 500 males. What percent of the measures in your simulation were greater than 341 millimeters? Compare your results with those found in Example 5.15 in this chapter.

Answers to Odd-Numbered Self-Testing Review Questions

Section 5.1

1. This is a binomial experiment with a trial consisting of a flu patient being given *Vantin.* A success is being cured, $n = 15$, $p = .77$ or 77 percent, $q = .23$ or 23 percent, and $r = 0, 1, 2, \ldots, 15$.

3. This is a binomial experiment with seventh graders being asked if they ever had an alcoholic drink. A "success" (?) is that they respond yes, $n = 5$, $p = .76$ or 76 percent, $q = .24$ or 24 percent, and $r = 0, 1, 2, \ldots, 5$.

5. This is a binomial experiment. A trial is the determination of whether a frequent flier had an income of over $35,000 a year. A success is that income exceeds $35,000, and $n = 12$, $p = .67$ or 67 percent, $q = .33$ or 33 percent, and $r = 0, 1, 2, \ldots, 12$.

Section 5.2

1. For questions 1 to 4, $n = 12$ and $p = .80$ or 80 percent. To use the binomial table in Appendix 1, your answers must be found in the $p = 1.00 - .80$ or .20 column. So .20 or 20 percent is the rate of "noncure," and the number of noncures $= 12 - r$. For $n = 12$, $p = .80$, $r = 8$, look in Appendix 1 under $n = 12$, $p = .20$, and $r = 4$ to find that the probability of exactly 8 cures (4 noncures) out of 12 is .1329.

3. If fewer than 5 have been cured, this means that 0, 1, 2, 3, or 4 have been cured. This, in turn, means that more than 7—that is, 12, 11, 10, 9, or 8—have been noncures. The answer is .0006.

5. With $n = 10$, $p = .10$ or 10 percent, and $r = 1$, $P(r = 1) = .0914$.

7. This event is the complement to the one in problem 6. So $1 - 1.000 = 0$.

9. For at least 5 being alive, calculate the probability of 5 or 6 being alive and then add the two probabilities. That is, ${}_6C_5(.83)^5(.17)^1 = 6(.83)^5(.17)^1 = .4018$, and ${}_6C_6(.83)^6(.17)^0 = 1(.83)^6(.17)^0 = .3269$. So $.4018 + .3269 = .7287$.

11. Problems 11 to 13 require using the binomial formula since the probability of .67 or 67 percent is not found in Appendix 1. For the probability of exactly 10 with income over $35,000: ${}_{12}C_{10}(.67)^{10}(.33)^2 = 66(.67)^{10}(.33)^2 = .1310$.

13. Using the complement rule and the results from problem 12, $1 - .1876 = .8124$.

15. With $r = 5$, the probability is .2188.

17. With $p = .40$, the probability of exactly 2 non-Democrats is .3110.

19. Adding the probabilities of 0, 1, or 2 non-Democrats $= .0467 + .3110 + .1866 = .5443$.

21. .0105

23. .0064

25. .0188

27. .0112

29. .1008

31. .1470

33. .6296

35. $E = 12(.8) = 9.6$ or 10 patients

37. $E = 185.5$

39. $E = .5$

41. $E = 4.98$

43. $E = 64$

Section 5.3

1. With $\mu = 1.2$ and $x = 3$, the probability is .0867.

3. With $\mu = 20$ and $x = 30$, the probability is .0083.

5. Use the complement rule. The answer is .9892.

7. .0498

9. $1 - .4060 = .5940$

11. .0916

13. .2240

15. .0293

Section 5.4

1. $.4940 + .4738 = .9678$

3. $.5 - .3340 = .1660$

5. $.4990 - .4943 = .0047$

7. $.5 - .4997 = .0023$

9. .4564

11. $.4857 + .4838 = .9695$

13. $.4686 - .2734 = .1952$

15. $.5 - .4982 = .0018$

17. $P(z > 0) = .5$

19. $.3944 - .1915 = .2029$

21. $.4525 + .2486 = .7011$

23. $P(z > 0) = .5$

25. $.1554 + .2881 = .4435$

27. .5461

29. .0026

31. $.1554 + .3849 = .5403$

33. $.5 - .4861 = .0139$

35. $.4251 + .4545 = .8796$

37. .0455

39. $.5 - .4999683 = .0000317$

41. .3360

43. .7967

45. .9495

47. .1949

49. .3707

51. .3748

53. .1190

55. .5375

57. .9987

59. .5705

61. .3791

63. .0045

65. .1685

67. If .90 or 90 percent lies to the right of the unknown z score, then .40 lies between the mean and the required z score. The area closest to .40 is .3997, and this corresponds to a z score of -1.28.

69. Since $.50 - .25 = .25$, and the closest area to .25 is .2486, the z score from Appendix 2 is .67. So the answer is $z = -.67$.

71. This is equivalent to .4950 or 49.50 percent being between the mean and the unknown z scores. The areas of .4949 and .4951 are equally distant from .4950, so we use $z = \pm 2.575$.

73. The required z score is 1.645. So $x = 3.5 + 1.645(.4) = 4.16$ years or about 50 months.

CHAPTER 6

Sampling Concepts

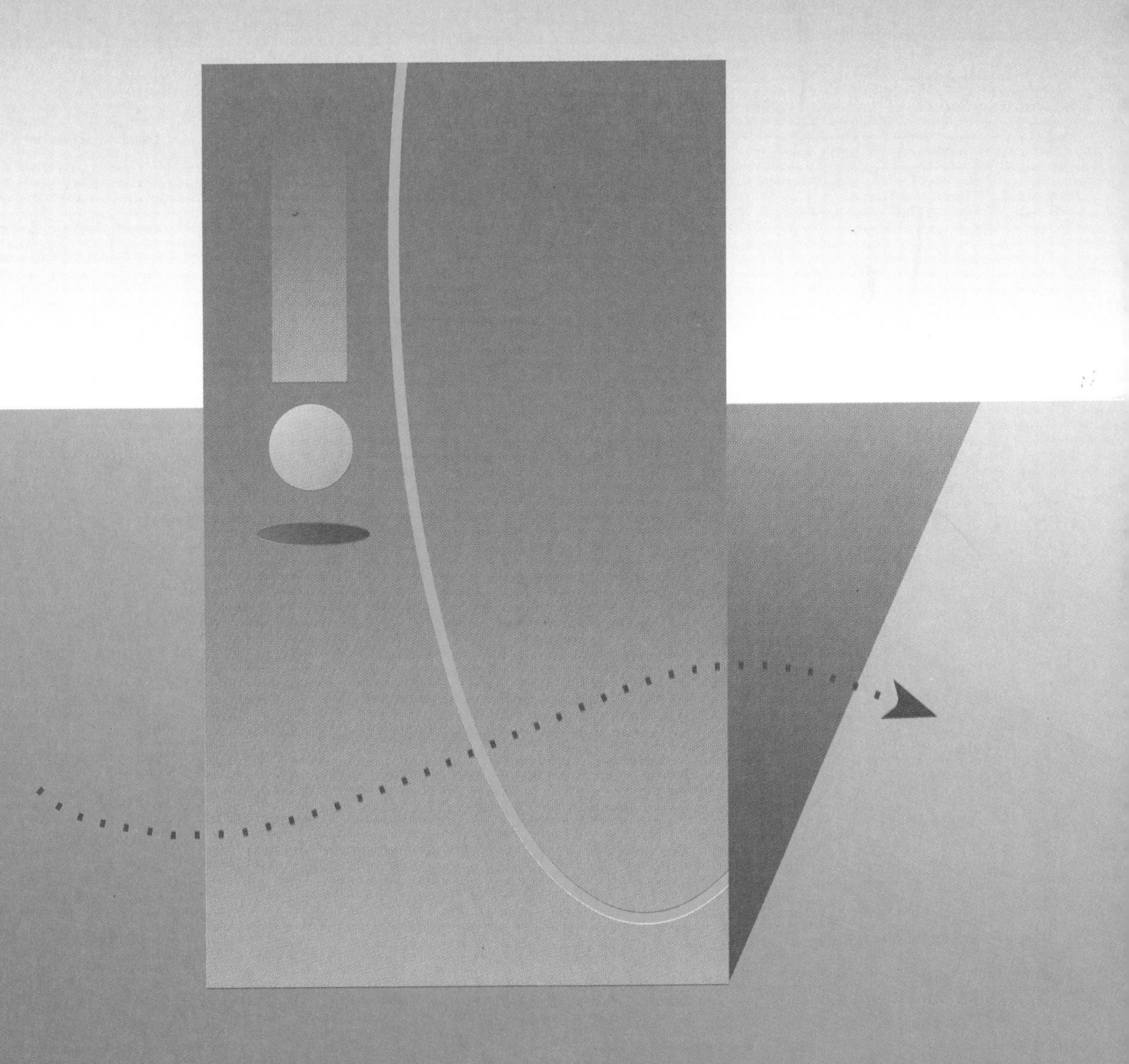

LOOKING AHEAD

Probability samples—the simple random, systematic, stratified, and cluster variations—were introduced and defined in Chapter 1. And Chapters 4 and 5 have focused on probability concepts and probability distributions. Now it's time to see why it is valid to use probability sample results to make inferences about population characteristics. Therefore, this chapter presents the theoretical and intuitive bases for using probability samples to estimate population values and to test hypotheses about those values.

Thus, after studying this chapter, you should be able to:

- Appreciate the need for sampling and the advantages that sampling may provide.
- Trace through the steps that are required to (*a*) produce a sampling distribution of sample means, (*b*) compute the mean of this sampling distribution, and (*c*) compute the standard deviation of this sampling distribution.
- Define the Central Limit Theorem and explain the relationship that exists between the standard error of the mean and the size of the sample.
- Trace through the steps necessary to (*a*) produce a sampling distribution of sample percentages, (*b*) compute the mean of this sampling distribution, and (*c*) compute the standard deviation of this sampling distribution,

6.1 Sampling: The Need and the Advantages

The Need

Sampling occurs frequently in the course of daily events and shouldn't be viewed as just a concept employed solely by statisticians. Although the samplings in daily life may not have the sophistication of formal statistical studies, they do serve a fundamental purpose of providing information for judgments. Here are a few examples:

1. A cook tastes a spoonful of soup to see if it has an acceptable flavor.
2. A prospective car buyer test-drives an automobile to compare it to others.
3. Pieces of ore are analyzed to learn the potential of a new mine.

You can undoubtedly add other similar examples to this list, but let's look now at the rationale for sampling.

Sampling is needed to provide sufficient information so that inferences may be made about the characteristics of a population. The population of interest may be *finite* or *infinite*:

> A **finite population** is one where the total number of members (items, measurements, and so on) is fixed and could be listed.
>
> An **infinite population** has an unlimited number of members.

Examples of finite populations are the computers installed in labs on your campus or the net weights of the 5,000 jars of jam filled in a production run. And to illustrate an infinite population, a computer could be programmed to generate the results produced by the rolling of a simulated pair of dice. If never turned off, the machine could produce an indefinitely large number of rolls of the dice.

Most of the time it's just not feasible to study an entire finite population to learn its true character, and, of course, it's impossible to consider all the elements in an infinite population. In the examples mentioned above, the cook can't taste the whole pot of soup to see if it's really acceptable, and the test driver can't drive the car for 3 years to find out if it will eventually be a lemon. But the data produced by sampling can be used to support judgments about the population.

The Advantages

Complete information acquired through a census is generally desirable. If every item in a population data set is examined, we can be confident in describing the population. But, as in many situations, what you want is not necessarily what you can get. Census data are a luxury in most situations and are usually not available for studying a population. Data gathering by sampling, rather than census taking, is the rule rather than the exception because of the *sampling advantages* discussed below.

Cost. Any data-gathering effort incurs costs for such things as mailings, interviews, and data tabulations. The more data to be handled, the higher the costs will likely be. Consider a consumer survey of the United States: If an attempt were made to poll every citizen, the cost would easily run into many millions of dollars. Any benefits derived from such census data would likely be negated by the cost. For example, a national food company might want to make a product change to improve sales. The company could survey every potential customer, but it's very likely that the costs of a census would wipe out any additional revenues generated from a changed product. Any time a sample can be taken with less expenditure than that required for a census, cost becomes an acceptable (although not sufficient) reason for sampling.

Time. Speed in decisions is often crucial. Let's assume you're the owner of a company and you've got an innovative idea for a better automobile antitheft device. You know that rival firms are also racing to create a similar product. Being the first to have a better product on the market may lead to high sales revenues, but do you actually have a better device? Will the public beat a path to your company's door for your idea, or will the new item fail to appeal to the public? Obviously, a census to help answer these questions requires too much time. The answer lies in sampling since it can produce adequate information about the public's response to your idea in a shorter period.

Accuracy of Sample Results. Sometimes a small sample provides information that's almost as accurate as the results obtained from a complete census. How is this possible? Remember that the object in sampling is to achieve representation of the population characteristics. There are sampling methods which produce samples that are highly representative of the population. In such cases, larger samples will not produce results that are *significantly* more accurate. Consider that soup again. If the cook has stirred it well before sampling, one or two sips should be sufficient to make a judgment about the entire pot. Any additional sips will only serve to decrease the volume of soup available for supper.

STATISTICS IN ACTION

Do You Still Want to Conduct a Census?

When the first U.S. census was carried out in 1790, a total of 17 marshals and 600 assistants traveled around the country counting all the inhabitants. Much of that work was done in door-to-door visits. In 1990, most of the census taking was done by mail, but 400,000 temporary government employees helped with the effort. A total of 248.7 million people were counted in the 1990 census. The total cost to carry out that constitutionally mandated task: $2.51 billion. Thus, it cost over $10 per person to count each person in the United States in 1990.

Focus Groups and Politicians

To determine campaign strategies, politicians hire pollsters to assemble people who are encouraged to say exactly what's on their minds. This assembly is called a *focus group*. One such group of about 30 voters from the Chicago suburbs had supported George Bush in 1988 but were undecided in 1992. Each member of this group was given a hand-held *perception analyzer*—a simple dial wired to a computer—and was instructed to move the dial to the right when President Bush said something they agreed with and to move the dial to the left when he said something they disagreed with. A high agreement level was reached in response to the statement that "this government is too big and spends too much." Other group agreements: Barbara Bush was a major asset, Bill Clinton needed a new haircut, and George Bush had lost touch with ordinary people.

Other Advantages. Destructive tests are often used to judge product quality. For example, a production manager may want to know the tensile strength of a truckload of iron bars she has received. To test their tensile strength, the bars are subjected to pressure until they break. All the bars can be tested, but only if the manager wants a truckload of broken bars. Since that's certainly not the case, a sample of bars must be used.

Sometimes the resources may be available for a census, but the nature of the population requires a sample. Suppose we're interested in the number of humpback whales left in the world. Environmental organizations may be willing to sponsor our count, but migration movements, births, and deaths prevent a complete count. One approach to the problem is to sample a small area of the ocean and use the results to make a projection.

Self-Testing Review 6.1

1. Give three illustrations of how sampling is used in daily life.

2. What is the difference between a finite and an infinite population?

3. What is the purpose in sampling?

4. Describe some advantages in using sample data to study population parameters.

5–11. Describe an appropriate population for each of the samples listed below:

5. Ten students from your statistics class.

6. A group of 37 psychology majors at your school.

7. Forty students from your school.

8. A group of 200 Geo Prisms assembled this week.

9. Stock prices today for 54 stocks listed on the New York Stock Exchange.

10. Fifteen schizophrenic patients undergoing an experimental treatment.

11. Seventy-two patients undergoing a new reconstructive knee surgery technique.

12–15. The freshman class at Casestudy College has 367 students. The Dean of Admissions collected data on 27 of them and found their mean score on one of the SAT tests was 517. The mean for the entire freshman class was therefore estimated to be approximately 517 on this test. A subsequent computer analysis of all freshmen showed the true mean to be 512.

12. What is the population?

13. What is the sample?

14. What number is a parameter?

15. What number is a statistic?

16. If sampling is done correctly, there are never any errors in using the sample information to describe the population from which the sample was drawn. True or false?

17. If the population remains the same, then all samples from that population will produce the same information. True or false?

6.2 Sampling Distribution of Means: A Pattern of Behavior

The sample mean seldom has exactly the same value as the population mean. Suppose the average income of a probability sample of city residents is $21,251. We could venture to say that the estimate of the population mean for all city residents is $21,251. But we intuitively know that the chances are slim that the sample mean *exactly* equals the population mean. A different sample of residents would most likely yield a different mean, such as $21,282, while another sample might produce a mean of $21,244.

In stating that a sample mean is an approximation of the population mean, we've made an assumption that the sample mean is related in some manner to the population mean. We intuitively assume that the value of the sample mean *tends toward* the value of the population mean. As we shall see later in this chapter, our intuition is correct: The population characteristics determine the range of values a sample mean may take.

Let's assume that we have a *population* of 15 cards. You'll remember from Chapter 3 (and Table 3.9) that the population size is identified by N, so $N = 15$ in this case. These cards (and population values) are numbered from 0 through 14. And let's further assume that we want to select random samples (without replacement) of size 6 from this population. You'll also remember from Chapter 3 that the sample size is identified by n. The number of *possible* samples that *could* be selected is a combinations problem of the type discussed in Chapter 5. We've been using ${}_nC_r$ to represent the combination of n things taken r at a time, but in this case it might be clearer to use ${}_NC_n$ to represent the combination of N items in a population taken n at a time in a sample. Anyway:

$${}_{15}C_6 = \frac{15!}{6!9!} = 5{,}005 \text{ possible samples}$$

One of these 5,005 possible samples consists of the cards numbered 2, 4, 6, 8, 10, and 12; a *second* possible sample selection is made up of the cards numbered 1, 4, 3, 7, 8, and 13; and a *third* possible sample comprises the cards numbered 14, 0, 7, 10, 9, and 8. (You can figure out the other 5,002 possible samples next summer at your leisure.) The arithmetic *means* of these 3 possible samples, in the order presented, are 7, 6, and 8. If there are 5,005 possible samples, there are, of course, 5,005 possible sample means. And if we were to select all the 5,005 possible samples, compute the mean of each of these samples, and arrange the 5,005 sample means in a frequency distribution, this distribution would be called a *sampling distribution of means*:

> A **sampling distribution of means** is the distribution of the arithmetic means of all the possible random samples of size n that could be selected from a given population.

If we're not careful at this point, we can run into some difficulties with definitions. Let's pause here to consider the three fundamental types of distributions in Figure 6.1. There's nothing new about Figure 6.1*a*; it's merely the frequency distribution of whatever population happens to be under study and could, of course, take many shapes. The mean and standard deviation of the *population distribution*—the symbols, you'll recall, are μ and σ—were discussed in Chapter 3 and are familiar to us. And there's really nothing very new about the distributions in Figure 6.1*b*; they are simply the

FIGURE 6.1 Educational schematic of three fundamental types of distributions.

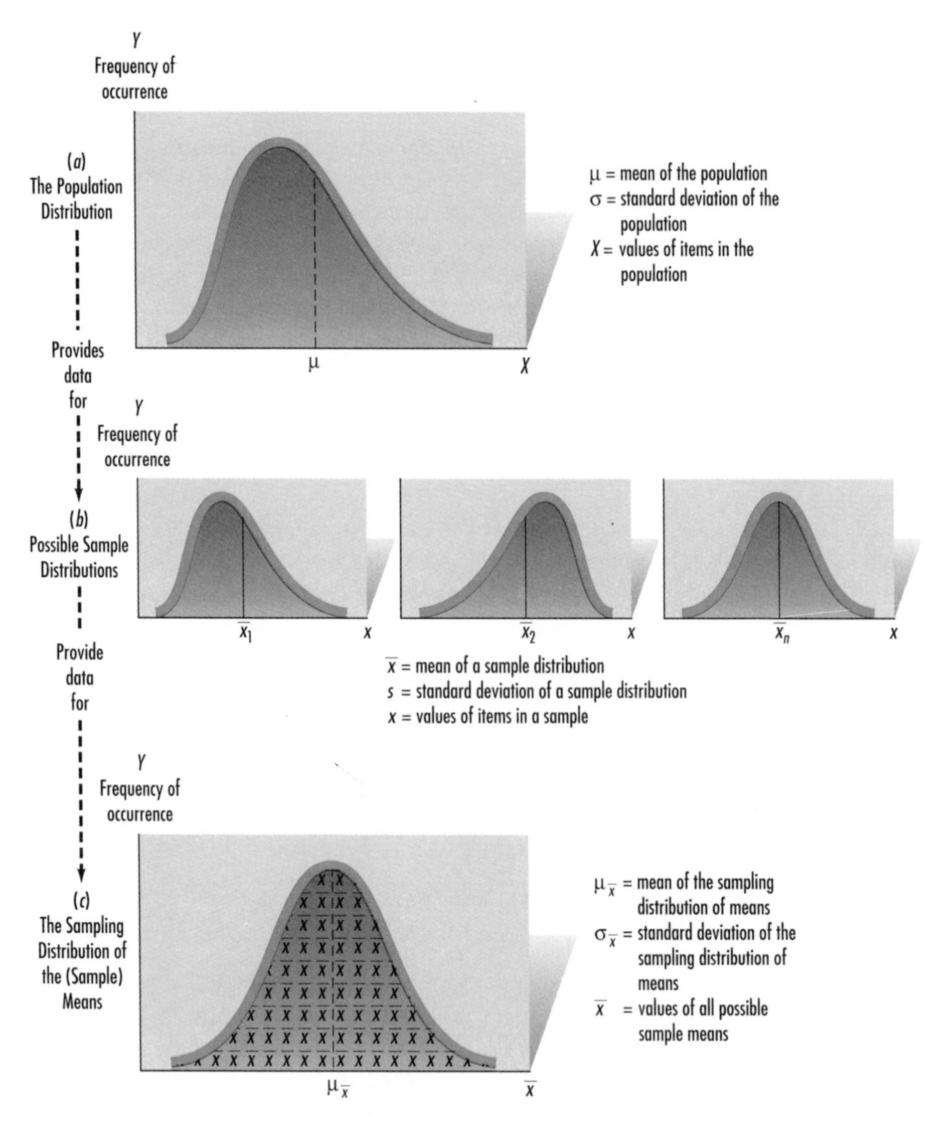

frequency distributions of several possible samples that could be selected from the population that is shown in Figure 6.1*a*. *Sample distributions* can have any shape. Each will have its own mean (remember that the symbol for the sample mean is $\bar{x}$) and its own standard deviation (that symbol is s). Of course, there are as many sample distributions as there are possible samples. Thus, for each of the 5,005 samples consisting of 6 cards from the population of 15 cards, there are 5,005 possible sample distributions and 5,005 possible sample means.

Finally, in Figure 6.1*c*, we come to the *sampling distribution of the means*—a distribution that *is* new to us and one that should *not* be confused with a *sample distribution* (even though the terms are confusingly similar). Thus, while we have 5,005 possible sample distributions—each with its own mean—there is but one sampling distribution of these 5,005 means. In the sampling distribution in Figure 6.1*c*, the possible sample mean values are distributed about the mean of the sampling distribution (sometimes called the *grand mean*). The mean of the sampling distribution is identified by the symbol $\mu_{\bar{x}}$ and is computed by adding up all the possible sample mean values and dividing by the number of samples. The standard deviation of the sampling distribution of the means measures the dispersion of the distribution and is identified by the symbol $\sigma_{\bar{x}}$.

Mean of the Sampling Distribution of Means

My center is giving way, my right is falling back, the situation is excellent. I attack.
—*Marshal Foch, Battle of the Marne, World War I*

Be alert and pay attention now because we are about to attack you with a very important fact: *The mean of the sampling distribution of means is equal to the population mean*—that is, $\mu_{\bar{x}} = \mu$. "What's that?" you said, "I have read some ridiculous statements in the past, and more than a few of them have come from this book, but . . ." Anticipating just such a skeptical attitude, we've prepared an example.

Let's assume the population is a small group of 5 students enrolled in a statistics course and an instructor wants to estimate the average amount of time spent by each student preparing for classes each week. Table 6.1 lists the amount of time each student spends per week preparing for class (but the instructor doesn't have access to this information). As we can see from Table 6.1, the population mean preparation time is 6 hours.

TABLE 6.1 POPULATION OF STUDENTS AND THEIR WEEKLY PREPARATION TIME

Student	Preparation Time (Hours)
A	7
B	3
C	6
D	10
E	4
	$\Sigma X = 30$

$$\mu = \frac{\Sigma X}{N} = \frac{30}{5} = 6$$

If the instructor takes a sample of 3 students, what are the possible values of the sample mean? How different from the true mean of 6 might a sample mean be? Table 6.2 provides the answer; it also provides us with the data needed to compute the mean of the sampling distribution using the following formula:

$$\mu_{\bar{x}} = \frac{\bar{x}_1 + \bar{x}_2 + \bar{x}_3 + \cdots + \bar{x}_{{}_NC_n}}{{}_NC_n} \tag{6.1}$$

$$= \frac{5.33 + 6.67 + 4.67 + \cdots + 6.67}{5!/(3!)(2!)} = \frac{60}{10} = 6$$

Dropout Factors

Roger McIntire, a professor at the University of Maryland, has conducted a study to identify factors that will help predict if a student will drop out of college. Using a sample of 910 Maryland students, he found that those who worked more than 21 hours a week, paid more than 30 percent of their own expenses, commuted 8 minutes or more from home to campus, spent less than 2 hours a week socializing on campus, and had fewer than two friends on campus were the ones most likely to quit school. McIntire concluded that campus jobs and affordable housing might help more students stay in school.

Thus, as you can see, the mean in Table 6.1 equals the mean in Table 6.2. That is, $\mu_{\bar{x}} = \mu$.

TABLE 6.2 SAMPLING DISTRIBUTION OF MEANS

Sample Combinations	Sample Data	Sample Means ($\bar{x}$)	$(\bar{x} - \mu_{\bar{x}})$	$(\bar{x} - \mu_{\bar{x}})^2$
1. A, B, C	7, 3, 6	5.33	−.67	.45
2. A, B, D	7, 3, 10	6.67	.67	.45
3. A, B, E	7, 3, 4	4.67	−1.33	1.77
4. A, C, D	7, 6, 10	7.67	1.67	2.79
5. A, C, E	7, 6, 4	5.67	−.33	.11
6. A, D, E	7, 10, 4	7.0	1.00	1.00
7. B, C, D	3, 6, 10	6.33	.33	.11
8. B, C, E	3, 6, 4	4.33	−1.67	2.79
9. B, D, E	3, 10, 4	5.67	−.33	.11
10. C, D, E	6, 10, 4	6.67	.67	.45
		60.0		10.0

$$\mu_{\bar{x}} = \frac{\Sigma(\bar{x}_1 + \bar{x}_2 + \bar{x}_3 + \cdots + \bar{x}_{N}C_n)}{{}_NC_n} = \frac{60}{10} = 6$$

where the numerator is the sum of all possible sample means, and the denominator is the number of possible samples.

This isn't an isolated case. Let's go back to our earlier example of the population of 15 cards numbered 0 to 14. The sum of the integers on the 15 cards is 105, and the mean of this population of 15 numbers is 7 (105/15). We didn't take those 5,000+ different samples to prove that $\mu_{\bar{x}}$ equals this μ of 7. But we did take 150 samples of 6 cards from our population of 15, and we then computed the sample mean for each of these 150 samples. Well, actually we didn't do this tedious task at all. Rather, we turned it over to a computer running the *Minitab* statistical package and promptly received the output shown in Figure 6.2.

In the first two lines of Figure 6.2, the program was told to put 150 samples into 150 rows, with each row or sample having 6 data items randomly selected from the integers 0 to 14. Line 3 of Figure 6.2 then instructed the program to (1) compute the mean of the 6 data items in each of the 150 rows and (2) store these 150 sample means in a separate column (C7). The next instruction calls for the program to compute the mean from the data set of 150 sample means. This value is 7.0144, very close to the population mean of 7.00. A stem-and-leaf display is then prepared to show the values of the 150 sample means. As you can see, the smallest of the 150 sample means has a value of about 3.0, and the largest sample mean is about 11.1. Finally, a box-and-whiskers display of our data set of 150 sample means is presented. You'll notice that the box representing the middle 50 percent of the sample means has a lower hinge of about 5.6 and an upper hinge of a little over 8.0.

A dozen more simulations, each with 150 random samples, could be processed by the computer in short order. All of these simulations would undoubtedly produce different overall averages from their data sets of 150 sample means. But these dozen values, like our mean of 7.0144, would be close to the population mean of 7 because ultimately, as we've seen, $\mu_{\bar{x}} = \mu$.

```
MTB > RANDOM 150 SAMPLES INTO C1-C6;
SUBC> INTEGERS 0 TO 14.
MTB > RMEAN C1-C6 INTO C7
MTB > MEAN C7
   MEAN     =        7.0144
MTB > STEM-AND-LEAF C7

Stem-and-leaf of C7            N  = 150
Leaf Unit = 0.10


    2    3 01
    4    3 56
    8    4 0013
   20    4 555566666888
   29    5 001111333
   41    5 555555566888
   57    6 0000001111133333
   73    6 5555566888888888
  (15)   7 000001111333333
   62    7 555666666668888888
   44    8 00001111133333
   30    8 555566
   24    9 0001111113
   14    9 55558
    9   10 033
    6   10 56688
    1   11 1

MTB > GBOXPLOT C7
```

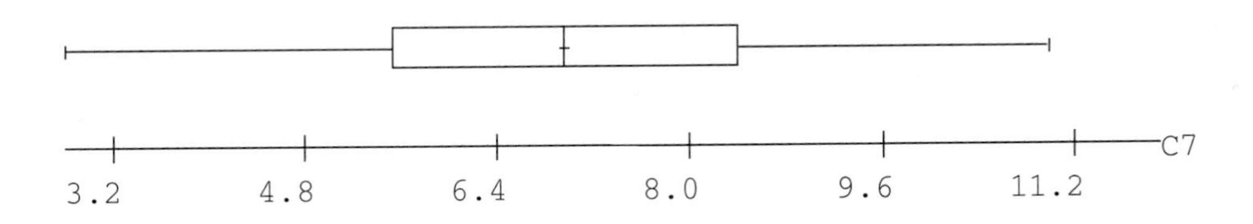

FIGURE 6.2 A computer simulation of 150 samples of 6 cards each, selected from a population of 15 integers numbered 0 to 14. The μ is 7, and the mean of this data set of 150 simulated sample means is 7.0144.

As the devil's advocate, we may say "So what?" to the fact that $\mu_{\bar{x}} = \mu$. No one in a realistic situation really takes all possible sample combinations and calculates the sample means. In practice only one sample is taken. What benefit is there in discussing the sampling distribution? Shouldn't we really be concerned with the proximity of a single sample mean to the population mean? In essence, the discussion of the sampling distribution *is* concerned with the proximity of a sample mean to the population mean.

You can see from Table 6.2 and Figure 6.2 that the possible values of the sample means *tend toward* the μ. Since these values have frequencies of occurrence, the sampling distribution is essentially a probability distribution. If the sample size is *sufficiently large* ($n > 30$), the sampling distribution approximates the *normal distribution whether or not the population is normally distributed.* And the sampling distribution is *normally distributed regardless of sample size if the population is normally distributed.* Figure 6.3 illustrates the sampling distribution as a normal distribution.

You'll recall that in a normal probability distribution the likelihood of an outcome is determined by the number of standard deviations from the mean of the distribution. Therefore, as you can see in Figure 6.3, there's a 68.3 percent chance that a sample selected at random will have a mean that lies within 1 standard deviation ($\sigma_{\bar{x}}$) of the population mean. Also, there's a 95.4 percent chance that the sample mean will lie within 2 standard deviations of the μ. Thus, a knowledge of the properties of the sampling distribution tells us the probable proximity of a sample mean outcome to the value of μ. With a knowledge of the sampling distribution, probability statements can be made about the range of possible values a sample mean may assume. This range of possible values can be calculated if a value for the standard deviation of the sampling distribution ($\sigma_{\bar{x}}$) is available. The computation of $\sigma_{\bar{x}}$ is shown in the next section.

Standard Deviation of the Sampling Distribution of Means

To gauge the extent to which a sample mean can differ from the population mean, we need some *measure of dispersion.* In other words, we must be able to compute the likely deviation of a sample mean from the mean of the sampling distribution. That is, we must be able to compute the *standard error of the mean:*

The **standard error of the mean** is the standard deviation of the sampling distribution of sample means and is represented by the symbol $\sigma_{\bar{x}}$.

For the data given in Table 6.2, the calculation of this measure is similar to the calculation of any other standard deviation about a mean:

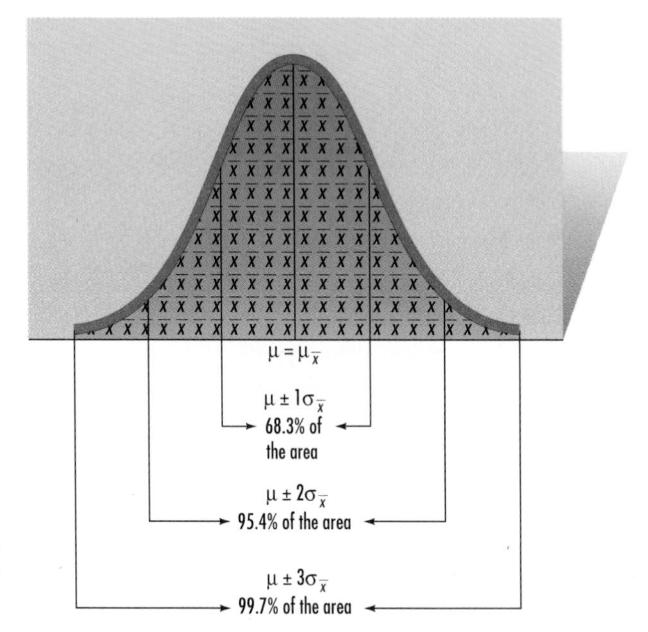

FIGURE 6.3 μ, $\mu_{\bar{x}}$, and $\sigma_{\bar{x}}$ for the areas under the sampling distribution of means.

$$\sigma_{\bar{x}} = \sqrt{\frac{\Sigma (\bar{x} - \mu_{\bar{x}})^2}{N}} \qquad (6.2)$$

where N = the total number of possible samples

$$\sigma_{\bar{x}} = \sqrt{\frac{10.00}{10}} = \sqrt{1.00} = 1.00$$

However, as pointed out in an earlier paragraph, no one (other than an eccentric statistician or a student with an assigned problem) ever deals with all the possible sample combinations. Therefore, an alternative method to compute $\sigma_{\bar{x}}$ must exist.

Since we've seen the relationship that exists between $\mu_{\bar{x}}$ and μ, we might intuitively assume that there is a relationship between the $\sigma_{\bar{x}}$ and the σ that will produce a shortcut method of computing the $\sigma_{\bar{x}}$. As a matter of fact, we are right, and the $\sigma_{\bar{x}}$ may be computed for a *finite population* with the following formula:

$$\sigma_{\bar{x}} = \frac{\sigma}{\sqrt{n}} \sqrt{\frac{N - n}{N - 1}} \qquad (6.3)$$

where σ = the population standard deviation
N = the population size
n = the sample size
$\sqrt{\frac{(N - n)}{(N - 1)}}$ = the finite population correction factor

Note that a **finite population correction factor** is used when a finite population of size N is involved.

From the data in Table 6.1, the σ is computed as follows:

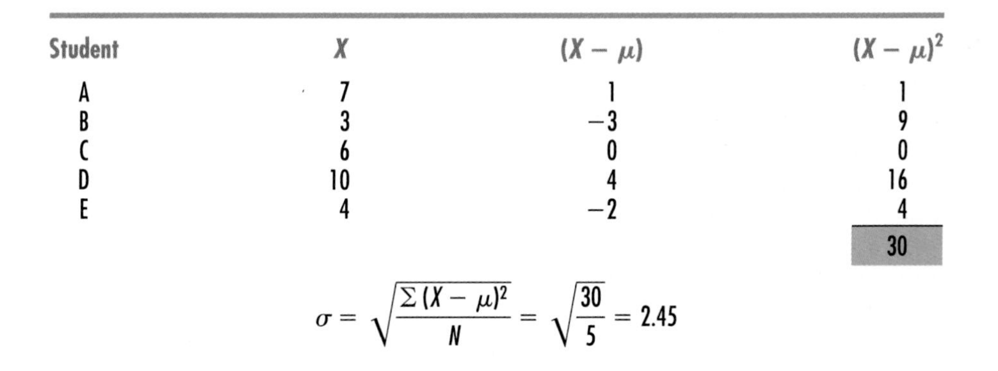

Student	X	$(X - \mu)$	$(X - \mu)^2$
A	7	1	1
B	3	−3	9
C	6	0	0
D	10	4	16
E	4	−2	4
			30

$$\sigma = \sqrt{\frac{\Sigma (X - \mu)^2}{N}} = \sqrt{\frac{30}{5}} = 2.45$$

Therefore, the standard error for the data given in Table 6.1 is computed as follows:

$$\sigma_{\bar{x}} = \frac{2.45}{\sqrt{3}} \sqrt{\frac{5 - 3}{5 - 1}} = 1.4145(.7071) = 1.00$$

We can now see that the results of the computations based on formulas 6.2 and 6.3 are equal. Thus, $\sigma_{\bar{x}}$ may be determined with a knowledge of the σ, the sample size, and the population size.

If the population is *infinite in size*, as, for example, in the case of items from an assembly-line operation, the standard error does not require a finite correction factor and may be computed as follows:

$$\sigma_{\bar{x}} = \frac{\sigma}{\sqrt{n}} \tag{6.4}$$

If the population is infinite, there's no need for the correction factor. However, a finite population doesn't necessarily mean that the correction has to be used. At this point the last statement may cause you to wince since there's an apparent contradiction. Let's look at the following example and, we hope, relieve your frustration.

Suppose we have a finite population of approximately 200 million and we have taken a sample of 2,000. If we followed the rules strictly, we would have:

$$\begin{aligned} \sigma_{\bar{x}} &= \frac{\sigma}{\sqrt{n}} \sqrt{\frac{N - n}{N - 1}} \\ &= \frac{\sigma}{\sqrt{n}} \sqrt{\frac{200{,}000{,}000 - 2{,}000}{200{,}000{,}000 - 1}} \\ &= \frac{\sigma}{\sqrt{n}} (.99999) \end{aligned}$$

However, the size of the population is so large that the finite correction factor for all practical purposes is 1. If the population size is extremely large compared to the sample size, formula 6.4 may be used to calculate the standard error of a finite population. *A general rule followed by many statisticians is to use the correction factor only if the sample size represents more than 5 percent of the population size.*

The Relationship between n and $\sigma_{\bar{x}}$

The $\sigma_{\bar{x}}$ is, of course, *a measure of the dispersion* of sample means about the μ. If the degree of dispersion *decreases*, the range of probable values a sample mean may assume also *decreases*, meaning the value of any single *sample mean* will probably be closer to the value of the *population mean* as the standard error decreases. And with formula 6.3 or 6.4, the value of $\sigma_{\bar{x}}$ obviously must decrease as the size of n increases. That is:

$$\downarrow \sigma_{\bar{x}} = \frac{\sigma}{\sqrt{n}\uparrow}$$

To add meaning to this mathematical manipulation, let's look at the relationship between n and $\sigma_{\bar{x}}$ intuitively. Let's assume we wish to estimate some parameter of a population of 100 and we are initially entertaining the thought of taking a sample of 10 items. Ten items may provide adequate information, but it's clear that more information could be obtained from a larger sample such as 20. More information provides a

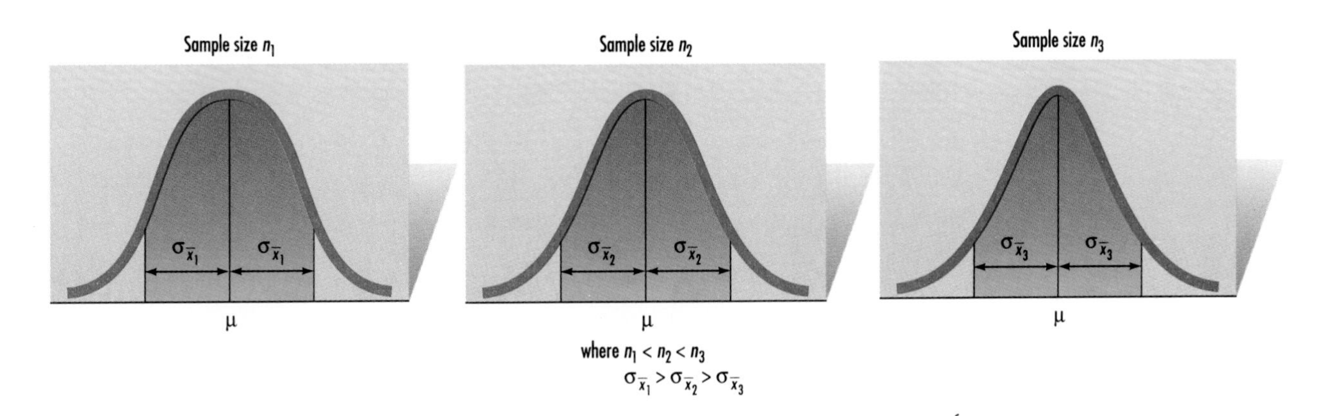

FIGURE 6.4 The relationship between n and $\sigma_{\bar{x}}$.

more precise estimate of the population parameter. As a matter of fact, a sample of 50 or 60 items would provide even more information and thus a more precise estimate. The ultimate option is that we could sample the entire population and obtain complete information, and thus there would be no difference between the sample statistic and the population parameter. In our example, if we were to estimate the μ, a sample of 100 would have the following $\sigma_{\bar{x}}$:

$$\sigma_{\bar{x}} = \frac{\sigma}{\sqrt{n}}\sqrt{\frac{N-n}{N-1}}$$
$$= \frac{\sigma}{\sqrt{n}}\sqrt{\frac{100-100}{99}} = 0$$

The general principle is that *as n increases, $\sigma_{\bar{x}}$ decreases.* As the sample size increases, we have more information with which to estimate the μ, and thus the probable difference between the true value and any sample outcome decreases. Figure 6.4 summarizes the points made in this section.

The Central Limit Theorem

Up to this point we've explained the concept of the sampling distribution of means in a rather intuitive way. Now, we're ready to formalize the concepts developed in previous sections and attribute the properties of the sampling distribution to the *Central Limit Theorem.* A detailed and mathematical discussion of the Central Limit Theorem is beyond the scope of this book, but the following summary is possible:

The Central Limit Theorem

The **Central Limit Theorem** basically states that for the distribution of the means of all samples of size *n*:

- The mean of the sampling distribution of means is equal to the population mean.

- The standard deviation of the sampling distribution of means ($\sigma_{\bar{x}}$) is equal to $\sigma/\sqrt{n}$ for an infinite population, and it's equal to $\sigma/\sqrt{n}\sqrt{(N-n)/(N-1)}$ for a finite population.
- If the sample size (n) is sufficiently large—statisticians often use a value of more than (>) 30 here—the sampling distribution approximates the normal probability distribution.
- If the population is normally distributed, the sampling distribution is normal regardless of sample size.

A considerable amount of important theoretical material has been presented in the last few pages. Let's look now at some example problems to illustrate certain basic concepts.

Example Problems

Example 6.1. The Bigg Truck Company has a fleet of 5 trucks which have monthly maintenance costs of \$200, \$175, \$185, \$210, and \$190. An estimate of the average monthly cost for a truck is to be obtained from a simple random sample of 3. What are the mean and the standard deviation of this sampling distribution of size 3?

The population mean (μ) = (\$200 + \$175 + \$185 + \$210 + \$190)/5 = \$192. Since the population mean is equal to the mean of the sampling distribution, $\mu_{\bar{x}}$ = \$192. The σ is:

$$\sigma = \sqrt{\frac{\Sigma\,(X-\mu)^2}{N}} = \sqrt{\frac{(8)^2 + (-17)^2 + (-7)^2 + (18)^2 + (-2)^2}{5}} = \$12.08$$

$$\text{So, } \sigma_{\bar{x}} = \frac{\sigma}{\sqrt{n}}\sqrt{\frac{N-n}{N-1}} = \frac{12.08}{\sqrt{3}}\sqrt{\frac{5-3}{5-1}} = \$4.93$$

Example 6.2. Sam and Janet Evening want to estimate the average dollar amount of the orders filled by their South Pacific Catering Company. They obtain their estimate by selecting a simple random sample of 49 orders. Sam and Janet don't know it, but their orders are normally distributed with a μ of \$120 and a σ of \$21. Now, what's the value of the standard error? What's the chance that the sample mean will fall between $\mu - 1\sigma_{\bar{x}}$ and $\mu + 1\sigma_{\bar{x}}$? What's the probability that the sample mean will lie between \$116.50 and \$124.00? Within what range of values does the $\bar{x}$ have a 95.4 percent chance of falling?

Well, $\sigma_{\bar{x}}$ is $\sigma/\sqrt{n}$ = \$21/7 = \$3.00. The chance that $\bar{x}$ will lie between \$120.00 ± \$3.00, or between \$117.00 and \$123.00, is 68.3 percent. And we follow the *approach* used in Chapter 5 to find the probability that a sample mean falls within a given range of values under a normal curve. In Chapter 5, a z value needed to solve a problem involving a single normally distributed variable was computed with this formula:

$$z = \frac{x-\mu}{\sigma}$$

But now we're dealing with a *sampling situation*, and the formula becomes:

$$z = \frac{\bar{x} - \mu}{\sigma_{\bar{x}}}$$

So the chance that the $\bar{x}$ falls between \$116.50 and the μ of \$120 is:

$$z_1 = \frac{\bar{x} - \mu}{\sigma_{\bar{x}}} = \frac{116.50 - 120.00}{3.00} = \frac{-3.50}{3.00} = -1.17, \text{ or a probability of } .3790$$

And the chance that the $\bar{x}$ lies between the μ of \$120 and a value of \$124.00 is:

$$z_2 = \frac{\bar{x} - \mu}{\sigma_{\bar{x}}} = \frac{124.00 - 120.00}{3.00} = \frac{4.00}{3.00} = 1.33, \text{ or a probability of } .4082$$

So the chance that the $\bar{x}$ falls between \$116.50 and \$124.00 is .3790 + .4082 or .7872. And finally, there's a 95.4 percent chance that $\bar{x}$ will fall between $\mu - 2\sigma_{\bar{x}}$ and $\mu + 2\sigma_{\bar{x}}$, or between \$114.00 and \$126.00.

Let's carry out another computer simulation to test the reasonableness of some of these figures. The first two lines of Figure 6.5 show that this simulation is based on 100 random samples of size 49 drawn from a normal population with a mean of \$120.00 and a standard deviation of \$21.00. The stem-and-leaf display shows the 100 sample means produced in this simulation, the histogram plots these 100 sample means, and the DESCRIBE command instructs the program to generate the descriptive values shown in Figure 6.5. As you can see, the mean of the 100 sample means—\$120.32 in this simulation—is close to the population mean of \$120.00, as it must be. The smallest sample mean produced in this simulation is \$112.80, and the largest sample mean is \$128.46. The middle 50 percent of the sample means lie between \$118.31 and \$122.42 (the Q_1 and Q_3 values), but what sample mean values bound the middle 68 percent of the data set?

Our previous analysis showed that *about* the middle two-thirds of the sample means should fall between \$117.00 and \$123.00. Do they in this simulation? Well, the middle 68 percent of the sample means fall between sample number 16 in the stem-and-leaf display and sample number 84 in that display. The value of sample mean number 16 is about \$117.60, and the value of sample mean number 84 is about \$123.30, so the simulation and the theory yield similar results. And 96 percent of the sample means in our simulation fall between \$114.40 and \$125.90—very close to our previously calculated 95.4 percent chance that a sample mean would lie between \$114.00 and \$126.00.

Notice, too, in the descriptive measures computed for our data set of 100 sample means at the bottom of Figure 6.5 that the standard deviation (STDEV) value is \$2.99. This measure of dispersion for our group of 100 sample means is very close to the measure of dispersion we've computed for *all* possible sample means. That is, the \$2.99 figure is close to the standard error value of \$3.00, as it should be.

As noted in our earlier simulation example, if we were to conduct another simulation and instruct the statistical package to randomly select another 100 samples of size 49, we would get different sample items, different sample means, a different stem-and-leaf display, a different histogram, and different descriptive values. But the mean (and other values) of this next simulated data set will be close to the values we've just examined, and they'll be close to the population values of interest.

Example 6.3. A manager at the Write-On Pen Company wants to estimate the average number of pens sold per month on the basis of the mean of a sample of 100 months. If the true population mean is 5,650 pens per month and the population

```
MTB > RANDOM 100 SAMPLES INTO C1-C49;
SUBC> NORMAL MU = 120     SIGMA = 21.
MTB > RMEAN C1-49 INTO C50
MTB > STEM-AND-LEAF C50

Stem-and-leaf of C50      N = 100
Leaf Unit = 0.10

   1  112 8
   2  113 3
   3  114 4
   8  115 22679
  13  116 23377
  21  117 03667889
  32  118 01133333689
  47  119 122223444568899
 (12) 120 011233455689
  41  121 11223356777899
  27  122 244556789
  18  123 0344578
  11  124 0129
   7  125 33899
   2  126 6
   1  127 9

MTB > GHISTOGRAM C50
```

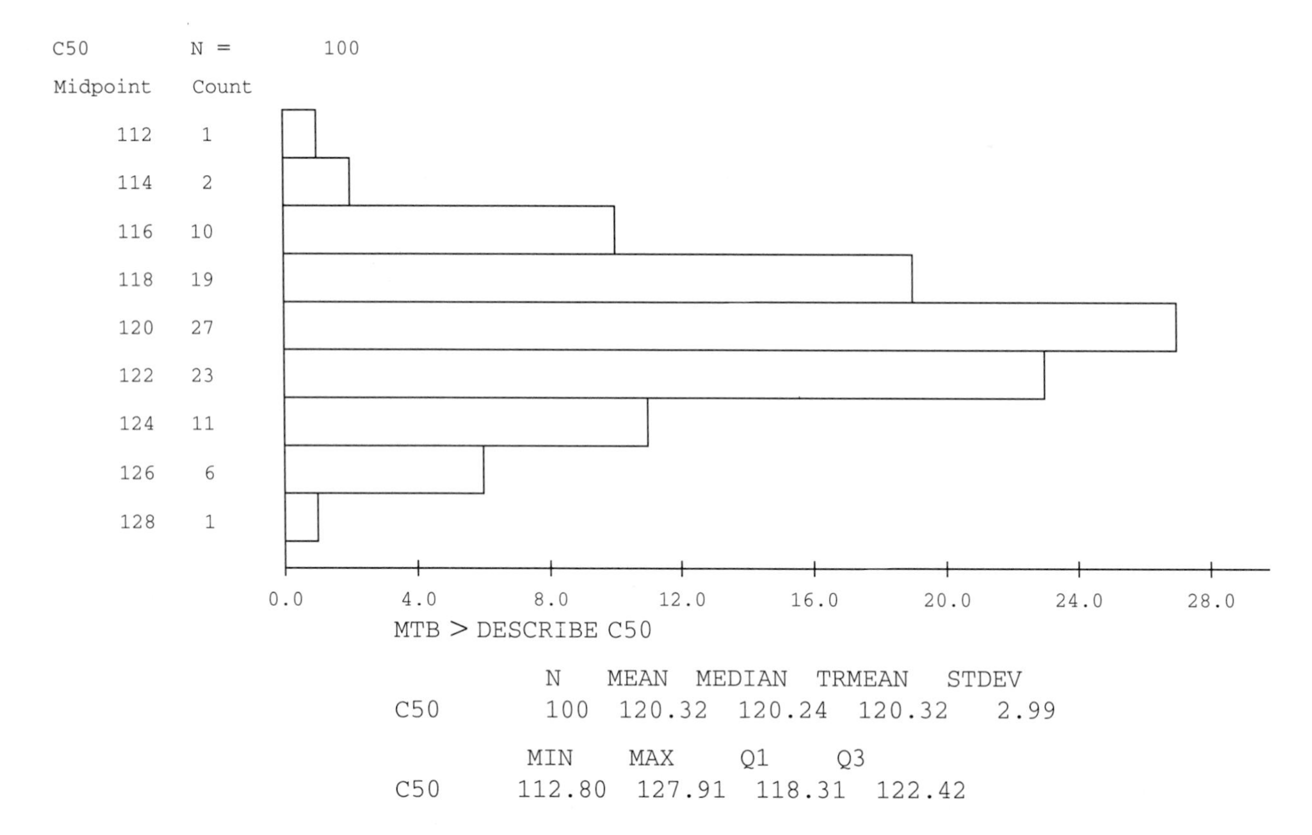


```
MTB > DESCRIBE C50

            N    MEAN   MEDIAN  TRMEAN   STDEV
C50       100  120.32   120.24  120.32    2.99

          MIN     MAX      Q1      Q3
C50    112.80  127.91  118.31  122.42
```

FIGURE 6.5 A computer simulation of 100 samples of size 49 taken from a normal distribution with a μ of \$120.00 and a σ of \$21.00. The sample means of the 100 samples are shown in the stem-and-leaf display and are plotted in the histogram.

standard deviation is 700 pens, what are the chances that the mean of a random sample will have a value within 200 pens of the true mean?

The problem basically asks, "What are the chances that the sample mean will lie between $\mu - 200$ and $\mu + 200$?" The manager is trying to find the chances that the $\bar{x}$ will fall within 5,650 ± 200, or between 5,450 and 5,850. Since the distribution of sample means is normal, we can calculate these chances by first finding the standard (z) scores for 5,450 and 5,850 and then using the z table in Appendix 2 to find the probabilities. You'll recall from Chapter 5 that the z value represented the number of standard deviations a value lies from the population mean. In that chapter we computed z as follows:

$$z = \frac{x - \mu}{\sigma}$$

And now when we work with a sampling distribution that is normal, we can compute the z score for the value of a sample mean by:

$$z = \frac{\bar{x} - \mu_{\bar{x}}}{\sigma_{\bar{x}}}$$

So to calculate the z scores of interest to the Write-On manager, we first find that the value of $\sigma_{\bar{x}} = \sigma/\sqrt{n} = 700/\sqrt{100} = 70$. Then:

$$z = \frac{\bar{x} - \mu_{\bar{x}}}{\sigma_{\bar{x}}} = \frac{5{,}850 - 5{,}650}{70} = 2.86$$

We can substitute the $\bar{x}$ of 5,450 for 5,850 and see that the other z score is -2.86. Consulting Appendix 2, we see that the probability that a sample mean will fall within 2.86 standard errors to one side of a true mean is 49.79 percent. Since we're concerned with 2.86 standard errors to both sides of the mean, the total likelihood of a sample mean value between 5,650 ± 200 is 99.58 percent.

Self-Testing Review 6.2

1. What is meant by the sampling distribution of means?
2. Discuss the difference between the mean of a sample distribution and the mean of the sampling distribution.

3–7. For each of the following, determine if the finite population correction factor is necessary to compute the standard error of the mean:

3. Cards are numbered 1 to 10. Five cards are selected at random with each card being replaced after it is selected.
4. Cards are numbered 1 to 10. Five different cards are selected at random.
5. A corporation has 200 employees, and each is given a number from 001 to 200. A random sample of 40 different employees is to be selected for evaluation.
6. A corporation has 200 employees, and each is given a number from 001 to 200. A sample of 40 is to be randomly selected to receive 40 prizes. Employees can receive more than 1 prize.

7. A college in the Bay Area has 7,496 registered students. A random sample of 30 is to be selected to test a new advisement program.

8–12. A population of 1,500 students took a math placement test. The population mean raw score was 37.1, and the population standard deviation was 5.2. For each of the following, compute the finite population correction factor and the standard error of the mean:

8. $n = 300$.

9. $n = 200$.

10. $n = 50$.

11. $n = 20$.

12. $n = 5$.

13. Examine the results of problems 8 to 12. What happens to the correction factor as the sample size gets smaller relative to the population size?

14–19. There is a population of 3 accountants (Andrews, Bartolini, and Cortez) working for Taxes-R-Us. The time each has worked for the company is as follows: Andrews, 1 year, Bartolini, 3 years, and Cortez, 5 years.

14. Compute the mean number of years at the company for this population.

15. Compute the standard deviation for the population.

16. List all the possible samples of $n = 2$ accountants formed without replacement, and show the numbers of years employed for the members of each sample.

17. Compute the mean years employed for each of the samples.

18. Compute the mean of the sample means found in problem 17. Compare this to your answer in problem 14.

19. Compute the standard deviation of all sample means found in problem 17. Compare this to your answer for problem 15.

20–25. Suppose there is a large number of accountants who will be taking a proficiency exam in the state of California after having worked 1 year, 3 years, and 5 years. Let A = those who have worked 1 year, B = those who have worked 3 years, and C = those with 5 years' experience, and assume that there is an equal number of accountants in each of these 3 categories:

20. Compute the population mean for the number of years worked.

21. Compute the population standard deviation.

22. List all the possible samples of $n = 2$ formed with replacement and the corresponding work experience values. (*Hint:* There are 9 possible samples—some are *AA*, *AB*, *AC*, *BA*, *BB*, and so on.)

23. Compute the mean of each of the samples.

24. Compute the mean of the sample means found in problem 23. Compare this to your answer in problem 20.

25. Compute the standard deviation of all sample means found in problem 23. Compare this to your answer for problem 21.

26–31. There is a population of 4 patients in a stress control program. The patients are

Aguilar, Brodski, Cho, and Davis. Each is asked for the number of anxiety attacks they have had during the previous week. The responses are: Aguilar 1, Brodski 2, Cho 8, and Davis 9:

26. Compute the mean number of anxiety attacks for the week for this population.

27. Compute the population standard deviation.

28. List all the possible samples of $n = 2$ formed without replacement and the corresponding values for the number of anxiety attacks. (You should have ${}_4C_2 = 6$ possible samples.)

29. Compute the mean of each of the samples.

30. Compute the mean of the sample means found in problem 29. Compare this to your answer in problem 26.

31. Compute the standard deviation of all sample means found in problem 29. Compare this to your answer for problem 27.

32–33. A food packing company fills sacks of cereal using automated machinery. The population mean weight of the sacks is 2 pounds with a standard deviation of 0.1 pound:

32. One of these sacks is selected at random. What is the probability that it weighs more than 2.08 pounds?

33. Twelve sacks are randomly selected and placed in a box for shipment. What's the probability that the mean weight of the sample of 12 sacks in the box exceeds 2.08 pounds?

34–36. There is a population of 5 students who are employed as chemistry lab assistants. The number of hours they work each week is as follows:

Student	Hours
a	7
b	16
c	20
d	12
e	22

34. Calculate the population mean and standard deviation for the number of hours worked per week.

35. A random sample of 3 different students is taken to estimate the population mean. What is the mean of the sampling distribution?

36. Calculate the standard deviation of the sampling distribution for samples of 3 students.

37–40. The Tite Wire Company makes wires for circus acts. The population mean thickness for the wires is 0.45 inches, and the population standard deviation is 0.03 inches. A sample of 100 pieces of wire is selected randomly and the thickness of each sample member is measured:

37. What is the mean and standard deviation of the sampling distribution?

38. What may be said about the shape of the sampling distribution? Why?

39. Within what range of values does the sample mean have a 68.3 percent chance of falling?

40. Within what range of values does the sample mean have a 95.4 percent chance of falling?

41–44. The mean of all credit card balances for I.O.U. Credit Corporation is $200, and the standard deviation is $15. Within what range of values will the mean of a random sample have a 95.4 percent chance of falling if the sample consists of:

41. 36 accounts?

42. 49 accounts?

43. 64 accounts?

44. What relationship do you observe between the sample size and the dispersion of the sampling distribution?

45. At the Keypon Trucking Company, the population average tonnage of freight handled per month is 225 tons, and the population standard deviation is 30 tons. What are the chances that in a random sample of 36 months the sample mean tonnage will have a value within 7 tons of the true mean?

46. The average verbal SAT score of all students attending McGuire College is 540, and the population standard deviation is 30 SAT points. What's the probability that a random sample of 36 McGuire students will have a mean verbal SAT score that is greater than 535?

47. The values of the accounts receivable of the Rice Corporation are normally distributed with a mean of $10,000 and a standard deviation of $2,000. If a random sample of 400 accounts is selected, what's the probability that the sample mean will be between $10,100 and $10,200?

48. The average gasoline consumption of all families in Rocktown is 16.9 gallons per week with a population standard deviation of 3.2 gallons per week. What's the probability that the mean of a random sample of 50 families exceeds 17.5 gallons per week?

49. A population of 500 children has a mean IQ score of 100 points and a standard deviation of 20 points. If a random sample of 30 children is selected, what's the probability that the mean IQ of the group exceeds 110?

6.3 Sampling Distribution of Percentages

We're often interested in estimating population percentages.* For example, a company might want to estimate the percentage of defective items produced by a machine,

* Many texts deal with the material in the discussion that follows in terms of *proportions* rather than percentages. We prefer to use percentages because many students seem to find the arithmetic easier and because percentages are more frequently used in everyday discussion. Of course, if you prefer to use proportions, you can simply move the decimal two places to the left in all computations. The ultimate results are identical.

the percent who intend to enter college. Suppose the true percentage for this population is 60 percent.

16. What is the mean of the sampling distribution of percentages? Why?

17. Is the finite population correction factor needed to calculate σ_p? If so, what's the value of the finite correction factor?

18. Calculate σ_p.

19–21. Vanity Press wants to estimate the percentage of books printed by an incompetent vendor that are defective and cannot be sold. If the true percentage of defective books is 8.5 percent, what's the probability that the sample percentage will be within 1 percent of the population percentage for a random sample of:

19. 100 books?

20. 1,000 books?

21. 5,000 books?

22. It's known that 64 percent of the population of Jamesburg favor Wanda A. Rounde for dogcatcher. What's the probability that a random sample of 100 voters will have a sample percentage favoring Ms. Rounde of between 60 and 68 percent?

23. Fifty-five percent of a television viewing population watched a fantastically popular program called *Name That Variable* with host Bart Hart last Thursday evening. What's the probability that, in a random sample of 100 viewers, less than 50 percent watched the program?

24–25. A population of 6 politicians and their ages are listed below:

Politician	Age
A	30
B	50
C	60
D	34
E	33
F	29

To serve as President of the United States, a politician must be 35 years of age or older:

24. What percentage of the above population is eligible to serve as President?

25. If simple random samples of size 2 (without replacement) are taken from the above population, what's the value of the standard error of percentage?

26. Suppose that 53 percent of a voting population favors Phil A. Buster in an election. Let's assume, though, that only 50 people turn out to vote. If we can consider the voters to be a random sample of the electorate, what's the probability that Phil loses (gets less than 50 percent of the vote)?

LOOKING BACK

1. A finite population is one where all members could be listed, while an infinite population is unlimited in size. A sample is a portion of the population selected for study. You've seen in this chapter that if a sample statistic is representative of a population parameter, it's possible to make an inference about the population measure from the sample figure. Although complete information, in the form of a census, may be desirable, sampling advantages such as reduced cost, faster response to questions, and acceptable accuracy often outweigh the disadvantage of sampling error.

2. Although sampling variation exists so that a statistic will not provide an exact value of a parameter, it's adequate for decision-making purposes to know that sample values are governed by population characteristics. When a simple random sample is used, the mean of a sampling distribution of either sample means or sample percentages is equal to the population parameter being sought—that is, the μ or the π. The standard deviation of the sampling distribution of sample means—the standard error of the mean ($\sigma_{\bar{x}}$)—is determined by σ and the sample size. And the standard deviation of the sampling distribution of percentages (σ_p)—the standard error of percentage—may be computed with knowledge of the π, population size, and sample size. Different samples that can be taken from a population will have different values, as we've seen in two computer simulations involving 150 and 100 samples. But the mean of these sample values tend toward the population mean value being sought.

3. The $\sigma_{\bar{x}}$ is a measure of the dispersion of sample means about the μ. If this value decreases, the range of probable values a sample mean may assume also decreases, meaning that the value of any single sample will probably be closer to the value of the unknown μ as the standard error decreases. How can this desirable result be achieved? Unfortunately, reducing the standard error requires that we increase the sample size (and the sample cost, and the time required to take the sample . . .).

4. The Central Limit Theorem summarizes several important facts presented in this chapter: The mean of the sampling distribution of means is equal to the population mean; if the sample size is sufficiently large, the sampling distribution approximates the normal probability distribution; and, if the population is normally distributed, the sampling distribution is normal regardless of sample size. Given this normality property, it's possible to make probability statements concerning the possible values a statistic may assume. The value of a sample statistic will tend toward the parameter value, and a probability statement can be made about the proximity of the statistic to the parameter.

5. The characteristics of the two sampling distributions (means and percentages) presented in Chapter 6 are summarized in Table 6.5.

TABLE 6.5 PROPERTIES OF SAMPLING DISTRIBUTIONS

Population	Sampling Distribution of Means ($\bar{x}$)	Sampling Distribution of Percentages (p)
Finite	$\mu_{\bar{x}} = \mu$	$\mu_p = \pi$
	$\sigma_{\bar{x}} = \frac{\sigma}{\sqrt{n}} \sqrt{\frac{N-n}{N-1}}$	$\sigma_p = \sqrt{\frac{\pi(100-\pi)}{n}} \sqrt{\frac{N-n}{N-1}}$
Infinite	$\mu_{\bar{x}} = \mu$	$\mu_p = \pi$
	$\sigma_{\bar{x}} = \frac{\sigma}{\sqrt{n}}$	$\sigma_p = \sqrt{\frac{\pi(100-\pi)}{n}}$

* Both sampling distributions approximate the normal probability distribution if the simple random sample size is sufficiently large. As a general rule of thumb, "sufficiently large" is over 30. In the case of percentages, "sufficiently large" is when $(n)(p) \geq 500$ and when $(n)(100 - p) \geq 500$.

Review Exercises

1–4. The fuel capacity for all Kawasaki Zephyrs is normally distributed with a mean of 4.5 gallons and a standard deviation of .7 gallons.

1. A Kawasaki Zephyr is chosen at random from the production line. What's the probability that it has a fuel capacity of more than 4.7 gallons?

2. A quality control officer tests a sample of 36 Kawasaki Zephyrs picked at random. What's the probability that the mean fuel capacity of these 36 Zephyrs is more than 4.7 gallons?

3. A quality control officer tests a sample of 100 Zephyrs picked at random. What's the probability that the mean fuel capacity of these 100 Zephyrs is more than 4.7 gallons?

4. Examine your answers to problems 1 to 3. What happens to the probability of the fuel capacity being more than 4.7 gallons as the sample size increases?

5–8. An educational research journal reported that a population of preservice teachers in their sophomore year thought an average of 10.43 hours should be spent in testing each week. The population standard deviation was 3.72 hours, and the responses were normally distributed:

5. One of these sophomores is questioned. What's the probability that he or she thinks that less than 9 hours a week should be spent testing?

6. A random sample of 20 sophomore preservice teachers is questioned. What's the probability that the mean time of the sample is less than 9 hours a week?

7. A random sample of 32 sophomore preservice teachers is questioned. What's the probability that the mean time of the sample is less than 9 hours a week?

8. Compare your answers for problems 5 to 7. What happens as the sample size is increased?

9–11. The First National Bank of Yourtown has a total of 5 offices. The number of transactions in a day is recorded for each branch as follows:

Branch	Transactions
Allwood Avenue	186
Broad Street	382
Central Office	841
Downtown Branch	729
Eastern Branch	336

A simple random sample of 3 branches is taken to estimate the population mean number of branch transactions in a day:

9. Calculate the population mean and the population standard deviation.

10. What is the mean of the sampling distribution ($n = 3$)?

11. What is the standard deviation of the sampling distribution?

12–14. There is a population of 5 chronic schizophrenic patients in a day hospital program run by the department of psychiatry at Bigstate University. Each patient is assessed by a trained psychiatrist using the Scale for Assessment of Negative Symptoms. The results of these tests are as follows:

Patient	Score
Abernathy	24
Benson	33
Cromwell	28
Davis	26
Eberly	38

A simple random sample of 3 patients is taken to estimate the population mean score on the Scale for Assessment of Negative Symptoms:

12. Calculate the population mean and the population standard deviation.

13. What is the mean of the sampling distribution?

14. What is the standard deviation of the sampling distribution?

15–17. *The American Journal of Public Health* recently reported that the mean amount of lead found in the printed sections of a population of soft plastic bread wrappers was 26 milligrams with a standard deviation of 6 milligrams. A sample of 100 soft plastic bread wrappers is randomly selected:

15. What is the mean and the standard deviation of the sampling distribution?

16. Within what range of values does the sample mean have a 68.3 percent chance of falling?

17. Within what range of values does the sample mean have a 95.4 percent chance of falling?

18–20. *The New England Journal of Medicine* recently published the results of a study of a population of patients who were treated by angioplasty after a heart attack. The number of days between the heart attack and a successful angioplasty procedure was normally distributed with a mean of 12 days and a standard deviation of 2 days. If a sample of 15 patients is selected from this population:

18. What is the mean and the standard deviation of the sampling distribution?

19. Within what range of values does the sample mean have a 68.3 percent chance of falling?

20. Within what range of values does the sample mean have a 95.4 percent chance of falling?

21–24. Knee flexion, a measure of the knee range of motion, is a normally distributed variable. *Physical Therapy* recently reported that the population mean number of degrees of knee flexion found in the First National Health and Nutrition Exam Survey (NHANES I) was 132 degrees with a standard deviation of 10 degrees. Within what range of values will the mean of a random sample have a 95.4 percent chance of falling if we have a sample of:

21. 10 individuals?

22. 100 individuals?

23. 1,000 individuals?

24. As the sample size increases, what happens to the dispersion of the sampling distribution?

25–28. The First National Health and Nutrition Exam Survey reported that the average body mass index (BMI) in the population of white women aged 25 to 39 was found to be 24 with a standard deviation of 5. Within what range of values will the mean of a random sample of white women in this age group have a 99.7 percent chance of falling if we have a sample of:

25. 25 women?

26. 49 women?

27. 81 women?

28. As the sample size increases, what happens to the dispersion of the sampling distribution?

29. The average length of all female babies born at Stork Memorial Hospital last year was 19.73 inches with a standard deviation of 0.16 inches. What are the chances that the mean of a random sample of 50 female babies born at Stork last year will be within .05 inches of the population mean?

30. The average weight of all male babies born at Stork Memorial Hospital last year was 6.95 pounds with a standard deviation of 1.02 pounds. What are the chances that the mean of a random sample of 64 male babies born at Stork last year will be within .3 pounds of the population mean?

31. The breaking strength of a material is normally distributed with a population mean of 85 pounds and a standard deviation of 18 pounds. What is the probability that the mean breaking strength for a sample of 12 pieces of this material will be less than 90 pounds?

32. *U.S. News & World Report* recently reported that the average Graduate Management Aptitude Test (GMAT) score for all students entering the Graduate School of Business at the University of Texas at Austin was 631 with a standard deviation of 80. What's the probability that the mean GMAT score of a random sample of 40 of these graduate business students at UT-Austin will be between 600 and 650?

33–35. A population of 15 college students has the flu. We randomly select a sample of 4 students to estimate the percentage whose symptoms are relieved after they take Cold-Ex. Let's assume the true percentage for the population of 15 is 80 percent.

33. What is the mean of the sampling distribution of percentages?

34. Is the finite population correction factor needed to calculate σ_p? If so, what is the value of the finite correction factor?

35. What is the value of σ_p?

36–38. There are 12 members of a board of trustees. We wish to estimate the true percentage of board members who will vote yes on the proposed budget, and a sample of 5 members is polled. If we assume that the true population percentage is 60 percent:

36. What's the mean of the sampling distribution of percentages?

37. Is the finite population correction factor needed to calculate σ_p? If so, what's the value of the finite correction factor?

38. What's the value of σ_p?

39–42. Fatzgon Dietary Products distributes 6 microwavable dinners. The dinners and their population caloric count are listed as follows:

Meal	Calories
Artichoke Chicken	364
Broiled Flounder	337
Chicken Cacciatore	428
Duck a l'Orange	326
Eggplant Parmigiana	297
Fish Medley	259

A simple random sample of 2 different meals is taken to estimate the population mean number of calories.

39. Calculate the population mean and the population standard deviation.

40. How many different samples of size 2 can be taken from this population?

41. What is the mean of the sampling distribution?

42. Calculate the standard deviation of the sampling distribution.

43–46. The city council has 6 members. The ages of this population are listed below:

Council Member	Age
Alvarez	36
Brown	53
Chang	42
Dougherty	47
Epstein	64
Ferraro	72

A simple random sample of 2 council members is taken to estimate the population mean age:

43. Calculate the population mean and the population standard deviation.

44. How many different samples of size 2 can be taken from this population?

45. What is the mean of the sampling distribution?

46. Calculate the standard deviation of the sampling distribution.

47. Up Up and Away Airlines wants to estimate the percentage of frequent fliers (those who made at least 3 domestic flights or 1 foreign trip in the past year) who have an average income of more than $35,000 a year. If the true population percentage is 67 percent, what are the chances that a simple random sample of 174 frequent fliers will be within 2 percent of the population percentage?

48. The manager of a car dealership wishes to estimate the percentage of new cars sold in the $13,500 to $17,400 price range. A recent *J. D. Power* report gives the true population percentage as 30 percent. What are the chances that a simple random sample of 46 cars will be within 5 percent of the population percentage?

49. The mean change in total exercise time for all coronary patients taking NORVASC is 62 seconds with a standard deviation of 17 seconds. Find the probability that a sample of 37 coronary patients taking NORVASC would have a mean change in exercise time between 60 and 70 seconds.

50. *The New York Times* recently reported that 45 percent of the population of Manhattan District 5 households participated in recycling. What's the probability that a random sample of 35 households in this district will have a sample percentage between 40 percent and 50 percent?

51. A study shows that 46 percent of all employees of TGB Corporation are less willing to give up free time today than they were 5 years ago. What's the probability that in a random sample of 25 employees, more than 50 percent will say they are less willing to give up free time today than they were 5 years ago?

52. *The New England Journal of Medicine* recently reported that the population mean and standard deviation figures for the cost per case for psychiatric evaluation services for self-referred patients are $3,222 and $1,451, respectively. What's the probability that a random sample of 67 such patients will have a mean cost of more than $3,500?

53. *The New England Journal of Medicine* also reported that the population mean and standard deviation figures for the cost per case for patients with medical back problems are $406 and $98, respectively. What's the probability that the mean cost per case for a random sample of 45 patients with medical back problems is between $375 and $450?

54. Another recent article in *The New England Journal of Medicine* reported that the number of days between a heart attack and successful angioplasty in a population of coronary patients is normally distributed with a mean of 12 days and a standard deviation of 2 days. What's the probability that the mean time between a heart attack and successful angioplasty treatment for a sample of 17 patients is between 11 and 13 days?

55. *The Journal of Abnormal Psychology* recently reported that the mean age at the onset of depression for a population of dysfunctional families is 30.6 years, and the standard deviation is 13.7 years. A sample of 42 depressed people from this population of dysfunctional families is selected at random. What's the probability that the mean age of this sample is between 25 and 35 years?

56. The population mean length of stay for patients at Keepum Memorial Hospital is 23.3 days and the standard deviation is 11.0 days. What's the probability that the mean length of stay for a sample of 31 patients selected at random is between 25 and 27 days?

57. The February 20, 1993, issue of *Advertising Age* reports that the mean salary for a population of creative directors working for ad agencies in a recent year was $85,000, and the population standard deviation was $5,000. If random samples consisting of income records for 32 creative directors are examined, what's the salary range within which 95.4 percent of the sample means will fall?

58. The mean length of time a population of sixth graders watch TV each day is found to be 118.3 minutes, and the standard deviation is 57.3 minutes. If samples of 46 sixth graders are selected at random, determine the range of time that will include 99.7 percent of the sample means.

59–60. A list of finalists for a high-level corporate position is handed to the CEO, who is interested in finding the percentage of females in this population. The finalists are:

Candidate	Gender
Adams	M
Benson	F
Clemens	F
Dvorak	M
Esposito	M

59. What percentage of the above population is female?

60. If simple random samples of size 2 are taken from this population, what's the value of σ_p?

Topics for Review and Discussion

1. Discuss the importance of sampling. What are some reasons a decision maker would choose to take a sample rather than a census?

2. What is the difference between a parameter and a statistic? Identify three of each, and give the symbols for each.

3. What is the difference between a sample distribution and a sampling distribution?

4. Discuss the consequences of the Central Limit Theorem. Under what conditions does it apply? Why is it important in statistical studies?

5. What is the standard error of the mean? Explain the relationship that exists between the standard error of the mean and the size of the sample.

6. Describe a situation in which the students in your statistics class comprise a population and a situation in which they are a sample.

7. In what situations is it not necessary to use the finite population correction factor to compute the standard error of the mean?

8. What is the sampling distribution of percentages? How does the Central Limit Theorem apply to this distribution?

9. What information is needed to compute the standard error of percentages?

Projects/Issues to Consider

1. Locate a population data set that is of interest to you. You may use a periodical from the library or distribute a questionnaire to your entire class with an appropriate question. After you've gathered your population data:

   a) Compute the population mean and standard deviation.

   b) Use the random number table in Appendix 3 at the back of this book, and select a random sample from your data set.

   c) Compute the mean of your sample and the standard error of the mean.

   d) Analyze the relationship between your sample mean and the population mean.

2. Locate a population with a small number of values, say, 5 or 6:

   a) Compute the population mean and standard deviation.

   b) Form all possible samples without replacement of $n = 3$.

   c) Compute the mean for each of your samples.

   d) Compute the mean of the set of means you've produced in part *b*, and compare your answer to the population mean found in part *a*.

   e) Compute the standard deviation for the set of values you've produced in part *c*. How does this value compare to the standard deviation of the population?

   f) Discuss the findings of this project.

3. Locate a population data set that will allow you to examine the percentage of a variable that interests you:

   a) Compute the population percentage.

   b) Form a random sample using the random number table in Appendix 3.

   c) Compute the sample percentage.

   d) Compute the standard deviation for the sampling distribution of percentages.

Computer Exercises

1–2. Repeat the simulation experiment shown in Figure 6.2 and discussed in this chapter:

1. What was the mean of your data set of 150 simulated sample means?

2. Explain why your result differed from the value of 7.0144 shown in the chapter. (It would be very unusual if your value *did* equal 7.0144.)

3–6. Repeat the simulation experiment shown in Figure 6.5 and discussed in this chapter:

3. What was the value of the smallest sample mean produced in your simulation?

4. What was the value of the largest sample mean?

5. What was the mean of your data set of 100 simulated sample means?

6. Explain why your result differed (it quite probably did) from the value of $120.32 shown in the chapter.

Answers to Odd-Numbered Self-Testing Review Questions

Section 6.1

1. Answers will vary here, of course, but anyone who has taken a taste of food, performed a chemistry experiment, had a blood test, or looked at a swatch of wallpaper has engaged in sampling.

3. The purpose of sampling is to investigate properties of a population of interest by studying part of that population.

5. All students in your class

7. All students from your school

9. All stocks listed on the NYSE

11. All knee surgery patients

13. The 27 freshmen whose data are collected

15. 517

17. False. Information will vary from sample to sample.

Section 6.2

1. If all possible samples of a given size (n) are taken from a given population and then for each sample the mean is computed, the distribution of these means is the sampling distribution of means.

3. The finite population correction (fpc) isn't needed. Since each card is replaced before the next is selected, the population is, in effect, infinite.

5. The fpc is needed. The population is finite and the sample is more than 5 percent of the population.

7. The fpc isn't needed. Although the population is finite, the sample size is less than 5 percent of the population size.

9. fpc = .9313, $\sigma_{\bar{x}} = .3424$

11. fpc = .9936, $\sigma_{\bar{x}} = 1.155$

13. As the sample size gets smaller (relative to the population size), the fpc gets closer to 1. For this reason, even though a population might be finite, we generally agree that the fpc isn't necessary when the sample size is less than 5 percent of the population.

15. $\sigma\sqrt{[(1-3)^2 + (3-3)^2 + (5-3)^2]/3} = 1.63$

17. The means of the samples are 2, 3, and 4.

19. $\sigma_{\bar{x}} = \sqrt{[(2-3)^2 + (3-3)^2 + (4-3)^2]/3} = .816$. The standard error of the means equals the population standard deviation divided by the square root of the sample size and multiplied by the correction factor using $N = 3$ and $n = 2$. That is, $.816 = 1.63/\sqrt{2} \cdot \sqrt{1/2}$.

21. $\sigma = \sqrt{[(1-3)^2 + (3-3)^2 + (5-3)^2]/3} = 1.633$

23. The means of the samples are 1, 2, 3, 2, 3, 4, 3, 4, 5.

25. $\sigma_{\bar{x}} = \sqrt{[(1-3)^2 + (2-3)^2 + (3-3)^2 + (2-3)^2 + (3-3)^2 + (4-3)^2 + (3-3)^2 + (4-3)^2 + (5-3)^2]/9} =$ 1.1547. You can verify that $1.1547 = 1.633/\sqrt{2}$.

27. $\sigma = \sqrt{[(1-5)^2 + (2-5)^2 + (8-5)^2 + (9-5)^2]/4} = 3.5355$

29. The means of the samples are 1.5, 4.5, 5, 5, 5.5, and 8.5.

31. $\sigma_{\bar{x}} = \sqrt{[(1.5-5)^2 + (4.5-5)^2 + (5-5)^2 + (5-5)^2 + (5.5-5)^2 + (8.5-5)^2]/6} =$ 2.0412. You can verify that $2.0412 = 3.5355/\sqrt{2} \cdot \sqrt{2/3}$.

33. This *is a sampling situation*, so:

$$z = \frac{\bar{x} - \mu}{\sigma_{\bar{x}}} = \frac{2.08 - 2.0}{.1/\sqrt{12}}$$

$$= \frac{.08}{.0289} = 2.77$$

The area under the normal curve corresponding to a z value of 2.77 is .4972. The probability that a sample mean lies 2.77 standard errors beyond the population mean is .5000 − .4972 or .0028. As these examples show, it's quite possible that you could have a single sack that weighed more than 2.08 pounds, but it's quite unlikely that a *sample of 12 sacks* would have a mean weight of more than 2.08 pounds.

35. The mean of the sampling distribution is the same as the population mean or 15.4 hours.

37. The mean of the sampling distribution, being equal to the population mean, is .45 inches. The standard error is $.03/\sqrt{100} = .003$ inches.

39. The range is .45 ± (1.00)(.003), or .447 to .453 inches.

41. There's a 95.4 percent chance that the sample mean will be located between the population mean ± 2(standard errors). And since the standard error is equal to $15/\sqrt{36}$ or 2.5, the range is 200 ± 2(2.5) or 195 to 205.

43. If the sample size is 64, the standard error is $15/\sqrt{64}$ or 1.88, and the range is 200 ± 2(1.88) or 196.24 to 203.76.

45. The problem, in effect, asks the question: "What is the chance that the sample mean will be located in the interval $\mu \pm z(\sigma_{\bar{x}})$, where $z(\sigma_{\bar{x}})$ is equal to ± 7 tons?" Since the standard error is $30/\sqrt{36}$ or 5 tons, the z value is $z(5) = \pm 7$ tons. Solving for z, we get $z = \pm 7/5 = \pm 1.4$. And since a z value of 1.4 corresponds to a normal curve area of .4192, the chance that a sample mean would be within ±7 tons of the population mean is 2(.4192) or .8384.

47.

$$z_1 = \frac{10{,}100 - 10{,}000}{2{,}000/\sqrt{400}} = 1.00$$

which corresponds to a table value of .3413

$$z_2 = \frac{10{,}200 - 10{,}000}{2{,}000/\sqrt{400}} = 2.00 \text{ which}$$

corresponds to a table value of .4772

And .4772 − .3414 = a probability of .1359.

49.

$$z = \frac{110 - 100}{(20/\sqrt{30})(\sqrt{470/499})}$$

$$= 10/3.5438 = 2.82$$

which yields a table value of .4976. The probability that the mean IQ of the sample of 30 exceeds 110 is thus .5000 − .4976 or .0024.

Section 6.3

1. It's the standard deviation of the distribution of all possible sample percentages for samples of size n that could be taken from a given population.

3.

Sample Members	CPA?	Sample Proportion
A, B, C	No, yes, no	1/3
A, B, D	No, yes, no	1/3
A, B, E	No, yes, yes	2/3
A, C, D	No, no, no	0
A, C, E	No, no, yes	1/3
A, D, E	No, no, yes	1/3
B, C, D	Yes, no, no	1/3
B, C, E	Yes, no, yes	2/3
B, D, E	Yes, no, yes	2/3
C, D, E	No, no, yes	1/3

5. $\sigma_p = 20$ percent. You can verify that 20 percent $= \sqrt{(40)(60)/3}\,\sqrt{2/4}$.

7. $\sigma_p = \sqrt{[(57)(43)]/100} = 4.95$

9. When the sample size increases, the dispersion of the sampling distribution of percentages decreases.

11. $\sigma_p = \sqrt{[(30)(70)]/100} = 4.58$

13. $\sigma_p = \sqrt{[(70)(30)]/100} = 4.58$

15. The greatest amount of dispersion occurs when $p = 50$ percent. The dispersion decreases as the values of p get closer to 0 or to 100 percent. Note the symmetry.

17. Since the population is finite and the sample is more than 5 percent of the population, the correction factor is needed. Its value is $\sqrt{(20-5)/(20-1)} = .89$.

19. Since the general form of the interval is $\pi \pm z(\sigma_p)$, the value of $z(\sigma_p)$ must equal 1 percent. The standard error is $\sqrt{(8.5)(91.5)/100} = 2.79$ percent. Therefore, $z(2.79 \text{ percent}) = 1$ percent, and $z = 1/2.79 = .358$ or .36. The normal curve area corresponding to a z value of .36 is .1406, and so the chance that a sample percentage will have a value within 1 percent of the population parameter is 2(.1406) or .2812.

21. The standard error is .394 percent, and the z score is 2.54. The probability of being within 1 percent of the population percentage is 2(.4945) or .989.

23.
$$z = \frac{50 - 55}{\sqrt{\dfrac{(55)(45)}{100}}} = \frac{-5}{4.97} = -1.01$$

which gives a table value of .3438. And .5000 − .3438 = the probability of .1562.

25. The $\sigma_p = \sqrt{(33.33)(66.67)/2}\,\sqrt{4/5} = 29.81$ percent.

CHAPTER 7

Estimating Parameters

Plant It Right There

Doctor Keith Aaronson of the University of Pennsylvania reported in a study that 29 percent of women patients referred to his hospital for transplant evaluation decided against it, but only 7 percent of the men refused. According to Dr. Aaronson, men get 81 percent of the heart transplants.

LOOKING AHEAD

You'll recall from Chapter 1 that statistical inference is the process of arriving at a conclusion about an unknown population parameter on the basis of information obtained from a sample statistic. In the last three chapters we've been looking at the probability and sampling concepts that underlie this process. Now, in this chapter, we'll see how statistical inference concepts enable us to use sample data to estimate the value of an unknown population mean, percentage, or variance. (In the next chapters, we'll see how these same concepts permit us to make tests to see if assumptions about one or more unknown population parameters are likely to be acceptable.)

Thus, after studying this chapter, you should be able to:

- Explain the basic concepts underlying the estimation of population means, percentages, and variances.
- Compute estimates of the population mean at different levels of confidence when the population standard deviation is unknown as well as when it's available.
- Compute estimates of the population percentage at different levels of confidence.
- Compute estimates of the population variance at different levels of confidence.
- Understand when and how to use the appropriate probability distributions needed for estimation purposes.
- Determine the appropriate sample size to use to estimate the population mean or percentage at different levels of confidence.

7.1 Estimate, Estimation, Estimator, Et Cetera

To guess is cheap,
To guess wrongly is expensive.—***An old Chinese proverb***

As the proverb implies, a guess is easily made. Anyone can offer an opinion. The naive as well as the expert can produce a value if asked to do so. José Q. Public or the famous economist Marge Propensity can give a figure for next year's gross domestic product. But the task in estimation *isn't* just to produce a figure; the challenge is to produce one that has a reasonable degree of accuracy.

Need for Accuracy

Let's assume that a sales manager who must make a sales forecast for the next period asks you, her trusted assistant, to estimate the average dollar purchase made by a typical customer. Perhaps you base your estimate on a method such as rolling a pair of dice, drawing from a deck of cards, or reading a cup of soggy tea leaves. Now maybe your luck holds and your hunch doesn't cause later embarrassment. But consider the alternative: On the basis of your wild guess, your boss makes a grossly inaccurate sales forecast that leads to the loss of hundreds of thousands of dollars. This result brings the wrath of the company president down on your boss, and she, in turn. . . . Well, you get the picture.

In this chapter we'll see how to estimate the population mean, population percentage, and population variance with some degree of confidence that the resulting figures approximate the true values. Of course, our methods won't produce exact population values. Some error is inevitable in estimation (as the Roman poet Ovid once wrote, "The judgment of man is fallible"*), but, as you'll see, the amount of error can be objectively assessed and controlled.

Some Terms to Consider

Although no formal definitions of the words "estimate" and "estimation" have yet appeared, you probably have a good grasp of their meanings. In case you don't, though, let's clarify some of the terms we'll be using.

Suppose the Adam and Eve Apple Orchards want to estimate the average dollar sales per day, and a sample of days has produced a sample mean of \$800. In this case the statistic ($\bar{x}$, the sample mean) may be used to estimate the parameter (μ, the population mean). The *sample value of* $\bar{x}$ = *\$800* is an *estimate* of the population value, μ:

An **estimate** is a specific value or quantity obtained for a statistic such as the sample mean, sample percentage, or sample variance.

Note, though, that an estimate isn't the same thing as an *estimator*:

An **estimator** is any statistic (sample mean, sample percentage, sample variance) that is used to estimate a parameter.

Thus, the sample mean ($\bar{x}$) is an estimator of the population mean (μ), the sample percentage (p) is an estimator of the population percentage (π), and the sample variance (s^2) is an estimator of the population variance (σ^2).

There are several reasons for selecting a particular statistic to be an estimator. A complete discussion of all the reasons is beyond the scope of this book, but we'll mention one important criterion here. The sample mean is selected as an estimator of the population mean, the sample percentage is an estimator of the population percentage, and the sample variance is an estimator of the population variance because these statistics are *unbiased.*

The concept of *expected value* was introduced in Chapter 5. If we make a study of any statistic such as the sample mean, sample percentage, or sample variance, take many samples of the same size that are picked in the same way, compute the value of the statistic under study for each sample selected, and then compute the mean of all the values of the statistic, this mean of all the statistic values is the *expected value* of the statistic. That is, the mean of the sampling distribution of a statistic is the expected value of that statistic. And if the expected value of the statistic equals the corresponding population parameter, we say that the statistic is an *unbiased estimator*:

---

* *Fasti*, chap. V, line 191. (There is nothing like a little Ovid to mollify the poets who are required to read this text.)

> An **unbiased estimator** is one that produces a sampling distribution that has a mean that's equal to the population parameter to be estimated.

We've already seen in Chapter 6 that the mean of the sampling distribution of sample means equals the population mean, and the mean of the sampling distribution of sample percentages equals the population percentage. Similarly, the mean of the sampling distribution of sample variances equals the population variance when the sample variances are computed with this formula:

$$s^2 = \frac{\Sigma (x - \overline{x})^2}{n - 1}$$

Thus, sample means, percentages, and variances are unbiased estimators. This unbiased tendency of an estimator is highly desirable.

We're now ready to define *estimation*:

> **Estimation** is the entire process of using an estimator to produce an estimate of the parameter.

There are two estimation types—point and interval.

Point Estimation. As you might expect:

> A **point estimate** is a single number used to estimate a population parameter, and the process of estimating with a single number is known as **point estimation.**

Thus, the sample mean of $800 in our previous example is a point estimate because the value is only 1 point along a scale of possible values. But is it likely that a single numeric estimate will be correct? This question brings us to the concept of interval estimation.

Interval Estimation. A parameter is usually estimated to be within a spread of values—that is, within an interval bounded by 2 values—rather than as a single number. It's unlikely that any particular sample mean will be exactly equal to the population mean, so allowances must be made for sampling error. Thus:

> An **interval estimate** is a spread of values used to estimate a population parameter, and the process of estimating with a spread of values is known as **interval estimation.**

With two methods of estimation, which one should be used? To answer this, let's look at the precision of a point estimate. You've seen in the computer simulations in Chapter 6 (and in Figures 6.2 and 6.5) that the means of different random samples taken from a population have different values. Sample means are likely to be close to the population mean, but it's highly unlikely that the value of a sample mean will be exactly equal to the mean of the population from which the sample was taken. A point estimate is not only likely to be wrong, it also doesn't allow us to evaluate the precision of the estimate.

The precision of an estimate is set by the degree of sampling error. And although a point estimate will likely be off the mark, this doesn't prevent us from placing considerable confidence in the estimate that the parameter will be within a given spread of values. For example, we may say that the daily average sales for the orchards is likely to be between \$785 and \$815 instead of simply saying that the true mean may be \$800. Thus, the parameter is estimated to be within a certain spread or interval. And since allowance is made for sampling error, the precision of the estimate can be assessed. Of course, an interval estimate may be wrong. But in contrast to a point estimate, the probability of error for an interval estimator can be known.

Don't get the impression, though, that a point estimate is of little value in the estimation process. As you'll see in the following pages, the interval estimate is actually based on the point estimate. In fact, the *point estimate is adjusted for sampling error to produce an interval estimate.* The following pages deal with interval estimates of the population mean, population percentage, and population variance, and the accuracy of these estimates can be determined with some degree of confidence.

A Link between Genius and Madness?

A 1-year study of the behavior of painters, novelists, playwrights, poets, and sculptors was carried out by Dr. Kay Jamison, a professor of psychiatry at UCLA. The conclusion was that episodes of intense creativity may be partially attributed to manic-depression. While severe depression and manic-depressive illness appear in about 6 percent of the general population, these illnesses appeared in over 50 percent of the people in the study, as evident from their having received medication for mania or depression. This study established a strong statistical link between creative genius and madness.

Self-Testing Review 7.1

1–7. Match the following numbered clues with the lettered responses given in the column on the right:

1. Estimators of population parameters.	a. Point estimate
2. Point estimate for μ, the population mean.	b. Unbiased estimator
3. Point estimate for π, the population percent.	c. The sample percent
4. Point estimate for σ^2, the population variance.	d. An interval estimate
5. If the expected value of the statistic equals the corresponding population parameter, we say the statistic is a(n) ______?	e. Sample statistics
6. A range of values used to estimate a population parameter.	f. The sample mean
7. A single value used as an estimator for a population parameter.	g. The sample variance

8. Why is an unbiased estimate desirable?

9. The difference between the population mean and the mean of a sample is most likely to be ______.

   a) Zero *b)* A large value *c)* A small value

10. The population mean will ______ fall within the interval estimate.

    a) Always *b)* Probably *c)* Occasionally *d)* Never

7.2 Interval Estimation of the Population Mean: Some Basic Concepts

In practice, only one sample of a population is taken, the sample statistic (mean, percentage, variance) is calculated, and an estimate of the population parameter is made. Although this section focuses on basic concepts involved in estimating the

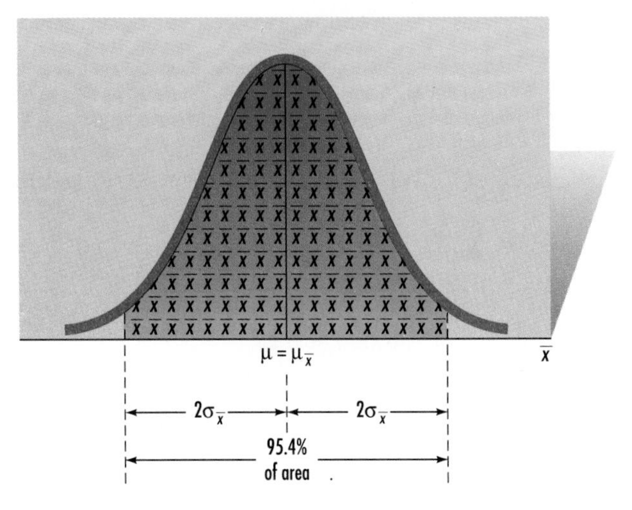

FIGURE 7.1 An educational schematic of the sampling distribution of the means when the sample size is large. There is a 95.4 percent chance that an $\bar{x}$ will have a value between $\mu \pm 2\sigma_{\bar{x}}$.

population mean, many of the points raised also apply to estimating the population percentage or variance. To estimate the population mean, though, we must know something about its relationship to the sample means.

The Sampling Distribution—Again

A quick review of the concepts of the sampling distribution of means shows the theoretical basis for the interval estimation of the population mean. Suppose we have a sample size that is sufficiently large so that the sampling distribution is approximately normal. Figure 7.1 shows that 95.4 percent of the possible outcomes of $\bar{x}$ are within $2\sigma_{\bar{x}}$ to each side of the mean of the sampling distribution. This means that if a mad statistician takes 1,000 samples of the same size from a population, about 954 of the sample means will fall within 2 standard errors to both sides of the population mean. (If this last sentence—and Figure 7.1—puzzles you, review Chapter 6 again.)

Interval Width Considerations

If 95.4 percent of the possible values of the sample mean fall within 2 standard errors of the population mean as shown in Figure 7.1, then obviously μ will not be farther than $2\sigma_{\bar{x}}$ from 95.4 percent of the possible values of $\bar{x}$. Now let's show in nonstatistical terms the logic of the preceding sentence. Let's assume we have 1,000 towns located at various distances from the city of Boston and it happens that 95.4 percent of these towns are within a 50-mile radius of Boston. If 954 towns are within a 50-mile radius of Boston, then logically Boston must fall within a 50-mile radius of each of these 954 towns. If Hingham is within 50 miles of Boston, then Boston will certainly be no farther than 50 miles from Hingham (see Figure 7.2). If we randomly select a large number of towns from the 1,000, Boston will be within a radius of 50 miles of 95.4 percent of all the towns selected. If all this appears simple and trite, then we've accomplished something.

In returning to the statistical world, substitute the population mean for Boston, let the possible sample means be the towns, and use $2\sigma_{\bar{x}}$ in place of the 50-mile radius. To

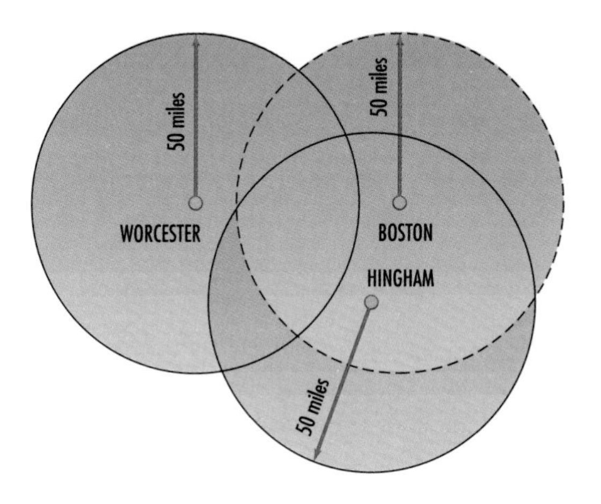

FIGURE 7.2 An illustration of distance relationships.

repeat, then, if 95.4 percent of the sample means are within $2\sigma_{\bar{x}}$ of the μ, then certainly the μ must be within $2\sigma_{\bar{x}}$ of 95.4 percent of the sample means. Thus, if we use the method of $\bar{x} \pm 2\sigma_{\bar{x}}$ to estimate the population mean and if we construct a large number of intervals, 95.4 percent of the interval estimates will include μ.

Now let's assume that we have 1,000 possible samples and thus 1,000 sample means, 3 of which are shown in Figure 7.3. The population mean will be located within 95.4 percent of the 1,000 possible intervals that could be constructed using $\bar{x} \pm 2\sigma_{\bar{x}}$. Any specific single interval may or may not contain μ (note that in Figure 7.3 the intervals produced using $\bar{x}_1$ and $\bar{x}_2$ do include μ, but the interval constructed using $\bar{x}_3$ fails to reach μ), but the method employed assures that if a large number of intervals are constructed, the μ will be included in 95.4 percent of them.

We'll not be limited to using a 95.4 percent probability of estimating the population mean for the rest of this book. Thus, we need to generalize what has been discussed so far so that we can apply different interval estimates to a variety of situations. If the

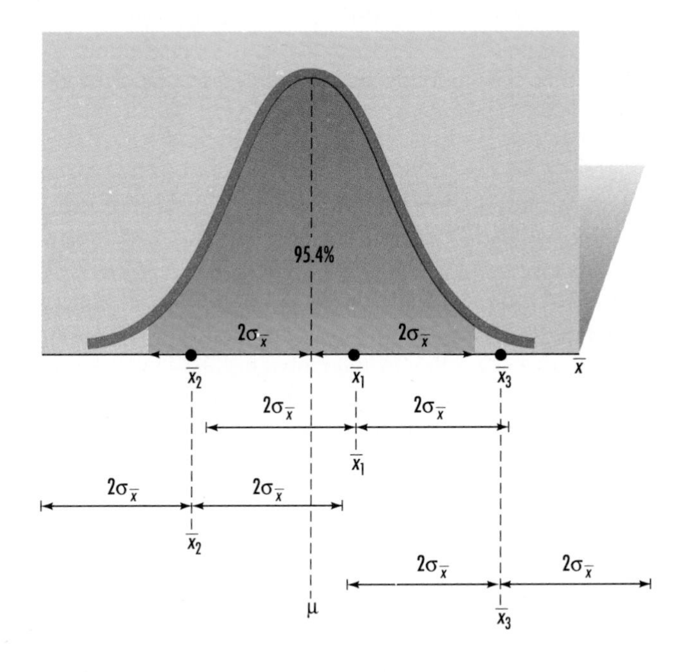

FIGURE 7.3 An illustration of the interval relationship between μ and $\bar{x}$.

sampling distribution is normal, an interval estimate of μ may be constructed in the following manner:

$$\underbrace{\bar{x} - z\sigma_{\bar{x}}}_{\text{lower limit of estimate}} < \mu < \underbrace{\bar{x} + z\sigma_{\bar{x}}}_{\text{upper limit of estimate}} \quad (7.1)$$

where $\bar{x}$ = the sample mean (and point estimator of μ)
$\sigma_{\bar{x}}$ = the standard error of the mean
z = the standard normal value determined by the probability associated with the interval estimate—that is, the value associated with a certain likelihood that μ will be *included* in a large number of interval estimates

The Level of Confidence

Statisticians assign a *level of confidence* to the interval estimates they produce.

The **level of confidence** (or the **confidence coefficient**) refers to the probability of correctly including the population parameter being estimated in the interval that is produced.

The word "confidence" is used because the probability value shows the likelihood that the computed spread of values will include the population mean. That is, the higher the level of probability associated with an interval estimator, the more confidence there is that the method of estimation will produce a result that contains the population mean.

In practice, the confidence level is generally identified before estimation. Thus, a 90 percent level of confidence might be specified, and this means that an analyst wants to be 90 percent sure that the population mean is included in the interval obtained. The analyst knows that if the sample mean selected to produce the estimate falls within the middle 90 percent of the area under the sampling distribution of means, then the desired result will be achieved.

What's the appropriate z value to use in formula 7.1 to construct an interval estimate that will include the population mean 90 percent of the time? The answer is the z value that separates the middle 90 percent of the area under the normal sampling distribution curve from the remaining 10 percent. Thus, to use the table for the standard normal distribution found in Appendix 2 at the back of the book, we need to know the z value that divides 45 percent of the area in each half of the normal distribution from the remaining 5 percent. Looking in the body of that table, we find the figure nearest to .4500 (or 45 percent). The z value corresponding to the table figure of .4495 is 1.64, and the z value for the table figure of .4505 is 1.65. The most accurate z value is thus 1.645. That is, the z value corresponding to an area of .45 (or 45 percent) is 1.645. Thus, the interval estimate of μ using a 90 percent confidence level is:

$$\bar{x} - 1.645\sigma_{\bar{x}} < \mu < \bar{x} + 1.645\sigma_{\bar{x}}$$

The confidence levels generally used in interval estimation are 90, 95, and 99 percent. The z values and the general forms of the interval estimates associated with these confidence levels are shown in Table 7.1. To summarize:

Confidence intervals are those interval estimates based on specified confidence levels, and the upper and lower limits of the intervals are known as **confidence limits.**

TABLE 7.1 COMMONLY USED CONFIDENCE COEFFICIENTS AND CONFIDENCE INTERVALS FOR A LARGE SAMPLE

Confidence Coefficient	z value	General Form of the Interval Estimate
90	1.645	$\bar{x} - 1.645\sigma_{\bar{x}} < \mu < \bar{x} + 1.645\sigma_{\bar{x}}$
95	1.96	$\bar{x} - 1.96\sigma_{\bar{x}} < \mu < \bar{x} + 1.96\sigma_{\bar{x}}$
99	2.575	$\bar{x} - 2.575\sigma_{\bar{x}} < \mu < \bar{x} + 2.575\sigma_{\bar{x}}$

The sampling error in estimating a population parameter is the distance between the true population parameter and the sample statistic. Thus, if we are estimating a population mean and if we have a sample size that is sufficiently large so that the sampling distribution is approximately normal, then this sampling error or **error of estimate** (often designated E) is likely to be less than $z(\sigma_{\bar{x}})$. You can see from this relationship that if the z value or confidence level is increased, then the size of the error of estimate is also increased. And you can also see from formula 7.1 that the confidence interval *width* for a given level of confidence is equal to *twice the value of the error of estimate.* For example, if we have a normal sampling distribution and we are seeking to produce a 90 percent confidence interval, then the error of estimate is $1.645(\sigma_{\bar{x}})$. And since this error of estimate can be to the right of (added to) or to the left of (subtracted from) the sample mean point estimate, the confidence interval width is twice the value of the error of estimate or $\pm\ 1.645(\sigma_{\bar{x}})$.

At this point you may wonder why it's necessary to have various confidence levels when it seems logical that the highest level of confidence is desirable in estimating μ. You might be thinking: "If I'm required to give an accurate estimate of the true mean, why shouldn't I always use a 99 percent confidence level? It makes sense to have as much confidence as possible in my estimate!"

There's no question that it's highly desirable to have as much confidence as possible in the estimate. But if more confidence is desired, you must allow for more sampling error—that is, you must accept a larger error of estimate. The width of the interval thus increases, and the estimate loses some precision. Table 7.1 shows this relationship between the confidence coefficient and the interval width. If the interval estimate is too wide, the estimate will have no utility.

Let's assume that you are an advertising manager and you must submit an advertising budget request to the finance department. You have been told that the advertising expenditures for the next period should be 10 percent of sales. A random sample has produced a mean sales figure of $\bar{x} = \$250,000$ with a $\sigma_{\bar{x}}$ of \$2,000. Using the general forms of the interval estimates shown in Table 7.1, you construct intervals for both the 90 and 99 percent levels of confidence. If you base your advertising budget on the estimate with a coefficient of 90 percent, you tell the finance department you'll need between \$21,710 and \$28,290 [\$25,000 ± 1.645(\$2,000)]. But if your advertising budget is based on an estimate with a 99 percent confidence level, you allow more room for error and tell the finance department to provide between \$19,850 and \$30,150 [\$25,000 ± 2.575(\$2,000)]. As the range in your budget request increases, the planning in the

What About 51.5 to 44.5?

A few years ago, Everett Carll Ladd wrote the following comments in *The Wall Street Journal*: "In a July statement, Louis Harris proclaimed that 'the selection of Rep. Geraldine Ferraro from New York as the vice presidential choice of Walter Mondale increases the Democratic chances of winning November's election. When paired with Mondale on the ticket, Ferraro narrows a 52–44% Reagan lead to 51–45%.' One doesn't know whether to laugh or cry. Polls can never, repeat *never*, achieve a measure of precision to such an extent that one percentage point means anything at all."

finance department becomes more uncertain. Financial people may then be forced to tie up more funds than necessary. In short, an increase in the confidence level might produce an estimate that isn't useful.

One caution should be mentioned here. The confidence coefficient should be stated *before* the interval estimation. Sometimes a novice researcher calculates a number of interval estimates on the basis of a single sample while varying the confidence level. After obtaining these estimates, he or she then selects the one that seems most suitable. Such an approach is really manipulating data so that the results of a sample are the way a researcher would like to see them. This approach introduces the researcher's bias into the study, and it should be avoided.

Self-Testing Review 7.2

1. What is the standard error of a sampling distribution?

2. If a mad statistician (or a sane one with a statistical software program) takes 10,000 samples of the same sufficiently large size from a population, how many of the corresponding sample means would be expected to fall within 1 standard error of the population mean? How many would fall within 2 standard errors of the population mean? How many would likely lie within 3 standard errors of the population mean?

3–6. For a sampling distribution with $n = 32$, within how many standard errors:

3. Will 80 percent of the sample means fall?

4. Will 90 percent of the sample means fall?

5. Will 95 percent of the sample means fall?

6. As the level of confidence increases, what do you notice about the number of standard errors required for the confidence interval?

7–9. For a sampling distribution with $n = 40$, within how many standard errors:

7. Will 86 percent of the sample means fall?

8. Will 92 percent of the sample means fall?

9. Will 98 percent of the sample means fall?

10. What role does the level of confidence play in an interval estimate?

11. What happens to the width of an interval estimate as the level of confidence increases?

12. Why would someone use a 90 percent level of confidence instead of a 99 percent level if the 99 percent level has a greater chance of including the population mean?

13. The sampling error (the error of estimate) is the distance between the population mean and the sample mean. If the confidence level is increased, what happens to the error of estimate?

14. Discuss the following statement: "The width of a confidence interval is equal to twice the value of the error of estimate."

15. In your own words, discuss what a 95 percent confidence interval means.

16. We are 90 percent sure that the population mean falls within the interval from

84.1 − 5.3 to 84.1 + 5.3. What is the level of confidence, the population parameter, the point estimate, and error of estimate for this situation?

17. We are 99 percent sure that the population mean falls within the interval from 3.5 − .02 to 3.5 + .02. What is the level of confidence, the population parameter, the point estimate, and error of estimate for this situation?

7.3 Estimating the Population Mean

Now that we've considered the general form of (and the theoretical basis for) an interval estimate, let's look first at the approach used to estimate the population mean when the population standard deviation is known and the sample size (n) exceeds 30. (Later, we'll consider situations where σ is unknown and n is 30 or less).

Estimating the Mean When the σ Is Known and $n > 30$

When the *population standard deviation* (σ) *is known*, we may directly compute the standard error of the mean as follows:

$$\sigma_{\bar{x}} = \frac{\sigma}{\sqrt{n}} \quad \text{for an infinite population}$$

or

$$\sigma_{\bar{x}} = \frac{\sigma}{\sqrt{n}}\sqrt{\frac{N-n}{N-1}} \quad \text{for a finite population}$$

And the interval estimate may then be constructed in the following manner:

$$\underbrace{\bar{x} - z\sigma_{\bar{x}}}_{\text{lower confidence limit}} < \mu < \underbrace{\bar{x} + z\sigma_{\bar{x}}}_{\text{upper confidence limit}}$$

So let's now use this estimation procedure to solve some example problems.

Example 7.1. A manager at the Papyrus Paper Company wants to estimate the mean time required for a new machine to produce a ream of paper. A random sample of 36 reams required an average machine time of 1.5 minutes for each ream. Assuming $\sigma = 0.30$ minute, construct an interval estimate with a confidence level of 95 percent.

We have the following data: $\bar{x} = 1.5$ minutes, $\sigma = .30$, $n = 36$, and confidence level = 95 percent. The $\sigma_{\bar{x}}$ is computed as follows:

$$\sigma_{\bar{x}} = \frac{\sigma}{\sqrt{n}} = \frac{.30}{\sqrt{36}} = .05$$

With a 95 percent confidence coefficient, the z value equals 1.96. Thus, the interval estimate of the true mean time (μ) is constructed as follows:

$$\bar{x} - z\sigma_{\bar{x}} < \mu < \bar{x} + z\sigma_{\bar{x}}$$
$$1.5 - 1.96(.05) < \mu < 1.5 + 1.96(.05)$$
$$1.402 \text{ minutes} < \mu < 1.598 \text{ minutes}$$

We can use the data in this example, along with *Minitab* computer simulations, to verify some of the concepts we've now considered. Let's assume in our Papyrus Paper Company example that the true population mean time for the machine to produce a ream is actually 1.48 minutes. The σ, we've seen, is .30 minute. If we were to take 15 random samples, with each sample containing the time data required to produce 36 reams, about how many of the 15 sample confidence intervals would include the μ of 1.48 minutes if we use the 95 percent confidence level? We could answer this question without much hesitation if we were taking 1,000 random samples (the answer would be about 950 of them), but with just 15 samples we can't always be sure.

Figure 7.4*a* shows the results obtained when the statistical package simulates taking 15 random samples, with 36 sample items in each sample. The first line in Figure 7.4*a* instructs the program to randomly select the data items and to then put the data for each sample in a separate column. The first sample data goes in column 1 (C1), the second in C2, and so on. The second line in Figure 7.4*a* gives the package the values of μ and σ. A ZINTERVAL command is then used in line 3 to produce a 95 percent confidence interval for each of the 15 samples. The means of each of the simulated samples are listed in the program output. The SE MEAN gives the standard error figured earlier in this example, and the lower and upper confidence limits are given for each of the 15 intervals produced.

You'll see that 14 of the 15 samples have intervals that *include* the μ of 1.48 minutes. Only the interval in C2 has failed to include this parameter. Notice, too, that the sample in C14 has duplicated the results we noted earlier (this is just a coincidence). All intervals, of course, have the same width, but they have different lower and upper limits, as you would expect.

To graphically demonstrate how these lower and upper limits vary, Figure 7.4*b* shows *another simulation* of 15 samples. You can ignore the program commands at the top of Figure 7.4*b* and concentrate on the plot. The vertical scale shows the average time required to produce a ream of paper. A solid line is drawn horizontally through the middle of the plot to show the position of the μ of 1.48 minutes. The baseline in this case is simply a count of the samples. Each letter *A* represents the location of the *lower limit* of the confidence interval for one sample, and if you draw a vertical line up to the corresponding letter *B*, you'll have located the position of the *upper limit* of the sample's confidence interval. With one exception, all *A*'s fall below the population mean line of 1.48, and all *B*'s are scattered above that line. In the one case (sample 7) where both *A* and *B* are above the μ line, the interval fails to include the μ. Thus, in this second simulation, our samples have different means, different lower limits, and different upper limits, but 14 of the 15 intervals include the population mean. Don't think that every simulation of 15 samples will always produce one sample that fails to include the μ in its intervals. The next simulation might easily have no "failures" or 2 failures.

Example 7.2. The Ledd Pipe Company has received a shipment of 100 lengths of pipe, and a quality control inspector wants to estimate the average diameter of the pipes to see if they meet minimum standards. She takes a random sample of 50 pipes, and the sample produces an average diameter of 2.55 inches. In the past, the population standard deviation has been 0.07 inch. Construct an interval estimate with a 99 percent level of confidence.

We have the following data from the problem: $\bar{x} = 2.55$, $\sigma = .07$, $n = 50$, $N = 100$, and confidence level = 99 percent. The standard error of the mean is:

$$\sigma_{\bar{x}} = \frac{\sigma}{\sqrt{n}} \sqrt{\frac{N-n}{N-1}} = \frac{.07}{\sqrt{50}} \sqrt{\frac{100-50}{100-1}} = .007$$

```
MTB > RANDOM 36 SAMPLE ITEMS INTO C1-C15;
SUBC> NORMAL MU = 1.48, SIGMA = .30.
MTB > ZINTERVAL 95 PERCENT, SIGMA = .30, DATA IN C1-C15

THE ASSUMED SIGMA =0.300

          N      MEAN     STDEV   SE MEAN    95.0 PERCENT C.I.
C1       36    1.4602    0.2641    0.0500   (  1.3621,   1.5584)
C2       36    1.3522    0.2893    0.0500   (  1.2541,   1.4503)
C3       36    1.4846    0.3192    0.0500   (  1.3865,   1.5828)
C4       36    1.5185    0.3065    0.0500   (  1.4203,   1.6166)
C5       36    1.4471    0.2573    0.0500   (  1.3490,   1.5453)
C6       36    1.5056    0.3172    0.0500   (  1.4075,   1.6037)
C7       36    1.5157    0.3037    0.0500   (  1.4176,   1.6139)
C8       36    1.5327    0.3038    0.0500   (  1.4346,   1.6309)
C9       36    1.4700    0.2783    0.0500   (  1.3718,   1.5681)
C10      36    1.4726    0.3246    0.0500   (  1.3745,   1.5708)
C11      36    1.4092    0.3084    0.0500   (  1.3110,   1.5073)
C12      36    1.4958    0.2891    0.0500   (  1.3976,   1.5939)
C13      36    1.4691    0.2818    0.0500   (  1.3710,   1.5673)
C14      36    1.5007    0.3578    0.0500   (  1.4026,   1.5988)
C15      36    1.4271    0.2573    0.0500   (  1.3290,   1.5253)
```

(*a*)

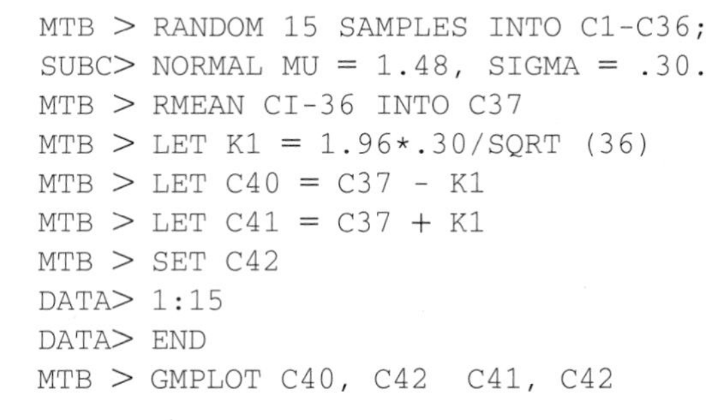

```
MTB > RANDOM 15 SAMPLES INTO C1-C36;
SUBC> NORMAL MU = 1.48, SIGMA = .30.
MTB > RMEAN CI-36 INTO C37
MTB > LET K1 = 1.96*.30/SQRT (36)
MTB > LET C40 = C37 - K1
MTB > LET C41 = C37 + K1
MTB > SET C42
DATA> 1:15
DATA> END
MTB > GMPLOT C40, C42  C41, C42
```

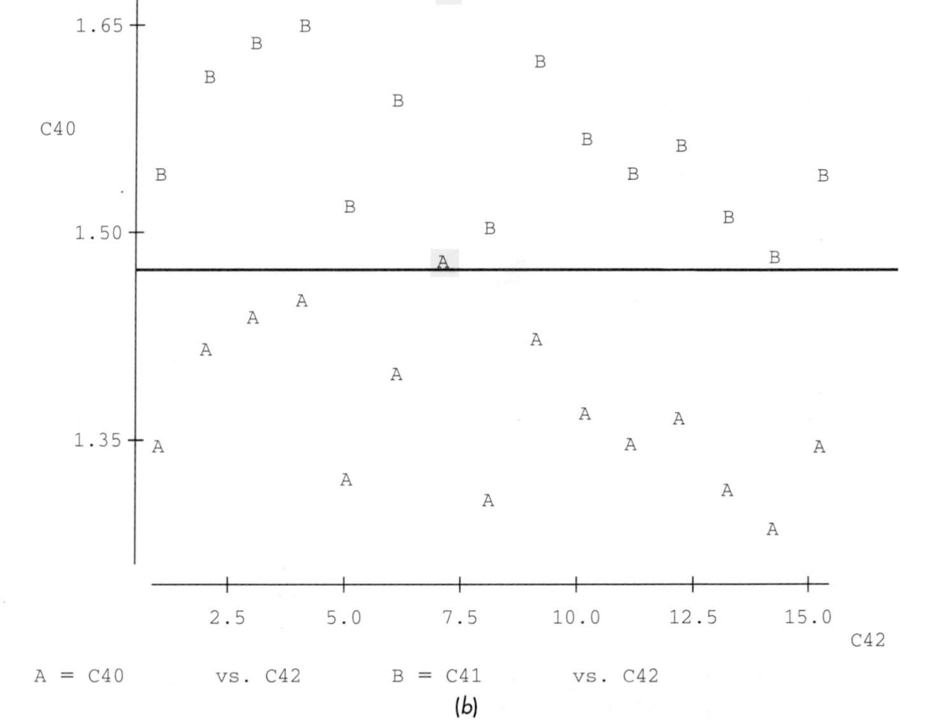

(*b*)

FIGURE 7.4 (*a*) The 95 percent confidence intervals produced when 15 random samples of size 36 are simulated using the Papyrus Paper Company data introduced in Example 7.1. (*b*) A plot of the lower (*A*) and upper (*B*) values of 15 more simulated random samples of size 36.

With a 99 percent confidence coefficient, the z value is 2.575. Therefore, the interval estimate of μ, the true average diameter of the shipment of pipes, is found as follows:

$$\bar{x} - z\sigma_{\bar{x}} < \mu < \bar{x} + z\sigma_{\bar{x}}$$
$$2.55 - 2.575(.007) < \mu < 2.55 + 2.575(.007)$$
$$2.532 \text{ inches} < \mu < 2.568 \text{ inches}$$

Note that the preceding examples are cases in which σ *is known (or can be identified), the sample size exceeds 30, and the sampling distribution is thus normally distributed.* The general procedure for interval estimation under such conditions is shown in Figure 7.5.

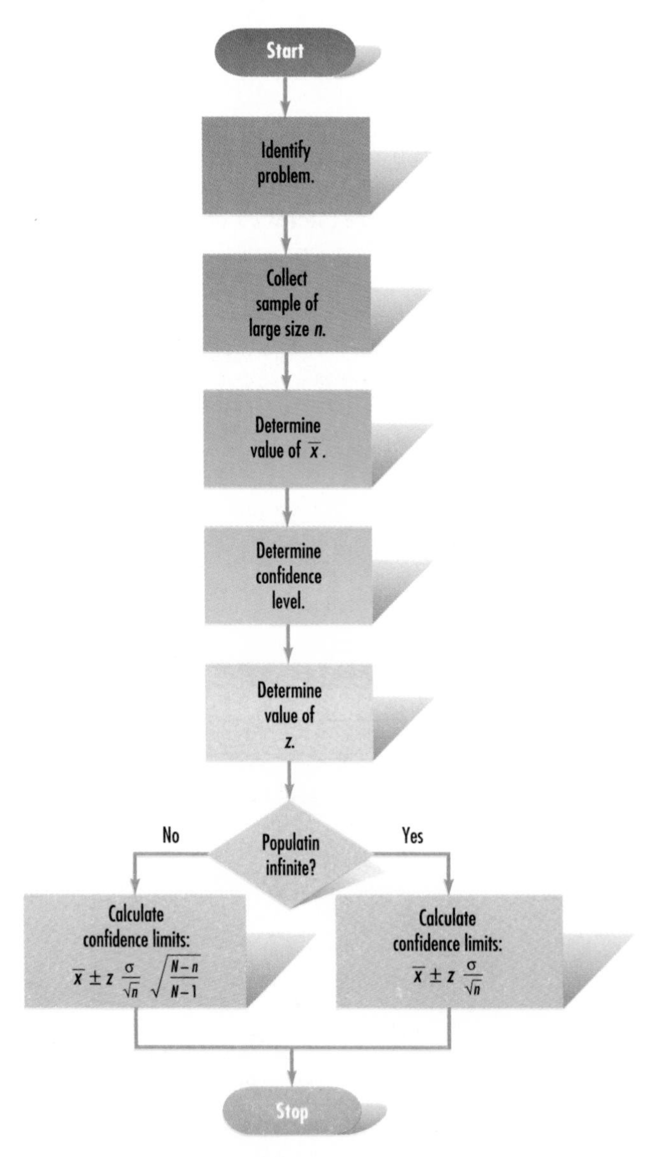

FIGURE 7.5 Procedure for the interval estimation of μ with σ known and $n > 30$.

Estimating the Mean When σ Is Unknown and $n > 30$

In most situations, not only is the population mean unknown but the population standard deviation is also unknown. In fact, it's only in isolated cases that the σ is known, so it usually must be estimated along with the population mean.

The Estimator of σ. You saw in Chapter 3 that the formula needed to compute the standard deviation for a set of *population* values is:

$$\sigma = \sqrt{\frac{\Sigma\,(x - \mu)^2}{N}}$$

And you also saw in Chapter 3 that the formula needed to compute the standard deviation for *a sample* is:

$$s = \sqrt{\frac{\Sigma\,(x - \bar{x})^2}{n - 1}}$$

The only basic difference in these formulas is that a denominator of $n - 1$ (rather than n) is used to compute the sample measure. Let's briefly look at why we use a denominator of $n - 1$ rather than n to compute a sample standard deviation.

If we were computing the sample standard deviation for its own sake, we *could* use a denominator of n in the formula. But we seldom are interested only in s. Rather, s is almost always computed to provide an estimate for an unknown σ. And if we used a denominator of n, we would run into a *bias* problem. Why is that? Well, you've seen earlier that a sample mean is an unbiased estimator of μ because the mean of the sampling distribution of means from which $\bar{x}$ is taken equals the population mean—that is, because $\mu_{\bar{x}} = \mu$. But *if* we were to use a denominator of n to compute all the s values needed to produce a sampling distribution of standard deviations, the mean of that sampling distribution *wouldn't* equal σ. Instead, the mean of the sampling distribution of standard deviations would be less than the population standard deviation, and so there's a tendency for a sample standard deviation computed with a denominator of n to also be less than σ. That is, the computed result would be biased toward understating σ, and so this tendency must be removed. That's where a denominator of $n - 1$ comes in because the use of $n - 1$ has the effect of slightly increasing the value of s and this reduces the bias. (Using $n - 1$ wasn't someone's lucky hunch; its selection can be shown mathematically, but that's beyond the scope of this book.) And so it is that a denominator of $n - 1$ is used to compute the sample standard deviation and to thus arrive at a useful estimator of σ that needs no further manipulation.

Later in this chapter we'll also be estimating the population variance (σ^2). Our estimator of that parameter is the sample variance (s^2). As you might expect by now, we'll use this formula to arrive at a value for s^2:

$$s^2 = \frac{\Sigma\,(x - \bar{x})^2}{n - 1}$$

Computed in this way, the sample variance is an unbiased estimator of the population variance.

Now that we've considered these details, let's look at how we obtain the estimated standard error needed to approximate the population mean when σ is unknown. The formula for this estimated standard error is:

Hi-Yo, Hi-Yo, . . .

Doctor John Forest of Baylor University College of Medicine in Houston studied a sample of 497 adults and found that those who gained or lost no more than 5 pounds over a 1-year period enjoyed the greatest psychological well-being. Forest concluded that "yo-yo" dieting not only endangered physical health but also put mental health at risk.

$$\hat{\sigma}_{\bar{x}} = \frac{s}{\sqrt{n}} \quad \text{for an infinite population} \tag{7.2}$$

or

$$\hat{\sigma}_{\bar{x}} = \frac{s}{\sqrt{n}}\sqrt{\frac{N-n}{N-1}} \quad \text{for a finite population} \tag{7.3}$$

You'll notice that the mark (ˆ) over the standard error symbol (or any other symbol) means that we have an *estimated value*. And you'll also notice that the calculation of $\hat{\sigma}_{\bar{x}}$ is the same as the computation of $\sigma_{\bar{x}}$ except that the population standard deviation is replaced by the sample standard deviation (s).

If the population standard deviation is unknown, the sampling distribution of means can be assumed to be approximately normal *only when the sample size is relatively large (over 30)*. With *estimated values* for σ (remember that $s = \hat{\sigma}$) and $\sigma_{\bar{x}}$, the general form of the interval estimate *for large samples* is altered slightly so that it appears as follows:

$$\underbrace{\bar{x} - z\hat{\sigma}_{\bar{x}}}_{\text{lower confidence limit}} < \mu < \underbrace{\bar{x} + z\hat{\sigma}_{\bar{x}}}_{\text{upper confidence limit}} \tag{7.4}$$

As you can see, we merely substituted the estimated value for the true value of the standard error of the mean. Again, let's look at some examples to illustrate the above points.

Example 7.3. Sam, owner of Sam's Convenience Store, wants to estimate the average dollar purchase per customer. A sample of 100 customers produces a mean spending figure of \$3.50, with a *sample* standard deviation of \$0.75. Estimate the true mean expenditure with a 90 percent confidence level.

We have the following data from the problem situation: $\bar{x} = \$3.50$, $s = .75$, $n = 100$, confidence level = 90 percent. Therefore, the estimate of $\hat{\sigma}_{\bar{x}}$ is computed as follows:

$$\hat{\sigma}_{\bar{x}} = \frac{s}{\sqrt{n}} = \frac{.75}{\sqrt{100}} = .075$$

With a 90 percent confidence level, and with a sample size of over 30, we use the z value of 1.645. The interval estimate is thus:

$$\bar{x} - z\hat{\sigma}_{\bar{x}} < \mu < \bar{x} + z\hat{\sigma}_{\bar{x}}$$
$$\$3.50 - 1.645(.075) < \mu < \$3.50 + 1.645(.075)$$
$$\$3.38 < \mu < \$3.62$$

Example 7.4. The Rogers Poultry Company receives a shipment of 100 hens, and the manager wants to estimate the true mean weight of the hens to see if they meet Rogers' standards. A sample of 36 hens yields a mean weight of 3.6 pounds with a *sample* standard deviation of .6 pound. Construct an interval estimate of the true mean weight with a 99 percent confidence level.

We have the following data: $\bar{x} = 3.6$, $s = .6$, $n = 36$, confidence level = 99 percent, $N = 100$. The estimate of the standard error is computed as follows:

$$\hat{\sigma}_{\bar{x}} = \frac{s}{\sqrt{n}} \sqrt{\frac{N-n}{N-1}} = \frac{.6}{\sqrt{36}} \sqrt{\frac{100-36}{100-1}} = .08$$

With a 99 percent confidence level, the z value is 2.575. Therefore, the interval estimate is:

$$\bar{x} - z\hat{\sigma}_{\bar{x}} < \mu < \bar{x} + z\hat{\sigma}_{\bar{x}}$$
$$3.6 - 2.575(.08) < \mu < 3.6 + 2.575(.08)$$
$$3.394 \text{ pounds} < \mu < 3.806 \text{ pounds}$$

When σ is unknown and when the *sample size is large,* the sampling distribution is approximately normally distributed. But if $\sigma_{\bar{x}}$ must be estimated and if the sample size is *30 or less,* the sampling distribution will not be normally distributed, and therefore the interval estimate *cannot* be calculated with the use of the z distribution. What distribution should then be used? The following section provides us with the answer.

Estimating the Mean Using the *t Distribution*

The Central Limit Theorem in Chapter 6 told us that if the sample size is sufficiently large (>30), the sampling distribution of means approximates the normal (z) probability distribution regardless of the shape of the population distribution. And it also told us that the sampling distribution is a normal distribution regardless of sample size if the population is normally distributed. But *what if, in sampling from a normal population, the sample size is 30 or less and we have to use* s as an estimator of an unknown σ? In that case, the appropriate sampling distribution of means follows a *t distribution* rather than a z distribution.

t Distribution

A **t distribution** (often called the ***Student t distribution***) uses the test statistic:

$$t = \frac{\bar{x} - \mu}{s/\sqrt{n}}$$

rather than:

$$z = \frac{\bar{x} - \mu}{\sigma/\sqrt{n}}$$

and has the following properties:

- It is similar to a z distribution with a zero mean and a symmetrical (bell) shape about that mean.

> - But its shape depends on the sample size (the t distribution is really a family of distributions, and there's a different one for each sample size).
> - With a small sample, the shape of the corresponding t distribution is less peaked than the z distribution, but as the sample size increases and approaches 30, the shapes of the t distributions lose their flatness and approximate the shape of the z distribution. (Thus, when $n > 30$, we can use z values.)

Figure 7.6 shows the effect of sample size on the shape of the t distributions, and Figure 7.7 summarizes the conditions under which the t or z distributions are used for estimation purposes. You can see in Figure 7.7 that it's possible to still use z values when the sample size is 30 or less *if* the σ is known, and *if* it's also known that the items in the population are normally distributed. But knowing both these facts is unlikely, and so we'll consider more plausible situations in the next few pages.

If σ is unknown, and if the sample size is small, the interval estimate of the population mean has the following form:

$$\underbrace{\bar{x} - t_{\alpha/2}\,\hat{\sigma}_{\bar{x}}}_{\text{lower confidence limit}} < \mu < \underbrace{\bar{x} + t_{\alpha/2}\,\hat{\sigma}_{\bar{x}}}_{\text{upper confidence limit}} \quad (7.5)$$

Like the z value, the value of t depends on the confidence level.

Appendix 4 at the back of the book is a table of t-distribution values. But the t table doesn't look like the z table. If, for example, we are interested in making an estimate at the 95 percent confidence level, the t-table format is not designed to emphasize the 95 percent chance of including μ in the estimate; rather, the presentation focuses attention on the 5 percent chance of *not including* μ. *This chance of error is labeled* α *(alpha) and in decimal form equals 1.00 minus the confidence coefficient.* For example, if the confidence coefficient is .95 (or 95 percent), then α is $1.00 - .95$ or .05. Since α represents the *total* chance of error—that is, the chance of not including μ—and since the particular t distribution being used is symmetrical, the total error is divided evenly between the chance of overestimation and the chance of underestimation. As shown by the shaded portion in the figure at the top of Appendix 4, however, the *t table only deals with areas to one side of the distribution.* Consequently, the subscript $\alpha/2$ follows t

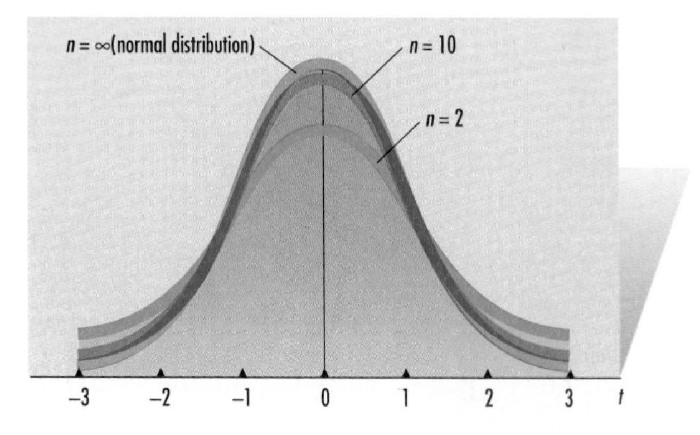

FIGURE 7.6 The effect of sample size on the shape of the t distributions.

distribution is needed for the construction of a 90 percent confidence interval. Find the appropriate t or z value:

7. $n = 13$, σ is unknown, and population is normal.

8. $n = 11$, σ is known, and population is normal.

9. $n = 34$, σ is unknown.

10. $n = 17$, σ is unknown, and population is normal.

11–15. Given sample statistics of $\bar{x} = 75$, $s = 12$, $n = 58$, and a desired 95 percent confidence interval:

11. Find the point estimate of the population mean.

12. Estimate the standard error of the mean. $\frac{s}{\sqrt{n}}$

13. Determine the appropriate t or z value to use.

14. Compute the error of estimate.

15. Construct the 95 percent confidence interval for μ.

16–20. Given sample statistics of $\bar{x} = 24.7$, $s = 5.3$, $n = 18$, and a desired 99 percent confidence interval (assume the population values are normally distributed):

16. Find the point estimate of the population mean.

17. Estimate the standard error of the mean.

18. Determine the appropriate t or z value to use.

19. Compute the error of estimate.

20. Construct the 99 percent confidence interval for μ.

21–25. If it is known that $\sigma = 3.81$, the population is normally distributed, and the sample statistics are $\bar{x} = 62.7$ and $n = 17$:

21. Find the point estimate for the population mean.

22. Estimate the standard error of the mean.

23. Determine the appropriate t or z value to use to get a 95 percent confidence interval.

24. Compute the error of estimate at the 95 percent confidence level.

25. Construct the 95 percent confidence interval for estimating the μ.

26–30. Given a normally distributed finite population of $N = 80$, and sample statistics of $\bar{x} = 769.53$, $s = 15.72$, and $n = 14$:

26. Find the point estimate for the population mean.

27. Estimate the standard error of the mean.

28. Determine the appropriate t or z value to use to get a 90 percent confidence interval.

29. Compute the error of estimate at the 90 percent confidence level.

30. Construct the 90 percent confidence interval for μ.

31–35. We have a normally distributed finite population of $N = 38$, and sample statistics of $\bar{x} = 346.8$, $s = 18.9$, and $n = 10$. If we want to construct a 98 percent confidence interval:

31. Find the point estimate for the population mean.
32. Estimate the standard error of the mean.
33. Determine the appropriate t or z value.
34. Compute the error of estimate.
35. Construct the 98 percent confidence interval.

36. *The New England Journal of Medicine* recently reported that in a sample of 155 self-referred psychiatric patients, the mean cost for services was $3,222 with a standard deviation of $1,450. Assuming that this is a random sample, form a 95 percent confidence interval for the population mean.

37. Professor Al Jebrah is a first-year teacher of mathematics who has a total of 150 students. To estimate the mean time his students spend on their homework each night, he randomly selects 20 of his students and finds the mean and standard deviation of this sample to be 2.3 hours and .7 hours, respectively. Assuming the population study time is normally distributed, construct a 95 percent confidence interval for the mean amount of time spent each night for all of Professor Jebrah's students.

38. A study to determine factors that would predict divorce was recently reported in the *Journal of Personality and Social Psychology.* A random sample of 222 stable couples reported the mean and standard deviation for the time they knew each other were 57.07 months and 43.46 months, respectively. Construct a 95 percent confidence interval for the mean time of all such couples.

39. A random sample of 15 patients with chronic lower back pain who participated in a recent study reported a mean duration of back pain of 17.6 months with a standard deviation of 5.0 months. Assuming the duration of back pain in the population is normally distributed, construct a 90 percent confidence interval for the mean duration of back pain for the population of such patients.

40. *Career Development Quarterly* recently reported on evaluations of career development variables for several ethnic groups. In an assessment of career decision-making attitudes, a random sample of 58 Asian-Americans had a mean score of 12.71 with a standard deviation of 2.6 on an assessment scale. Form a 90 percent confidence interval for the population of Asian-Americans.

41. In a recent study published in the *Journal of Educational Research,* a random sample of 84 sophomores enrolled in a teacher preparation program were asked what they thought was the appropriate number of hours to spend in testing per week. The mean of the responses was 10.43 hours, and the standard deviation was 6.72 hours. Form a 99 percent confidence interval for the mean number of hours all sophomores in such a program believe should be spent testing each week.

42. A senior manager of a large accounting firm wants to find the mean number of hours it takes for all current first-year accountants to complete an in-house training program. There is a population of 45 such accountants currently employed at the firm who have completed the training. She randomly selects a sample of 7 of these employees and records their times as follows: 25, 19, 15,

25, 12, 20, and 12. Find the sample mean and standard deviation, and construct a 95 percent confidence interval for the population mean training time required. (Assume the population values are normally distributed.)

43. Wing-and-a-Prayer Airlines wants to estimate the mean time it takes to fly from San Francisco to Los Angeles. In a random sample of 8 flights, the mean time is 1.2 hours, and the standard deviation is 0.4 hours. Assuming population values are normally distributed, form a 95 percent confidence interval for the population mean length of flight time for this route.

44. In a recent *American Journal of Public Health* survey, data were reported on a random sample of 123 caregivers who helped patients with dementia who live at home. The sample mean years of education for these caregivers was 13.5, with a standard deviation of 3.2 years. Form a 99 percent confidence interval for the number of years of education for the population of these caregivers.

45. Western Trucking Company has 42 vehicles in its fleet. To estimate the average mileage in its entire fleet, the dispatcher took a random sample of 6 trucks and found the sample mean and sample standard deviation to be 57,393 miles and 12,300 miles, respectively. Assuming the population mileage figures are normally distributed, form a 99 percent confidence interval for the mileage on all company trucks.

46. It was recently reported in the *Journal of Personality and Social Psychology* that a random sample of 71 patients completed the Center for Epidemiological Studies Depression Scale (CES-D), a self-report scale used to measure levels of depression. The mean score of the sample on the variable "perceived control" was 30.1 and standard deviation was 4.8. Form a 90 percent confidence interval for the mean score of all such patients.

47. A random sample of 36 chronic schizophrenic patients was selected, and these patients were given the Wisconsin Card Sorting Test. The sample mean was 43.0 errors with a standard deviation of 25.8 errors. Assuming a normally distributed population, construct a 95 percent confidence interval for the mean score on this test that would be achieved by all such patients.

48. A random sample of 126 electronic assemblies is selected to determine the life expectancy of a certain component. The sample mean life expectancy was 648 hours, and the standard deviation was 58 hours. At the 90 percent level of confidence, estimate the true life expectancy of the population of such components.

But 20 and 39 Percent . . .
Martin Zweig, an investor and operator of mutual funds, has carried out statistical studies of how certain monetary factors affect stock prices. In one example, reported in *Winning on Wall Street*, he analyzed various buy and sell "signals" by comparing the prime rate indicator with Standard & Poor's 500 index from 1945 to 1988. His analysis showed that 80 percent of the time his sell signals were correct and 61 percent of the time his buy signals were correct.

7.4 Estimating the Population Percentage

Since the mean of the sampling distribution of percentages is equal to the population percentage, the *sample* percentage (p) is an unbiased estimator of the population percentage (π). If the sample size is sufficiently large—that is, if $n \cdot p$ [the *sample size* (n) times the *sample percentage* (p)] is ≥ 500 and if $n(100 - p)$ is also ≥ 500—then the sampling distribution approximates the normal distribution. Thus, we are able to make probability statements about the interval estimates of π that are based on sample percentages. In this section we'll discuss only the large-sample case in the interval estimation of π since the small-sample approach is beyond the scope of this book.

We use p, the sample percentage, as the point estimate for π and construct the interval estimate of the population percentage as follows:

$$\underbrace{p - z\hat{\sigma}_p}_{\text{lower confidence limit}} < \pi < \underbrace{p + z\hat{\sigma}_p}_{\text{upper confidence limit}} \tag{7.6}$$

A z value is used here in exactly the same way it was used to estimate a population mean. And when we compute a value for $\hat{\sigma}_p$, we have produced an *estimate* of the standard deviation of the sampling distribution of percentages—i.e., an estimate of the standard error of percentage. An unbiased estimate of the standard error of percentage may thus be found in this way:

$$\hat{\sigma}_p = \sqrt{\frac{p(100 - p)}{n}}\sqrt{\frac{N - n}{N - 1}} \quad \text{for a finite population} \tag{7.7}$$

or:

$$\hat{\sigma}_p = \sqrt{\frac{p(100 - p)}{n}} \quad \text{for an infinite population} \tag{7.8}$$

The estimate of the standard error is *always used* in the construction of an interval estimate. Why? Because the true standard error cannot be computed for an interval estimate of π. This fact is obvious from the following formula:

$$\sigma_p = \sqrt{\frac{\pi(100 - \pi)}{n}}$$

where, as you can see, the calculation of σ_p requires knowledge of π. Yet π is what we are trying to estimate! To resolve this dilemma, we must use formulas 7.7 and 7.8.

We are now ready to show the similarity of the procedures used to estimate population means and percentages by considering the following example problems.

Example 7.8. An accountant at the Highland Fling Scottish Boomerang Company wants to estimate the percentage of credit customers who have bought boomerangs with bad checks. A random sample of 150 accounts showed that 15 customers had passed bad checks. Estimate at the 95 percent confidence level the true percentage of credit customers who have written bad checks.

We have the following data: $p = 15/150 = 10$ percent, $n = 150$, and the confidence coefficient is 95 percent. The *estimate* of σ_p is computed as follows:

$$\hat{\sigma}_p = \sqrt{\frac{p(100 - p)}{n}} = \sqrt{\frac{10(90)}{150}} = 2.45 \text{ percent}$$

With a confidence coefficient of 95 percent, the z value is 1.96. So the interval estimate of the true percentage of credit customers who pass bad checks is:

$$p - z\hat{\sigma}_p < \pi < p + z\hat{\sigma}_p$$
$$10 \text{ percent} - 1.96(2.45 \text{ percent}) < \pi < 10 \text{ percent} + 1.96(2.45 \text{ percent})$$
$$5.20 \text{ percent} < \pi < 14.80 \text{ percent}$$

Example 7.9. A high school counselor is interested in the proportion of male students who would volunteer for military service. From 600 male students she randomly samples 50 and finds that 15 of them would like to enlist. Use a 99 percent confidence coefficient to estimate the true percentage.

The data are: $p = 15/50 = 30$ percent, $n = 50$, confidence coefficient $= 99$ percent. The estimate of the standard error is computed as follows:

$$\hat{\sigma}_p = \sqrt{\frac{p(100-p)}{n}}\sqrt{\frac{N-n}{N-1}}$$
$$= \sqrt{\frac{30(70)}{50}}\sqrt{\frac{600-50}{600-1}} = 6.21 \text{ percent}$$

With a confidence coefficient of 99 percent, the z value is 2.575. Therefore, the interval estimate is:

$$p - z\hat{\sigma}_p < \pi < p + z\hat{\sigma}_p$$
$$30 \text{ percent} - 2.575(6.21 \text{ percent}) < \pi < 30 \text{ percent} + 2.575(6.21 \text{ percent})$$
$$14.01 \text{ percent} < \pi < 45.99 \text{ percent}$$

(You'll notice that this estimate may not be of much help to the counselor. Its large width results from the high level of confidence specified and the relatively small sample size.)

The general procedure for constructing an interval estimate of π in the large-sample case is summarized in Figure 7.10.

Example 7.10. Political polls represent one of the major uses of interval estimation of π. Let's assume that Senator Phil A. Buster faces a tough reelection campaign and orders a poll to learn how the voters view his candidacy. A random sample of 1,200 voters reveals that 532 are likely to vote for Senator Buster, while the others polled prefer his opponent or are undecided. At the 95 percent level of confidence, what's the population percentage of voters who express a preference for the Senator?

The data are: $p = 532/1{,}200 = 44.33$ percent, $n = 1{,}200$, confidence coefficient $=$ 95 percent. The estimate of the standard error is:

$$\hat{\sigma}_p = \sqrt{\frac{p(100-p)}{n}} = \sqrt{\frac{44.33(55.67)}{1{,}200}} = 1.43 \text{ percent}$$

With a confidence coefficient of 95 percent, the z value is 1.96. Thus, the interval estimate is:

$$p - z\hat{\sigma}_p < \pi < p + z\hat{\sigma}_p$$
$$44.33 - 1.96(1.43) < \pi < 44.33 + 1.96(1.43)$$
$$41.53 \text{ percent} < \pi < 47.13 \text{ percent}$$

Senator Buster had better try to swing those undecided voters into his camp!

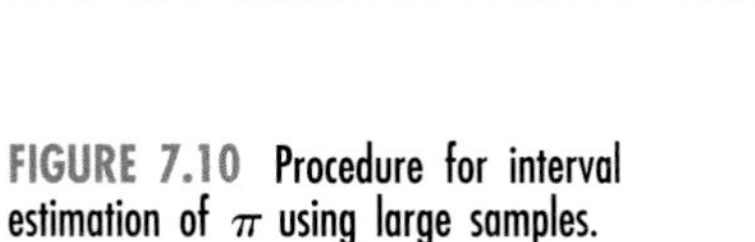

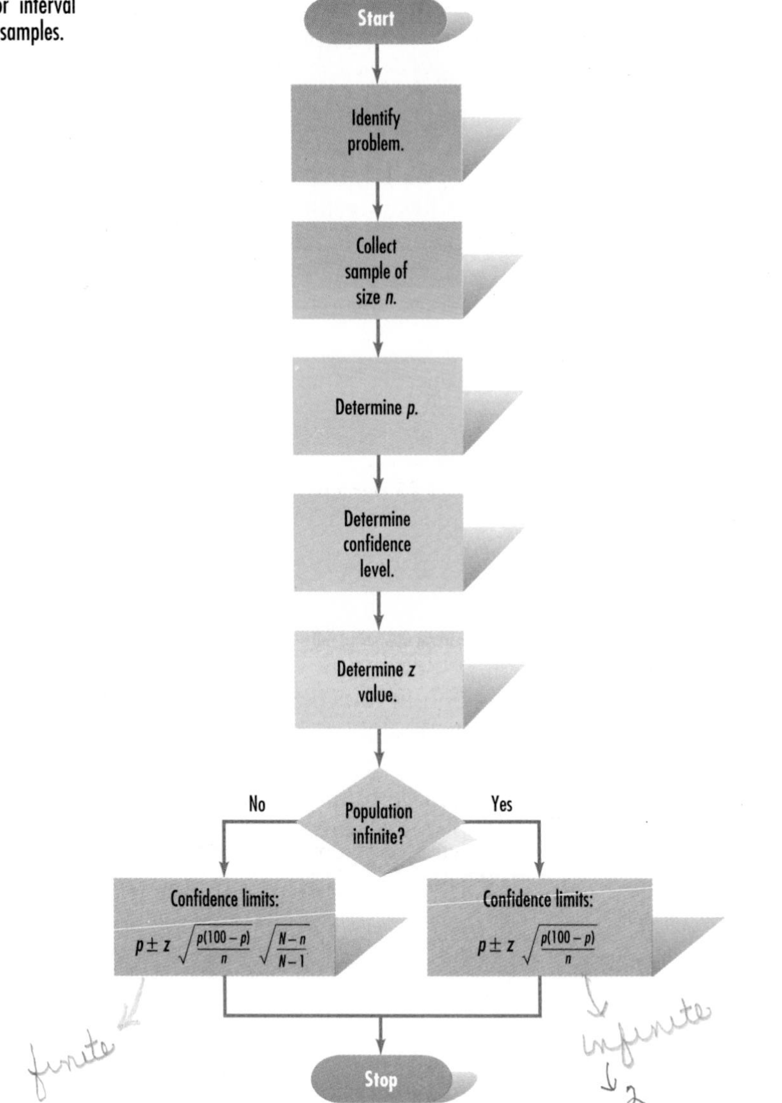

FIGURE 7.10 Procedure for interval estimation of π using large samples.

Self-Testing Review 7.4

1. A random sample of 319 male Vietnam veterans who suffered from posttraumatic stress disorder (PTSD) were selected to participate in a study reported in *The Journal of Applied Psychology*. Of these 319 veterans, 43 were unemployed. Construct a 95 percent confidence interval for the percentage of all such veterans who suffer from PTSD who are unemployed.

2. A random sample of 864 requests for magnetic resonance imaging (MRI) scans made in 1991 were evaluated in a study reported in *The New England Journal of Medicine*. In 502 of these sample cases, the requesting physician had an ownership interest in the imaging facility. Form a 90 percent confidence interval for the population percent of physicians who requested MRI scans and who had an ownership interest in the imaging facility.

3. The *American Journal of Psychiatry* recently reported the findings of a study of HIV-related risk behaviors among adolescents. Of the 76 adolescents hospitalized for psychiatric reasons in the random sample, 40 reported that they were sexually active. Form a 95 percent confidence interval for the population percent of such hospitalized adolescents that are sexually active.

4. A recent article in *Focus on Critical Care AACN* reported on a fellowship program in the Washington (D.C.) Hospital Center to train graduate nurses to work in intensive care units (ICUs). If a random sample of 46 nurses begins such a training program and if 27 of these nurses are currently working in an ICU, what's the 95 percent confidence interval for the population percent of nurses who begin such a program and then later work in an ICU?

5. An article in *The New England Journal of Medicine* evaluated a syringe-exchange program in New Haven, Connecticut. In a test of a random sample of 160 needles that were employed by injection-drug users before the exchange program was started, 108 were found to be HIV-positive. Form a 99 percent confidence interval of the population percentage of such preexchange needles of injection-drug users that are HIV-positive.

6. *The New England Journal of Medicine* has published a study that seeks to determine if vitamin use decreases the incidence of neural-tube defects in newborns. Vitamin supplements were not used in a random sample of 2,052 pregnancies, and there were 6 cases of neural-tube defects. Form a 95 percent confidence interval for the population percent of neural-tube defects in pregnancies when vitamin supplements aren't used.

7. Three-color lenses are glued to bicycles using a new process. A quality control officer wants to determine the percent of the lenses that don't adhere properly. From a random sample of 135 bicycles, 17 had lenses that did not adhere properly. Form a 90 percent confidence interval for the population percent of lenses that are not glued properly.

8. The results of a study to determine patterns of drug use in adolescents was recently published in *The Journal of Abnormal Psychology.* Of the 4,145 seventh graders in a random sample, 124 reported using alcohol on a weekly basis. Form a 90 percent confidence interval for the percent of weekly alcohol use among all seventh graders.

9. An article published in the *American Journal of Public Health* investigated the percent of the elderly who use mental health services in nursing homes. A random sample of 4,646 nursing home residents found that 109 of these residents used a mental health specialist. Construct a 99 percent confidence interval for the percentage of all elderly in nursing homes who use mental health specialists.

10. There are 241 residents of a dormitory. A vote is needed to change the policy of "quiet hours." In a random sample of 34 residents, 23 say they will vote in favor of the change. Construct a 95 percent confidence interval for the population percentage favoring the change in policy.

11. There are 219 pages in an issue of a magazine. The art director wants to estimate the percent of pages that have color. He randomly chooses 15 pages and finds that 6 have color. Form a 90 percent confidence interval for the population percentage of pages with color in this issue.

Phone Book Errors

The Reuben H. Donnelley Corporation publishes hundreds of telephone directories for hundreds of cities in the United States. About 400,000 businesses advertise in these directories. The Donnelley Corporation uses a statistical software package and its statistical process control graphics features to plot the average number of errors per 100 documents and to plot the percentage of ads that fail to conform to a client's original specifications. This helps the Donnelley Corporation monitor the advertising production process to ensure that advertisers receive the best possible service, and it alerts Donnelley managers when this process may be drifting into an "out-of-control" situation.

7.5 Estimating the Population Variance

We've now seen examples that show why estimates of population means and percentages are needed and how those estimates are computed. But why would an estimate of a population variance (σ^2) be needed? As you know, the variance and standard deviation are measures that show the amount of spread or scatter that exists in a data set. And it's often desirable to know the extent of this scatter so that steps can be taken to control it. It's not enough that a tire maker knows that the mean life expectancy of a line of tires is 40,000 miles. The manufacturer also wants to be sure that the tires produced are of a consistent quality so that there isn't a wide spread in tire mileage results to alienate customers. And a drug supplier has to be focused on the potency of the tablets produced. It's not enough to make tablets that have an acceptable mean strength if some tablets are unduly weak while others produce overdoses. Thus, it's understandable that manufacturers who want to produce consistent quality results will want to keep close tabs on the population variance or standard deviation.

The Chi-Square Distribution

Assuming—and this is an important caveat—that the population values are *normally distributed,* we can estimate the population variance (and thus the population standard deviation) by using a sampling distribution that is new to us.

To explain this new distribution, let's assume that we take a random sample of 20 drug tablets from a production line. Let's further assume for the moment that we know the value of the population variance (σ^2). We then compute the sample variance (s^2) and find the value for the following variable:

$$\frac{(\text{df})s^2}{\sigma^2} = \frac{(n-1)s^2}{\sigma^2} = \frac{(20-1)s^2}{\sigma^2}$$

Note that in this case, df, or degrees of freedom $= n - 1$.

If we record the result of this computation, repeat the same procedure a very large number of times (a distressing thought!), and put all the trial results in a frequency distribution, the theoretical result would be a distribution called a *chi-square* (χ^2) *distribution.* (The symbol for the Greek letter chi is χ, and is pronounced like the first two letters in the word "kind.")

Chi-Square Distribution

A **chi-square distribution** is the sampling distribution for the variable

$$\frac{(\text{df})s^2}{\sigma^2}$$

and it has the following properties:

- Since values are obtained by using squared numbers, all χ^2 values are zero or positive—a property that's not found with z and t distributions. Thus, the

population percentage of pre-exchange-program needles that are HIV-positive.

9–10. *The Journal of Applied Psychology* recently published an article about male Vietnam veterans who suffered from posttraumatic stress disorder (PTSD). A psychologist wishes to estimate the population percentage of these vets who suffer from PTSD who are unemployed:

9. If no preliminary study is available, how large a sample size is needed to be 95 percent confident the estimate is within 3 percent of π?

10. In a preliminary random sample, 43 of 319 male Vietnam veterans who suffered from PTSD were unemployed. Using this preliminary study, how large a sample is needed to construct a 95 percent confidence interval within 3 percent of π?

11. In a preliminary study reported in *Physical Therapy*, the sample standard deviation for the duration of a particular back pain suffered by patients was 18.0 months. How large a random sample is needed to construct a 90 percent confidence interval for the duration of such back pain for the population of back patients so that the estimate is within 2 months of the actual duration?

12. An accounting firm wishes to form a 90 percent confidence interval for the population mean tax refund for its clients who receive refunds. How large a random sample is needed to be within $10 of the actual amount if a preliminary study finds the standard deviation to be $42.67?

13. A preliminary sample of 512 Denver County employees was questioned in a study in *The American Journal of Public Health* to assess the prevalence of symptoms attributed to the work environment. And 45 employees reported experiencing eye irritation. How large a random sample is needed to be 90 percent confident of being within 3 percent of the population percent of those who experience eye irritation?

14. A clinical test on a random sample of 58 cardiac patients was performed to assess the increase in exercise time after 3 weeks of using the medication NORVASC. The mean and standard deviation for the increase were 62 seconds and 17 seconds, respectively. How large a sample is needed to be 99 percent confident that the sample mean is within 2 seconds of the population mean?

LOOKING BACK

1. Any specific value of a statistic is an estimate, and any statistic used to estimate a parameter is an estimator. An unbiased estimator is one that produces a sampling distribution that has a mean that's equal to the population parameter to be estimated. Thus, sample means, sample percentages, and sample variances are unbiased estimators of population means, percentages, and variances. The entire process of using an estimator to produce an estimate of the parameter is known as *estimation.*

2. A single number used to estimate a population parameter is called a *point estimate,* and the process of estimating with a single number is known as *point estimation.* Although an unbiased estimator tends toward the value of the population parame-

ter, it's unlikely that the value of a point estimate will be exactly equal to the parameter value. Thus, an *interval estimate*—one that uses a spread of values to estimate a parameter—is desired over a point estimate because allowances are made for sampling error. As you've seen in this chapter, a point estimate is adjusted for sampling error to produce an interval estimate.

3. On the basis of the properties of their sampling distributions, it's possible to construct an interval estimate of μ, π, or σ^2 with some degree of certainty. The width of the interval estimate increases as the level of confidence increases since allowance must be made for more sampling error. The *level of confidence* (or *confidence coefficient*) refers to the probability of correctly including the population parameter being estimated in the interval that is produced. The confidence levels generally used in interval estimation are 90, 95, and 99 percent. The interval estimates based on specified confidence levels are known as *confidence intervals*, and the upper and lower limits of the intervals are known as *confidence limits.*

4. The formulas needed to estimate the population mean, when σ is known or unknown and when the population is finite or infinite, are summarized in Table 7.3.

TABLE 7.3 SUMMARY OF INTERVAL ESTIMATION UNDER VARIOUS CONDITIONS

	Estimating μ				Estimating π
	σ Known		σ Unknown		(When $np \geq 500$ and
Population	$n \leq 30$*	$n > 30$	$n \leq 30$*	$n > 30$	$n(100 - p) \geq 500$)
Finite	$\bar{x} \pm z\frac{\sigma}{\sqrt{n}}\sqrt{\frac{N-n}{N-1}}$	$\bar{x} \pm z\frac{\sigma}{\sqrt{n}}\sqrt{\frac{N-n}{N-1}}$	$\bar{x} \pm t_{\alpha/2}\frac{s}{\sqrt{n}}\sqrt{\frac{N-n}{N-1}}$	$\bar{x} \pm z\frac{s}{\sqrt{n}}\sqrt{\frac{N-n}{N-1}}$	$p \pm z\sqrt{\frac{p(100-p)}{n}}\sqrt{\frac{N-n}{N-1}}$
Infinite	$\bar{x} \pm z\frac{\sigma}{\sqrt{n}}$	$\bar{x} \pm z\frac{\sigma}{\sqrt{n}}$	$\bar{x} \pm t_{\alpha/2}\frac{s}{\sqrt{n}}$	$\bar{x} \pm z\frac{s}{\sqrt{n}}$	$p \pm z\sqrt{\frac{p(100-p)}{n}}$

*Population values are assumed to be normally distributed.

You've seen that the sample size must be considered in estimation since it affects the width of the confidence interval. If the sample size is sufficiently large (over 30), the sampling distribution of means approximates the normal distribution and z values are used in the estimation process. On rare occasions, z values can also be used with smaller samples if it's known that the population is normally distributed and the σ is also known. In most cases with smaller samples, though, if we know that the population is normally distributed but we *don't* know the value of σ, the sampling distribution approximates a t distribution. The shape of a t distribution is determined by the sample size.

5. The formulas needed to estimate the population percentage are also summarized in Table 7.3. These formulas are valid when $np \geq 500$ and when $n(100 - p)$ is also ≥ 500. The procedure for the interval estimation of π is similar to the procedure for estimating μ. Interval estimates of π values can be found almost every week in the polls published by newspapers and magazines.

6. It's often desirable to know the extent of the spread or scatter that exists in a data set so that steps can be taken to control it. When that's the case, an estimate of the

population variance may be needed. If population values are normally distributed, the σ^2 can be estimated by using a chi-square distribution. The interval estimate of σ^2 takes a familiar form:

lower confidence limit $< \sigma^2 <$ upper confidence limit

Or more specifically:

$$\frac{(n-1)s^2}{\chi^2_{\alpha/2}} < \sigma^2 < \frac{(n-1)s^2}{\chi^2_{1-\alpha/2}}$$

7. Discussions of how to determine the appropriate sample size when estimating the population mean or population percentage have been covered in this chapter. Determining the sample size for an interval estimate of μ requires an assumption about the value of σ. Likewise, determining the sample size to estimate the value of π requires an approximate value of π. If the researcher has absolutely no idea of the approximate value of π, then that value is assumed to be 50 percent.

Review Exercises

1–4. Find the z value corresponding to each of the following confidence coefficients:

1. 90 percent.

2. 85 percent.

3. 95 percent.

4. 98 percent.

5–8. Find the t value corresponding to each of the following conditions:

5. $n = 21$, 95 percent confidence level.

6. $n = 6$, 99 percent confidence level.

7. $n = 14$, 90 percent confidence level.

8. $n = 4$, 95 percent confidence level.

9–11. In each of the following exercises, a set of conditions is given. To construct a 95 percent confidence interval for the population mean, would you use the z distribution, a t distribution, or neither? Where appropriate, determine the z or t value.:

9. $n = 47$, $\bar{x} = 83.2$, $s = 5.5$, population distribution shape unknown, σ unknown.

10. $n = 21$, $\bar{x} = 23.1$, $s = 7.5$, population distribution shape normal, $\sigma = 7.9$.

11. $n = 17$, $\bar{x} = 3.2$, $s = 1.4$, population distribution shape normal, σ unknown.

12–14. In each of the following exercises, a set of conditions is given. To construct a 90 percent confidence interval for the population mean, would you use the z distribution, a t distribution, or neither? Where appropriate, determine the z or t value:

12. $n = 16$, $\bar{x} = 733.2$, $s = 45.5$, population distribution shape normal, $\sigma = 41.7$.

13. $n = 15$, $\bar{x} = 23.1$, $s = 7.5$, population distribution shape unknown, σ unknown.

14. $n = 37$, $\bar{x} = 8.1$, $s = 3.9$, population distribution shape unknown, σ unknown.

15–16. Determine the values of χ^2 needed to form the confidence intervals for the population variance for each of the following conditions:

15. 90 percent level of confidence, $n = 22$.

16. 95 percent level of confidence, $n = 13$.

17–20. Construct a 95 percent confidence interval for the population mean if $\bar{x} = 76.1$, $s = 14.2$, and the sample size is:

17. $n = 36$.

18. $n = 49$.

19. $n = 64$.

20. Examine the results of problems 17 to 19. What happens to the width of a confidence interval as the sample size increases but all other conditions remain the same?

21–24. If $\bar{x} = 364.1$, $s = 61.7$, and $n = 100$, construct a confidence interval for μ using each of the following confidence coefficients:

21. 90 percent.

22. 95 percent.

23. 99 percent.

24. Using the results of problems 21 to 23, what happens to the width of the confidence interval as the level of confidence increases but all other conditions remain the same?

25. Construct a 95 percent confidence interval for the σ^2 if $n = 17$ and $s^2 = 31.8$.

26. Construct a 99 percent confidence interval for the σ^2 if $n = 24$ and $s^2 = 2.9$.

27–29. In 1919 a study of the four blood groups was conducted for various populations. In a random sample of 116 residents of Bougainville Island, it was observed that 74 had type A blood.

27. What is the point estimate of the percent of Bougainville Island residents who had type A blood?

28. Estimate the standard error.

29. Construct a 90 percent confidence interval for the percent of Bougainville Island residents who had type A blood.

30–32. A recent article in *The Journal of Clinical Psychology* reported that in a random sample of 1,190 male Vietnam veterans, 319 stated that they had suffered from posttraumatic stress disorder (PTSD).

30. What is the point estimate of the percent of male Vietnam veterans who suffered from PTSD?

31. Estimate the standard error.

32. Construct a 95 percent confidence interval for the population percentage of male Vietnam veterans who suffered from PTSD.

33–36. The American Journal of Psychiatry recently published the results of a study to determine if there was a possible type of brain dysfunction associated with infantile autism. Each child in the study was given a behavioral test and scored on a scale from 0 to 116, where 0 = absence of symptoms and 116 = maximum severity. (Assume the scores on this test are normally distributed.) The scores of the random sample of 21 children in the study were:

27 35 65 67 47 46 63 44 34
51 17 40 41 60 24 48 29 73
60 41 27

33. What is the value of the point estimate you would use to estimate the mean score of *all* children with symptoms of infantile autism?

34. Calculate the estimated standard error.

35. Construct a 90 percent confidence interval for the mean score of all such children (assume a normally distributed population).

36. Construct a 90 percent confidence interval for the population variance.

37–40. Each year, *Money* magazine consultants compile a list of 50 top blue chip stocks, analyze them, and suggest which ones to buy, which ones to hold, and which ones to sell. The following is a random sample of 9 of the stock prices on this list:

$31.25 33.50 104.75 25.50 53.25
57.25 30.75 35.00 58.50

37. What is the value of the point estimate of the mean price of the 50 top blue chip stocks?

38. Calculate the estimated standard error.

39. Assuming population values are normally distributed, construct a 95 percent confidence interval for the population mean price of the 50 top blue chip stocks.

40. Construct a 95 percent confidence interval for the population standard deviation.

41–44. A recent issue of *The Lancet* described

a study that dealt with the effects of increased inspired oxygen concentrations on exercise performance in chronic heart failure patients. A random sample of 12 such patients were included in the study. The oxygen consumption during the exercise test (in milliliters per minute per kilogram) was:

9.7 21.0 14.3 15.2 12.8 8.6 10.9
8.3 19.1 7.0 19.5 12.5

41. What is the value of the point estimate of the mean amount of oxygen consumed by all such patients?

42. Calculate the estimated standard error of the mean.

43. Construct a 99 percent confidence interval for the population mean oxygen consumption. (Assume population values are normally distributed.)

44. Construct a 99 percent confidence interval for the population standard deviation.

45–48. *Nation's Business* recently published results of a study to determine an employer's health care costs when using various plans. The following is the data on the mean cost for an HMO in a random sample of 12 major cities (we'll assume the cost data are normally distributed):

Atlanta	$3,259
Chicago	3,133
Cleveland	3,465
Dallas-Fort Worth	2,963
Houston	3,295
Los Angeles	3,025
Minneapolis-St. Paul	2,673
New York Metro	3,254
Philadelphia	2,882
Richmond	2,448
San Francisco	2,939
Seattle	2,624

45. Calculate a point estimate for the mean cost of an HMO in a major city.

46. Determine the estimated standard error of the mean.

47. Construct a 95 percent confidence interval for the population mean cost of an HMO in a major city.

48. Construct a 95 percent confidence interval for the population variance.

49. To study the characteristics of first-episode schizophrenic patients, data were collected on a random sample of 32 such patients, and these facts were reported in a recent issue of *The American Journal of Psychiatry.* The mean and standard deviation for the number of years of education for the patients in this sample were 12.4 years and 3.0 years, respectively. Construct a 95 percent confidence interval for the mean number of years of education for all such patients.

50. The same issue of *The American Journal of Psychiatry* reported that an IQ test was administered to a random sample of 26 chronic schizophrenic patients and the mean was found to be 97.6. For this test, the population values are known to be normally distributed with a standard deviation of 15. Construct a 90 percent confidence interval for the mean IQ of the population of chronic schizophrenic patients.

51. A recent issue of *SAM Advanced Management Journal* described a study that was conducted to examine the attitudes and perceptions of undergraduate business students. Of 431 business students who were in the random sample, 99 expressed an interest in international business. Form a 90 percent confidence interval for the percentage of all business students interested in international business.

52. Using the information in problem 51 to form a preliminary estimate, how large a sample would be necessary to be 90 percent confident that the sample mean is within 2 percent of the population mean?

53. A study to show expenditures in caring for patients with dementia who live at home was recently published in *The American Journal of Public Health.* Surveys were returned by a random sample of 141 caregivers. The caregivers reported the average number of care-giving hours a week to be 106.9 with a standard deviation of 68.2 hours. Form a confidence interval at the 90 percent level for the population mean number of caregiver hours.

54. Frieda Paine, a physical therapist, is testing a new technique on patients who are recovering from sports injuries. To estimate the average length of time she spent with each of these patients last week, she randomly selects a sample of 10. The mean and standard

deviation of her sample were 147.8 and 31.2 minutes, respectively. Assuming population values are normally distributed, construct a 95 percent confidence interval for the time spent with the population of patients during the week.

55. Form a 95 percent confidence interval for the population standard deviation for Ms. Paine's patients. (See problem 54.)

56. A study in *The American Journal of Psychiatry* has found that of a random sample of 36 World War II POW survivors, 24 reported that they had recurrent distressing dreams. Form a 90 percent confidence interval for the population percentage of such POW survivors who have recurrent distressing dreams.

57. How large a sample is needed to form a 90 percent confidence interval for problem 56 and be within 5 percent of the population percentage?

58. *The Journal of Educational Research* has published a study in which a random sample of 281 college students took a word recall test. The mean and standard deviation scores were 58.3 and 20.5, respectively. Construct a 95 percent confidence interval for the population mean score.

59. Use the study in problem 58 for a preliminary estimate of σ. How large a sample will be necessary to be 95 percent confident of being within 2 units of the population mean?

60. An article in *The Journal of Applied Psychology* reported that in a random sample of 871 male Vietnam veterans who did not suffer from posttraumatic stress disorder (PTSD), 22 were unemployed. Form a 95 percent confidence interval for the population percentage of those who don't suffer from PTSD and are unemployed.

61. Anna Liszt, a stock broker, manages 75 accounts. To estimate the mean earnings for her accounts in the past year, she randomly chooses 32 accounts and finds the mean and standard deviation to be $5,295 and $869, respectively. Construct a 99 percent confidence interval for the average earnings of all Anna's clients in the past year.

62. *The Career Development Quarterly* has reported on the results of a test designed to assess career decision-making skills. A random sample of 122 middle-class high school students had a mean score of 13.07 with a standard deviation of 2.4. Construct a 99 percent confidence interval for the mean score of all middle-class high school students.

63. According to a recent report in *The American Journal of Public Health*, a random sample of 512 Denver County employees was questioned in a study to assess the prevalence of symptoms attributed to the workplace. Forty-five employees reported experiencing eye irritation. Find the 90 percent confidence interval for the percent of all Denver County workers experiencing eye irritation.

64. A clinical test on a random sample of 58 cardiac patients was performed to assess the increase in exercise time after 3 weeks of using the medication NORVASC. The mean and standard deviation for the increase were 62 and 17 seconds, respectively. Construct a 99 percent confidence interval for the mean time increase of the population of all such patients.

65. In a random sample of 1,625 patients with artificial joints, 1,003 were 65 years old or over. Construct a 95 percent confidence interval for the percent of all patients with artificial joints that are 65 or older.

66. A random sample of 64 unstable couples was part of a study (reported in the *Journal of Personality and Social Psychology*) to determine factors that would predict divorce. On a test for extroversion, the husbands had a mean and standard deviation of 24.70 and 5.14, respectively. Form a 99 percent confidence interval for the mean score on this test of all husbands in unstable marriages.

67. According to a study in *The New England Journal of Medicine*, the mean cost per case for physical therapy in a random sample of 1,017 self-referred patients was $404 with a standard deviation of $102. Form a 99 percent confidence interval for the population mean cost in such cases.

68. To assess the effectiveness of a new marketing campaign, a week's sales figures were compiled for 14 randomly selected employees at Smears Department store. The mean and standard deviation of the weekly sales were $12,386 and $3,352, respectively. Assuming a normally distributed population, form a 90 percent confidence interval for the weekly sales of all Smears employees during this marketing campaign.

69. A study reported in *Physical Therapy* was made to compare the effect of varied training frequencies on the development of isometric lumbar torque (strength). A random sample of 10 subjects with lower back pain received training every other week during this study. The mean and standard deviation for the beginning weight of this sample were 24.7 and 3.6 kilograms, respectively. Assuming that population values are normally distributed,

construct a 95 percent confidence interval for the beginning population mean weight.

70. Using the study in problem 69 for a preliminary estimate of σ, how large a sample would be needed to be 95 percent confident of being within 1 kilogram of the population mean?

71. Senator E. Z. Wynn is running for re-election and wants to estimate the percent of voters who plan to vote for him. He instructs his staff to conduct a poll so that they are 95 percent sure their estimate will be accurate within ± 3 percent. What is the minimum number of voters that must be used in the poll if no preliminary study is available?

72. An industrial engineer wishes to estimate the mean life of a calculator battery (in hours) to within 2 hours of the true value. Past experience has shown that the standard deviation is 15.2 hours. How large a sample should she select to be within 5 hours 90 percent of the time.

73. A child psychologist wishes to estimate the mean length of time 6-year-old children spend with their parents each day. Past experience has shown the standard deviation to be 127 minutes. How large a sample should he select to be within 15 minutes 99 percent of the time?

74. A plant manager wants to form a 99 percent confidence interval to estimate the percent of defective products from the production line. She wants the estimate to be accurate within ± 5 percent. What is the minimum number she needs for her sample if no preliminary study is available?

Topics for Review and Discussion

1. What is an unbiased estimate? Why is such an estimate desirable? What are the unbiased estimators for each of the parameters μ, σ, σ^2, and π?

2. Although a population parameter will seldom be exactly equal to a point estimate, discuss how a point estimate is used to estimate a population parameter.

3. Discuss, in your own words, the meaning of a 90 percent confidence interval.

4. What effect does an increase in the confidence level have on the width of the confidence interval?

5. What effect does an increase in the sample size have on the width of the confidence interval?

6. Under what conditions does the sampling distribution of the mean approximate a normal distribution?

7. Under what conditions is a t distribution used to approximate the distribution of sample means?

8. Under what conditions can we estimate the value of the population variance? What distribution is used in this estimate?

9. Under what conditions must the finite correction factor be used to approximate the standard error of the mean?

10. It is known that in using the highest confidence interval, there is a greater probability of including the population mean in the interval. What are the disadvantages in using the highest confidence interval?

11. What should a researcher do if the interval estimate is too wide?

12. We can use the parameter σ_x in an interval estimate for μ. Why isn't it possible to use the parameter σ_p in an interval estimate for π?

Projects/Issues to Consider

1. Locate a population data set (you may use the one found for the projects/issues section in Chapter 6) from which you can determine the parameters μ and σ. Select a random sample of size n from this population and compute $\overline{x}$ and s.

a) Construct a 90 percent confidence interval estimate of the population mean. Use the known value of σ. Your sample size and the shape of the distribution will determine if you use the z or a t distribution. Do you need a finite correction factor?

b) Compare the interval estimate from part a with the known population mean. Did your population mean fall inside the interval you constructed? If not, why not?

2. Identify a current controversial topic, and formulate a question regarding opinions on this topic. Your question may be at the local or national level, but be sure it is not stated in a biased manner. Identify the population for your question. You will form an interval estimate for the percent of the population that responds yes to your question, but first determine the level of precision you want. Find the minimum sample size to achieve this level of precision, and collect a sufficient number of responses. Now, form a 90 percent confidence interval for the population percentage.

3. Locate a percentage estimate achieved by polling in a recent newspaper or periodical. (These poll results are often reported as a point estimate with a statement to the effect that the results are within ±3 percentage points of the population percentage.) Use this estimate to produce a confidence interval.

Computer Exercises

1–2. Refer to Example 7.1 in this chapter, and repeat the computer simulation experiment shown in Figure 7.4*a*:

> *1.* How many of your 15 samples produce 95 percent confidence intervals that include the population mean of 1.48 minutes?
>
> *2.* Why do your confidence intervals differ from those shown in Figure 7.4*a*?

3–4. Refer again to Example 7.1 in this chapter, and repeat the simulation experiment shown in Figure 7.4*b*:

> *3.* How many of your 15 plots produce 95 percent confidence intervals that include the population mean of 1.48 minutes?
>
> *4.* Why do your plots differ from those shown in Figure 7.4*b*?

5–7. Locate the billable hours for the sample of 50 workers that Bill Alott used to prepare his GOTCHA report in STR 3.2, page 49 (you may have saved this data set after doing the computer exercises in Chapter 3, and we'll use it again in the computer exercises section of Chapter 8):

> *5.* Enter (or retrieve) the 50 data items in this large sample, and use your software to calculate a 95 percent confidence interval for the mean number of billable hours per week for all Global Technologies consultants.
>
> *6.* Now calculate a 99 percent confidence interval.
>
> *7.* Discuss the outputs produced at the two levels of confidence.

8–10. Cerebral vascular accident (CVA)—otherwise known as stroke—is an interruption of the flow of blood to the brain. Many who suffer from CVA must undergo occupational therapy to rehabilitate paralyzed limbs. The following data represent the number of weeks that a random sample of 26 stroke patients participated in an occupational therapy program (we'll use this data set again in the computer exercises section of Chapter 8):

8 21 6 9 4 15 10 9 7 9 6
17 8 9 9 2 8 8 3 10 16 13
5 3 2 1

Assuming that the weeks of therapy in the population are normally distributed, enter the sample data and use your software to:

> *8.* Produce a 95 percent confidence interval for the mean number of weeks in therapy for all stroke patients in this program.
>
> *9.* Now produce a 99 percent confidence interval.
>
> *10.* Discuss the outputs produced at the two levels of confidence.

Answers to Odd-Numbered Self-Testing Review Questions

Section 7.1

1. e 3. c 5. b 7. a 9. c

Section 7.2

1. It is the standard deviation of a sampling distribution.

3. 80 percent of the sample means will fall within 1.28 standard errors of $\mu_{\bar{x}}$.

5. 95 percent of the sample means will fall within 1.96 standard errors of $\mu_{\bar{x}}$.

7. 86 percent of the sample means will fall within 1.48 standard errors of $\mu_{\bar{x}}$.

9. 98 percent of the sample means will fall within 2.33 standard errors of $\mu_{\bar{x}}$.

11. As the level of confidence increases, the width of the interval increases (when all other factors remain the same).

13. It is also increased.

15. If we computed a large number of these intervals, then 95 percent of them would span the population mean. (But, of course, 5 percent would not.)

17. The level of confidence is 99 percent. The parameter is the population mean (which is unknown). The point estimate is 3.5 (the known sample mean), and the error of estimate is .02.

Section 7.3

1. Since the areas entered in the table in Appendix 2 represent the area between the mean and the z score and the curve is symmetric about the mean, we must first take half of the 86 percent area. We next look for an area entry (found in the body of the table) that is closest to half of .86 or .43. The nearest area entry is .4306. The z value corresponding to this area is 1.48. So 43 percent of the area under the curve falls between vertical lines drawn at the mean and at $z = 1.48$. This means that 86 percent of the area under the curve lies between vertical lines erected at $z = -1.48$ and $z = +1.48$.

3. Looking up an area nearest to .49, we find .4901. This corresponds to a z score of 2.33. So 98 percent of the area under the normal curve lies between vertical lines erected at $z = \pm 2.33$.

5. The area in the right tail is $(100 - 99)/2$ or .005 and df = 6, so $t = \pm 3.707$.

7. This is a small sample with σ unknown so we use a t distribution with 12 df. The corresponding area in the right tail is .05, so $t = \pm 1.782$.

9. This is a large sample so we use the normal distribution with $z = \pm 1.645$.

11. The point estimate is the sample mean of 75.

13. Use $z = 1.96$ since this is a large sample estimate.

15. The 95 percent confidence interval is 75 ± 3.088 or 71.912 to 78.088.

17. The estimated standard error of the mean is $5.3/\sqrt{18} = 1.249$.

19. The maximum error of estimate is $t(\hat{\sigma}_{\bar{x}}) = 2.898(1.249) = 3.620$.

21. The point estimate is the sample mean of 62.7.

23. Use $z = 1.96$ since we know the population standard deviation and the population distribution is normal.

25. The 95 percent confidence interval is 62.7 ± 1.811 or 60.889 to 64.511.

27. This is a finite population so the correction factor must be used. The estimated standard error is $4.201(\sqrt{66/79}) = 3.840$.

29. The maximum error of estimate is $1.771(3.840) = 6.801$.

31. The point estimate is the sample mean of 346.8.

33. Use $t = 2.821$ since this is a small sample estimate.

35. The 98 percent confidence interval is 346.8 ± 14.667 or 332.133 to 361.467.

37. The population here is finite, and the sample size exceeds 5 percent of the population so the estimated standard error of the mean = $(.7/\sqrt{20})\sqrt{(150 - 20)/(150 - 1)} = .157(.934) = .147$. So the 95 percent confidence interval is $2.3 \pm 2.093(.147)$, which is 1.992 to 2.608 hours per night.

39. The estimated standard error is $5/\sqrt{15} = 1.291$. The confidence interval is $17.6 \pm 1.761(1.291)$ or 15.327 to 19.873 months.

41. The estimated standard error is $6.72/\sqrt{84} = .733$. The confidence interval is $10.43 \pm 2.575(.733)$ or 8.542 to 12.318 hours.

43. The estimated standard error is $.4/\sqrt{8} = .141$. The confidence interval is $1.2 \pm 2.365(.141)$ or .867 to 1.533 hours, which is about 52 to 92 minutes.

45. The estimated standard error is $(12{,}300/\sqrt{6})\ \sqrt{36/41} = 5{,}021.4(.937) = 4{,}705.1$. The confidence interval is $57{,}393 \pm 4.032(4{,}705.1)$ or 38,422.04 to 76,363.96.

47. The estimated standard error is $25.8/\sqrt{36} = 4.3$. The confidence interval is $43.0 \pm 1.96(4.3)$ or 34.572 to 51.428.

Section 7.4

1. The sample percentage = 43/319 = .1348 or 13.48 percent. The estimated standard error ($\hat{\sigma}_p$) is $\sqrt{[(13.48)(100 - 13.48)]/319} = 1.91$ percent. The confidence interval is 13.48 ± 1.96(1.91) or 9.733 to 17.227 percent.

3. $p = 40/76 = 52.63$ percent. $\hat{\sigma}_p = \sqrt{[(52.63)(100 - 52.63)]/76} = 5.727$. The confidence interval is 52.63 ± 1.96(5.727) or 41.40 to 63.85 percent.

5. $p = 108/160 = 67.5$ percent. $\hat{\sigma}_p = 3.70$. The confidence interval is 67.5 ± 2.575(3.70) or 57.97 to 77.03 percent.

7. $p = 17/135 = 12.59$ percent. $\hat{\sigma}_p = 2.86$. The confidence interval is 12.59 ± 1.645(2.86) or 7.885 to 17.295 percent.

9. $p = 109/4{,}646 = 2.35$ percent. $\hat{\sigma}_p = .22$. The confidence interval is 2.35 ± 2.575(.22) or 1.78 to 2.92 percent.

11. $p = 6/15 = 40$ percent. $\hat{\sigma}_p = 12.65\sqrt{204/218} = 12.237$ percent. The confidence interval is 40 ± 1.645(12.237) or 19.870 to 60.130 percent.

Section 7.5

1. The χ^2 value for the lower limit with 11 degrees of freedom is 19.68, and the χ^2 value for the upper limit is 4.57.

3. Using the χ^2 distribution with 26 df, the χ^2 values are 38.9 and 15.38. The lower limit for the confidence interval is 26(63.2)/38.9 = 42.24, and the upper limit is 26(63.2)/15.38 = 106.84. The confidence interval is $42.24 < \sigma^2 < 106.84$.

5. Using the χ^2 distribution with 15 df, the limits for the confidence interval are 15(26.8)/32.8 and 15(26.8)/4.60. The confidence interval for the population variance is 12.256 to 87.39.

7. With 13 df, the limits are 13(1,204.09)/29.8 and 13(1,204.09)/3.57. The confidence interval for the population variance is 525.27 to 4,384.64.

9. With 24 df, the limits for the confidence interval are 24(17.4)/39.4 and 24(17.4)/12.40, or 10.60 to 33.68.

11. With 22 df, the limits are 22(.27)/42.8 and 22(.27)/8.64, or 0.139 to 0.69.

13. With 26 df, the limits are 26(196)/48.3 and 26(196)/11.16, or 105.51 to 456.63.

Section 7.6

1. We want to be within 5 hours of the population mean, and the form for the confidence interval is $\bar{x} \pm z\sigma_{\bar{x}}$), so $z(\sigma_{\bar{x}}) = 5$. Since the confidence interval is at the 95 percent level, $z = 1.96$. Therefore, $1.96\ (\sigma_{\bar{x}}) = 5$. And $\sigma_{\bar{x}} = 5/1.96 = 2.55$. Since the finite correction factor is not needed, $\sigma_{\bar{x}} = \sigma/\sqrt{n}$. Solving for n we get $n = \sigma^2/\sigma_{\bar{x}}^2$. Assume the population standard deviation is 68.2, so $n = (68.2)^2/(2.55)^2 = 715.3$. Go to the next whole number (since the sample size must be a whole number). If the sample mean is to be within 5 hours of the population mean, the sample must be at least 716. Since the preliminary study yielded information from 100 patients, it will be necessary to obtain data from another 616 patients if we want to be 95 percent confident that the sample mean is within 5 hours of the population mean.

3. As the amount of tolerance (allowable error) decreases, the size of the sample must increase for the required accuracy.

5. If the confidence level is changed to 95 percent, we must change the value of the z score to 1.96. Then the minimum value for n is $(4.346)^2/(.5/1.96)^2$ or 290.24. Now the necessary sample size is 291.

7. With no preliminary sample, we assume that $\pi = 50$ percent. Then, $n = [(50)(50)]\ /\ (5/1.96)^2 = 2{,}500/6.5077 = 384.16$. We would need at least 385 in the sample.

9. With no preliminary sample we use $\pi = 50$ percent, so $n = [(50)(50)]/(3/1.96)^2 = 1{,}067.12$. The minimum sample size would be 1,068.

11. We want to be within 2 months of the actual time, so $E = 2$. We assume that $\sigma = 18.0$ months, and since we are using a 90 percent interval, $z = 1.645$. The minimum value for $n = (18.0)^2/(2/1.645)^2 = 219.19$ or 220.

13. We can use the preliminary study and estimate π with $p = 45/512 = 8.79$ percent. So $n = [(8.79)(100 - 8.79)]/(3/1.645)^2 = 241.06$ or 242. Since the preliminary study had 512 employees in the sample, there is already sufficient data to construct a confidence interval within the required tolerance.

Testing Hypotheses: One–Sample Procedures

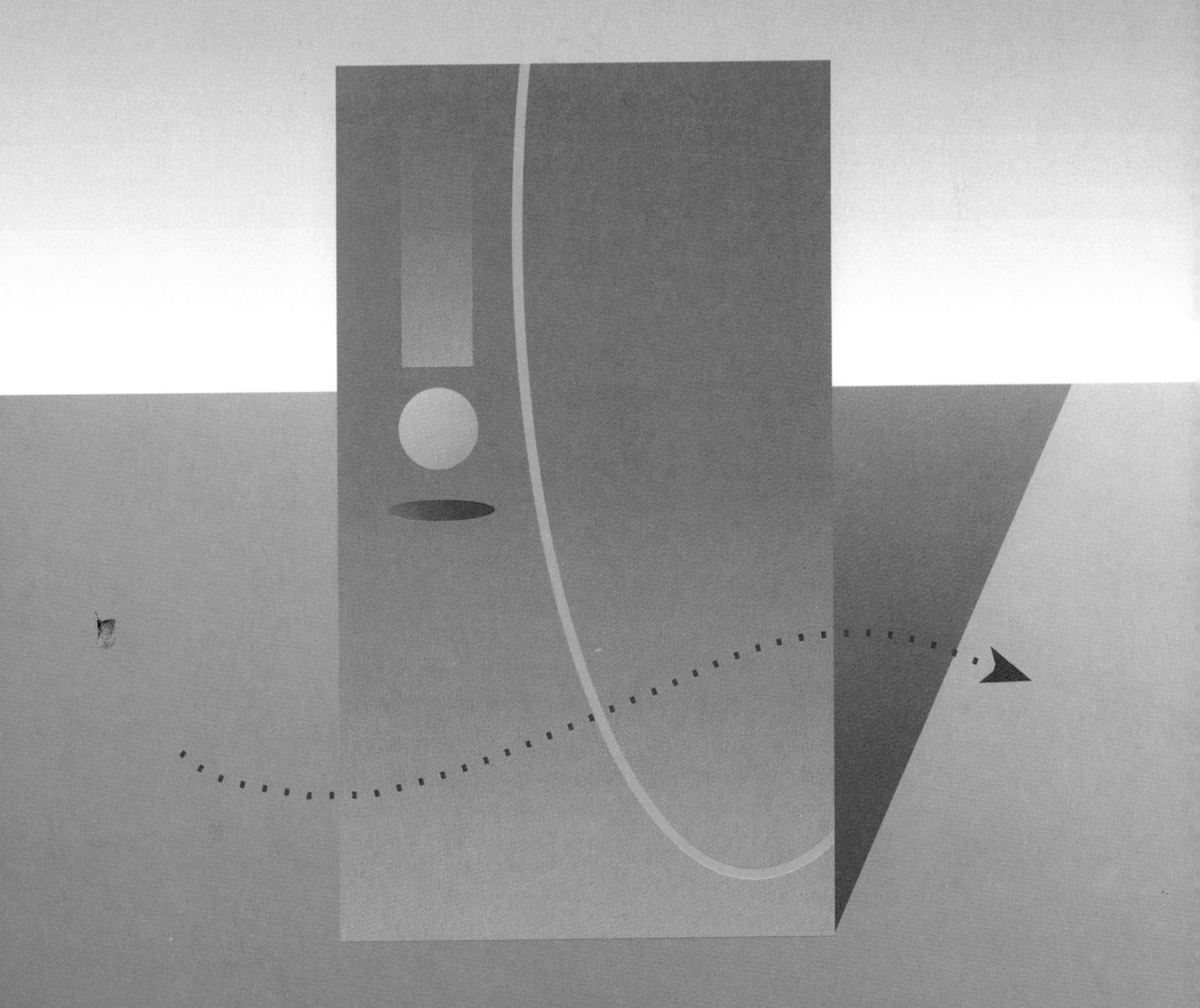

Incentive

A Census Bureau study has found that on the average people with bachelor's degrees earn $2,116 a month in the United States. The same study reported that high school graduates earn an average of $1,077 a month.

LOOKING AHEAD

You'll learn procedures in this chapter to help you decide if sample results *support* a hypothesis about a parameter value or if the results show that the hypothesis should be *rejected.* That is, you'll learn that if the sample results differ from the hypothetical population value being tested by an amount that *exceeds* what might be expected because of sampling variation, the difference is called a *statistically significant difference,* and this difference is the basis for rejecting the hypothesis being tested.

But we're getting ahead of ourselves. Before the chapter asks you to test hypotheses and make decisions, it outlines a general hypothesis-testing procedure for you to use to conduct one-sample hypothesis tests of means, percentages, and variances. (In later chapters, we'll look at hypothesis-testing procedures that involve two or more samples.)

Thus, after studying this chapter, you should be able to:

- Explain the necessary steps in the general hypothesis-testing procedure.
- Compute one-sample hypothesis tests of means (both one- and two-tailed versions) when the population standard deviation is known as well as when it's not available.
- Compute one-sample hypothesis tests of percentages for both one- and two-tailed testing situations.
- Compute one-sample hypothesis tests of variances for both one- and two-tailed testing situations.

8.1 The Hypothesis-Testing Procedure in General

We saw in the last chapter that when we have an estimating situation, the value of a population parameter is unknown and sample results are manipulated to provide some insight about the true value. In this chapter, though, the sample results are used for a different purpose. Although the exact value of a parameter may still be unknown, there's often some hunch or hypothesis about its true value. Sample results may bolster the hypothesis, or they may indicate that the assumption is untenable. For example, Dean Maria Santoro may state that the average IQ of the students at her university is 130. This statement may be an assumption on her part, and there should be some way of testing her claim. One possible method of validation involves sampling. If a random sample of these students produces an average IQ of 104, it's easy to reject the assumption that the true average is 130 because of the large discrepancy between the sample mean and the assumed value of the population mean. Similarly, if the sample mean is 131, it's reasonable to accept the dean's statement. Unfortunately, life's decisions are not always as clear-cut as this. Usually, the difference between the value of the sample statistic and the assumed parameter is neither too large nor too small, and thus obvious decisions are rare. Suppose, for example, the average IQ of a sample is 136; or suppose it's 122. Does either value warrant rejection of the statement that $\mu = 130$? Obviously, the decision process must be based on some sort of criterion.

Before we present the formal steps in the hypothesis-testing procedure, let's consider another example. Suppose the mayor of a town states that the average per-capita income of the town's citizens is $30,000, and you have a statistician friend—Stan Strate—who is hired by the town council to verify or discredit the mayor's claim. Obviously, Stan's knowledge of sampling variation tells him that even if the true mean is $30,000 as stated, a sample mean will *most likely not equal* the parameter value. Stan realizes there will probably be a difference between the sample mean and the assumed population value. The immediate question confronting him is how large or *significant* should the difference between the $\overline{x}$ and the assumed population value be to provide sufficient reason to dismiss the mayor's claim? Is a difference in values of $100 significant? Is a difference of $1,000 significant? Well, the significant differences can be found through statistical techniques.

Still Another Look at the Sampling Distribution of Means

Let's look at Figure 8.1 and assume that we have a sampling distribution of means where (1) the true mean (μ) is actually equal to the hypothesized value (μ_{H_0}) of $30,000 and (2) the standard error is equal to $200. *In other words, we are assuming that the mayor is actually correct and* μ *is indeed $30,000.* (Of course, Stan and the town council aren't aware of this fact.) Suppose further that Stan takes a sample of townspeople, with the result that the sample mean per-capita income is equal to $30,200. Is it reasonable for Stan to expect this result with a μ_{H_0} of $30,000 and a $\sigma_{\overline{x}}$ of $200? How likely is it that an $\overline{x}$ of $30,200 will occur in this situation? As a more general question, what are the chances of Stan's getting a sample mean that differs from the μ_{H_0} of $30,000 by ±$200?

Since the sampling distribution in Figure 8.1 is approximately normal, Stan can check the likelihood that a sample mean equals $30,200 or $29,800 by seeing how many standard errors from the μ of $30,000 a difference of $200 represents. How many z values does $30,200 or $29,800 lie from the true and assumed population mean of $30,000—that is, what's the standardized difference or the number of *standard units*? Stan can calculate the standard units in this way:

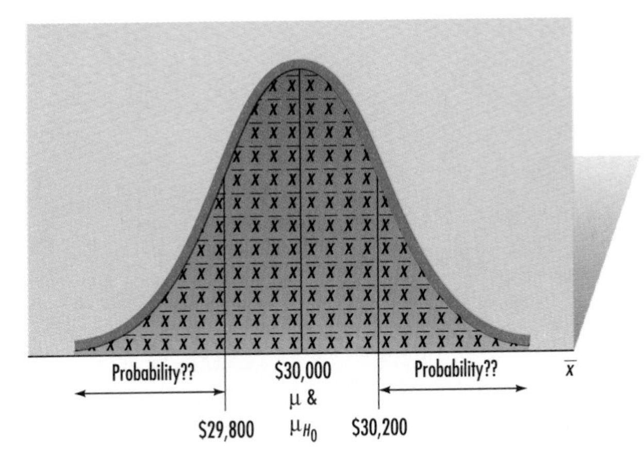

FIGURE 8.1 Educational schematic of a normally shaped sampling distribution where the assumed mean and the true mean happen to be of equal value.

$$z = \frac{\bar{x} - \mu_{H_0}}{\sigma_{\bar{x}}}$$

$$z = \frac{\$30{,}200 - \$30{,}000}{\$200} = 1.00 \quad \text{and} \quad z = \frac{\$29{,}800 - \$30{,}000}{\$200} = -1.00$$

Thus, we can see that if a sample mean in our example differs from the assumed value by \$200, it differs by 1 standard unit or 1 standard error. Consulting the z table in Appendix 2, we note that the area *under one side* of the distribution that corresponds to a z value of 1.00 is .3413 and the total area between $z \pm 1.00$ is .6826. This means that there's a .1587 chance that the $\bar{x}$ may be *less than* the population mean by *1 or more standard errors.* All this is demonstrated in Figure 8.2, where it's shown that there's a total chance of 31.74 percent that $\bar{x}$ will differ from μ by 1 standard unit or more. Consequently, Stan could report to the town council that a sample mean of \$30,200 is likely to occur with a population mean of \$30,000 and a \$200 difference is not sufficiently significant for him to reject the mayor's claim.

Suppose Stan's sample mean is \$30,400 instead of \$30,200. Would he reject the mayor's claim with this sample result? (Remember, he really doesn't know the true value of the population mean.) Converting this \$400 difference between $\bar{x}$ and μ_{H_0} into standard units, we get:

$$z = \frac{\bar{x} - \mu_{H_0}}{\sigma_{\bar{x}}} = \frac{\$30{,}400 - \$30{,}000}{\$200} = 2.00$$

Thus, the total chance that an $\bar{x}$ will differ from our true mean of \$30,000 by 2 or more standard errors is only approximately 4.6 percent, as shown in Figure 8.3. Given such a low chance of obtaining a sample mean of \$30,400, Stan would likely be justified in *rejecting* the mayor's claim. Now there's sufficient statistical evidence for him to conclude that the mayor's claim is incorrect.

As you can see, the difference between the value of an obtained sample mean and an

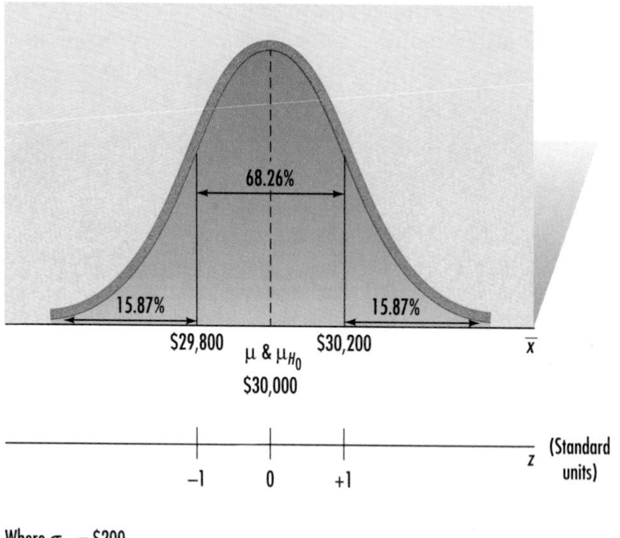

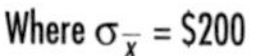

FIGURE 8.2 Illustration of the likelihood of obtaining an $\bar{x}$ that differs from the true mean by 1 standard error or more.

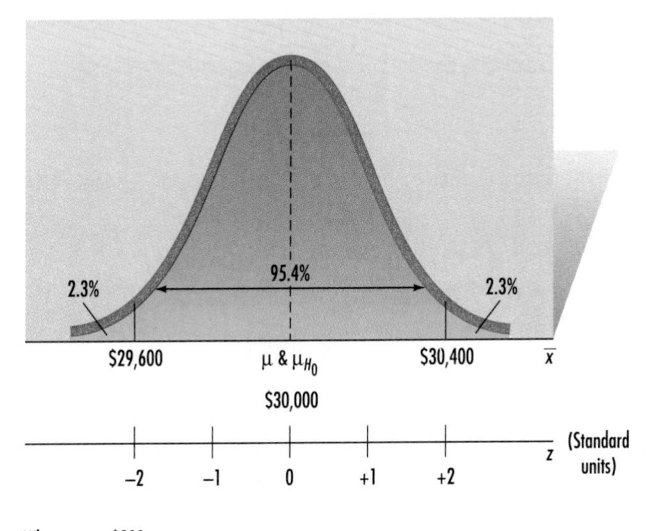

FIGURE 8.3 Illustration of the likelihood of obtaining an $\bar{x}$ that differs from the true mean by 2 or more standard errors.

assumed value of a hypothetical population mean is considered significantly large to warrant rejection of the hypothesis if the likelihood of the value of a sample mean is too low. The criterion of "too low" varies with the standards of researchers. For now, it's sufficient to state that all hypothesis tests must have some established rule that rejects a hypothesis if the likelihood of a value of $\bar{x}$ falls below a minimum acceptable probability level.

Unfortunately, since Stan doesn't know that the true population mean is indeed \$30,000, he may justifiably but erroneously reject the mayor's claim if he obtains a sample mean of \$30,400. As a matter of fact, if he establishes a rule that any sample mean value that differs from the assumed mean of \$30,000 by 2 or more standard errors in either direction of the sampling distribution will cause the hypothesis to be rejected and if the true mean is indeed \$30,000, he will erroneously reject the mayor's claim 4.6 percent of the time if he makes a large number of tests. In other words, a particular sample mean may be a part of a sampling distribution in which the value of the true mean happens to be equal to the assumed value, but the likelihood of that particular sample mean occurring may be so low that there is sufficient reason not to accept the hypothesized value as the true value. In short, the minimum acceptable likelihood of a sample mean is also the *risk of rejecting a statement that is actually true.*

With this basic example in mind, we're now ready to study the formal steps in the *classical* (or *traditional*) hypothesis-testing procedure.

Stressed Out

Psychologist Sheri Johnson of Brown University studied a sample of 304 teens and found that although both sexes face a similar amount of stress in the environment, females report greater personal distress. She hypothesized that "masculine" strategies such as pursuing diverting activities to gain a fresh perspective on problems seem to make teenage boys better able to recover from stress, while "feminine" strategies of dwelling on the cause of the problem tend to magnify the emotions and prolong the negative reaction to stress.

Steps in the Classical Hypothesis-Testing Procedure

There are seven steps in this testing procedure, and we'll encounter these same seven steps repeatedly in the remainder of this book.

Step 1: State the Null and Alternative Hypotheses The first step in traditional hypothesis testing is to specifically note the assumed value of the parameter *before* sampling:

This assumption to be tested is known as the **null hypothesis**, and the symbol for the null hypothesis is H_0.

Suppose we want to test the hypothesis that the population mean is equal to 100. The format of this hypothesis is:

H_0: $\mu = 100$

As we've seen earlier, the hypothesized value of the population mean when used in calculations is identified by the symbol μ_{H_0}.

It's important to note here that the null hypothesis is the one that contains the *equality* relation—that is, the H_0 states that some parameter (mean, percentage, variance) is *equal to* a specified value. A test is often carried out for the purpose of trying to show that the H_0 *isn't* true. For example, a researcher may hope that the population mean life expectancy of a new type of automobile battery will exceed the 60-month mean life expectancy of a currently sold product. But in conducting a test of the new battery, the H_0 is that the population mean life expectancy is equal to 60 months. The researcher's hope in this test is that the sample results will show a higher mean life expectancy and thus *won't* support the H_0.

If the sample results don't support the null hypothesis, we must obviously conclude something else:

The conclusion that is accepted contingent on the rejection of the null hypothesis is known as the **alternative hypothesis**, and the symbol for the alternative hypothesis is H_1.

There are three possible alternative hypotheses to the null hypothesis stated above:

H_1: $\mu \neq 100$
H_1: $\mu > 100$
H_1: $\mu < 100$

The selection of an alternative hypothesis depends on the nature of the problem, and later sections of this chapter discuss these alternative hypotheses. (The researcher carrying out the test of the new type of automobile battery would like to have reason to reject the H_0 that the $\mu = 60$ months and accept the H_1 that the μ is >60 months.) As with the null hypothesis, the alternative hypothesis should be stated *prior to* actual sampling.

Step 2: Select the Level of Significance Having noted the null and alternative hypotheses, the second step is to establish a criterion for rejecting or accepting the null hypothesis. If the true mean is actually the assumed value, we know that the probability of the differences between sample means and the μ_{H_0} diminishes as the size of the difference increases. That is, extremely large differences are unlikely. We must state, *prior to* sampling, the minimum acceptable probability of occurrence for a difference between $\bar{x}$ and μ_{H_0}. In our previous example involving the mayor's claim, a difference between $\bar{x}$ and μ_{H_0} with a likelihood of only 4.6 percent or less was considered unlikely, and so Stan felt there was sufficient reason to reject the hypothesis. In that case, a 4.6 percent chance of occurrence was the minimum acceptable probability level.

As noted earlier, if the true mean is indeed equal to the assumed value, the minimum acceptable probability level is also the risk of *erroneously* rejecting the null hypothesis when that hypothesis is *true.* Therefore, the next step in the hypothesis-testing procedure is to state the level of risk of rejecting a true null hypothesis:

This risk of erroneous rejection of the H_0 is known as the **level of significance,** which is denoted by the Greek letter α (alpha).

Of course, the costlier it is to mistakenly reject a true hypothesis—maybe because you might be sued and lose a lot of money—the smaller α should be. Thus, the value of α is small, usually .01 or .05. Technically, α is known as the risk of a **Type I error**—that is, the risk that a true hypothesis will be rejected. When a *false* hypothesis is erroneously *accepted* as true, it's known as a **Type II error**. When we decrease the probability of a Type I error, we raise the probability of a Type II error. We'll concentrate on the role of the Type I error; a thorough treatment of Type II errors is beyond the scope of this book. (Some university students were unkind enough to suggest to the author a few years ago that registering for his statistics course was known on campus as a Type III error.)

Step 3: Determine the Test Distribution to Use Once the level of significance is chosen, it's then necessary to select the correct probability distribution to use for the particular test. In this chapter, as in Chapter 7, we'll focus on the normal (z), t, and chi-square (χ^2) distributions. In later chapters we'll see that other probability distribution options are possible, but for now we'll use the z distribution in hypothesis tests of *means and percentages* when the sample size (n) is sufficiently large. We can also use the z distribution in tests of means when we have a smaller sample *if* two conditions are met: (1) It's known that the population values are normally distributed, and (2) the value of the population standard deviation is known. (This is just a restatement of the rules graphically presented in Figure 7.7, page 245.) If n is ≤ 30, the population values are known to be normally distributed, but the value of σ is *unknown*, then a t distribution is needed. And when we are concerned with one-sample hypothesis tests of *variances*, we use the chi-square distributions introduced in Chapter 7.

Step 4: Define the Rejection or Critical Regions Once the appropriate test distribution is determined, it's then possible to move to the next step. Suppose in a test using the z distribution that it's known that the level of significance (the acceptable risk of erroneous rejection of the null hypothesis) is $\alpha = .05$. This means that the null hypothesis will not be accepted if the difference expressed in standard units between $\bar{x}$ and μ_{H_0} has only a 5 percent or less chance of occurring. Since in many cases the null hypothesis can be rejected if the $\bar{x}$ is either too high or too low, we may want a .025 chance of erroneous rejection in each tail of the sampling distribution if the true mean is equal to the assumed value. In this situation, an α value of .05 represents the *total* risk of error. Figure 8.4 shows how the normal curve is partitioned. With .025 in each tail, the remaining area *in each half* of the sampling distribution is .4750 (.5000 − .0250). Appendix 2 shows that the z value for an area figure of .4750 is 1.96.

What does the partitioning of the normal curve in Figure 8.4 mean? Figure 8.5 shows that if a sample mean differs from the hypothetical mean by 1.96 or more standard errors in either direction, there's sufficient reason to reject the null hypothesis at the .05 level of significance. Thus, a z value of 1.96 represents the level in standard units at which the difference between $\bar{x}$ and μ_{H_0} becomes significant enough to raise doubt that

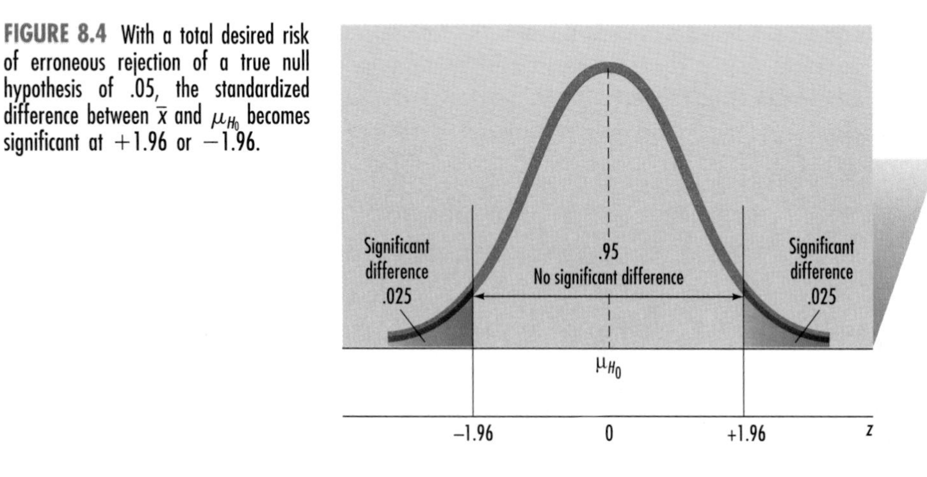

FIGURE 8.4 With a total desired risk of erroneous rejection of a true null hypothesis of .05, the standardized difference between $\bar{x}$ and μ_{H_0} becomes significant at +1.96 or −1.96.

$\mu_{H_0} = \mu$. A **significant difference** is a difference between $\bar{x}$ and μ_{H_0} that leads to the rejection of the null hypothesis:

> The **rejection region** (or **critical region**) is thus that part of the sampling distribution—equal in total area to the level of significance—that's specified as being unlikely to contain a sample statistic if the H_0 is true. (There may be more than one rejection region.)

The **acceptance region**, of course, is the remainder of the sampling distribution under consideration.

After the level of significance is stated and the proper test distribution is selected, the fourth step in our procedure is to find the boundaries for the rejection region (or regions) of the sampling distribution, which is represented in standard units. If the difference between an obtained $\bar{x}$ and the assumed μ_{H_0} has a value that falls into a rejection region, the null hypothesis is rejected. (If the difference doesn't fall into a critical region, of course, there's no statistical reason to doubt the hypothesis.) But a word of caution concerning conclusions about the validity of a null hypothesis is needed here. A test *never proves* that a null hypothesis is true. Rather, a test merely provides statistical evidence for not rejecting a null hypothesis. The only standard of truth is the population parameter, and since the true value of that parameter is un-

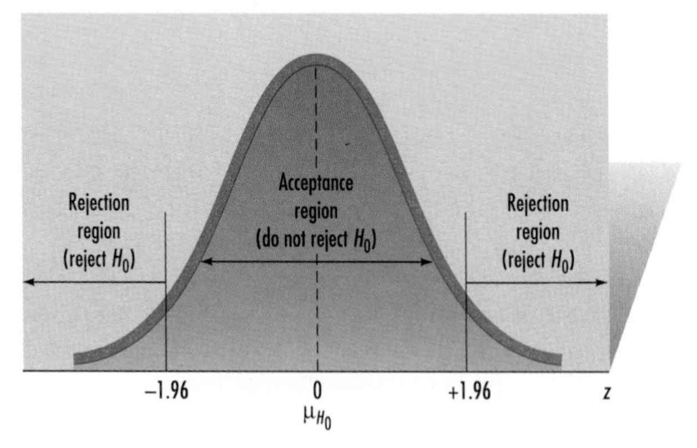

FIGURE 8.5 Construction of acceptance and rejection regions with a significant level of .05.

known, the assumption can never be proven. Thus, when we say in this chapter that a null hypothesis is "accepted," we merely mean that there's no statistically valid reason to reject the assumed parameter value. A similar analogy can be found in a criminal trial when a jury renders a "not-guilty" verdict. The jurors may not know for sure if the defendant is innocent of the charges, but there isn't enough evidence to cause them to convict the defendant.

Step 5: State the Decision Rule After we've stated the hypotheses, selected the level of significance, determined the test distribution to use, and defined the rejection region(s), the fifth step is to prepare a *decision rule*:

A **decision rule** is a formal statement of the appropriate conclusion to be reached about the null hypothesis based on sample results.

The general format of a decision rule is:

Reject H_0 if the standardized difference between $\bar{x}$ and μ_{H_0} falls into a rejection region. Otherwise, accept H_0.

Step 6: Make the Necessary Computations After all the ground rules have been laid out for the test, the next step is the actual data analysis. A sample of items is collected, the sample statistic(s) are computed, and an estimate of the parameter is calculated. Assuming that we're testing a hypothesis about the value of the population mean, we first calculate the value of a sample mean. To convert the difference between $\bar{x}$ and μ_{H_0} into a standardized value, it's then necessary to compute the standard error of the mean. The standardized difference between the statistic and the assumed parameter is called the *test ratio*:

The **test ratio (TR)** is the standardized difference between the statistic and the assumed parameter that is the basis for determining if we accept or reject the null hypothesis.

The test ratio for a hypothesis test of a population mean might be determined as follows:

$$TR = \frac{\bar{x} - \mu_{H_0}}{\sigma_{\bar{x}}} \tag{8.1}$$

Step 7: Make a Statistical Decision *If the value of the test ratio falls into a rejection region, the null hypothesis is rejected.* For example, Figure 8.6 shows the rejection regions of a normal curve with $\alpha = .01$. Referring to the z table, a *total* risk of 1 percent corresponds to z values of -2.575 and $+2.575$. Suppose a sample produced a test ratio of 2.60. Since the TR falls into a rejection region, there's sufficient reason to reject the null hypothesis, and the risk of erroneous rejection is only 1 percent.

At this point, your head may be dizzy with definitions and procedural steps. To help you sort out your head, Figure 8.7 summarizes the general seven-step procedure for a classical hypothesis test.

STATISTICS IN ACTION

Take This Cube and Shove It

In a University of Wisconsin study involving 115 incarcerated delinquent boys and 39 nondelinquent boys, it was found that there's no support for the common belief that sugary foods can provoke aggression or other behavior problems in delinquent and nondelinquent boys. Laboratory tests were used to assess the behavior of the subjects after meals, and it was concluded that no differences in behavior occurred in either group when their meals contained sugar or when they contained a sugar substitute.

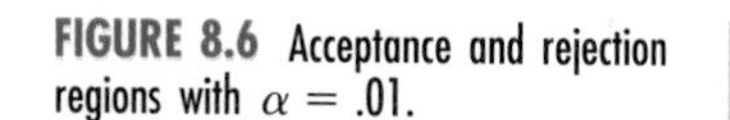

FIGURE 8.6 Acceptance and rejection regions with $\alpha = .01$.

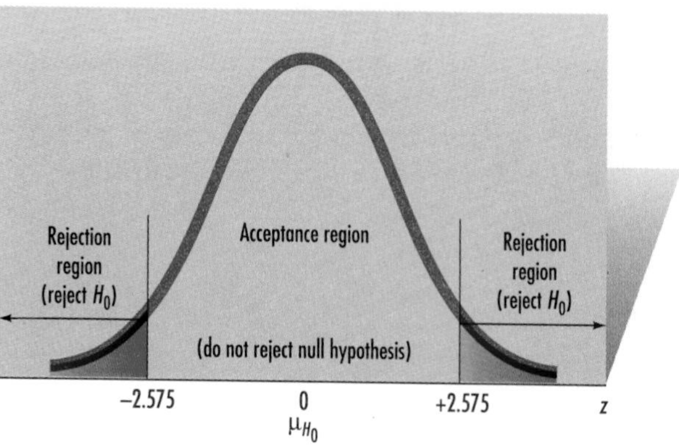

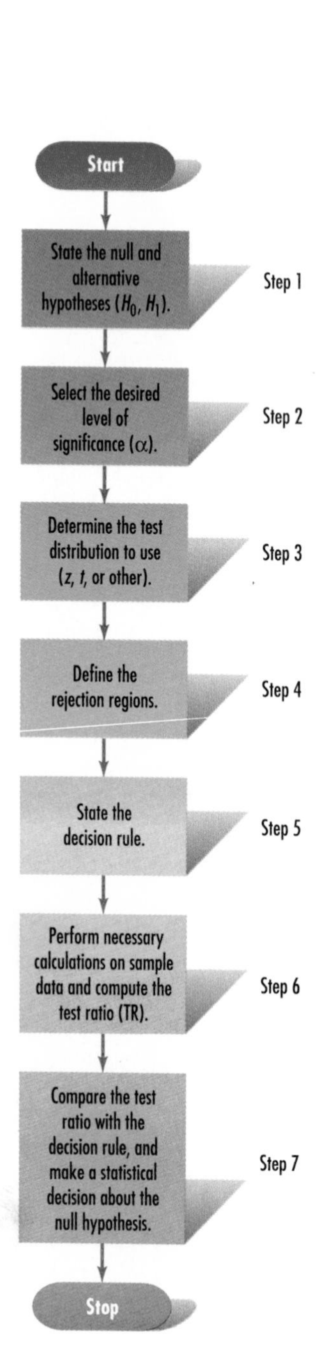

FIGURE 8.7 Classical seven-step hypothesis-testing procedure.

Managerial Decisions and Statistical Decisions: A Caution

Let's conclude this section on a nonstatistical note. Although statistical laws give us objective ways to assess hypotheses, a statistical conclusion doesn't represent the final word in decision making. Consumers of statistical reports use quantitative results as one form of input in a complex network of factors that affect an ultimate decision. But decision making is full of uncertainty, and although statistical results serve to reduce and control some of this uncertainty, they don't completely eliminate doubt. Problems may be quantified and a result obtained, but the solution is only as good as the input that has gone into structuring the problem.

Thus, statistical results, although objectively determined, shouldn't be blindly accepted. Other situational factors must also be considered. For example, a statistical test may tell a production manager that a machine probably isn't producing as much as she had assumed. But this result doesn't tell her what action to take. She may replace the machine, fix it, or leave it in its present condition. The ultimate decision is made by considering the available money for replacement, the repair record of the machine, the availability of new machines, and so on. Thus, the statistical conclusion is not necessarily the managerial conclusion; it's simply one factor that must be considered in the context of the whole problem.

Self-Testing Review 8.1

1. If a sample has 57 members, what is the total chance that the mean of this sample will fall 1.96 or more standard errors from the true population mean?
2. If a sample has 14 members and the population is normally distributed but we don't know the population standard deviation, what is the total chance that the mean of this sample will fall 2.65 or more standard errors from the true population mean?
3. If a sample has 22 members and we know the population is normally distributed with a standard deviation of 62.9, what's the total chance that the mean of this sample will fall 2.33 or more standard errors from the true population mean?
4. What is a Type I error?
5. What is a Type II error?

Now, you compute the test ratio (*Step 6*):

$$TR = \frac{\bar{x} - \mu_{H_0}}{\sigma_{\bar{x}}} = \frac{\bar{x} + \mu_{H_0}}{\sigma/\sqrt{n}} = \frac{14.1333 - 14.00}{.15/\sqrt{6}} = \frac{.1333}{.0612} = 2.18$$

Conclusion (Step 7):

Since your TR falls between ±2.575, you accept the null hypothesis that the machine is operating properly at the .01 level of significance.

A *p*-Value Approach to Hypothesis Testing

We've now looked at three examples that have followed the classical hypothesis-testing procedure. But there are other ways to conduct such tests. For example, one procedure uses a *p-value* (or *probability-value*) testing approach. Most of the steps used in a *p*-value hypothesis test are the same as those we've been following in using the classical procedure, but there are also a few differences. Let's consider the problem situation in Example 8.3 again, but this time we'll use a *p*-value test to see if the bolt-making machine is properly adjusted.

Steps in the p-Value Procedure

Step 1: State the Null and Alternative Hypotheses In our example problem:

H_0: $\mu = 14.00$ millimeters
H_1: $\mu \neq 14.00$ millimeters

As you can see, this first step is the same in the classical and *p*-value procedures.

Step 2: Select the Level of Significance In our example, α is .01, and, again, this second step is the same in traditional and *p*-value tests. (We'll see in a moment, though, that some researchers who use the *p*-value approach may omit this step.)

Step 3: Determine the Test Distribution to Use The z distribution is correct for our example, and there is still no difference between classical and *p*-value tests.

Step 4: State the Decision Rule The classical step of defining the rejection (or critical) region is omitted in a *p*-value test, and a decision rule *may* be formulated next. The decision rule may be stated as follows:

Reject H_0 if the *p*-value is less than α. Otherwise, accept H_0.

Thus, in our Example 8.3 problem, we'll accept the H_0 if the p value is equal to or greater than .01. Note, however, that some researchers who prefer to use a *p*-value hypothesis test don't select a level of significance and they don't specify a decision rule. Rather, they merely report the p value(s) produced by their studies and leave the interpretations to those who read their reports in professional journals. When that approach is used, a published p value of less than .01 is considered to be a very strong statistical argument *against* accepting the H_0, and a p value of from .01 to .05 may give readers sufficient reason to doubt the validity of the H_0.

Step 5: Compute the Test Ratio There's no difference between classical and p-value methods here. We've seen that the test ratio for our Example 8.3 problem is:

$$\text{TR} = \frac{\bar{x} - \mu_{H_0}}{\sigma_{\bar{x}}} = \frac{\bar{x} - \mu_{H_0}}{\sigma/\sqrt{n}} = \frac{14.1333 - 14.00}{.15/\sqrt{6}} = \frac{.1333}{.0612} = 2.18$$

Step 6: Compute the Appropriate p Value It's helpful here to sketch the correct probability distribution—the z distribution in this example—and locate the TR value on the sketch. For our Example 8.3 problem, which is concerned with both tails of the probability distribution, such a sketch is shown in Figure 8.10. As you can see, the p value in *each tail* of the probability distribution is the area that falls *beyond* the TR value of 2.18. The area between the μ_{H_0} and the TR value of +2.18 is found to be .4854 in the z table in Appendix 2. And the same area, of course, is found between μ_{H_0} and a TR value of −2.18. So in the *left tail*:

$$p \text{ value} = \text{probability that } z < -2.18 \text{ if } H_0 \text{ is true} = .5000 - .4854 = .0146$$

And in the *right tail*:

$$p \text{ value} = \text{probability that } z > +2.18 \text{ if } H_0 \text{ is true} = .5000 - .4584 = .0146$$

What meanings can we attach to these values of .0146? Well, as you've seen in earlier discussions, the probability that a sample mean will fall 2.18 or more standard errors to the *right* of a population mean is .0146, and, of course, the probability that a sample mean will be 2.18 or more standard errors to the *left* of a population mean is also .0146. Thus, to compute the probability or p value that a sample mean will fall into *either* the right-tail *or* left-tail areas beyond 2.18 standard errors in a *two-tailed* test using the normal probability distribution, we employ this formula:

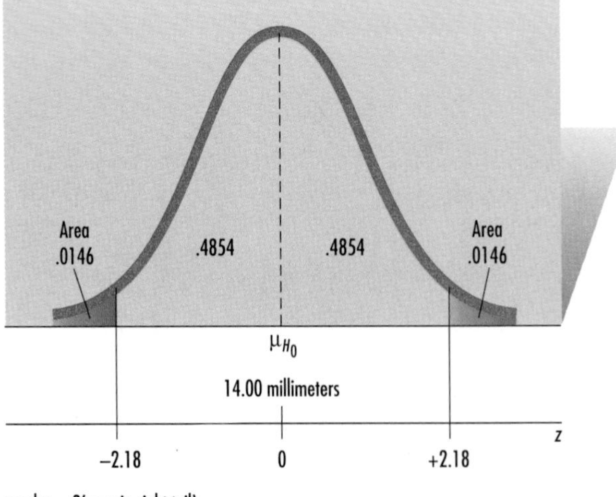

FIGURE 8.10 Illustration of p values for a two-tailed test using the data in Example 8.3.

$$p \text{ value (two-tailed)} = P(z < -|\text{TR}|) + P(z > |TR|) \tag{8.2}$$

where $|\text{TR}|$ = the absolute value (ignore the signs) of the test ratio

And in our Example 8.3 problem, the p value we are looking for is:

$$\begin{aligned} p \text{ value (two-tailed)} &= P(z < -|\text{TR}|) + P(z > |\text{TR}|) \\ &= .0146 + .0146 = .0292 \text{ or } .03 \end{aligned}$$

To summarize, then:

The **p value** of a hypothesis test is the probability of obtaining a difference between the sample statistic and the hypothetical population parameter that is at least as extreme as the one actually observed assuming the H_0 is true. The smaller the p value, the stronger the case against the H_0.

Step 7: Make a Statistical Decision Since the computed p value of .03 is greater than the level of significance of .01, we accept the null hypothesis. In our example, the p value of .03 tells us that a sample with a mean of 14.1333 or larger could be expected to happen 3 percent of the time when the μ is 14.00 millimeters. As noted earlier, though, if we hadn't selected a level of significance in advance, a p value of .03 might be interpreted by some as sufficient reason to doubt the validity of our hypothesis that the population mean is 14.00 millimeters.

You've spent some time now following through the computations needed to conduct classical and p-value hypothesis tests of the sample bolt data presented in Example 8.3. But computer statistical packages can easily handle such tasks. As you can see in Figure 8.11, a package user may simply key in the sample data, use a ZTEST command, supply the values of μ_{H_0} and σ, and receive a printout giving the same values you've studied. The TR value of 2.18 is shown under the column labeled Z in the printout, and the p value of .030 is produced.

```
MTB > SET C2
MTB > END
MTB > PRINT C2

C2
   14.15    13.85    13.95    14.20    14.30    14.35

MTB > ZTEST  MU = 14,   SIGMA = .15, DATA IN C2

TEST OF MU = 14.0000 VS MU N.E. 14.0000
THE ASSUMED SIGMA = 0.150

                N        MEAN      STDEV    SE MEAN       Z     P VALUE
C2              6     14.1333     0.1966     0.0612    2.18       0.030
```

FIGURE 8.11 A *Minitab* computer printout of the two-tailed z test for Example 8.3.

Classical One-Tailed Tests When σ Is Known

Many times it's not enough to simply conclude that the true value is *not equal* to the assumed value. If the null hypothesis isn't tenable, we often want to know if the rejection occurs because the true value is *probably higher* or *probably lower* than the assumed value. In other words, is the hypothesis rejected because the true value is likely to be greater than or less than the assumed value? In such situations, the null hypothesis is still:

H_0: μ = assumed value

But the *alternative hypothesis* is one of the following:

H_1: $\mu >$ assumed value

or

H_1: $\mu <$ assumed value

Scientists in federal, state, and local governments use statistical data to evaluate and control the levels of pollution they encounter. (*Comstock*)

The nature of either alternative hypothesis indicates a *one-tailed test*:

In a **one-tailed test**, there's only one rejection region, and the null hypothesis is rejected only if the value of a sample statistic falls into this single rejection region. If the single rejection region is in the right tail of the sampling distribution so that H_1: parameter > assumed value, then the one-tailed test is also known as a **right-tailed test**; but if the single rejection region is in the left tail so that H_1: parameter < assumed value, then the one-tailed test is a **left-tailed test.**

Right-Tailed Tests. When the alternative hypothesis is:

H_1: $\mu >$ assumed value

then the rejection region is in the right tail of the sampling distribution, and the null hypothesis is rejected *only* if the value of a sample mean is *significantly high*. If the value of a sample mean is low compared to the assumed value, the null hypothesis isn't rejected. The *major pitfall with a right-tailed test* is that the true value may be less than the assumed value, but because of the structure of the right-tailed test, the H_0 isn't rejected. In such a test, the attention is focused on rejecting H_0 solely on the basis that the true value might be greater than the assumed value.

If you're confused by the above paragraph, consider this analogy. Suppose you and a friend are guessing a third person's age. Your friend hypothesizes that the third party is 23 years old, but you believe that he is older. Finally, you ask the third person, "Are you more than 23 years old?" He says no. As a result, you cannot reject your friend's assertion, but the opinion may be wrong because you didn't ask the person if he was less than 23 years old. The nonrejection of H_0 in a right-tailed test is similar to this analogy.

Left-Tailed Tests. When the alternative hypothesis is:

H_1: $\mu <$ assumed value

then we're interested in seeing if the true value is *less* than the assumed value. In this case, the H_0 is rejected only if the value of a sample mean is *significantly low*. In a left-tailed test, the H_0 isn't rejected if the true value is likely to be more than the assumed value.

The distinctions between left- and right-tailed tests are shown in Figure 8.12.

Level of Significance Considerations. The level of significance (α) is the *total risk* of erroneously rejecting H_0 when it's actually true. In a two-tailed test, the total risk is evenly divided between each tail. But in a one-tailed test (since there's only one rejection region), *an area in the single tail is assigned the total risk or α*. If the z distribution is applicable, the correct z value is thus determined by the one-tailed probability of $.5000 - \alpha$. For example, if the level of significance is .05 for a left-tailed test, the boundary of the rejection region is a z value of -1.64 (see Figure 8.13).

Decision Rule Statements. If a z distribution is to be used, the decision rule for a *left-tailed test* takes this form (remember the rules of first-year algebra here):

Reject H_0 and accept H_1 if TR $< -z$ value. Otherwise, accept H_0.

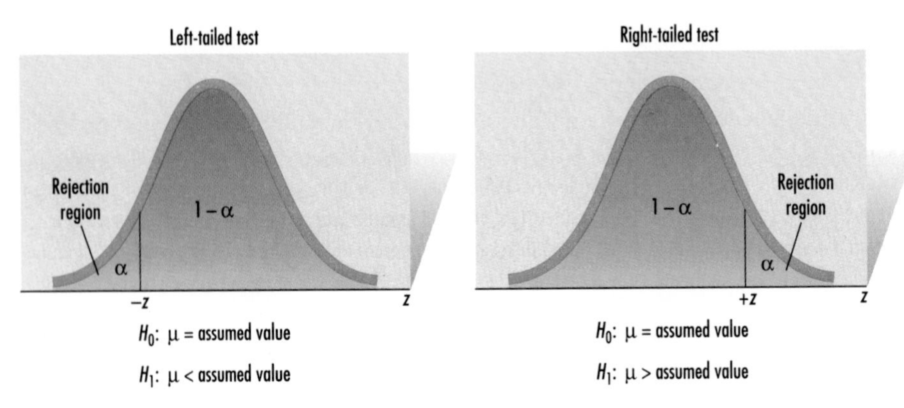

FIGURE 8.12 Illustration of the rejection regions for left-tailed and right-tailed tests.

The decision rule for a *right-tailed test* is:

Reject H_0 and accept H_1 if TR $> z$ value. Otherwise, accept H_0.

The test ratio is computed the same way for a one-tailed test as for a two-tailed test.

Let's look now at the following examples, where a one-tailed test is applicable and where the σ is known.

Example 8.4. Juanita Lopez, a production supervisor at a chemical company, wants to be sure that the Super-Duper can is filled with an average of 16 ounces of product. If the mean volume is significantly less than 16 ounces, customers (and regulatory agencies) will likely complain, prompting undesirable publicity. The physical size of the can doesn't allow a mean volume significantly above 16 ounces. A random sample of 36 cans shows a sample mean of 15.7 ounces. Assuming σ is 0.2 ounce, conduct a hypothesis test with $\alpha = .01$.

Hypotheses (Step 1):

H_0: $\mu = 16$ ounces
H_1: $\mu < 16$ ounces

The nature of the problem is such that if the null hypothesis is rejected, Juanita will conclude that the sample mean is significantly low.

We know that $\alpha = .01$ *(Step 2)*, and with $n > 30$ and with σ known, the z distribution is used *(Step 3)*. Thus, with $\alpha = .01$, and with a left-tailed test, the rejection region begins at a z-tail value beyond -2.33 *(Step 4)*.

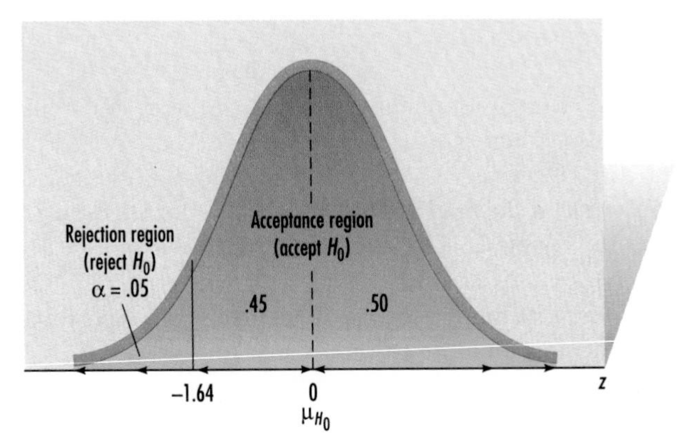

FIGURE 8.13 Rejection for a left-tailed test at the .05 level of significance.

Decision Rule (Step 5):

Reject H_0 and accept H_1 if TR < -2.33 (remember that $-2.34 < -2.33$). Otherwise, accept H_0.

Test Ratio (Step 6):

$$\text{TR} = \frac{\bar{x} - \mu_{H_0}}{\sigma_{\bar{x}}} = \frac{\bar{x} - \mu_{H_0}}{\sigma/\sqrt{n}} = \frac{15.7 - 16}{.2/\sqrt{36}} = -9.00$$

Conclusion (Step 7):

Since TR < -2.33, Juanita must reject H_0 and *rush* to correct the filling process. It's virtually impossible that a sample selected from a sampling distribution that has a true mean of 16 ounces will have a sample mean located 9.00 standard errors to the left of the true mean!

Example 8.5. Frank N. Cence, a perfume distributor, believes that the mean cost to process a sales order is \$13.25. Minnie Mize, cost controller, fears that the average cost of processing is more than that. She is interested in taking action if costs are high, but she can accept the situation if the actual mean cost is below the assumed value. A random sample of 100 orders has a sample mean of \$13.35. Assuming the σ is \$0.50, conduct a test at the .01 level of significance.

Hypotheses (Step 1):

H_0: $\mu = \$13.25$ cost
H_1: $\mu > \$13.25$ cost

This is a right-tailed test because only a significantly high sample mean value will lead Minnie to reject the null hypothesis. With $\alpha = .01$ (*Step 2*), and with $n > 30$, the z distribution is applicable (*Step 3*), and the correct z value is 2.33 (*Step 4*).

Decision Rule (Step 5):

Reject H_0 and accept H_1 if TR > 2.33. Otherwise, accept H_0.

Test Ratio (Step 6):

$$\text{TR} = \frac{\bar{x} - \mu_{H_0}}{\sigma_{\bar{x}}} = \frac{\bar{x} - \mu_{H_0}}{\sigma/\sqrt{n}} = \frac{\$13.35 - \$13.25}{\$0.50/\sqrt{100}} = 2.00$$

Conclusion (Step 7):

Since TR < 2.33, Ms. Mize has no reason to reject Mr. Cence's statement at the .01 level of significance.

A *p*-Value One-Tailed Test

A two-tailed *p*-value hypothesis test was presented a few pages earlier. Let's use the data in Example 8.5 now to show how a one-tailed *p*-value test is conducted.

The first steps to conduct the test in Example 8.5 don't change. The *hypotheses* for this right-tailed test are still:

H_0: $\mu = \$13.25$ cost
H_1: $\mu > \$13.25$ cost

The *level of significance* (α) remains .01, and we'll still use the z distribution to carry out this test. Our *decision rule* now is:

Reject H_0 if the p value is less than .01. Otherwise, accept H_0.

The computation of the *test ratio* doesn't change:

$$\text{TR} = \frac{\bar{x} - \mu_{H_0}}{\sigma_{\bar{x}}} = \frac{\bar{x} - \mu_{H_0}}{\sigma/\sqrt{n}} = \frac{\$13.35 - \$13.25}{\$0.50/\sqrt{100}} = 2.00$$

To compute the appropriate p value for a one-tailed test when the z distribution is used, we select *one* of these formulas:

$$p \text{ value (right-tailed test)} = P(z > TR) \quad (8.3)$$

$$p \text{ value (left-tailed test)} = P(z < TR) \quad (8.4)$$

Since this is a right-tailed test, our p value is:

p value (right-tailed test) $= P(z > \text{TR}) = P(z > 2.00)$
$= .5000 - .4772$ (Use Appendix 2 to find that the area between μ_{H_0} and a z value of 2.00 is .4772.)
$= .0228$

Since the computed p value of .0228 is greater than the level of significance of .01, we accept the H_0, and Ms. Mize should accept Mr. Cence's statement that the mean cost to process a sales order is $13.25. Our p value of .0228 tells us that a sample with a mean of $13.35 or larger could be expected to happen 2.28 percent of the time when the μ is $13.25. Of course, if this test had been conducted with $\alpha = .05$, then we would have rejected the H_0.

Classical Two-Tailed Tests When σ Is Unknown

Up to now, we've been working hypothesis tests with σ known. But as you saw in Chapter 7, knowledge of σ is rare. Usually, the sample standard deviation (s) is used in the testing procedure.

With σ unknown, the following aspects of the traditional hypothesis-testing procedure are affected:

1. The correct sampling distribution can no longer be *assumed* to be approximately normally shaped if n is 30 or less.
2. In the computation of the test ratio (TR), an estimated standard error—$\hat{\sigma}_{\bar{x}}$—must be used instead of $\sigma_{\bar{x}}$.

Thus, when σ is unknown, the z distribution (and Appendix 2) can be used only to find the rejection regions when the sample size exceeds 30. But if σ is unknown, if population values are known to be normally distributed, and if the sample size is 30 or less, then the sampling distribution takes the shape of a t distribution. The t value used to find the boundary of a rejection region depends on the level of significance and the degrees of freedom (which are $n - 1$ for the tests in this chapter). For example, suppose you are making a *two-tailed test* at the .05 level of significance with a sample size of 16. In the t table in Appendix 4, the t value with 15 degrees of freedom is 2.131. That is, $t_{\alpha/2}$ or $t_{.025} = 2.131$. The t table is set up to show the rejection region in *one tail.*

As we shall see in the following examples, the seven-step testing procedure is the same with an unknown σ as with a given σ, with the exceptions that (1) the proper test distribution (z or t) must be used and (2) the correct method of calculating the test ratio must be followed.

Example 8.6. John Drinkwater, co-owner of the J-T Pub, believes his business sells an average of 17 pints of Border Ale daily. His partner, Tess Totaler, thinks this estimate is wrong. A random sample of 36 days shows a mean sales of 15 pints and a sample standard deviation (s) of 4 pints. Test the accuracy of Drinkwater's opinion at the .10 level of significance.

Hypotheses (Step 1):

H_0: $\mu = 17$ pints
H_1: $\mu \neq 17$ pints

This is a two-tailed test because we wish to assess only the validity of Drinkwater's belief. The level of significance (*Step 2*) is .10. With $n = 36$, the sampling distribution approximates the normal distribution, and thus the *z distribution* applies (*Step 3*).

Since this is a two-tailed test and $\alpha = .10$, the risk of error in each tail is .05. The z value corresponding to $.5000 - .05 = .45$ is 1.645 (*Step 4*).

Decision Rule (Step 5):

Reject H_0 and accept H_1 if $TR < -1.645$ or $TR > +1.645$. Otherwise, accept H_0.

With $s = 4$ and $n = 36$, $\hat{\sigma}_{\bar{x}}$ is estimated in the following manner:

$$\hat{\sigma}_{\bar{x}} = \frac{s}{\sqrt{n}} = \frac{4}{\sqrt{36}} = .667$$

Test Ratio (Step 6):

$$TR = \frac{\bar{x} - \mu_{H_0}}{\hat{\sigma}_{\bar{x}}} \tag{8.5}$$

$$= \frac{15 - 17}{.667} = -3.00$$

Conclusion (Step 7):

Since $TR < -1.645$, it's necessary to reject Drinkwater's claim at the .10 level of significance.

Example 8.7. The height of female adults in Biglandia is normally distributed, and a journal article claims that the mean height of these females is 64 inches. To test this claim, a sociologist takes a random sample of 16 Biglandia women and finds that the mean is 62.9 inches and the standard deviation is 2.5 inches. Can the claim made in the article be accepted at the .05 level of significance?

Hypotheses (Step 1):

H_0: $\mu = 64$ inches
H_1: $\mu \neq 64$ inches

The nature of the problem indicates a *two-tailed test*, with a risk of erroneous rejection of .025 in each tail. We'll use the .05 level of significance (*Step 2*), and the *t distribution* is applicable because the population heights are normally distributed and the sample size is only 16 (*Step 3*). With 15 degrees of freedom and a .025 risk in each tail, $t_{.025} = 2.131$ (*Step 4*).

Decision Rule (Step 5):

Reject H_0 and accept H_1 if TR < -2.131 or TR $> +2.131$. Otherwise, accept H_0.

Test Ratio (Step 6):

$$\text{TR} = \frac{\bar{x} - \mu_{H_0}}{\hat{\sigma}_{\bar{x}}} = \frac{\bar{x} - \mu_{H_0}}{s/\sqrt{n}} = \frac{62.9 - 64.0}{2.5/\sqrt{16}} = \frac{-1.1}{.625} = -1.76$$

Conclusion (Step 7):

Since TR falls between ± 2.131, there is no reason to reject the article's statement at the .05 level of significance.

Example 8.8. Let's go back to the problem in Example 8.3 and change some assumptions. Let's suppose that the bolts made by the automatic machine are normally distributed and should still have a μ diameter of 14.00 millimeters. If bolts vary significantly in either direction from this standard, they aren't suitable for their intended use. But let's assume now that the σ isn't known, and we must instead calculate a sample standard deviation. As before, the latest sample has 6 bolts with the following diameters (in millimeters): 14.15, 13.85, 13.95, 14.20, 14.30, and 14.35. Our previous decision was that the bolt machine was operating properly. Is that decision still the correct one at the .01 level?

Hypotheses (Step 1):

H_0: $\mu = 14.00$ millimeters
H_1: $\mu \neq 14.00$ millimeters

This is still a two-tailed test, and $\alpha = .01$ (*Step 2*), but we must now use the *t* distribution (*Step 3*) because the sample size is only 6 and we don't know σ. With 5 degrees of freedom and a .005 risk in each tail, $t_{.005} = 4.032$ (*Step 4*).

Decision Rule (Step 5):

Reject H_0 and accept H_1 if TR < -4.032 or TR $> +4.032$. Otherwise, accept H_0.

The value of s must be computed as follows (remember from Example 8.3 that $\bar{x} = 14.1333$):

This is a *left-tailed test* because only a sample mean that is significantly low will cause rejection of the null hypothesis. And since this is a one-tailed test, the area in the single rejection region is equal to the significance level of .05 *(Step 2)*.

With $n = 25$, the *t* distribution applies *(Step 3)*, there are 24 degrees of freedom, and $t_{.05} = 1.711$ *(Step 4)*.

Decision Rule (Step 5):

Reject H_0 and accept H_1 if $\text{TR} < -1.711$. Otherwise, accept H_0.

Test Ratio (Step 6):

$$\text{TR} = \frac{\bar{x} - \mu_{H_0}}{\hat{\sigma}_{\bar{x}}} = \frac{\bar{x} - \mu_{H_0}}{s/\sqrt{n}} = \frac{4{,}460 - 4{,}500}{250\ \sqrt{25}} = \frac{-40}{50} = -.80$$

Conclusion (Step 7):

Since $\text{TR} > -1.711$, there's no significant reason for Mr. Stone to doubt the manager's claim. It's quite possible that a sample could be selected with a mean located only $-.80$ standard error from a true mean.

Example 8.10. Hiram N. Fyrem, owner of the HNF Employment Agency, believes that the agency receives an average of 16 complaints per month from companies that hire the agency's people. Mollie DeGree, an interviewer, is concerned that the true mean is higher than Hiram believes. If Hiram's hypothesis is an understatement, something must be done about the agency's employee screening procedures. A sample of 10 months yields an average of 18 complaints with a standard deviation of 3 complaints. Conduct a test at the .01 level.

Hypotheses (Step 1):

H_0: $\mu = 16$ complaints per month
H_1: $\mu > 16$ complaints per month

If the statistical evidence cannot support the H_0, Mollie wants to conclude that the parameter value is more than assumed. Thus, this is a *right-tailed test.*

With $n = 10$ and $\alpha = .01$ *(Step 2)*, the *t* value *(Step 3)* with 9 degrees of freedom is $t_{.01} = 2.821$ *(Step 4)*.

Decision Rule (Step 5):

Reject H_0 and accept H_1 if $\text{TR} > 2.821$. Otherwise, accept H_0.

Test Ratio (Step 6):

$$\text{TR} = \frac{\bar{x} - \mu_{H_0}}{\hat{\sigma}_{\bar{x}}} = \frac{\bar{x} - \mu_{H_0}}{s/\sqrt{n}} = \frac{18 - 16}{3/\sqrt{10}} = \frac{2}{.95} = 2.11$$

Conclusion (Step 7):

Since $\text{TR} < 2.821$, there's no sufficient reason at the .01 level of significance to reject Hiram's hypothesis.

Example 8.11. We've seen how the *Minitab* statistical program processes two-tailed *z* and *t* tests in earlier examples. But one-tailed tests are also handled with ease. Let's suppose that Mollie isn't satisfied with the results obtained in the preceding example and decides to sample another 10-month period. The sample gave the following complaint data: 20, 14, 12, 24, 17, 22, 13, 16, 15, and 19.

Hypotheses (Step 1):

H_0: $\mu = 16$ complaints per month
H_1: $\mu > 16$ complaints per month

And at the .01 level (*Step 2*), a t value (*Step 3*) of 2.821 (*Step 4*) is still needed.

Decision Rule (Step 5):

Reject H_0 and accept H_1 if TR > 2.821. Otherwise, accept H_0.

The sample mean and sample standard deviation are found as follows:

x	$(x - \bar{x})$	$(x - \bar{x})^2$
20	2.8	7.84
14	−3.2	10.24
12	−5.2	27.04
24	6.8	46.24
17	−.2	.04
22	4.8	23.04
13	−4.2	17.64
16	−1.2	1.44
15	−2.2	4.84
19	1.8	3.24
172		141.60

$$\bar{x} = \frac{\Sigma x}{n} = \frac{172}{10} = 17.2$$

$$s = \sqrt{\frac{\Sigma (x - \bar{x})^2}{n - 1}} = \sqrt{\frac{141.60}{9}} = 3.967$$

With $s = 3.967$, we can now compute the test ratio:

Test Ratio (Step 6):

$$\text{TR} = \frac{\bar{x} - \mu_{H_0}}{\hat{\sigma}_{\bar{x}}} = \frac{\bar{x} - \mu_{H_0}}{s/\sqrt{n}} = \frac{17.2 - 16}{3.967/\sqrt{10}} = \frac{1.2}{1.254} = .96$$

Conclusion (Step 7):

Since the TR is <2.821, we accept the H_0. There's still no reason to doubt Hiram's claim.

All of these results are duplicated in the *Minitab* computer output shown in Figure 8.15. The sample data are entered, a TTEST command is used, the μ_{H_0} of 16 complaints is recorded, and an ALTERNATIVE subcommand is specified. The ALTERNATIVE = +1 subcommand tells the program that it's to perform a right-tailed test. (If

```
MTB > SET C1
MTB > END
MTB > PRINT C1

C1
   20    14    12    24    17    22    13    16    15    19

MTB > TTEST  MU = 16,  DATA IN C1;
SUBC> ALTERNATIVE = +1.

TEST OF MU = 16.000 VS MU G.T. 16.000

              N      MEAN     STDEV   SE MEAN       T    P VALUE
C1           10    17.200     3.967     1.254    0.96       0.18
```

FIGURE 8.15 A *Minitab* computer printout of the one-tailed t test for Example 8.11.

we had wanted a left-tailed test, the proper subcommand would have been ALTERNATIVE = −1.) As you can see, the printout also produces the p value obtained when this formula is used:

$$p \text{ value (right-tailed test)} = P(t > \text{TR}) = P(t > .96)$$

Again, we can't find the exact p value for this example from Appendix 4, but Figure 8.15 tells us it is .18. And since $.18 > .01$, we know that Mollie should accept the H_0.

We've now discussed one-sample hypothesis tests of means under various conditions. The *classical testing procedure* is similar under all conditions, and the testing differences that exist are reflected in the differences that appear in the decision rules. These differences are summarized in Table 8.1. The *p-value testing procedure* uses the same decision rules in all tests (reject H_0 if the p value is $<\alpha$. Otherwise, accept H_0).

TABLE 8.1 DECISION RULES UNDER VARIOUS CONDITIONS WITH CLASSICAL HYPOTHESIS TESTING OF MEANS

$n > 30$, or σ Known and Population Values Known to Be Normally Distributed	$n \leq 30$, σ Unknown, and Population Values Known to Be Normally Distributed
TWO-TAILED TEST	
Reject H_0 and accept H_1 if TR $> +z$ value or TR $< -z$ value. Otherwise, acccept H_0.	Reject H_0 and accept H_1 if TR $> +t_{\alpha/2}$ value or TR $< -t_{\alpha/2}$ value. Otherwise, accept H_0.
LEFT-TAILED TEST	
Reject H_0 and accept H_1 if TR $< -z$ value. Otherwise, accept H_0.	Reject H_0 and accept H_1 if TR $< -t_{\alpha}$ value. Otherwise, accept H_0.
RIGHT-TAILED TEST	
Reject H_0 and accept H_1 if TR $> +z$ value. Otherwise, accept H_0.	Reject H_0 and accept H_1 if TR $> +t_{\alpha}$ value. Otherwise, accept H_0.

Self-Testing Review 8.2

1–3. Assume $n > 30$ or σ is known and the population distribution is normal. Determine the critical z value for each of the following types of hypothesis tests:

1. Right-tailed test, $\alpha = .01$.

2. Two-tailed test, $\alpha = .05$.

3. Left-tailed test, $\alpha = .05$.

4–6. Determine the critical t value to be used for the following tests:

4. $n = 13$, left-tailed test, $\alpha = .05$.

5. $n = 21$, two-tailed test, $\alpha = .01$.

6. $n = 9$, right-tailed test, $\alpha = .05$.

7–10. For each of the following, determine if the use of a t or z distribution is appropriate for the hypothesis test. Find the critical t or z value.

7. $n = 19$, σ is unknown and population distribution is normal, left-tailed test, $\alpha = .05$.

8. $n = 11$, σ is known and population distribution is normal, right-tailed test, $\alpha = .01$.

9. $n = 34$, σ is unknown, two-tailed test, $\alpha = .01$.

10. $n = 12$, σ is unknown and population distribution is normal, left-tailed test, $\alpha = .05$.

11. What is the meaning of the term "p value" for a right-tailed hypothesis test? What is the meaning for a left-tailed test? For a two-tailed test?

12. How does a p-value procedure differ from the classical hypothesis-testing procedure?

13. The level of significance for a test is .05, and the p value is .07. What is the decision? (Do you reject or fail to reject (accept) the null hypothesis?)

14. The level of significance for a test is .01, and the p value is .004. What is the decision?

15. The p value for a test is .23. What decision is made at the .05 level of significance? At the .01 level?

16. The p value for a test is .002. What decision is made at the .05 level of significance? At the .01 level?

17. The p value for a test is .02. What decision is made at the .05 level of significance? At the .01 level?

18–20. For each of the following, use the seven-step classical hypothesis-testing procedure (and where necessary, assume the population is normally distributed):

18. Test the claim that μ is equal to 50 at the .05 level of significance. Use sample data of $n = 45$, $\bar{x} = 48.3$, $\sigma = 5.1$.

from the hypothetical mean than the specific test ratio. The "further away" direction depends on whether we have a right-tailed, left-tailed, or two-tailed test.

13. Since the *p* value is $>\alpha$, you accept the H_0. (The test ratio does not fall in the critical region.)

15. Since .23 is > both .01 and .05, you fail to reject (accept) the H_0 at both levels of significance.

17. Since .02 is < .05, the H_0 is rejected at the .05 level. Since .02 is >.01, you accept the H_0.

19. Step 1: State the null and alternative hypotheses. H_0: $\mu = 231$, and H_1: $\mu > 231$. *Step 2: Select the level of significance.* $\alpha = .01$. *Step 3: Determine the test distribution to use.* We use a *t* distribution with 17 df. *Step 4: Define the critical or rejection region(s).* The value for a one-tailed test at the .01 level is $t = 2.567$. *Step 5: State the decision rule.* Reject H_0 and accept H_1 if the TR is >2.567. Otherwise, accept H_0. *Step 6: Compute the test ratio (TR).* The standard error is $15.7/\sqrt{18} = 3.70$, so the TR = (235.3 − 231)/3.70 = 1.162. *Step 7: Make the statistical decision.* Since the TR value of 1.162 does not fall in the rejection region, we accept the H_0.

21. Step 1: State the null and alternative hypotheses. H_0: $\mu = 32$, and H_1: $\mu \neq 32$. *Step 2: Select the level of significance.* $\alpha = .05$. *Step 3: Determine the test distribution to use.* The *z* distribution is used. *Step 4: Define the critical or rejection region(s).* The values for a two-tailed test at the .05 level are $z = -1.96$ and $z = +1.96$. *Step 5: State the decision rule.* Reject H_0 and accept H_1 if the TR is < −1.96 or > +1.96. Otherwise, accept H_0. *Step 6: Compute the test ratio (TR).* The TR = (34.4 − 32)/1.314 = 1.83. *Step 7: Make the statistical decision.* Since the TR value of 1.83 does not fall in the rejection region, we accept the H_0 that mean score of patients with dementia is 32.

23. Step 1: H_0: $\mu = 15.08$, and H_1: $\mu < 15.08$. *Step 2:* $\alpha = .01$. *Step 3:* Use the *t* distribution with 26 df. *Step 4:* The value for this left-tailed test at the .01 level is $t = -2.479$. *Step 5:* Reject H_0 and accept H_1 if the TR is < −2.479. Otherwise, accept H_0. *Step 6:* The TR = (14.44 − 15.08)/.639 = −1.00. *Step 7:* Since the TR value of −1.00 does not fall in the rejection region, we accept the H_0 that women rate advancement at the same level as do men.

25. Step 1: H_0: $\mu = 3$, and H_1: $\mu \neq 3$. *Step 2:* $\alpha = .05$. *Step 3:* The *z* distribution is used. *Step 4:* The values for a two-tailed test at the .05 level are $z = \pm 1.96$. *Step 5:* Reject H_0 and accept H_1 if the TR is < −1.96 or > +1.96. Otherwise, accept H_0. *Step 6:* The TR = (3.07 − 3.0)/.09699 = .72. *Step 7:* Since the TR value of .72 does not fall in the rejection region, we accept the H_0 and decide that the mean rating is 3.

27. Step 1: H_0: $\mu = \$300$, and H_1: $\mu > \$300$. *Step 2:* $\alpha = .05$. *Step 3:* Use the *t* distribution with 18 df. *Step 4:* The value for a right-tailed test at the .05 level is $t = 1.734$. *Step 5:* Reject H_0 and accept H_1 if the TR is > 1.734. Otherwise, accept H_0. *Step 6:* The TR = (338.2 − 300)/36.9 = 1.03. *Step 7:* Since the TR value of 1.03 does not fall in the rejection region, we accept the H_0 and decide that cellular phones cost an average of $300 at the time of the *Consumer Reports* survey.

29. Step 1: H_0: $\mu = 2.2$, and H_1: $\mu > 2.2$. *Step 2:* $\alpha = .01$. *Step 3:* Use the *z* distribution. *Step 4:* The value for a right-tailed test at the .01 level is $z = 2.33$. *Step 5:* Reject H_0 and accept H_1 if the TR is > 2.33. Otherwise, accept H_0. *Step 6:* The TR = (4.9 − 2.2)/.2182 = 12.37. *Step 7:* Since the TR value of 12.37 falls far out in the rejection region, we reject the H_0 and conclude that the surgery improves preoperative functioning.

31. Step 1: H_0: $\mu = 84.2$, and H_1: $\mu < 84.2$. *Step 2:* $\alpha = .01$. *Step 3:* Use the *t* distribution with 28 df. *Step 4:* The value for a left-tailed test at the .01 level is $t = -2.467$. *Step 5:* Reject H_0 and accept H_1 if the TR is < −2.467. Otherwise, accept H_0. *Step 6:* The TR = (69.1 − 84.2)/2.90 = −5.21. *Step 7:* Since the TR value of −5.21 falls in the rejection region, we reject the H_0 and decide that patients who take *Captopril* have lower blood pressure than those who do not.

33. Step 1: State the null and alternative hypotheses. H_0: $\mu = 19.43$, and H_1: $\mu > 19.43$. *Step 2: Select the level of significance.* $\alpha = .01$. *Step 3: Determine the test distribution to use.* The *z* distribution is used. *Step 4: State the decision rule.* Reject H_0 and accept H_1 if the *p*-value is <.01. Otherwise, accept H_0. *Step 5: Compute the test ratio (TR).* The TR = (29.85 − 19.43)/3.021 = 3.44. *Step 6: Compute the appropriate p value.* Although the corresponding area does not appear in the *z* table, we know that the area to the right of $z = 3.44$ is less than the area to the right of $z = 3.09$ which is on the table and is equal to .5 − .4990 = .0010. Thus, $p < .0010$. *Step 7: Make the statistical decision.* The *p* value for this test is <.0010, which in turn is $<\alpha = .01$. Following the decision rule, we reject the H_0.

35. Step 1: H_0: $\mu = 46.3$, and H_1: $\mu \neq 46.3$. *Step 2:* $\alpha = .05$. *Step 3:* Use the *z* distribution.

Step 4: Reject H_0 and accept H_1 if the p value is $<.05$. Otherwise, accept H_0. *Step 5*: The TR = $(32.78 - 46.3)/1.9092 = -7.08$. *Step 6*: Although the corresponding area does not appear in the table, we know that twice the area to the left of $z = -7.08$ is <twice the area to the left of $z = -3.09$ [which is in the table and is equal to $.5 - .4990$ or $.0010$, and $2(.0010) = .0020$]. Thus, $p < .0020$. *Step 7*: The p value for this test is $<.0020$, which is < the .05 level of significance. Following the decision rule, we reject the H_0.

Section 8.3

1. *Step 1: State null and alternative hypotheses.* H_0: $\pi = 35$ percent, H_1: $\pi > 35$ percent. *Step 2: Select level of significance.* $\alpha = .01$. *Step 3: Determine test distribution to use.* Use the z distribution. *Step 4: Define the critical or rejection region(s).* The value for a right-tailed test at the .01 level is $z = 2.33$. *Step 5: State the decision rule.* Reject H_0 and accept H_1 if the TR > 2.33. Otherwise, accept H_0. *Step 6: Compute the test ratio (TR).* The $\sigma_p = \sqrt{[(35)(65)]/114} = 4.4672$, and the TR = $(36.8421 - 35)/4.4672 = .4124$. *Step 7: Make the statistical decision.* Since the TR value of .4124 does not fall in the rejection region, we accept the H_0.

3. *Step 1: State null and alternative hypotheses.* H_0: $\pi = 50$ percent, H_1: $\pi < 50$ percent. *Step 2: Select level of significance.* $\alpha = .05$. *Step 3: Determine test distribution to use.* Use the z distribution. *Step 4: Define the critical or rejection region(s).* The value for a left-tailed test at the .05 level is $z = -2.33$. *Step 5: State the decision rule.* Reject H_0 and accept H_1 if the TR is < -2.33. Otherwise, accept H_0. *Step 6: Compute the test ratio (TR).* The TR = $(48.02 - 50)/1.759 = -1.1256$. *Step 7: Make the statistical decision.* Since the TR value of -1.1256 does not fall in the rejection region, we accept the H_0.

5. *Step 1*: H_0: $\pi = 34$ percent, H_1: $\pi > 34$ percent. *Step 2*: $\alpha = .05$. *Step 3*:. Use the z distribution. *Step 4*: The value for a right-tailed test at the .05 level is $z = 1.645$. *Step 5*: Reject H_0 and accept H_1 if the TR is >1.645. Otherwise, accept H_0. *Step 6*:. The TR = $(45.99 - 34)/.5818 = 20.61$. *Step 7*: Since the TR value of 20.61 falls way out in the rejection region, we reject the H_0 that the population percentage is 34 percent.

7. *Step 1*: H_0: $\pi = 6$ percent, H_1: $\pi > 6$ percent. *Step 2*: $\alpha = .01$. *Step 3*:. Use the z distribution. *Step 4*: The value for a right-tailed test at the .01 level is $z = 2.33$. *Step 5*: Reject H_0 and accept H_1 if the TR is >2.33. Otherwise, accept H_0. *Step 6*:. The TR = $(8.772 - 6)/.5743 = 4.8267$. *Step 7*: Since the TR value of 4.8267 falls in the rejection region, we reject the H_0 that the population percentage is 6 percent.

9. *Step 1*: H_0: $\pi = 80$ percent, H_1: $\pi > 80$ percent. *Step 2*: $\alpha = .05$. *Step 3*:. Use the z distribution. *Step 4*: The value for a right-tailed test at the .05 level is $z = 1.645$. *Step 5*: Reject H_0 and accept H_1 if the TR is >1.645. Otherwise, accept H_0. *Step 6*:. The TR = $(85.0146 - 80)/1.08148 = 4.6368$. *Step 7*: Since the TR value of 4.6368 falls in the rejection region, we reject the H_0. It appears that more than 80 percent are in favor of the waiting period.

11. *Step 1*: H_0: $\pi = 40$ percent, H_1: $\pi > 40$ percent. *Step 2*: $\alpha = .05$. *Step 3*:. Use the z distribution. *Step 4*: The value for a right-tailed test at the .05 level is $z = 1.645$. *Step 5*: Reject H_0 and accept H_1 if the TR is >1.645. Otherwise, accept H_0. *Step 6*:. The TR = $(47.9167 - 40)/5 = 1.5833$. *Step 7*: Since the TR value of 1.5833 does not fall in the rejection region, we accept the H_0.

13. *Step 1*: H_0: $\pi = 15$ percent, H_1: $\pi \neq 15$ percent. *Step 2*: $\alpha = .01$. *Step 3*:. Use the z distribution. *Step 4*: The values for a two-tailed test at the .01 level are $z = \pm 2.575$. *Step 5*: Reject H_0 and accept H_1 if the TR is < -2.575 or $> +2.575$. Otherwise, accept H_0. *Step 6*:. The TR = $(14.0502 - 15)/.9560 = -0.9935$. *Step 7*: Since the TR value of $-.9935$ does not fall in the rejection region, we accept the H_0.

Section 8.4

1. Use χ^2 with 12 − 1 or 11 df. Since a .05 area must lie to the left of the critical value, .95 must lie to the right. Look under the column headed .95 to find $\chi^2 = 4.57$.

3. Use χ^2 with 24 − 1 or 23 df. Since a .01 area must lie to the right, read directly under the .01 column to see that $\chi^2 = 41.6$.

5. *Step 1: State null and alternative hypotheses.* H_0: $\sigma = 4.26$, and H_1: $\sigma > 4.26$. *Step 2: Select level of significance.* $\alpha = .05$. *Step 3: Determine test distribution to use.* Use the χ^2 distribution with 23 df. *Step 4: Define the critical or rejection region(s).* For a right-tailed test, look under the .05 column to find $\chi^2 = 35.2$. *Step 5: State the decision rule.* Reject H_0 and accept H_1 if the TR is >35.2. Otherwise, accept H_0. *Step 6: Compute the test ratio (TR).* The TR = $23(5.317)/(4.26)^2 = 6.739$. *Step 7*:

Make the statistical decision. Since the TR of 6.739 does not fall in the rejection region, we accept the H_0.

7. *Step 1: State null and alternative hypotheses.* H_0: $\sigma^2 = 4$, H_1: $\sigma^2 < 4$. *Step 2: Select level of significance.* $\alpha = .01$. *Step 3: Determine test distribution to use.* Use the χ^2 distribution with 13 df. *Step 4: Define the critical or rejection region(s).* For a left-tailed test, look under the .99 column to find $\chi^2 = 4.11$. *Step 5: State the decision rule.* Reject H_0 and accept H_1 if the TR is <4.11. Otherwise, accept H_0. *Step 6: Compute the test ratio (TR).* The TR = 13(3.92)/4 = 12.74. *Step 7: Make the statistical decision.* Since the TR of 12.74 does not fall in the rejection region, we accept the H_0.

9. *Step 1*: H_0: $\sigma^2 = 0.5$, and H_1: $\sigma^2 < 0.5$. *Step 2*: $\alpha = .01$. *Step 3*: Use the χ^2 distribution with 15 df. *Step 4*: $\chi^2 = 5.23$. *Step 5*: Reject H_0 and accept H_1 if the TR is <5.23. Otherwise, accept H_0. *Step 6*: The TR = 15(.35)/0.5 = 10.5. *Step 7*: Since the TR of 10.5 does not fall in the rejection region, we accept the H_0.

11. *Step 1*: H_0: $\sigma = 90$, and H_1: $\sigma > 90$. *Step 2*: $\alpha = .05$. *Step 3*: Use the χ^2 distribution with 23 df. *Step 4*: $\chi^2 = 35.2$. *Step 5*: Reject H_0 and accept H_1 if the TR is >35.2. Otherwise, accept H_0. *Step 6*: The TR = $23(113)^2/(90)^2$ = 36.2577. *Step 7*: Since the TR of 36.2577 falls in the rejection region, we reject the H_0.

13. *Step 1*: H_0: $\sigma = 15$, and H_1: $\sigma > 15$. *Step 2*: $\alpha = .01$. *Step 3*: Use the χ^2 distribution with 25 df. *Step 4*: $\chi^2 = 44.3$. *Step 5*: Reject H_0 and accept H_1 if the TR is >44.3. Otherwise, accept H_0. *Step 6*: The TR = $25(16.1)^2/(15)^2$ = 28.8011. *Step 7*: Since the TR of 28.8011 does not fall in the rejection region, we accept the H_0.

CHAPTER 9

Testing Hypotheses: Two-Sample Procedures

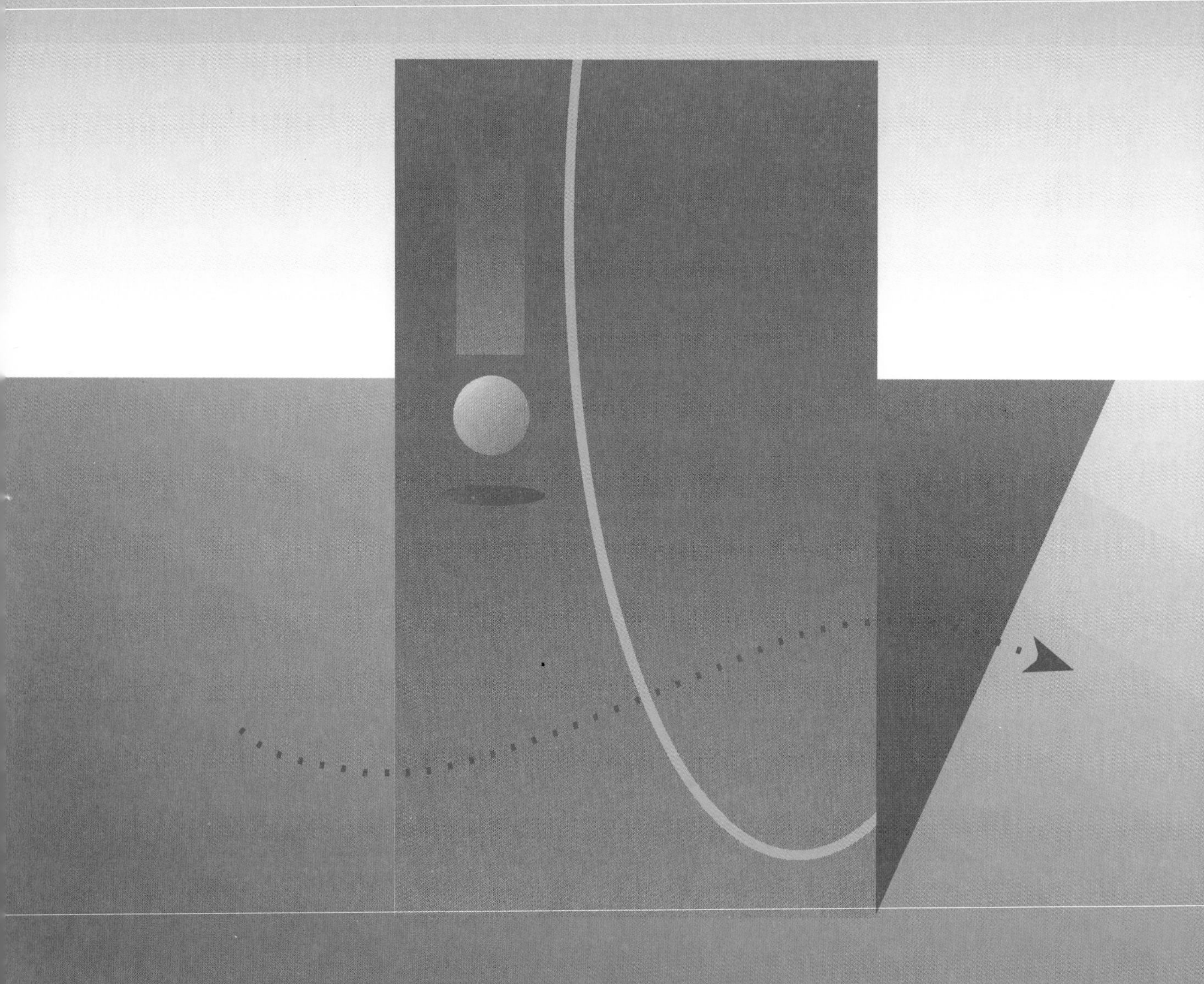

LOOKING AHEAD

In this chapter we'll use the data obtained from two samples taken from different populations. And the hypothesis-testing concepts introduced in Chapter 8 are used to make relative comparisons between (1) two population variances, (2) two population means, and (3) two population percentages. (Since the results obtained in comparing two variances may be needed during an analysis of two means, we'll consider variances first in this chapter.)

Thus, after studying this chapter, you should be able to:

- Explain the purpose of two-sample hypothesis tests of variances, means, and percentages.
- Understand the procedures to be followed in conducting these tests of variances, means, and percentages.
- Perform the necessary computations and make the appropriate statistical decisions in two-sample hypothesis-testing situations.

9.1 Hypothesis Tests of Two Variances

Some General Thoughts

Decision makers often want to see if two populations are similar or different with respect to some characteristic. For example, an instructor may want to know if male professors receive higher salaries than female professors for the same teaching load. Or a psychologist may want to see if an experimental group responds differently than a control group to an experimental stimulus. Or the purchasing agent of a firm that manufactures cooling towers may need to know if the cooling fan motors of one supplier are more durable than those of another vendor. In short, there are many situations that require groups to be compared on the basis of a given trait.

Chapter 8 showed us ways to test the validity of an assumed value of a parameter. The assumed value was a single quantity that was subjected to statistical testing. In this chapter, though, we're concerned with the parameters of two different populations, but we're not primarily interested in estimating the *absolute values* of the parameters. Rather, the topic of interest is the *relative values* of the parameters. That is, does one population appear to possess more or less of a trait than the other? Our purpose in this chapter, then, is to use the data from two samples obtained from two populations to see if there's likely to be a statistically significant difference between the parameters of two populations.

Two-Variance Testing: Purpose and Assumptions

And our *purpose in this section* is to use sample variance (s^2) data to arrive at conclusions about the corresponding population variance (σ^2) parameters. Thus, we'll take random samples from two populations, compute the variance for each sample

The Smiling Soothsayers

"Economists state their GNP growth projections to the nearest tenth of a percentage point to prove that they have a sense of humor."
—*E.R. Fielder*

"Give them a number or give them a date, but never both."
—*E.R. Fielder*

"It is surprising that one soothsayer can look at another without smiling."
—*Cicero*

"I have been increasingly moved to wonder whether my job is a job or a racket, whether economists . . . should cover their faces or burst into laughter when they meet on the street."
—*Frank H. Knight in an address to the American Economic Association*

data set, and use the results obtained as the basis for comparing the population variances (and standard deviations).

The following two assumptions must be valid if the testing procedure we'll follow is to produce usable results:

1. The data in the two populations we sample must be *normally distributed.*
2. The data *source* (persons, objects, and so on) in the first population must be *independent* of the data source in the second population. Thus, we assume that *independent samples* are used:

A sample selected from the first population is said to be an **independent sample** if it isn't related in some way to the data source found in the second population. If, however, the same (or related) data sources are used to generate the data sets for each population, then the samples taken from each population are said to be **dependent samples.**

If our populations consist of two groups of students in two separate sections of a statistics course, and if we randomly select samples from each population and test the knowledge in both classes, then we have independent samples. But if we test the knowledge of a sample of statistics students *before* they take an intensive review course, and retest the same students *after* the course is completed to evaluate the difference in scores for each student, then we have dependent samples. Such "before-and-after" tests usually involve dependent samples. There are ways to handle such situations (as we'll soon see), but the procedure to test two variances described in this section requires independent samples.

The Testing Procedure

The procedure used to test hypotheses about the variances of two populations follows the seven familiar steps outlined in Chapter 8. We'll focus on the *classical procedure* in this chapter, but remember that *p-value tests* produce the same results.

Step 1: State the Null and Alternative Hypothesis The *null hypothesis* in a test of two variances is that there's *no difference* in the variability of the two populations. That is:

$$H_0 : \sigma_1^2 = \sigma_2^2$$

The *alternative hypothesis* is either that there *is* a significant difference between the population variances or that one variance identified by the analyst is greater than the other. Thus, the alternative hypothesis takes one of these familiar forms:

$$H_1 : \sigma_1^2 \neq \sigma_2^2$$
$$H_1 : \sigma_1^2 > \sigma_2^2$$
$$H_1 : \sigma_1^2 < \sigma_2^2$$

Step 2: Select the Level of Significance The analyst must choose the value of α.

Step 3: Determine the Test Distribution to Use We've used the z, t, and χ^2 probability distributions at different times in Chapters 7 and 8. But now we must consider a new test distribution. To get a feel for this new test distribution, let's assume that the null hypothesis is *true*, and σ_1^2 does equal σ_2^2. In that case, the ratio $\sigma_1^2/\sigma_2^2 = 1$. And if the H_0 is true and if we compute variances for the samples taken from each population, then the ratio s_1^2/s_2^2 should also yield a result that doesn't deviate too far from a value of 1.

Of course, sampling variation can be expected to cause *some* disparity in the two sample variances even when the H_0 is true. But how much disparity can be accepted—that is, how far from 1 can the ratio stray—before it must be concluded that the H_0 *isn't* true? The answer to this question is found through the use of an *F distribution*:

F Distribution

An ***F* distribution** is the sampling distribution for the variable s_1^2/s_2^2, and it has the following properties:

- There are no negative values in an *F* distribution, so the scale of possible *F* values extends from 0 to the right in a positive direction.
- An *F* distribution isn't symmetrical like the *z* or *t* distributions; rather, it is skewed to the right like a χ^2 distribution.
- There are many *F* distributions, and each one is determined by the number of samples and the number of observations in the samples. Thus, the *F* distribution used to compare two samples of size 8 and 9 is different from the ones used to compare samples of 9 and 10, 6 and 7, 15 and 15, and so on.

Figure 9.1 shows the general shape of *F* distributions. The *F*-distribution values we'll need for our two-variance testing procedure are given in the tables in Appendix 5 at the back of the book. To use these tables to look up the critical *F* values we need, we must know three things:

1. The *level of significance* (specified in ***step 2*** of our testing procedure).
2. The *degrees of freedom* (df) for the sample used in the *numerator* of our test ratio s_1^2/s_2^2. To use the *F* tables in Appendix 5 if our alternative hypothesis is $H_1{:}\sigma_1^2 \neq \sigma_2^2$, we *specify that the sample with the largest sample variance is designated as sample 1*, and this larger variance is always used in the numerator of our test ratio. If we want a one-tailed test, the simplest way to run the test is to always designate the alterna-

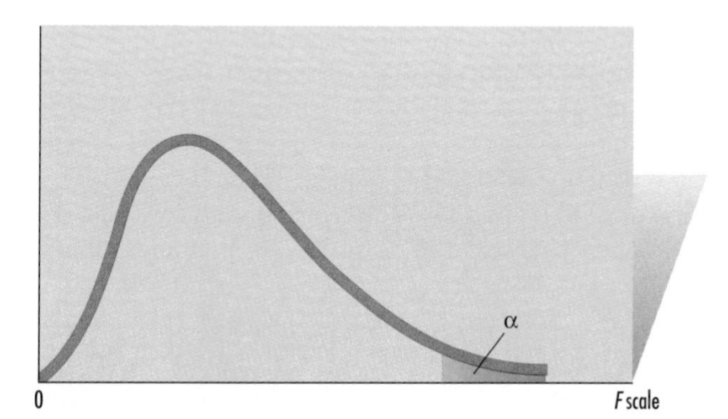

FIGURE 9.1 The general shape of *F* distributions.

tive hypothesis as $H_1{:}\sigma_1^2 > \sigma_2^2$ (this might require us to interchange sample 1 and sample 2). Then the test ratio will be s_1^2/s_2^2. In either case, the degrees of freedom for the *numerator*, then, is $n_1 - 1$.

3. The *degrees of freedom* (df) for the sample used in the *denominator* of our test ratio. This value is $n_2 - 1$.

Step 4: Define the Rejection or Critical Regions Let's assume we're conducting a two-tailed test at the .05 level of significance. There are 10 items in sample 1 and 11 items in sample 2. The .025 areas in each tail of the *F* distribution (a total of .05) are the rejection regions, and the remaining .95 area is the acceptance region. To find the critical *F* value we need in this case, we look in Appendix 5 and locate the "critical values" table for $\alpha = .025$ (these are one-tailed tables).

As you've seen, to use the *F*-distribution tables in Appendix 5, you must also know the df values for numerator and denominator. In our example, $\text{df}_{\text{num}} = 10 - 1$ or 9, and $\text{df}_{\text{den}} = 11 - 1$ or 10. You'll notice in Appendix 5 that the table *columns* correspond to degrees of freedom in the numerator, and the *rows* represent degrees of freedom in the denominator. So in the table for $\alpha = .025$, the critical *F* value found at the intersection of the column numbered 9 and the row numbered 10 is 3.78. Thus, a test ratio with a value that exceeds 3.78 in this case falls into the rejection region.

Step 5: State the Decision Rule We can now state the *decision rule* for this situation as follows:

Reject H_0 and accept H_1 if the test ratio (TR) is > 3.78. Otherwise, accept H_0.

Step 6: Compute the Test Ratio The formula needed to compute the test ratio (*F* value) is:

$$\text{TR} = \frac{s_1^2}{s_2^2} \tag{9.1}$$

Step 7: Make the Statistical Decision If the TR is less than 3.78 in this instance, the null hypothesis that $\sigma_1^2 = \sigma_2^2$ is accepted.

Now let's look at an example problem and follow the steps we've just outlined.

Example 9.1. Let's suppose that two experimental diets designed to add weight to malnourished third-world children are being tested. It's assumed that each diet will produce weight gains that are normally distributed. Although factors such as the average weight gain produced by each diet and each diet's cost per serving are obviously important, the dietitian in charge of the testing also wants to know if there is a significant difference in the variability of the weight gains produced by each diet.

To test this variability, the first diet (A) is given to 8 children, and the second diet (B) is supplied to 9 suffering from hunger. The weight gains for diet A (in pounds) after a 6-week period are:

4.1 4.3 6.0 5.6 8.5 7.9 5.1 4.9

And the gains (in pounds) made by those fed diet B during the same time are:

7.3 6.7 8.3 7.0 6.6 6.8 9.2 7.6 5.9

Table 9.1 presents these two sample data sets and shows the computation of the sample variance values that we'll soon need.

TABLE 9.1 WEIGHT GAIN (IN POUNDS)

Diet A			Diet B		
x	$(x-\bar{x})$	$(x-\bar{x})^2$	x	$(x-\bar{x})$	$(x-\bar{x})^2$
4.1	−1.7	2.89	7.3	.03	.0009
4.3	−1.5	2.25	6.7	−.57	.3249
6.0	.2	.04	8.3	1.03	1.0609
5.6	−.2	.04	7.0	−.27	.0729
8.5	2.7	7.29	6.6	−.67	.4489
7.9	2.1	4.41	6.8	−.47	.2209
5.1	−.7	.49	9.2	1.93	3.7249
4.9	−.9	.81	7.6	.33	.1089
46.4	0	18.22	5.9	−1.37	1.8769
			65.4	0	7.8401

$$\bar{x}_1 = \frac{\Sigma x}{n} = \frac{46.4}{8} = 5.80 \text{ pounds} \qquad \bar{x}_2 = \frac{\Sigma x}{n} = \frac{65.4}{9} = 7.27 \text{ pounds}$$

$$s_1^2 = \frac{\Sigma (x-\bar{x})^2}{n-1} = \frac{18.22}{8-1} = 2.6029 \qquad s_2^2 = \frac{\Sigma (x-\bar{x})^2}{n-1} = \frac{7.8401}{9-1} = .9800$$

The dietitian's interest now is to see if there's a significant difference in the variability of the weight gains produced by each diet. Thus, the following *Null Hypothesis* (*Step 1*) is that such a difference *doesn't* exist:

$$H_0 : \sigma_1^2 = \sigma_2^2$$

The *Alternative Hypothesis* is simply that there *is* a significant difference between the population variances. Thus, the alternative hypothesis takes the familiar *two-tailed test* form:

$$H_1 : \sigma_1^2 \neq \sigma_2^2$$

The dietitian specifies that $\alpha = .05$ (*Step 2*), and, of course, we'll use the F distribution in this test (*Step 3*). As you can see in Table 9.1, sample 1 with the larger variance of 2.6029 consists of the 8 children fed diet A, and sample 2 with the smaller variance of .9800 is the 9 children receiving diet B. Thus, the size of sample 1 is 8 ($n_1 = 8$), and the size of sample 2 is 9 ($n_2 = 9$). The degrees of freedom for the *numerator*, then, is $n_1 - 1$, or $8 - 1$, or 7, and the degrees of freedom for the *denominator* is $n_2 - 1$, or $9 - 1$, or 8.

Figure 9.2 shows the areas of acceptance and rejection of interest to us in our test involving diets A and B. With $\alpha = .05$ and with a two-tailed test (remember, $H_1 : \sigma_1^2 \neq \sigma_2^2$), we need to find the appropriate F value that separates the .025 area in the right tail from the rest of the distribution. Now you can see why we stipulated earlier that the sample with the largest variance is always labeled sample 1 and that the larger sample variance is always placed in the numerator of the test ratio. When set up in this way, the computed ratio will always be 1 or greater, and we need only find the F value that separates the 2.5 percent of the area in the right tail from the remaining 97.5 percent of the area. Thus, in the table for $\alpha = .025$, the critical F value (Step 4) found at the intersection of the column numbered 7 and the row numbered 8 is 4.53 (see Figure 9.2).

Decision Rule (Step 5):

Reject H_0 and accept H_1 if the TR is > 4.53. Otherwise, accept H_0.

And from Table 9.1 we see that $s_1^2 = 2.6029$ and $s_2^2 = .9800$.

Test Ratio (Step 6):

$$\text{TR} = \frac{s_1^2}{s_2^2} = \frac{2.6029}{.9800} = 2.656$$

Conclusion (Step 7):

Since the TR of 2.656 is less than 4.53, we accept the null hypothesis that $\sigma_1^2 = \sigma_2^2$. There doesn't appear to be a significant difference in the variability of the weight gains produced by the two experimental diets.

Suppose the dietitian wants to conduct a *one-tailed test* so that the alternative hypothesis reads:

$H_1 : \sigma_1^2 > \sigma_2^2$.

In this case the entire α value of .05 is in the right tail of the F distribution, and the decision rule is:

Reject H_0 and accept H_1 if TR > 3.50 (.05 table, $\text{df}_{\text{num}} = 7$, $\text{df}_{\text{den}} = 8$). Otherwise, accept H_0.

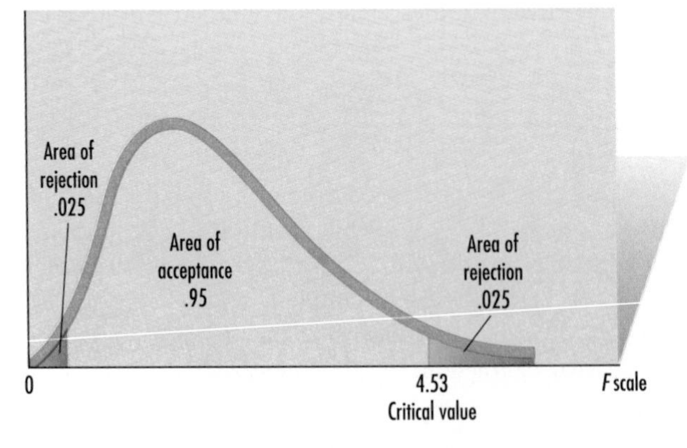

FIGURE 9.2 The critical F value of 4.53 defines the acceptance and rejection regions of the F distribution with 7 degrees of freedom in the numerator and 8 degrees of freedom in the denominator.

Since the TR of 2.656 is < 3.50, the decision is still to accept the H_0 in this one-tailed test at the .05 level.

Self-Testing Review 9.1

1. What are the two necessary conditions that must be satisfied if the testing procedure for equal variances is to produce usable results?
2. Use a right-tailed test and test the equality of population variances at the .01 level of significance given the following data: For sample A, $n = 21$, $s^2 = 35.43$, and $\bar{x} = 67.47$. And for sample B, $n = 13$, $s^2 = 48.91$, and $\bar{x} = 307.84$.
3. Use a two-tailed test and test the equality of population variances at the .05 level of significance given the following data: For sample A, $n = 16$, $s^2 = 457.21$, and $\bar{x} = 247.16$. And for sample B, $n = 7$, $s^2 = 99.08$, and $\bar{x} = 247.27$.
4. The length of time customers at Shop N Pay must wait in line before they can leave the checkout station with their purchases is measured. For a random sample of 7 customers, the sample variance is 17.84. The competition down the street, Buy Fair, uses a different type of checkout system. The length of time 25 customers wait in line at Buy Fair is measured, and the variance is 15.93. Test the hypothesis at the .05 level that there is less population variance at Buy Fair.
5. On-The-Ball, Inc., makes ball bearings that are used in tractors and other equipment. On the 8-to-4 shift, a random sample of 16 ball bearings is selected and the diameters are measured. The variance is 17.39. Later, a random sample of 13 ball bearings is selected from the 4-to-midnight shift, and the variance for the diameter measures is found to be 12.83. Test the hypothesis at the .05 level that the population variances for both shifts are equal.
6. Participants in a consumer product research study published in the *Journal of Consumer Research* were randomly divided into two groups, given 4 brands of orange juice to sample, and then asked to rate the quality of each brand. The 29 subjects in the first group were given 1 juice brand to drink at a time. About 5 minutes elapsed between consecutive samples, during which time the tasters were asked to eat a cracker. In the second group, the 29 subjects were given all 4 brands simultaneously and were instructed to go back and forth between the different brands and were not provided with crackers. The taste ratings for each brand are given below for each method of tasting:

Brand	One Brand at a Time	All Brands Simultaneously
A	6.31	7.38
D	5.53	5.04
B	3.34	2.92
C	1.88	.75

Test the hypothesis at the .05 level that the taste ratings will have equal population variances when subjects are given one brand to test at a time and when they are given all 4 brands simultaneously.

7. *Fortune* magazine recently listed "America's Most Admired Corporations." A random sample of senior executives was asked to rate the 10 largest companies in their own industry based on specified attributes. For the computers and office equipment industry, the companies and their ratings were:

Hewlett Packard	7.34	Apple Computer	6.96
Sun Microsystems	6.86	Compaq Computer	6.53
IBM	6.50	NCR	6.03
Digital Equipment	6.00	Pitney Bowes	5.82
Unisys	3.32	Wang Labs	3.17

For the top 10 in the electronics industry, the companies were rated by a sample of executives as follows:

General Electric	7.67	Motorola	7.42
Emerson Electric	7.04	Raytheon	6.48
Cooper Industries	6.33	Whirlpool	6.29
Texas Instruments	6.12	Rockwell Internatl.	6.05
TRW	6.04	Westinghouse Electric	5.65

Test the hypothesis at the .05 level that the population variance for ratings in the computer and office equipment industry is equal to the variance in the electronics industry.

8. A pharmaceutical company makes a *Weight-Away* tablet that is advertised to control appetite for 10 hours. It is in direct competition with the popular *Fatnomore*. The duration of time each tablet is effective was measured in clinical tests. A random sample of 14 people who tested *Weight-Away* had a variance of 3.92. For a similar sample of 7 people who tested *Fatnomore*, the variance was 7.21. Test the hypothesis at the .05 level that the population variances for both products are equal.

9. An experiment reported in the *American Journal of Public Health* was conducted to evaluate the effectiveness of a work-site health promotion program in reducing obesity as measured by a body mass index (BMI). A random sample of 16 work sites received classes on weight control combined with payroll incentives. Another sample of 16 work sites served as a control and received no classes or incentives. The following data represent the mean BMI for each site in the control group after 2 years:

26.06 26.40 25.53 26.28 25.39 25.69 26.12 26.24
26.53 26.37 26.22 26.42 25.57 24.94 25.95 26.47

The mean BMI for each site in the treatment group after 2 years is:

26.02 25.87 25.02 25.46 25.70 26.10 26.24 26.57
24.57 25.18 26.84 26.31 26.22 25.61 26.42 25.16

Test the hypothesis at the .05 level that after 2 years there was no difference between the population variance of the treatment group and the control group.

10. A machine is designed to fill boxes of cereal, and the boxes state that the contents weigh 18 ounces. When the conveyer belt that the boxes travel on was set at a speed of 2 inches per second, a random sample of 28 boxes had a variance of 1.3. The production manager decided to speed up the conveyor belt to 5 inches per second. When this was done, a random sample of 16 boxes had a variance of 4.9. Test the hypothesis at the .05 level that when the conveyor belt is set at the faster speed, the population variance in the amount of cereal in the boxes will be greater.

11. Participants in a consumer product research study were randomly divided into two groups, given 4 brands of grape juice to sample, and then asked to rate the quality of each brand. The 25 subjects in the first group were given 1 juice brand to drink at a time. About 5 minutes elapsed between consecutive samples, during which time the tasters were asked to eat a cracker. In the second group, the 25 subjects were given all 4 brands simultaneously and were instructed to go back and forth between the different brands and were not provided with crackers. The first group (one juice at a time) had a mean taste quality score for brand A of 6.31 (on a scale of 0 to 10) and a standard deviation of 1.2. The second group (all brands simultaneously) had a mean taste quality rating for brand A of 7.38 and a standard deviation of .97. At the .05 level, test the hypothesis that the population variances for the ratings of the two groups for brand A juice are equal.

12. *Macworld* has published results of tests that were made to compare the time needed by a sample of slide-making software packages to produce high-resolution and medium-resolution slides. The time (in seconds) needed by each package to produce a slide with a simple text and a logo image is as follows:

Software Package	High-Resolution Time	Medium-Resolution Time
Colorfast	340	167
FilmPrinter Turbo II	120	90
LFR	290	79
LFR Mark II	100	41
PCR II	178	94
Personal LFR	138	95
ProColor Premier	174	99
Spectra Star 450	642	184

Test the hypothesis at the .05 level that the population variance in speed is equal for the two resolution categories.

13. A study is made that involves individual measures of neuropsychological functioning for random samples of chronic schizophrenic patients and for normal subjects. The 25 schizophrenic patients had a mean of 93.2 on the Wide Range Achievement Test (WRAT) and a standard deviation of 16.8. The 25 normal subjects had a mean WRAT score of 107.8 with a standard deviation of 11.6. Test the hypothesis at the .05 level that there's no difference in the population variances of the two groups.

14. A study was published in the *American Journal of Psychology* that dealt with the effect of *Lorazepam* on brain glucose metabolism. In this study, there were random

It Was Worth a Try

A total of 2,235 women at four Latin American health centers were thought to have a higher-than-average risk of delivering a low-birth-weight infant. Of these women, 1,115 were randomly assigned to an intervention program that provided four to six home visits from a nurse or social worker in addition to routine prenatal care. The remaining 1,120 received only routine prenatal care. There was little difference in the weights of the children born to the two groups. The study concluded that interventions designed to provide psychological support and health education during high-risk pregnancies are unlikely to reduce the incidence of low birth weight among infants.

samples of 13 normal subjects and 10 alcoholic subjects. The verbal IQ for the normal subjects was 116 with a standard deviation of 23. The verbal IQ for the alcoholic sample was 109 with a standard deviation of 13. Test the hypothesis at the .05 level that the population variances of the two groups are equal.

15. In the study cited in problem 14 of the effect of *Lorazepam* on brain glucose metabolism, 5 of the alcoholic subjects were given *Lorazepam* and 5 were given a placebo. Brain metabolic values (μmol/100 g/min) were determined. At the thalamus (a part of the brain), the metabolic value for the subjects receiving a placebo was 40.7 and the standard deviation was 4. For the subjects receiving *Lorazepam*, the mean metabolic value was 34.5 and the standard deviation was 7. Test the hypothesis at the .05 level that variances for the metabolic rates measured at the thalamus are no different for alcoholics given a placebo and those given *Lorazepam.*

16. Many amputees participate in a variety of athletic activities, and so there's a demand for prosthetic legs that will improve their athletic performance. Different types of artificial limbs designed for those who have had one leg amputated below the knee were compared against each other and against an intact flesh-and-bone counterpart in a recent issue of *Physical Therapy.* In one segment of the study, the duration of single-limb support (SLS) while walking on a solid ankle cushion heel prosthetic leg was compared with the duration of SLS for a patient's intact leg. In 10 trials of the artificial leg, the mean duration of SLS was .32 second, and the standard deviation was .00027 second. Ten trials were also made on the natural leg, and the mean duration of SLS was .41 second, and the standard deviation was .00018 second. Test the hypothesis at the .05 level that the variance for the duration of SLS on the prosthetic leg is equal to the variance for the duration of SLS on the natural limb.

17. A measure of the right atrial pressure (mm Hg) was made for cardiac patients who were given *Captopril* and for those given *Nitroprusside.* The random sample of 21 patients given *Captopril* had a right atrial pressure of 7.8 mm Hg and a standard deviation of 1.4 mm Hg. For the 21 patients given *Nitroprusside,* the mean right atrial pressure was 6.6 mm Hg, and the standard deviation was 1.6 mm Hg. Test the hypothesis at the .05 level that there's no difference in the population variance of right atrial pressure between the two groups.

18. When offered a 20 percent discount, a random sample of 6 customers at the May Company reported a mean of 4.17 on a "change-of-intention-to-buy" scale (that goes from -9 to $+9$) with a standard deviation of .32. At the same store at the 30 percent discount level, a random sample of 7 customers reported a mean of 4.00 and standard deviation of .08 for their intention-to-buy scores. Test the hypothesis at the .05 level that at this May Company store there's no difference in variance in the intention-to-buy scale when a discount rate is 20 percent and when it is 30 percent.

9.2 Hypothesis Tests of Two Means

We've just looked at a *single* testing procedure that allowed us to arrive at a decision about two population variances. That procedure assumed that (1) the data in the two populations were normally distributed and (2) independent samples were used.

In considering hypothesis tests of two means, we'll continue to assume that the data in the two populations are normally distributed. But instead of a single procedure to follow, there are now several possible paths to take, as you can see in Figure 9.3. Of course, only one of the four procedures labeled in Figure 9.3 is used in any given test, and thus only one of the paths is followed to conduct that test. "Well fine," you may be thinking, "but how do I pick the right path through that maze in Figure 9.3?" The answer to your reasonable question is that you begin at the "Start" symbol at the top of

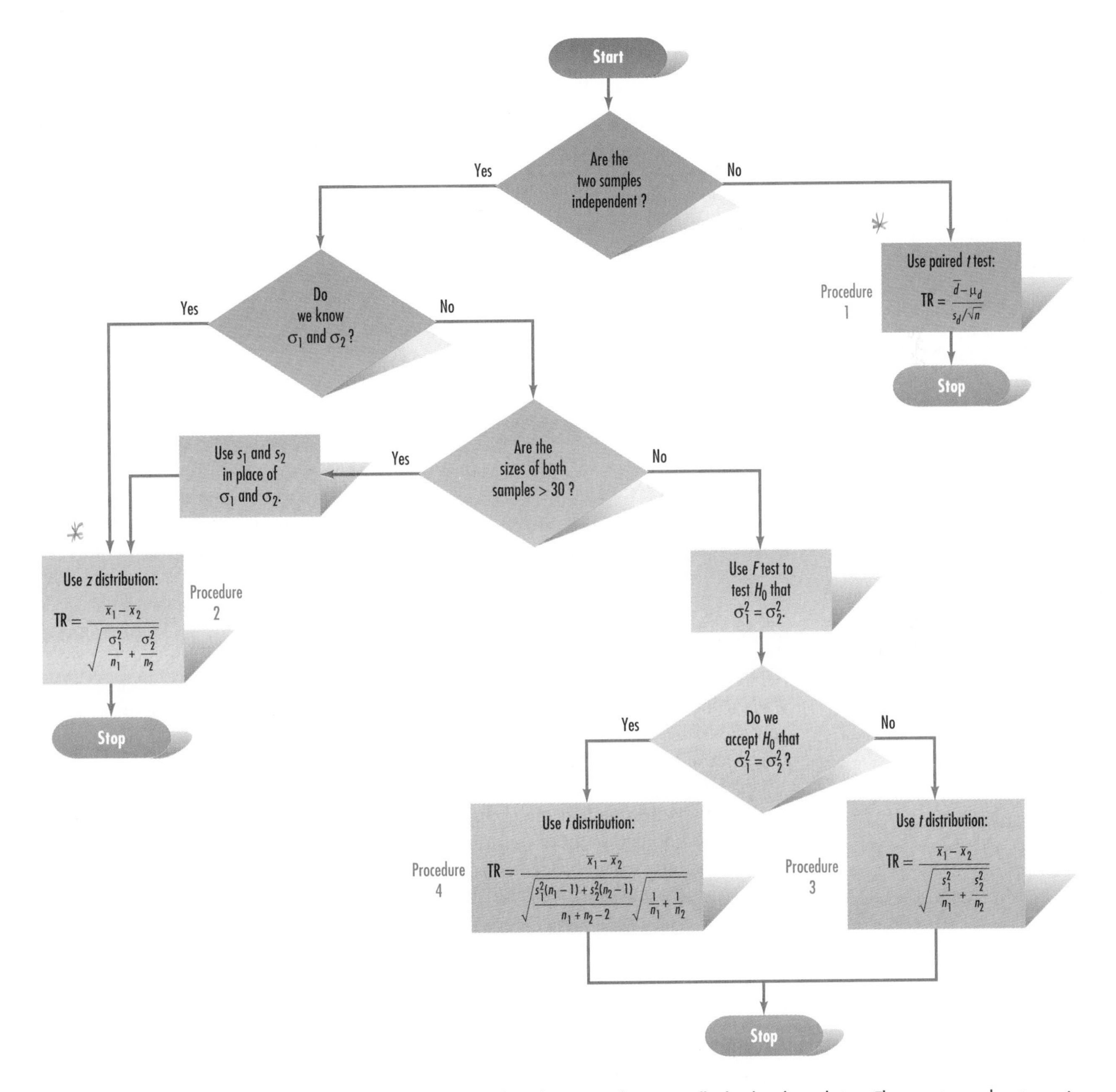

FIGURE 9.3 Four procedures followed to conduct hypothesis tests about the means of two normally distributed populations. The correct procedure to use in a given situation depends on the answers to the questions posed in the four diamond-shaped decision symbols.

Figure 9.3 and consider each of the questions raised in the four diamond-shaped decision symbols until you've arrived at the correct test procedure to use. Thus:

- You'll follow *Procedure 1* to conduct a paired t test if you have two *dependent* samples taken from normally distributed populations.
- You'll follow *Procedure 2* and use the z distribution to conduct the test if you have independent samples and normally distributed populations and if you know σ_1 and σ_2, or if you know that the sizes of both samples exceed 30.
- If σ_1 and σ_2 are unknown and the sample sizes are small, however, you first use the F test we've just studied to test the null hypothesis that $\sigma_1^2 = \sigma_2^2$. If the result of that F test is to *reject* the H_0 that $\sigma_1^2 = \sigma_2^2$, then you'll use the t distribution and *Procedure 3* to conduct the test.
- If the result of the F test is to *accept* the H_0 that $\sigma_1^2 = \sigma_2^2$, then you'll use the t distribution and *Procedure 4* to conduct the test.

Let's now consider two-sample hypothesis-testing examples using each of these four procedures. Although details vary, you'll be relieved to know that the same familiar seven steps are used in each testing procedure.

Procedure 1: The Paired t Test for Dependent Populations

Example 9.2. An industrial engineer is evaluating a new technique to assemble air compressors. If there's a difference in the number of compressors that can be assembled when the existing procedure is used, and when the new technique is followed, she will recommend that the company use the approach that results in the greatest worker productivity. A sample of 8 employees is selected at random, and the number of compressors they each produce in 1 week using the existing procedure is recorded. The same 8 workers are then trained to use the new technique, and their output for 1 week is then noted. (Since the same workers are used to produce each sample, the samples are dependent.) The engineer's data are given in the first three columns of Table 9.2.

The paired t test for dependent samples follows the same procedures used in Chapter 8 when the t distribution is employed in one-sample hypothesis tests of means. But now the test is applied to the *differences between paired values.* These differences form a single set of observations, and our testing effort follows a familiar path.

This path begins with a statement of the Null and Alternative Hypotheses (Step 1). The null hypothesis is that the population mean output using the existing methods is equal to the population mean output obtained with the new technique. In other words, the H_0 is that the mean difference (μ_d) between *before* and *after* production methods is zero. Thus, the *null hypothesis* can be stated in this way:

$$H_0 : \mu_d = 0$$

And the *alternative hypothesis* is that there *is* a mean difference between production methods:

$$H_0 : \mu_d \neq 0$$

TABLE 9.2 SAMPLE DATA BEFORE AND AFTER THE USE OF A NEW COMPRESSOR ASSEMBLY TECHNIQUE, AND CALCULATIONS NEEDED FOR A PAIRED t TEST

Employee	Production after New Method (x_1)	Production before New Method (x_2)	Difference ($x_1 - x_2$) (d)	($d - \bar{d}$)	($d - \bar{d}$)^2
A	85	80	5	3	9
B	84	88	−4	−6	36
C	80	76	4	2	4
D	93	90	3	1	1
E	83	74	9	7	49
F	71	70	1	−1	1
G	79	81	−2	−4	16
H	83	83	0	−2	4
			16	0	120

$$\bar{d} = \frac{\Sigma d}{n_{\text{pairs}}} = \frac{16}{8} = 2.00$$

$$s_d = \sqrt{\frac{\Sigma (d - \bar{d})^2}{n - 1}} = \sqrt{\frac{120}{8 - 1}} = \sqrt{17.143} = 4.14$$

The next move is to select the level of significance (*Step 2*), and the engineer wants to conduct this test at the .05 level of significance. *Step 3* is to determine the test distribution to use, and since the samples are small and the name of this test is the paired t test, it's no surprise that we'll use the t distribution. *Step 4* is to define the rejection or critical regions. With a *two-tailed test* here, there's a .025 rejection region in each tail of the t distribution. And the $t_{.025}$ value with $n - 1$, or $8 - 1$, or 7 degrees of freedom is found in Appendix 4 to be 2.365. *Note that n in a paired t test represents the number of pairs of data.*

Decision Rule (Step 5):

Reject H_0 and accept H_1 if TR < -2.365 or TR $> +2.365$. Otherwise, accept H_0.

Test Ratio (Step 6):

You'll recall from Chapter 8 that when the t distribution is used in one-sample hypothesis tests of means, the test ratio (or t value) is found with this formula:

$$\text{TR} = \frac{\bar{x} - \mu_{H_0}}{\hat{\sigma}_{\bar{x}}} = \frac{\bar{x} - \mu_{H_0}}{s / \sqrt{n}}$$

The difference between a sample mean and a hypothetical population mean is found, and this difference is then divided by an estimated standard error to get a standardized value. Except for the fact that we're dealing with paired differences now, the following test ratio formula follows exactly the same approach:

$$\text{TR} = \frac{\bar{d} - \mu_d}{s_d / \sqrt{n}} \tag{9.2}$$

Instead of a sample mean ($\bar{x}$) and hypothetical population mean (μ_{H_0}) in the numerator of the test ratio, we use the mean of the differences between the sample pairs ($\bar{d}$ which is equal to $\Sigma d/n_{\text{pairs}}$) and the hypothetical difference between the two population means (μ_d), which is equal to zero if the hypothesis is true. And in the denominator, we now use the standard deviation of the paired sample d values (s_d) and the square root of the *number of pairs* of data (n_{pairs}). The value of s_d is found with this formula:

$$s_d = \sqrt{\frac{\Sigma (d - \bar{d})^2}{n - 1}} \tag{9.3}$$

As you can see in Table 9.2, the value of $\bar{d}$ is 2.00, and the s_d value is 4.14. Thus:

$$\text{TR} = \frac{\bar{d} - \mu_d}{s_d / \sqrt{n}} = \frac{2.00 - 0}{4.14 / \sqrt{8}} = \frac{2.00}{1.464} = 1.37$$

Conclusion (Step 7):

Make the statistical decision. Since the TR of 1.37 is $<$ the t value of 2.365, we accept the null hypothesis that the mean difference in production methods is zero. The engineer can't conclude that one assembly method is better than the other.

Suppose we want to conduct a *one-tailed test* so that the alternative hypothesis reads

$H_1 : \mu_d > 0$

In this case, the entire α value is in the right tail of the t distribution. Thus, if we are making a paired t test with 7 paired observations at the .05 level, the $t_{.05}$ value with $n - 1$, or $7 - 1$, or 6 df is 1.943. And the decision rule is:

Reject H_0 and accept H_1 if the TR is $>$ 1.983. Otherwise, accept H_0.

We would then calculate the test ratio and make the statistical decision just as we did for a two-tail test.

Procedure 2: The z Test for Independent Populations

A z test is used when:

1. The samples are taken from two independent and normally distributed populations.
2. The values of σ_1 and σ_2 are known, or the size of both samples exceeds 30.

When these conditions are met, begin the testing procedure by (surprise!) formulating the *Null and Alternative Hypotheses (Step 1)*. The *null hypothesis* is:

$H_0 : \mu_1 = \mu_2$

This hypothesis states that the true mean of the first population is equal to the true mean of the second population. If the null hypothesis can't be supported, there are three possible alternative hypotheses that may be accepted:

$$H_1 : \mu_1 \neq \mu_2$$
$$H_1 : \mu_1 > \mu_2$$
$$H_1 : \mu_1 < \mu_2$$

When the null hypothesis states that the true mean of group 1 is equal to the true mean of group 2 ($\mu_1 = \mu_2$), it's essentially saying that the *difference between the parameters of the two groups is zero*—that is, $\mu_1 - \mu_2 = 0$. This idea makes it possible for us to visualize another type of sampling distribution.

The Sampling Distribution of the Differences between Sample Means. A conceptual schematic of this new sampling distribution is shown in Figure 9.4. Distribution A in Figure 9.4 is the sampling distribution of the sample means for population 1, and distribution B is the corresponding sampling distribution for population 2. Each of these theoretical distributions is developed from the means of all the possible samples of a given size that can be drawn from a population. Now if we were to select a single sample mean from distribution A and another sample mean from distribution B, we could subtract the value of the second mean from the value of the first mean and get a difference—that is, $\bar{x}_1 - \bar{x}_2$ = difference. This difference is either a negative or positive value, as shown in the examples between distributions A and B in Figure 9.4.

We could theoretically continue to select sample means from each population and continue to compute differences until we reached an advanced stage of senility. If we then constructed a frequency distribution of *all the sample differences,* we would have distribution C in Figure 9.4, which is the **sampling distribution of the differences between sample means.** And, as noted in Figure 9.4, *if* H_0 is true and *if* μ_1 is equal to

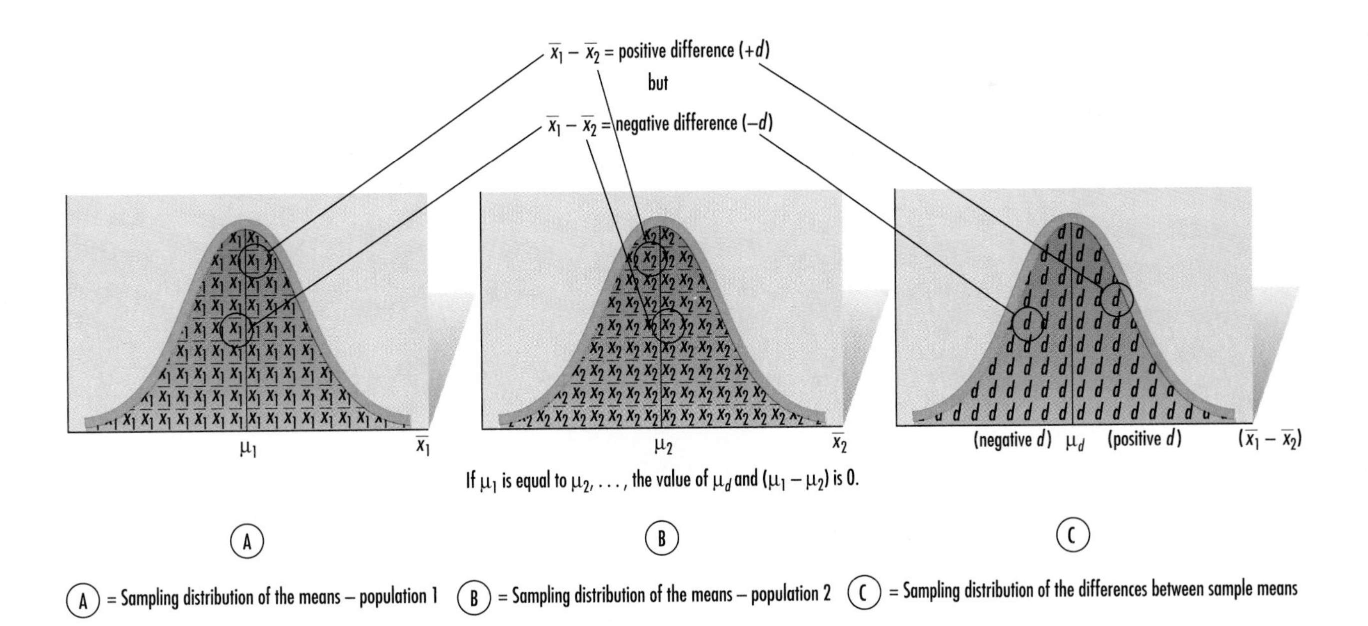

FIGURE 9.4 Conceptual schematic of the sampling distribution of the differences between sample means.

μ_2, *then* the value of the mean of the sampling distribution of the differences (μ_d) is equal to $\mu_1 - \mu_2$. That is, μ_d is zero. In short, the negative differences and the positive differences cancel, and the mean is zero.

Even if the mean of the sampling distribution of differences is zero, though, the characteristics of the sampling distribution of differences allows the value of $\bar{x}_1 - \bar{x}_2$ to deviate from zero. If the parameters are truly equal and if random samples are taken from the two populations, it's unlikely that the difference of $\bar{x}_1 - \bar{x}_2$ will equal zero; there will usually be some sampling variation. But if the true means are equal, the likelihood of an extremely large difference between $\bar{x}_1$ and $\bar{x}_2$ is small, especially in two large samples. Thus, if an extremely large difference occurs between $\bar{x}_1$ and $\bar{x}_2$, it's justifiable to conclude that the true means aren't equal. The immediate problem, of course, is to figure out when the difference between samples becomes significant so that the null hypothesis can be rejected.

If we know the standard deviation of each population or if we take a large sample from each population, the shape of the sampling distribution of the differences between means is approximately normal. Thus, the middle 68.26 percent of the differences in that sampling distribution are found within 1 standard deviation of the mean, and 2 standard deviations to either side of the mean accounts for 95.4 percent of the differences. The standard deviation of the sampling distribution of differences is called the **standard error of the difference between means** and is identified by the symbol $\sigma_{\bar{x}_1-\bar{x}_2}$ (see Figure 9.5).

Now that we've considered these theoretical concepts, we can briefly summarize the remaining six steps in our z test for independent samples. *Step 2* is to pick a level of significance, and we've seen that the test distribution to use (*Step 3*) is the z distribution. The process of establishing rejection regions (*Step 4*) and stating *Decision Rules* (*Step 5*) is exactly the same here as in Chapter 8, when the z distribution was applicable. For example, with a two-tailed test and a .05 level of significance, the z values for the boundaries of the rejection regions are ± 1.96. For a left-tailed test and a .05 level, the z value is -1.645.

Just as in other hypothesis tests, a *Test Ratio* (TR) (*Step 6*) must also be calculated for two-sample tests of means. You'll recall from Chapter 8 that one TR was found with this formula:

$$\text{TR} = \frac{\bar{x} - \mu_{H_0}}{\sigma_{\bar{x}}}$$

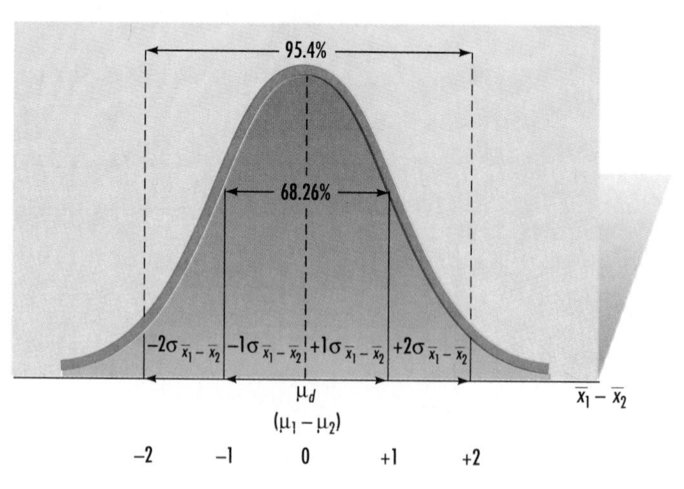

FIGURE 9.5 The sampling differences between means when σ_1 and σ_2 are known or when n_1 and n_2 are both > 30.

That is, the ratio was the difference between actual and hypothetical values expressed in number of standard errors. Now, our test ratio uses the standardized difference between $\bar{x}_1$ and $\bar{x}_2$ and is computed as follows:

$$\text{TR} = \frac{(\bar{x}_1 - \bar{x}_2) - (\mu_1 - \mu_2)}{\sigma_{\bar{x}_1-\bar{x}_2}} \quad \text{or} \quad \text{TR} = \frac{\bar{x}_1 - \bar{x}_2}{\sigma_{\bar{x}_1-\bar{x}_2}} \tag{9.4}$$

since $\mu_1 - \mu_2$ is assumed to be zero if H_0 is true. And (if it's known that we have two independently drawn random samples with known population standard deviations) the standard error of the difference is computed as follows:

$$\sigma_{\bar{x}_1-\bar{x}_2} = \sqrt{\frac{\sigma_1^2}{n_1} + \frac{\sigma_2^2}{n_2}} \tag{9.5}$$

Let's now look at the following examples, which illustrate two-tailed and one-tailed tests.

Example 9.3: Two-Tailed Testing When σ_1 and σ_2 Are Known. The Russ Trate Traffic Signal Company has decided to install new types of microcomputers in its traffic light assemblies so that these units can more efficiently monitor and control traffic flows. Microcomputers from two suppliers are judged to be suitable for the application. To have more than one source of supply, the Trate Company prefers to buy microcomputers from both suppliers, provided that there's no significant differ-

Highway engineers analyze traffic-pattern statistics as they plan for future roadways. (*Comstock*)

ence in durability. A random sample of 35 computer assemblies of brand A and a sample of 32 computers of brand B are tested. The mean time between failure (MTBF) for the brand A computers is found to be 2,800 hours, and the MTBF for the brand B units is found to be 2,750 hours. Information from industry sources indicates that the population standard deviation is 200 hours for brand A and 180 hours for brand B. At the .05 level of significance, is there a difference in durability?

Hypotheses (Step 1):

$H_0 : \mu_1 = \mu_2$
$H_1 : \mu_1 \neq \mu_2$

Since the Trate Company is interested in testing only for a significant difference, this is a two-tailed test. Also, the level of significance is specified at the .05 level (*Step 2*), and the knowledge of the population standard deviations (and the large sample sizes) enable us to use the z distribution (*Step 3*). Thus, the rejection regions are bounded by $z = \pm 1.96$ (*Step 4*).

Decision Rule (Step 5):

Reject H_0 and accept H_1 if TR < -1.96 or TR $> +1.96$. Otherwise, accept H_0.

With $\sigma_1 = 200$ hours, $n_1 = 35$, $\sigma_2 = 180$ hours, and $n_2 = 32$:

$$\sigma_{\bar{x}_1-\bar{x}_2} = \sqrt{\frac{\sigma_1^2}{n_1} + \frac{\sigma_2^2}{n_2}} = \sqrt{\frac{200^2}{35} + \frac{180^2}{32}} = 46.43 \text{ hours}$$

Test Ratio (Step 6):

$$\text{TR} = \frac{\bar{x}_1 - \bar{x}_2}{\sigma_{\bar{x}_1-\bar{x}_2}} = \frac{2{,}800 - 2{,}750}{46.43} = 1.08$$

Conclusion (Step 7):

Since the test ratio falls within the acceptance region of ± 1.96, we can conclude that there's no significant difference in the durability of the two microcomputer brands.

Example 9.4: One-Tailed Testing When σ_1 and σ_2 Are Known. Discount Stores Corporation owns outlet A and outlet B. For the past year outlet A has spent more dollars advertising casual slacks than outlet B. The corporation's advertising manager wants to see if the advertising has resulted in more sales for outlet A. A random sample of 36 days at outlet A had a mean of 170 slacks sold daily. A random sample of 36 days at outlet B had a mean sales of 165 slacks. Assuming $\sigma_1^2 = 36$ and $\sigma_2^2 = 25$, what can be concluded if a test is conducted at the .05 level of significance?

Hypotheses (Step 1):

$H_0 : \mu_1 = \mu_2$
$H_1 : \mu_1 > \mu_2$

This is a *right-tailed test* because the manager wants to see if the sales performance of outlet A is better than the performance of outlet B. With a .05 level (*Step 2*), the z value (*Step 3*) that bounds the rejection region is 1.645 (*Step 4*).

Decision Rule (Step 5):

Reject H_0 and accept H_1 if TR > 1.645. Otherwise, accept H_0.

Test Ratio (Step 6):

$$TR = \frac{\bar{x}_1 - \bar{x}_2}{\sigma_{\bar{x}_1 - \bar{x}_2}} = \frac{170 - 165}{\sqrt{36/36 + 25/36}} = \frac{5}{1.3017} = 3.84$$

Conclusion (Step 7):

Since TR is more than 1.645, there's sufficient reason to believe that outlet A has sold more slacks than outlet B.

Example 9.5: Two-Tailed Testing When σ_1 and σ_2 Are Unknown. When the population standard deviations are unknown—the usual situation—and when the samples from both populations are large (>30), then the sample standard deviations are used to estimate the population values. That is, $s_1 = \hat{\sigma}_1$, and $s_2 = \hat{\sigma}_2$.

An estimated standard error of the difference between means is then computed as follows:

$$\hat{\sigma}_{\bar{x}_1 - \bar{x}_2} = \sqrt{\frac{s_1^2}{n_1} + \frac{s_2^2}{n_2}} \qquad (9.6)$$

Let's now consider the following example.

Dr. I. M. Sain, a psychologist, administered IQ tests to see if there was a difference in the scores produced by a population of business majors and a population of psychology majors. A random sample of 40 business majors had a mean score of 131 with a standard deviation of 15. The random sample of 36 psychology majors had a mean of 126 and a standard deviation of 17. At the .01 level of significance, is there a difference?

Hypotheses (Step 1):

$H_0: \mu_1 = \mu_2$
$H_1: \mu_1 \neq \mu_2$

Since Dr. Sain is interested only in concluding equality or nonequality between groups, this is a two-tailed test. The .01 level is specified (*Step 2*), and since our sample sizes are large, the z distribution is used (*Step 3*). Thus, the rejection regions are bounded by $z = -2.575$ and $z = +2.575$ (*Step 4*).

Decision Rule (Step 5):

Reject H_0 and accept H_1 if TR < -2.575 or TR $> +2.575$. Otherwise, accept H_0.

Test Ratio (Step 6):

With $s_1 = 15$, $n_1 = 40$, $s_2 = 17$, and $n_2 = 36$, the test ratio (*Step* 6) is:

$$TR = \frac{\bar{x}_1 - \bar{x}_2}{\sqrt{(s_1^2/n_1) + (s_2^2/n_2)}} = \frac{131 - 126}{\sqrt{(15^2/40) + (17^2/36)}} = \frac{5}{3.695} = 1.35$$

Conclusion (Step 7):

Since the test ratio falls between ±2.575, we can conclude that there is no significant difference in the IQ scores of the two groups.

Example 9.6: One-Tailed Testing When σ_1 and σ_2 Are Unknown. A Chamber of Commerce manager is seeking to attract new industry to his area. One argument

Still A Problem

In a recent survey of college freshmen, it was found that 85.1 percent of those polled disagreed with the statement, "Racial discrimination is no longer a major problem in America." The preceding year the percentage disagreeing with the same statement was 79.7. And the survey found that an all-time high of 42 percent of the students agreed that "helping to promote racial understanding" was an essential or very important goal.

he has been using is that average wages paid for a particular type of job are lower than in other parts of the nation. A rather skeptical company president assigns his brother-in-law the task of testing this claim. A random sample of 60 workers (group 1) performing the particular job in the chamber manager's area is taken, and the sample mean is found to be \$9.75 per hour with a sample standard deviation of \$2.00 per hour. Another random sample of 50 workers (group 2) taken in region NE produced a sample mean of \$10.25 per hour with a sample standard deviation of \$1.25 per hour. At the .01 level, what report should the brother-in-law give to the president?

Hypotheses (Step 1):

$H_0 : \mu_1 = \mu_2$
$H_1 : \mu_1 < \mu_2$

This is a one-tailed test because the validity of the chamber manager's claim is being evaluated—i.e., that wages paid are less than in other areas. At the .01 level (*Step 2*), the z value (*Step 3*) that bounds the rejection region is -2.33 (*Step 4*).

Decision Rule (Step 5):

Reject H_0 and accept H_1 if TR < -2.33. Otherwise, accept H_0.

Test Ratio (Step 6):

With $s_1 = \$2.00$, $n_1 = 60$, $s_2 = \$1.25$, and $n_2 = 50$, the test ratio (*Step 6*) is:

$$\text{TR} = \frac{\bar{x}_1 - \bar{x}_2}{\sqrt{(s_1^2/n_1) + (s_2^2/n_2)}} = \frac{\$9.75 - \$10.25}{\sqrt{(\$2.00^2/60) + (\$1.25^2/50)}} = \frac{-\$0.50}{\$0.313} = -1.60$$

Conclusion (Step 7):

Since TR falls into the area of acceptance, the chamber manager's claim is not supported by the sample results at the .01 level. The brother-in-law is relieved to report that the test results tend to confirm the president's doubts.

Procedure 3: Small-Sample t Test for Independent Populations When $\sigma_1^2 \neq \sigma_2^2$

As you can see in Figure 9.3, the following conditions must be met before Procedure 3 can be applied:

1. The two samples are taken from two independent (and normally distributed) populations.
2. The values of σ_1 and σ_2 are unknown.
3. The size of n_1 or n_2 is small (≤ 30).
4. An F test leads to the conclusion that σ_1^2 and σ_2^2 are likely to be *different* (*unequal*).

When these conditions are met, the t-test steps followed in Procedure 3 closely parallel those we've just followed to conduct one- and two-tailed tests. Since that's the case, we need not consider as many examples in this section.

Example 9.7. An apartment rental agent tells the personnel manager of a firm thinking of building a plant in the agent's city that the mean rental rates for two-bedroom apartments are the same in sectors A and B of the city. To test this claim, the personnel manager randomly samples apartment complexes in each sector and obtains the following data:

Sector A	Sector B
$\bar{x}_1 = \$595$	$\bar{x}_2 = \$580$
$n_1 = 10$	$n_2 = 12$
$s_1 = \$62$	$s_2 = \$32$
$s_1^2 = 3{,}844$	$s_2^2 = 1{,}024$

What can the personnel manager conclude about the agent's claim at the .05 level?

Since the two samples are small and come from independent populations with unknown population standard deviations, we must first conduct an F test to see if σ_1^2 and σ_2^2 are likely to be different. Sector A has the largest variance, so that value becomes s_1^2, and the variance in sector B is identified as s_2^2. All other values in sectors A and B have similar subscripts.

Hypotheses (Step 1):

$H_0 : \sigma_1^2 = \sigma_2^2$
$H_1 : \sigma_1^2 \neq \sigma_2^2$

With $\alpha = .05$ (*Step 2*), and with $n_1 - 1$, or $10 - 1$, or 9 degrees of freedom in the numerator and with $n_2 - 1$, or $12 - 1$, or 11 degrees of freedom in the denominator, the critical F value (*Step 3*) in Appendix 5 is found in the table for $\alpha = .025$ at the intersection of the *column* numbered 9 and the *row* numbered 11. This critical F value is 3.59 (*Step 4*).

Decision Rule (Step 5):

Reject H_0 and accept H_1 if TR is $>$ 3.59. Otherwise, accept H_0.

Test Ratio (Step 6):

$$\text{TR} = \frac{s_1^2}{s_2^2} = \frac{3{,}844}{1{,}024} = 3.754$$

Conclusion (Step 7):

Since the TR of 3.754 is $>$ 3.59, we reject the null hypothesis and conclude that σ_1^2 and σ_2^2 are likely to be different.

The results of this preliminary F test clear the way for us to use Procedure 3 to test the rental agent's claim that the rental rates for two-bedroom apartments are the same in sectors A and B.

Hypotheses (Step 1):

$H_0 : \mu_1 = \mu_2$
$H_1 : \mu_1 \neq \mu_2$

This is a *two-tailed test* since we are merely trying to validate the agent's claim. The personnel manager specifies that $\alpha = .05$ (*Step 2*). With small samples and unknown population standard deviations, the t distribution must be used (*Step 3*). With $\alpha = .05$ and with a two-tailed test, we need $t_{.025}$ values for the boundaries of the rejection regions. To find those values in Appendix 4, though, we must know the degrees of freedom (df). In this Procedure 3 test, *the df figure is the smaller of* $n_1 - 1$ *or* $n_2 - 1$. Thus, the df figure we need is $n_1 - 1$, or $10 - 1$, or 9. And so the $t_{.025}$ values that bound the rejection regions are ± 2.262 (*Step 4*).

Decision Rule (Step 5):

Reject H_0 and accept H_1 if TR is < -2.262 or TR is $> +2.262$. Otherwise, accept H_0.

Test Ratio (Step 6):

The test ratio for this Procedure 3 test is computed in a familiar way, and it's frequently used, but it only produces approximate results. However, no better test has been devised to satisfy the conditions spelled out at the beginning of this section. The formula for this test ratio is:

$$\text{TR} = \frac{\bar{x}_1 - \bar{x}_2}{\sqrt{(s_1^2/n_1) + (s_2^2/n_2)}} \tag{9.7}$$

With $s_1^2 = 3{,}844$, $n_1 = 10$, $s_2^2 = 1{,}024$, and $n_2 = 12$, the test ratio is:

$$\text{TR} = \frac{\bar{x}_1 - \bar{x}_2}{\sqrt{(s_1^2/n_1) + (s_2^2/n_2)}} = \frac{\$595 - \$580}{\sqrt{(3{,}844/10) + (1{,}024/12)}} = \frac{\$15}{21.673} = .692$$

Conclusion (Step 7):

Since the TR value of .692 falls between ± 2.262, we accept the null hypothesis. The rental agent appears to know what she is talking about when she says that mean rates for two-bedroom apartments are the same in sectors A and B.

What if this had been a *one-tailed test*? What if the test had been used to see if rates are higher in sector A than in sector B? In that case, the alternative hypothesis is $H\colon \mu_1 > \mu_2$, and the *decision rule* is:

Reject H_0 and accept H_1 if TR > 1.833 ($t_{.05}$, df $= 9$). Otherwise, accept H_0.

The computed TR value remains .692 in a one-tailed test, and, in this case, the statistical decision is still to accept the null hypothesis because $.692 < 1.833$.

Procedure 4: Small-Sample t Test for Independent Populations When $\sigma_1^2 = \sigma_2^2$

As shown in Figure 9.3, the following conditions must be met if Procedure 4 is to be used:

1. The two samples are taken from two independent (and normally distributed) populations.
2. The values of σ_1 and σ_2 are unknown.
3. The size of n_1 or n_2 is small (≤ 30).

4. An F test leads to the conclusion that σ_1^2 and σ_2^2 are likely to be *equal.*

When these conditions are met, the seven t-test steps followed in Procedure 4 also parallel those we've already considered. The *Null Hypothesis (Step 1)* for Procedure 4 remains:

$H_0 : \mu_1 = \mu_2$

And the alternative is one of the following:

$H : \mu_1 \neq \mu_2$
$H : \mu_1 > \mu_2$
$H : \mu_1 < \mu_2$

A level of significance must be specified *(Step 2)*, and values from the t table are used *(Step 3)* when we define the rejection region(s). The degrees of freedom value needed in Procedure 4 to use the t table when two small samples are tested is found with this formula:

$\text{df} = n_1 + n_2 - 2$

Thus, if there are 14 items in sample 1 and 16 items in sample 2, our df value is (14 + 16 − 2) or 28. If we are making a *two-tailed test* at the .05 level of significance with these samples, then our t value is 2.048 *(Step 4)*.

Decision Rule (Step 5):

Reject H_0 and accept H_1 if the TR is < -2.048 or TR is $> +2.048$. Otherwise, accept H_0.

Before we can compute the test ratio, we must first calculate the *estimated* standard error of the difference between means. Since we assume that an F test has been made and the result supports the assumption that the unknown population variances (and standard deviations) are equal, an estimate of the population standard deviation is found by *pooling or combining* information from the two samples. The formula for the pooled standard deviation is:

$$s_{po} = \sqrt{\frac{s_1^2(n_1 - 1) + s_2^2(n_2 - 1)}{n_1 + n_2 - 2}}$$

And the formula for the estimated standard error is:

$$\hat{\sigma}_{\bar{x}_1 - \bar{x}_2} = s_{po}\sqrt{\frac{1}{n_1} + \frac{1}{n_2}}$$

Test Ratio (Step 6):

These formulas can be combined in the test ratio formula as follows:

$$\text{TR} = \frac{\bar{x}_1 - \bar{x}_2}{\sqrt{\frac{s_1^2(n_1 - 1) + s_2^2(n_2 - 1)}{n_1 + n_2 - 2}}\sqrt{\frac{1}{n_1} + \frac{1}{n_2}}} \tag{9.8}$$

Conclusion (Step 7):

A statistical decision is made.

Now let's look at an example.

Example 9.8. Let's revisit Example 9.1 and consider again the case of the two experimental diets designed to add weight to malnourished third-world children. Table 9.1, page 333, presents the weight gains made by the 8 children who were fed diet A and the 9 children who received diet B. As you can see in Table 9.1, the relevant data are:

Diet A	Diet B
$\bar{x}_1 = 5.8$ pounds	$\bar{x}_2 = 7.27$ pounds
$n_1 = 8$	$n_2 = 9$
$s_1^2 = 2.6029$	$s_2^2 = .9800$

We've already used the F test in Example 9.1 and accepted the hypothesis that $\sigma_1^2 = \sigma_2^2$. Now, let's suppose that the dietitian wants to conduct a test at the .05 level to see if there's a significant difference in the weight gained by the two groups. Following the steps in Procedure 4, the *Hypotheses (Step 1)* are:

$$H_0 : \mu_1 = \mu_2$$
$$H_1 : \mu_1 \neq \mu_2$$

We're interested in testing only for a significant difference in weight gain. A .05 level is specified *(Step 2)*, the t distribution must be used *(Step 3)*, and the rejection regions are bounded by a t value with df = 15 and $\alpha = .05$. This t value is 2.131 *(Step 4)*.

Decision Rule (Step 5):

Reject H_0 and accept H_1 if TR < -2.131 or TR $> +2.131$. Otherwise, accept H_0.

Test Ratio (Step 6):

$$\text{TR} = \frac{\bar{x}_1 - \bar{x}_2}{\sqrt{\dfrac{s_1^2(n_1 - 1) + s_2^2(n_2 - 1)}{n_1 + n_2 - 2}}\sqrt{\dfrac{1}{n_1} + \dfrac{1}{n_2}}}$$

$$= \frac{5.8 - 7.27}{\sqrt{\dfrac{2.6029(7) + .9800(8)}{15}}\sqrt{\dfrac{1}{8} + \dfrac{1}{9}}}$$

$$= \frac{-1.47}{1.32(.4859)} = \frac{-1.47}{.6414} = -2.29$$

Conclusion (Step 7):

Reject the null hypothesis; the TR is < -2.131. The population mean weight gain of diet A isn't equal to the gain of diet B at the .05 level.

A Computer Makes It Easy. By now you know that generating statistics from the raw sample data taken from two samples and then computing standard errors and test

```
MTB > SET C1
DATA> 4.1, 4.3, 6.0, 5.6, 8.5, 7.9, 5.1, 4.9
DATA> END
MTB > SET C2
DATA> 7.3, 6.7, 8.3, 7.0, 6.6, 6.8, 9.2, 7.6, 5.9
DATA> END
MTB > TWOSAMPLE T C1 VS C2;
SUBC> POOLED.

TWOSAMPLE T FOR C1 VS C2 .
     N       MEAN      STDEV   SE MEAN
C1   8       5.80       1.61      0.57
C2   9      7.267      0.990      0.33

95 PCT CI FOR MU C1 - MU C2: (-2.83, -0.10)

TTEST MU C1 = MU C2 (VS NE): T= -2.29  P=0.037  DF=  15

POOLED STDEV =          1.32
```

FIGURE 9.6 The *Minitab* output duplicating the results computed in Example 9.8.

ratios for those samples can be a tedious process. But you also know that statistical software packages can quickly do these tiresome calculations.

Figure 9.6 shows how the *Minitab* program processes the data for Examples 9.1 and 9.8. As you can see in the first program lines, weight-gain figures for the children fed diets A and B are entered into program storage areas C1 and C2. The program user then gives a TWOSAMPLE T command and specifies that the data to be used in the test are located in C1 and C2 (see line 7 of Figure 9.6). A POOLED subcommand in line 8 tells the program how to calculate the estimated population standard deviation. The remaining lines in Figure 9.6 are then generated by the program.

You'll notice that the package produces a 95 percent *confidence interval* for the *difference* between μ_1 and μ_2 that we don't need. The interval -2.83 to -0.10 tells us, though, that we can be 95 percent sure that the population mean weight gain of diet A is from 2.83 to 0.10 pounds *less than* the weight gain produced by diet B. The last two lines of computer output show the same results that we've painstakingly calculated. The MU C1 = MU C2 is the null hypothesis, and the (VS NE) is the "not-equal" alternative hypothesis. The pooled estimate of the population standard deviation is shown to be 1.32, and the TR (or t in this case) is -2.29.

The *p value* of .037 gives us another way to arrive at our statistical decision. As you saw in Chapter 8, the decision rule for a p-value test is:

Reject H_0 if the p value is $< \alpha$.   Otherwise, accept H_0.

Since the p value of .037 here is less than α of .05, we know to reject H_0.

Self-Testing Review 9.2

1. Four tests were discussed in this section: the paired t test, the z test, the t test for populations with equal variances, and the t test for populations with unequal variances. For each of these tests, what assumptions must be satisfied to produce useful results?

2. A recent *American Journal of Public Health* study dealt with the type of care given to the disabled elderly. After short-term hospital stays, a random sample of 110 disabled elderly subjects were assigned to an experimental group that offered physician-led primary home care on a 24-hour basis. Another sample of 73 disabled elderly subjects in a control group were offered ordinary care. After 6 months, the experimental group of 110 had a mean and standard deviation of 24.1 and 5.24, respectively, on the MMSE (mini-mental-state examination, where scores range from 0 to 30). The control group of 73 had a mean and standard deviation on the same test of 23.9 and 6.29, respectively. At the .05 level, test the hypothesis that a population receiving the experimental treatment would do significantly better on the MMSE exam than one receiving ordinary care.

3. The "perceived discount," or PD, is measured as follows: PD = (perceived regular price − perceived sale price)/perceived regular price. A study to see if the PD for "high-image" stores is higher than the PD for stores with generally lower prices was published in *The Journal of Consumer Research.* At Nordstrom's (a high-image store), the mean PD for a random sample of 44 customers for name-brand items on sale was 38.08, and the standard deviation was .84. At the May Company, the mean PD for a random sample of 47 customers for similar sale items was 35.60 with a standard deviation of .79. Test the hypothesis at the .01 level that the population PD at Nordstrom's is higher than the population PD at the May Company.

4. *Fortune* magazine recently listed "America's Most Admired Corporations." A random sample of senior executives was asked to rate the 10 largest companies in their own industry based on specified attributes. For the computers and office equipment industry, the companies and their ratings were:

Hewlett Packard	7.34	Apple Computer	6.96
Sun Microsystems	6.86	Compaq Computer	6.53
IBM	6.50	NCR	6.03
Digital Equipment	6.00	Pitney Bowes	5.82
Unisys	3.32	Wang Labs	3.17

For the top 10 in the electronics industry, the companies were rated by a sample of executives as follows:

General Electric	7.67	Motorola	7.42
Emerson Electric	7.04	Raytheon	6.48
Cooper Industries	6.33	Whirlpool	6.29
Texas Instruments	6.12	Rockwell Internatl.	6.05
TRW	6.04	Westinghouse Electric	5.65

Test the hypothesis at the .05 level that the population mean rating in the computer and office equipment industry is lower than in the electronics industry. (Assume both populations are normally distributed and verify from problem 7, STR 9.1, that the population variances are not equal.)

5. An experiment was conducted to evaluate the effectiveness of a work-site health promotion program in reducing obesity as measured by a body mass index (BMI). A random sample of 16 work sites received classes on weight control combined with payroll incentives. Another sample of 16 work sites served as a control and received no classes or incentives. The following data represent the mean BMI for the *control group* before and after the experiment. The mean BMI for each site in the control group before the experiment was:

   26.50 26.07 25.37 27.41 25.39 25.40 25.79 26.34
   26.52 26.08 26.45 25.90 25.51 25.67 25.44 27.04

   The mean BMI for each site in the control group after 2 years was:

   26.06 26.40 25.53 26.28 25.39 25.69 26.12 26.24
   26.53 26.37 26.22 26.42 25.57 24.94 25.95 26.47

   Test the hypothesis at the .01 level that there was no difference in BMI for the control group before and after the treatment.

6. A study reported in the *Journal of Consumer Research* dealt with the influence of touch on consumers' evaluations of service. Experimenters were trained as servers in a restaurant. In alternating hours, the servers either touched or did not touch the diners. The mean and standard deviation ratings given by a random sample of 32 female diners who were touched were 2.84 and 1.10, respectively. And a random sample of 35 female diners who were not touched gave mean and standard deviation ratings of 2.21 and .82, respectively. Test the hypothesis at the .05 level that the population mean rating given by diners who are touched is higher than the population mean rating given by those who are not touched.

7. A study in the *American Journal of Public Health* dealt with a randomly selected group of 264 primary caregivers. These caregivers were interviewed and asked to keep diaries for 6 months. Diaries were returned by 141 caregivers, and 123 did not return diaries. The mean and standard deviation of the time spent in working as a caregiver for the sample of caregivers who returned diaries was 42.0 and 34.7 months, respectively. Of the sample who didn't return diaries, the mean and standard deviation for the number of months working as a caregiver was 37.2 and 33.4. Test the hypothesis at the .05 level that the population mean time spent working as a caregiver is the same for the two groups.

8. Participants in a consumer product research study were randomly divided into two groups, given 4 brands of orange juice to sample, and then asked to rate the quality of each brand. The 29 subjects in the first group were given one juice brand to drink at a time. About 5 minutes elapsed between consecutive samples, during which time the tasters were asked to eat a cracker. In the second group, the 29 subjects were given all 4 brands simultaneously and were instructed to go back and forth between the different brands and were not provided with crackers. The first group (one juice at a time) had a mean taste quality score for brand A of 6.31 (on a scale of 0 to 10) and a standard deviation of 1.2. The second group (all brands simultaneously) had a mean taste quality rating for brand A of 7.38 and a standard

deviation of .97. At the .01 level, test the hypothesis that the population mean taste rating will be lower when one brand is tried at a time than when all brands are tried simultaneously.

9. *Macworld* has published results of tests that were made to compare the time needed by a sample of slide-making software packages to produce high-resolution and medium-resolution slides. The time (in seconds) needed by each package to produce a slide with a simple text and a logo image is as follows:

Software Package	High-Resolution Time	Medium-Resolution Time
Colorfast	340	167
FilmPrinter Turbo II	120	90
LFR	290	79
LFR Mark II	100	56
PCR II	178	94
Personal LFR	138	95
ProColor Premier	174	99
Spectra Star 450	642	184

Test the hypothesis at the .05 level that the population mean time needed to produce high-resolution slides is significantly greater than the mean time needed to produce medium-resolution slides.

10. In the study described in problem 8, the taste ratings for each brand are given below for each method of tasting:

Brand	One Brand at a Time	All Brands Simultaneously
A	6.31	7.38
D	5.53	5.04
B	3.34	2.92
C	1.88	.75

Use the paired t test to test the hypothesis at the .01 level that there's no difference in the taste ratings using the two methods.

11. An experiment reported in the *American Journal of Public Health* was conducted to evaluate the effectiveness of a work-site health promotion program in reducing obesity as measured by a body mass index (BMI). A random sample of 16 work sites received classes on weight control combined with payroll incentives. Another sample of 16 work sites served as a control and received no classes or incentives. The following data represent the mean BMI for each site in the *control group* after 2 years:

26.06	26.40	25.53	26.28	25.39	25.69	26.12	26.24
26.53	26.37	26.22	26.42	25.57	24.94	25.95	26.47

The mean BMI for each site in the *treatment group* after 2 years was:

26.02 25.87 25.02 25.46 25.70 26.10 26.24 26.57
24.57 25.18 26.84 26.31 26.22 25.61 26.42 25.16

Test the hypothesis at the .05 level that after 2 years there was no difference in the population mean BMIs between the treatment group and the control group. (Assume that both populations are normally distributed, and verify from problem 9, STR 9.1, that the population variances are equal.)

12. A study is made that involves individual measures of neuropsychological functioning for random samples of chronic schizophrenic patients and for normal subjects. The 25 schizophrenic patients had a mean of 93.2 on the Wide Range Achievement Test (WRAT) and a standard deviation of 16.8. The 25 normal subjects had a mean WRAT score of 107.8 with a standard deviation of 11.6. Test the hypothesis at the .01 level that there's no difference in the population WRAT scores for the two groups. (Assume both populations are normally distributed and verify from problem 13, STR 9.1, that the population variances are equal.)

13. A test reported in *Physical Therapy* was conducted to determine the effectiveness of using an anti-inflammatory cream on delayed-onset muscle soreness. A random sample of 10 patients is treated with the cream on one arm and with a placebo on the other (control) arm. After 4 days a measure of muscle soreness is then taken for each patient on each arm. The results are:

Control Arm	Treated Arm
46	2
22	32
10	30
14	3
26	14
29	32
29	2
47	39
20	18
13	2

Using a paired t procedure, test the hypothesis at the .01 level that there is less soreness in the treated arm.

14. Some research indicates that antioxidant vitamins may be effective in cancer prevention. A study of the effect of the Treatwell Intervention Program was reported in the *American Journal of Public Health.* A record of the vitamin intake in milligrams per day was recorded for a random sample of subjects in a control group before the program started and after it was completed. The following is the before and after vitamin-intake record for the control group (the group in which no intervention was employed):

Vitamin	Baseline (before) Intake	Final (after) Intake
Vitamin B_6	1.63	1.60
Vitamin B_{12}	5.06	5.35
Folate	217.12	213.19
Pantothenic acid	3.69	3.77
Riboflavin	1.59	1.53
Vitamin C	129.43	123.76

Using the paired t procedure, test the hypothesis at the .01 level that there is no difference in vitamin intake for the control group between the time the program started and the time it was completed.

15. How well can divorce be predicted by a wife's personality score? In a study reported in *The Journal of Personality and Social Psychology*, a follow-up was made of a random group of 286 couples 5 years after their marriage. It was found that 222 couples remained together and 64 had dissolved their marriages. As newlyweds, all wives in the study group (and husbands, too, but that's another question) were given a personality test and were then rated on "dysfunctional beliefs." For the wives in the sample of couples who remained together, the mean rating for dysfunctional beliefs was 49.53 and the standard deviation was 12.62. For the 64 divorced wives, the mean dysfunctional beliefs score was 53.73 with a standard deviation of 4.74. Test the hypothesis at the .05 level that there's no significant difference in the mean dysfunctional beliefs scores for the populations of women considered in the problem.

16. A study was published in the *American Journal of Psychology* that dealt with the effect of *Lorazepam* on brain glucose metabolism. In this study, there were random samples of 13 normal subjects and 10 alcoholic subjects. The verbal IQ for the normal subjects was 116 with a standard deviation of 23. The verbal IQ for the alcoholic sample was 109 with a standard deviation of 13. Test the hypothesis at the .05 level that the population mean IQ of alcoholics is lower than that of normal subjects. (Assume both populations are normally distributed and verify from problem 14, STR 9.1, that the population variances are equal.)

17. An experimental diet was followed by a random sample of 6 people. The cholesterol level for each was measured before and after the diet as follows:

Before	After
196	174
212	160
254	151
207	121
221	275
223	118

Test the hypothesis at the .01 level that there's a significant decrease in the population cholesterol level after the diet.

18. In the study cited in problem 16 of the effect of *Lorazepam* on brain glucose metabolism, 5 of the alcoholic subjects were given *Lorazepam,* and 5 were given a placebo. Brain metabolic values (μ mol/100 g/min) were determined. At the thalamus (a part of the brain), the mean metabolic value for the subjects receiving a placebo was 40.7, and the standard deviation was 4. For the subjects receiving *Lorazepam,* the mean metabolic value was 34.5 and the standard deviation was 7. Test the hypothesis at the .01 level that population mean metabolic rates measured at the thalamus are no different for alcoholics given a placebo and those given *Lorazepam.* (Assume both populations are normally distributed and verify from problem 15, STR 9.1, that the population variances are equal.)

19. Many amputees participate in a variety of athletic activities, and so there's a demand for prosthetic legs that will improve their athletic performance. Different types of artificial limbs designed for those who have had one leg amputated below the knee were compared against each other and against an intact flesh-and-bone counterpart in a recent issue of *Physical Therapy.* In one segment of the study, the duration of single-limb support (SLS) while walking on a solid ankle cushion heel prosthetic leg was compared with the duration of SLS for a patient's intact leg. In 10 trials of the artificial leg, the mean duration of SLS was .32 second, and the standard deviation was .00027 second. Ten trials were also made on the natural leg, and the mean duration of SLS was .41 second, and the standard deviation was .00018 second. Test the hypothesis at the .05 level that there's no difference in the population mean duration of SLS between the prosthetic leg and the natural leg. (Assume both populations are normally distributed and verify from problem 16, STR 9.1, that the population variances are equal.)

20. A measure of the right atrial pressure (mm Hg) was made for cardiac patients who were given *Captopril* and for those given *Nitroprusside.* The random sample of 21 patients given *Captopril* had a right atrial pressure of 7.8 mm Hg and a standard deviation of 1.4 mm Hg. For the 21 patients given *Nitroprusside,* the mean right atrial pressure was 6.6 mm Hg, and the standard deviation was 1.6 mm Hg. Test the hypothesis at the .01 level that there's no difference in the population mean right atrial pressure between the two groups. (Assume both populations are normally distributed, and verify from problem 17, STR 9.1, that the population variances are equal.)

21. Do men and women have different values when choosing a career? The following random sample data from a *Career Development Quarterly* study gives the result of responses by males and females to a Life Roles Inventory questionnaire. Use the paired t test at the .01 level to see if the ratings are different for men and women:

Value	Male Rating	Female Rating
Ability utilization	16.69	17.17
Achievement	16.76	17.17
Advancement	15.08	14.44
Aesthetics	13.31	13.75
Altruism	14.95	16.49
Authority	14.61	14.51
Autonomy	15.79	16.37
Creativity	14.89	14.82

(Continued)

Value	Male Rating	Female Rating
Economics	16.31	16.25
Lifestyle	14.09	14.45
Personal development	17.31	18.02
Physical activity	14.18	13.97
Prestige	15.34	15.79
Risk	10.34	9.47
Social interaction	12.65	13.82
Social relations	15.44	16.93
Variety	13.06	13.96
Working conditions	13.75	14.91
Cultural identity	12.76	13.48
Physical prowess	9.49	8.51

22. A study in *The New England Journal of Medicine* sought to learn if there was a difference between the mean cost per case for physical therapy for a self-referred group and for an independent-referral group. In a random sample of 1,017 self-referral cases, the mean cost per case was \$404, and the standard deviation was \$102. In a sample of 240 independently referred cases, the mean cost per case was \$440, and the standard deviation was \$167. Test the hypothesis at the .05 level that the population mean cost per case is lower for the self-referred group.

23. An article in *Criminal Justice and Behavior* claims that an intellectual imbalance of performance IQ (P) with verbal IQ (V)—such that $P > V$—is a useful predictor of the probability of becoming delinquent. A random sample of 157 "high $P > V$" delinquents had a mean delinquency score of 255.7 with a standard deviation of 30.7, and a random sample of 356 "low $P > V$" delinquents had a mean delinquency score of 195 with a standard deviation of 41.8. Test the hypothesis at the .01 level that there is no difference in the delinquency scores for these two populations.

24. When offered a 20 percent discount, a random sample of 6 customers at the May Company reported a mean of 4.17 on a "change-of-intention-to-buy" scale and a standard deviation of .32. When offered a 30 percent discount, a random sample of 7 customers reported a mean of 4.00 and a standard deviation of .08 for their change-of-intention-to-buy scores. Test the hypothesis at the .01 level that at this May Company store there's a different population mean value on the intention-to-buy scale when a 20 percent discount rate is offered and when the rate is 30 percent. (Assume both populations are normally distributed and verify from problem 18, STR 9.1, that the population variances are not equal.)

25. What do members of the American Bowling Congress (ABC) do to relax away from the office? The answer is that they bowl. *Bowling Magazine* has published team averages for this year and for the previous year as follows:

Team	This Year's Team Average	Last Year's Team Average
Duds	525	527
So What?	535	556
Vintage Rock and Roll	559	571
Up in Smoke	535	559
Two D's and a Sub	538	514
A Beautiful Thing	520	555
3UDSU	520	512
Zebkuhler	553	576
Jerry's Kids	572	599
Skid Row Hook	546	553
Tri-Right	543	560
Bo's Bach	474	479
Woulda Coulda Did	593	599
VHS	565	556

Use a paired t test at the .05 level to see if the team averages have changed significantly between this year and the previous year.

STATISTICS IN ACTION

Not Up My Nose You Don't . . .

A year-long study of smokers was conducted at the Maudsley Hospital in London. A group of 227 smokers received 4 weeks of therapy designed to help them give up their habit. Half of the group were given an experimental nasal spray containing nicotine, and the other half were given a spray containing a placebo. One-fourth of those using the experimental spray and one-tenth of those using the placebo spray gave up smoking during the trial.

9.3 Hypothesis Tests of Two Percentages

There are two assumptions that must be met to carry out the seven-step testing procedure described in this section:

1. The two samples are taken from two independent populations.
2. The samples taken from each population are sufficiently large. That is, for each sample $np \geq 500$ and $n(100 - p)$ is also ≥ 500.

The purpose of conducting two-sample hypothesis tests of percentages is to see, through the use of sample data, if there's likely to be a statistically significant difference between the percentages of two populations. The null hypothesis in such tests of the differences between percentages is:

$H_0 : \pi_1 = \pi_2$

If the null hypothesis cannot be supported, one of three possible alternative hypotheses is accepted:

$H_1 : \pi_1 \neq \pi_2$
$H_1 : \pi_1 > \pi_2$
$H_1 : \pi_1 < \pi_2$

The Sampling Distribution of the Differences between Sample Percentages

The **sampling distribution of the differences between sample percentages** is theoretically analogous to the sampling distribution of the differences between sample means. The mean of the sampling distribution of the differences between percentages is zero *if* the null hypothesis is true—that is, if $\pi_1 = \pi_2$. If the sample size of each group is large, the shape of the sampling distribution is approximately normal. The standard deviation of the sampling distribution of the differences between percentages, which is called the **standard error of the difference between percentages**, is calculated as follows:

$$\sigma_{p_1-p_2} = \sqrt{\frac{\pi_1(100 - \pi_1)}{n_1} + \frac{\pi_2(100 - \pi_2)}{n_2}}$$

Unfortunately, the computation of $\sigma_{p_1-p_2}$ requires a knowledge of the parameters. If these values were known in the first place, there would be no need to conduct a test! Therefore, *in a test of differences between percentages, the estimator* $\hat{\sigma}_{p_1-p_2}$ *must always be used:*

$$\hat{\sigma}_{p_1-p_2} = \sqrt{\frac{p_1(100 - p_1)}{n_1} + \frac{p_2(100 - p_2)}{n_2}} \tag{9.9}$$

And the *test ratio* is computed as follows:

$$\text{TR} = \frac{(p_1 - p_2) - (\pi_1 - \pi_2)}{\hat{\sigma}_{p_1-p_2}} \quad \text{or} \quad \text{TR} = \frac{p_1 - p_2}{\hat{\sigma}_{p_1-p_2}} \tag{9.10}$$

since $\pi_1 - \pi_2$ will equal zero if the null hypothesis is true.

The general seven-step procedure for testing the differences between percentages is not any different than the z test for differences between means. Consequently, there's little need here for an extended further discussion. The following is an example of a two-tailed test.

Example 9.9. Ken Kharisma, candidate for public office, feels that male voters as well as female voters have the same opinion of him. A random sample of 36 male voters showed that 12 of these voters favored his election. Thus, p_1 is 12/36 or 33 percent. And it was found in a random sample of 50 female voters that 18 were Ken supporters. So p_2 is 18/50 or 36 percent. Test the validity of Ken's assumption, using a significance level of .05.

Hypotheses (Step 1):

$H_0 : \pi_1 = \pi_2$
$H_1 : \pi_1 \neq \pi_2$

This is a two-tailed test because Ken is interested only in the equality or nonequality of opinions between the two groups. The .05 level is specified (*Step 2*), the z distribution is applicable (*Step 3*), and the rejection regions are bounded by $z = \pm 1.96$ (*Step 4*).

Decision Rule (Step 5):

Reject H_0 and accept H_1 if TR < -1.96 or TR $> +1.96$. Otherwise, accept H_0.

Test Ratio (Step 6):

$$TR = \frac{p_1 - p_2}{\sqrt{\dfrac{p_1(100 - p_1)}{n_1} + \dfrac{p_2(100 - p_2)}{n_2}}} = \frac{33 \text{ percent} - 36 \text{ percent}}{\sqrt{\dfrac{(33)(67)}{36} + \dfrac{(36)(64)}{50}}}$$

$$= \frac{-3.00 \text{ percent}}{10.368} = -.29$$

Conclusion (Step 7):

Since the test ratio is between ±1.96, there is no reason to reject Ken Kharisma's claim. Apparently, both genders have about the same *low opinion* of Ken!

Had this been a *one-tailed test* to see if male voters were significantly less responsive to Ken's message than female voters, the alternative hypothesis would be:

$H_1 : \pi_1 < \pi_2$

And the decision rule would be:

Reject H_0 and accept H_1 if TR < -1.645. Otherwise, accept H_0.

The TR would remain unchanged, and the conclusion would still be to accept the null hypothesis.

But I'm Compelled . . .

A *Journal of Consumer Research* study assessed the characteristics of compulsive buyers and compared them with the characteristics of general consumers. There was no discernible difference in the household incomes of these groups. But while gender was about equally represented in the study (48.3 percent men and 51.7 percent women), compulsive buyers were heavily skewed toward female respondents (92 percent).

Self-Testing Review 9.3

1. A study in the *American Journal of Public Health* reported that out of a random sample of 512 Denver employees, 292 reported they experienced headaches. And of a sample of 281 Adams County employees, 172 reported they experienced headaches. Test the hypothesis at the .01 level that there's no difference in the population percentage of employees who experience headaches at the two locations.

2. During the period from July 29, 1991, through November 1, 1991, a random sample of 185 patients at a public clinic in Maryland participated in a study. Of these patients, 57 reported having "one-night stands." On November 7, 1991, Earvin "Magic" Johnson announced that he was infected with the HIV virus. From November 11, 1991, through February 14, 1992, a second study was conducted, and in a random sample of 97 surveyed, 19 reported having one-night stands. Test the hypothesis at the .05 level that there was a significantly smaller percentage of high-risk behavior in this population after the Johnson announcement.

3. Humor is commonly used as an advertising tool in the United States, but researchers know little about its use in foreign markets. One type of humor found in ads is the showing of situations in which there are expected-unexpected contrasts. In a *Journal of Marketing* study, a random sample of 36 U.S. ads showed that 16 contained an expected-unexpected contrast. In Germany, a sample of 36 ads had 18 with expected-unexpected contrasts. Test the hypothesis at the .01 level that there is a different population percentage of ads that have expected-unexpected contrasts in the two nations.

4. Patients in an *American Journal of Sports Medicine* study were divided into two groups to see if having prior reconstructive knee surgery had an effect on the

success rate of later reconstructive surgery using a prosthetic Dacron ligament. A random sample of 50 patients in group 1 had isolated anterior cruciate laxity but had not undergone any previous reconstructive surgery. For this group, surgery was successful for 31 of the 50 patients. The random sample of 34 patients in group 2 had complex laxity and had received a previous unsuccessful reconstruction. For this group, 11 of the 34 had successful surgery. Test the hypothesis at the .01 level that the population percent of success is equal for the two groups.

5. A study of psychological disorders and how they affect high school students was published in *The Journal of Abnormal Psychology.* In a random sample of 891 females in the study, 16 had attention-deficit hyperactivity. And in the random sample of 819 males, 37 had attention-deficit hyperactivity. Test the hypothesis at the .01 level that males have a greater population percentage of attention-deficit hyperactivity.

6. In a study reported in the *American Journal of Psychiatry*, the characteristics of a random sample of 17 patients with major depression were compared with the characteristics of a random sample of 47 patients without depression. The study found that 8 of the patients with major depression had a family history of psychiatric disorders while 23 of those without depression had a family history of psychiatric disorders. Test the hypothesis at the .01 level that the population percentage of patients with a family history of psychiatric disorders is the same for patients with and without major depression.

7. A large brokerage firm wants to see if the percent of new accounts valued at over $50,000 has changed. A random sample of 900 accounts opened last year showed that 36 were over $50,000 in size. And a random sample of 1,000 accounts opened this year showed that 44 were over $50,000. At the .05 level, what should the brokerage company conclude?

8. *The Journal of Research in Crime and Delinquency* has published a study on the situational characteristics of crimes. In a random sample of 96 robberies, 36 were committed by people who acted alone. And in a random sample of 69 assaults, 39 were committed by people who acted alone. Test the hypothesis at the .05 level that the population percentage of crimes committed by people acting alone is the same for robbery and assault.

9. In a study reported in the *American Journal of Psychiatry*, the responses regarding HIV risk behavior of a random sample of 76 hospitalized adolescents in a psychiatric facility were compared with the responses of a random sample of 802 school-based adolescents in the same city. Seven of the hospitalized adolescents reported using injection drugs while 29 of the school group reported the same behavior. Test the hypothesis at the .01 level that injection-drug use among the population of psychiatrically hospitalized adolescents is greater than it is among the population of school-based adolescents.

10. A survey has been made of how *Internet* (an electronic information superhighway) is used. In a random sample of 211 teachers who use *Internet* for professional activities, 192 said they used it to send electronic mail. And in a random sample of 119 teachers who use *Internet* for student activities, 94 said they used it to send electronic mail. Test the hypothesis at the .01 level that there's a higher population percentage of electronic mail users among teachers who use *Internet* for professional activities than among those who use it for student activities.

11. A random sample of 150 men are polled, and 80 of them prefer brand X toothpaste over competitive products. A random sample of 230 women shows that 130 of them also prefer brand X. At the .01 level, is there a significant difference in the toothpaste preference of the population of men and women?

12. An *American Journal of Public Health* study showed that in a random sample of 2,536 elderly medicaid patients in nursing homes, 72 received mental health services. And in a random sample of 2,109 elderly patients in nursing homes who were not on medicaid, 45 received mental health services. Test the hypothesis at the .01 level that there is a difference in the population percentage of medicaid and nonmedicaid patients who receive mental health services.

13. Thirty years after Dr. Martin Luther King, Jr.'s, historic march on Washington, an *Associated Press* national telephone poll asked this question: "Do minorities generally get equal justice in this country today, or is getting equal justice still a major problem for minorities?" There were 892 whites and 111 blacks who responded. Of the whites, 446 thought that obtaining equal justice was still a problem. And of the blacks, 96 thought obtaining equal justice was still a problem. Test the hypothesis at the .05 level that a lower population percentage of whites think obtaining justice is still a problem.

14. A study in *The New England Journal of Medicine* sought to learn if the risk of neural-tube birth defects could be decreased by mothers' taking multivitamins in the period preceding delivery. It was found that in a random sample of 2,394 women who were given vitamins, there were 67 congenital malformations in the babies they delivered. There were 2,310 women in a control group who were given placebos, and among this group there were 109 congenital malformations. Test the hypothesis at the .01 level that the population percentage of birth defects is equal for the group given vitamins and the group not given vitamins.

15. The public relations manager of Tailspin Airlines is concerned about a recent increase in the number of customer claims of damage to luggage. A random sampling of the records at two terminals yields the following data: At the Bayburg terminal, 760 pieces of luggage were handled, and 44 items were damaged. At the Beantown terminal 830 pieces of luggage were handled, and 60 were damaged. At the .05 level is there a significant difference in damage claims between the two terminals?

16. A recent study in the *American Journal of Psychiatry* used random samples of 202 nicotine-dependent people and 192 nondependent people to see if nicotine-dependent people had a greater vulnerability to psychiatric disorders. In the nicotine-dependent group, 169 were currently smoking and in the nondependent group, 123 were smoking. Test the hypothesis at the .05 level that the population percentage of those who are still smoking is higher in the nicotine-dependent group.

17. In another *American Journal of Psychiatry* study of the current and lifetime psychiatric diagnoses of World War II combat veterans who were prisoners of war and those combat veterans who weren't POWs, there were 23 POW survivors and 28 combat veterans. The study found that 16 of the POWs currently suffer from posttraumatic stress disorder (PTSD), and 5 of the combat veterans also suffer from PTSD. Test the hypothesis at the .01 level that there is a higher population percentage of PTSD among POW survivors than among combat veterans who weren't POWs.

LOOKING BACK

1. You've seen in this chapter how to evaluate two populations to see if they're likely to be similar or different with respect to some characteristic. Our interest isn't in testing for the absolute values of parameters; rather, our concern is to see if one population appears to possess more or less of a trait than the other. The prerequisites to using the hypothesis tests of two variances, two means, and two percentages that are discussed in this chapter are spelled out when the tests are introduced.
2. We've used testing procedures to make relative comparisons between two population variances, two population means, and two population percentages. The null hypothesis to be tested in these situations is that the two parameters are equal. That is, there's no significant difference between the two parameters. The alternative hypothesis may be either one-tailed or two-tailed depending on the logic of the situation.
3. An F distribution is the sampling distribution used to make tests of two variances, z and t distributions are used in tests of means, and the z distribution is used in our tests of percentages. The same basic seven-step testing procedure (state the hypotheses, select the level of significance, determine the correct test distribution, define the critical regions, state the decision rule, compute the test ratio, and make the statistical decision) is used in all the tests outlined in this chapter.
4. Only one procedure is discussed for tests of two variances or tests of two percentages. But four procedures are used to conduct tests of means. The choice of the correct procedure to use depends on the answers to the questions raised in the four diamond-shaped decision symbols found in Figure 9.3. Examples of each of these procedures are given and discussed in the chapter.

Review Exercises

1. A study reported in the *Journal of Consumer Research* sought to learn the effect of touching on customer behavior. Shoppers were approached at random by 3 male and 3 female experimenters as they entered a store, and the shoppers were handed a catalog. During alternate 1-hour periods, the experimenter either touched the subject lightly on the upper arm or did not. There were 17 in the group who were touched and 16 in the no-touch group. The mean and standard deviation of the shopping time for the shoppers who were touched were 22.11 and 4.74 minutes, respectively. For the no-touch group, the mean and standard deviation times were 13.56 and 5.67, respectively. Test the hypothesis at the .05 level that the population variances are equal for both groups.

2. Using the statistics given in problem 1, test the hypothesis at the .05 level that the population mean time spent shopping by those who are touched is significantly greater than the similar time spent by those who are not touched.

3. A recent *American Journal of Public Health* study dealt with the type of care given to the disabled elderly. After short-term hospital stays, a random sample of 110 disabled elderly subjects were assigned to an experimental group that offered physician-led primary home care on a 24-hour basis. Another sample of 73 disabled elderly subjects in a control group were offered ordinary care. After 6 months, the experimental group of 110 used a mean of 4.2 drugs with a standard deviation of 2.10 drugs. The control group of 73 used a mean of 4.6 drugs with a standard deviation of 3.15. At the .01 level, test the hypothesis that

the population represented by the experimental group needed fewer drugs.

4. As reported in the *Archives of General Psychiatry*, a study that began in 1971 followed the educational progress of a random sample of 91 boys with attention-deficit hyperactivity disorder. The progress of another random sample of 95 boys without the disorder was also followed. The study found that 4 of the hyperactive boys went on to graduate school to study a profession and 21 of the control group went into professions. Test the hypothesis at the .05 level that a smaller population percentage of attention-deficit hyperactive boys go into professional fields.

5. On November 7, 1991, Earvin "Magic" Johnson announced that he was infected with the HIV virus and would be retiring from professional basketball. Fourteen weeks before the announcement, a study conducted at a public clinic in Maryland found that out of a random sample of 186 participants, 60 reported having at least 3 partners of the opposite sex. Fourteen weeks after Johnson made his announcement, a random sample of 97 were surveyed. Twenty of these 97 reported having at least 3 partners of the opposite sex. Test the hypothesis at the .01 level that Magic Johnson's announcement may have had a significant effect on sexual risk behavior.

6. A work-site weight-reducing health program was tested for its effectiveness in reducing obesity as measured by the body mass index (BMI). A random sample of 16 work sites received classes on weight control combined with payroll incentives. Another sample of 16 work sites served as a control and received no treatment. The mean BMI for each site in the *treatment group* (the one that received classes) before and after a 2-year period is as follows:

BMI before Treatment	BMI after 2 Years
26.97	26.02
25.64	25.87
25.12	25.02
25.57	25.46
26.09	25.70
26.17	26.10
25.92	26.24
25.68	26.57
25.07	24.57
25.70	25.18
26.61	26.84
26.34	26.31
26.34	26.22
25.70	25.61
26.30	26.42
25.84	25.16

Test the hypothesis at the .01 level that there's no difference between the BMI of the treatment group before and after the 2-year treatment period.

7. A random sample of 8 pairs of identical 12-year-old twins took part in a study to see if vitamins helped their attention spans. For each pair, twin A was given a placebo, and twin B received a special vitamin supplement. A psychologist then determined the length of time (in minutes) each remained with a puzzle. The results were:

Twin A	Twin B
34	29
18	42
39	33
31	40
28	38
26	40
28	27
22	15

Use a paired t procedure to test the hypothesis at the .05 level that the vitamin supplement gives recipients a longer attention span.

8. A study in the *American Journal of Public Health* dealt with a randomly selected group of 264 primary caregivers. These caregivers were interviewed and asked to keep diaries for 6 months. Diaries were returned by 141 caregivers, and 123 did not return diaries. Of those who returned diaries, the average years of education was 13.9, and the standard deviation was 2.9 years. Of those who did not return diaries, the average years of education was 13.5, and the standard deviation was 3.2 years. Test the hypothesis at the .05 level that there was no difference in the education of the population of caregivers who fell into these two groups.

9. A researcher claims that the percent of patients who show signs of cognitive impairment is different for patients who are restrained and those who are unrestrained. A random sample of 5,834 nursing home patients who are restrained finds that 2,519 show no signs of cognitive impairment. Another random sample of 4,110 patients who are unrestrained finds that 2,983 show no signs of cognitive impairment. Test the researcher's claim at the .05 level of significance.

10. An experiment was conducted over a 3-year period in the Minneapolis-St. Paul area to evaluate the effectiveness of a work-site health promotion program in reducing obesity. Before the trial was run, the mean and standard deviation of body mass index (BMI), a measure of obesity, were determined in each work site. For the random sample of 16 work sites in the treatment group (the group receiving classes on weight control), the mean and standard deviation for BMI (kg/m^2) were 25.58 and .81, respectively. For the random sample of 16 work sites in the control group (the group not receiving any weight control instruction), the mean and standard deviation were 25.80 and .99, respectively. Test the hypothesis at the .05 level that there was no difference in the population variance of the BMI between the groups of work sites before the trial was run.

11. Use the statistics from problem 10 to test the hypothesis at the .01 level that there's no difference in the population mean BMI between the two groups of work sites.

12. A random sample of 937 factory workers participated in the first round of an intervention program for weight loss. The mean and standard deviation for the weight loss in round 1 was 5.9 and 1.1 pounds, respectively. In a later round of the intervention program, a random sample of 333 workers participated with a mean and standard deviation weight loss of 3.7 and 1.5 pounds, respectively. Test the hypothesis at the .05 level that there was no difference in the population weight loss between the workers in round 1 and the workers in the later round.

13. Some research indicates that antioxidant vitamins may be effective in cancer prevention. A study of the effect of the Treatwell Intervention Program was reported in the *American Journal of Public Health.* A record of the vitamin intake in milligrams per day was recorded for a random sample of subjects in a treatment group (the group in which intervention was employed) before the program started and after it was completed. The following is the before and after vitamin-intake record for the treatment group:

Vitamin	Baseline (before) Intake	Final (after) Intake
Vitamin B_6	1.54	1.61
Vitamin B_{12}	5.01	5.50
Folate	211.99	220.70
Pantothenic acid	3.66	3.74
Riboflavin	1.63	1.61
Vitamin C	129.66	134.84

Using the paired *t* procedure, test the hypothesis at the .01 level that there is no difference in vitamin intake for the treatment group between the time the program started and the time it was completed.

14. A study reported in the *American Journal of Sports Medicine* was undertaken to find the optimum type of knee surgery for patients with anterior cruciate ligament injuries. Two types of surgeries were performed. For technique A (anatomic location of the graft through a drill hole on the femur), surgery was successful for 22 out of the random sample of 30 patients. For technique B (modified over-the-top reconstruction), 32 out of a random sample of 54 surgeries were successful. Test the hypothesis at the .01 level that the population success rates for the two techniques are the same.

15. A study to see how well marital dissolution was predicted by the personality scores of husbands and wives was recently published in *The Journal of Personality and Social Psychology.* A random selection of 286 newlywed couples was made. After 5 years, 222 couples remained together, and 64 had dissolved their marriages. The mean age of the husband in the 222 stable marriages was 30.47, and the standard deviation was 7.65. For the 64 unstable couples, the mean age of the husband was 29.64, and the standard deviation was 6.83. Test the hypothesis at the .01 level that there's no difference in the population mean age of the husband in the two types of couples.

16. A claim is made that GM-powered race cars are faster than those using Ford engines. The qualifying times for a sample of 11 Ford-powered cars at a raceway was 119.02 seconds, and the standard deviation was 1.76 seconds. For a similar sample of 11 GM-powered cars, the time was 118.50 seconds, and the standard deviation was 1.24 seconds. Test the hypothesis at the .05 level that there's no difference in the population variance of the qualifying times for the cars powered by the two companies.

17. Use the statistics from problem 16 to test the hypothesis that the GM-powered race cars are faster than those using Ford engines. (This means they take *less time* to complete the qualifying laps.) Use the .05 level of significance.

18. Some professors of education claim there's a significant difference of opinion about testing and test uses between preservice and inservice teachers. In a study reported in the *Journal of Educational Research,* random samples from both groups were asked their opinion on this statement: "Standardized tests serve a useful purpose." (The response scale ranged from 1 = strongly agree to 6 = strongly disagree.) The 84 preservice sophomore education majors had a mean response of 4.02 with a standard deviation of .83, while the 32 inservice teachers responded to the statement with a mean of 2.93 and a standard deviation of .97. Test the professors' claim at the .01 level.

19. A Project CALC course is quite different from a traditional calculus course because it emphasizes interactive computer laboratories, cooperative learning, and extensive student writing. A final exam was given to a random sample of 46 students in a traditional course, who scored a mean of 9.64 and a standard deviation of 5.21 on a portion of the exam involving concepts. For the same portion of the test, 41 Project CALC students scored a mean of 11.52 and a standard deviation of 7.07. Test the hypothesis at the .05 level that the population mean score for Project CALC students is higher than the mean score for the traditional students.

20. A random sample of 996 franchise operations (fast-food restaurants, convenience stores, and so on) on a national mailing list had a mean age of 10.07 years and a standard deviation of 10.10 years. A consultant conducted a study in one region of the country using a random sample of 100 franchise operations. The sample mean age was 17.78 years, and the standard deviation was 11.99 years. Test the hypothesis at the .05 level that there's a difference in the population mean ages for these two groups.

21. In an *American Journal of Psychology* study of the characteristics of alcoholics, there were random samples of 12 normal subjects and 10 alcoholic subjects. The normal subjects had a mean score of 1 with a standard deviation of 0.6 on the Hamilton Rating Scale for Depression. For the alcoholic subjects, the mean rating on the same scale was 6 with a standard deviation of 3. Assume that the population variances are not equal, and test the hypothesis at the .01 level that alcoholics have a higher population mean rating on the Hamilton scale than do normal subjects.

22. *Macworld* has published results of tests that were made to compare the time needed by a sample of slide-making software packages to produce high-resolution and medium-resolution slides. The time (in seconds) needed by each package to produce a slide with a simple text and a logo image is as follows:

Software Package	High-Resolution Time	Medium-Resolution Time
Colorfast	340	167
FilmPrinter Turbo II	120	90
LFR	290	79
LFR Mark II	100	56
PCR II	178	94
Personal LFR	138	95
ProColor Premier	174	99
Spectra Star 450	642	184

Use the paired *t* procedure to test the hypothesis at the .05 level that there's a significant difference between high-resolution and medium-resolution times for the same software.

23. An article in *Teaching of Psychology* compared attitudes toward nuclear disarmament (AND) scores of teachers who have discussed nuclear disarmament issues in any course against the AND scores of those teachers who have not discussed the issue. For the random sample of 72 teachers who have discussed this issue, the mean on the AND questionnaire was 77.04 with a standard deviation of 13.28. For the random sample of 55 teachers who have not discussed this issue, the mean AND score was 69.33 with a standard deviation of 13.52. Test the hypothesis at the .01 level that population attitudes toward nuclear disarmament are different for the two groups of teachers.

24. Is having CPA certification an important salary factor? According to a recent issue of *Management Accounting*, a random sample of 15 accountants with no certification had a mean salary of $49,287. A sample of 12 accountants with CPA certification had a mean salary of $61,936. Assume the standard deviation for both samples is $500, and test the hypothesis at the .05 level that accountants with CPAs have higher salaries.

25. A study compared the step length in meters of amputees using a solid ankle cushion heel (SACH) prosthesis with those using a Carbon Copy II (CC II) device. For the random sample of 10 using SACH, the mean step length was .764 meter, and the standard deviation was .0011 meter. For the sample of 10 using CC II, the mean step length was .766 meter, and the standard deviation was .00054 meter. Test the hypothesis at the .05 level that the population variances for the two types of prosthetic feet are equal.

26. Use the data from problem 25 to test the hypothesis that there's no difference in the population mean step length between the two types of prosthetic feet. Use the .01 level.

27. *The Journal of Research in Crime and Delinquency* published a study on the situational characteristics of crimes. In a random sample of 96 robberies, 46 were related in some way to drugs or alcohol. And in a random sample of 69 assaults, 43 were related to drugs or alcohol. Test the hypothesis at the .05 level that the population percent of robberies related to drugs or alcohol is lower than that for assaults.

28. In a study in the *Journal of Small Business Management*, retail buyers ranked the "quality of the product" and "product fit" as the first and second most important variables they use to select a product. A random sample of 312 buyers evaluated "the quality of the product" with a mean score of 4.756 on a scale and a standard deviation of .486. For product fit, the mean score from a sample of 310 buyers was 4.697, and the standard deviation was .532. Test the hypothesis at the .05 level that there's no significant difference between the population mean scores for the two variables.

29. The mean heart rate (beats per minute) for a random sample of 31 patients with congestive heart failure who were given *Captopril* was 75.2 beats per minute, and the standard deviation was 2.7 beats per minute. The random sample of 31 patients who were given *Nitroprusside* had a mean heart rate of 77.4 beats per minute, and a standard deviation of 2.4 beats per minute. Test the hypothesis at the .01 level that there's no difference in heart rate for the population of patients given *Captopril* and those given *Nitroprusside.*

30. A study on psychological disorders and how they affect high school students was recently published in *The Journal of Abnormal Psychology.* Out of a random sample of 891 females in the study, 12 had an eating disorder. Of the random sample of 819 males, 1 had an eating disorder. Test the hypothesis at the .01 level that eating disorders are greater for the population of females.

31. Random samples of 26 men and 27 women responded to a questionnaire designed to determine the value of various activities in their lives. A comparison of the mean scores on a scale for each activity is as follows:

Activity	Men	Women
Studying	34.34	34.85
Working	43.95	44.31
Community	30.58	31.92
Home and Family	44.80	46.55
Leisure	39.42	40.39

Use the paired *t* procedure to test the hypothesis at the .01 level that there's no difference in the values of these activities in men and women.

32. The purpose of a study in *The New England Journal of Medicine* was to see if there was a difference between the mean cost per

case for psychiatric evaluation services for a self-referred group and for an independent-referral group. In a random sample of 155 self-referral cases, the mean cost per case was $3,222 with a standard deviation of $1,450. In a sample of 65 independently referred cases, the mean cost per case was $2,549, and the standard deviation was $742. Test the hypothesis at the .01 level that the population mean cost per case is higher for the self-referred group.

33. Syringe-exchange programs represent one attempt to slow the spread of HIV infection among injection-drug users. In a study conducted in New Haven, Connecticut, and reported in *The New England Journal of Medicine*, a random sample of 160 needles was tested for the HIV virus, and 108 were found to be positive. A year later, after the introduction of a syringe-exchange program, a sample of 338 needles was tested, and 164 were found to be HIV positive. Test the hypothesis at the .05 level that the population percentage of HIV-positive needles used by injection-drug users in New Haven was lower after the exchange program was in operation.

34. An article in *Criminal Justice and Behavior* claims that an intellectual imbalance of performance IQ (P) with verbal IQ (V)—such that $P > V$—is a useful predictor of the probability of becoming delinquent. A random sample of 269 male juvenile delinquents were classified in the low $P > V$ group, and a sample of 253 were classified in the high $P > V$ group. In a full-scale IQ test, the boys in the low $P > V$ group had a mean of 90.57 with a standard deviation of 16.1. For the high $P > V$ group, the mean IQ was 96.51 with a standard deviation of 11.4. Test the hypothesis at the .05 level that there's no difference between the population mean IQ scores for these groups.

35. A random sample of 21 sophomores who stayed home and commuted to their schools had a mean GPA of 2.685 with a standard deviation of .792. A random sample of 21 sophomores who went away to school had a mean GPA of 2.480 and a standard deviation of .689. At the .05 level, are the population variances for the two groups equal?

36. Do students who stay at home and commute to college have better GPAs than their friends who go away to school? Test the hypothesis at the .05 level that students who stay at home have a higher population mean GPA than those who go away. Use the data from problem 35.

37. Patients with non-Hodgkin's lymphoma are divided into two treatment groups. A random sample of 127 in the first group are put on a regimen called COPA. In the second group, a sample of 122 are put on a regimen that includes an interferon called I-COPA. There are 79 2-year survivors in the COPA group and 87 such survivors in the I-COPA group. Test the hypothesis at the .01 level that the 2-year survival rates are equal for the two treatments.

38. Transcutaneous electrical nerve stimulation (TENS) devices are frequently used in the management of acute and chronic pain conditions. An important component of the TENS system is the skin electrode. A study in *Physical Therapy* was done to determine conductive differences among the electrodes used with TENS devices. A sample of 11 electrodes from a low-impedance group were tested in two different trials. The results were:

Electrode Number	Trial 1 Impedance (Ohms)	Trial 2 Impedance (Ohms)
8	1200	1900
9	1200	1100
13	1000	1000
14	1600	1600
15	1400	1600
16	1400	1400
17	1200	1100
20	1700	1400
21	1600	1800
23	1300	1400
25	1600	1400

Use the paired t procedure to test the hypothesis at the .05 level that there is no significant difference between the impedance measurements in the two trials.

39. As noted in problem 38, an important component of the TENS system is the skin electrode. A test was carried out to determine the conductive differences among electrodes used with TENS devices. For the first trial of a random sample of 11 electrodes in the low-impedance group, the impedance measures (in ohms) were:

1200 1200 1000 1600 1400 1400
1200 1700 1600 1300 1600

And for the first trial of a random sample of 12 electrodes in the medium-impedance group, the measures (in ohms) were:

3100 3100 2900 2700 2500 3600
3100 3100 2100 4300 3000 2600

Test the hypothesis at the .05 level that the population variances for the two impedance groups are equal.

40. Use the statistics from problem 39 to test the hypothesis at the .05 level that the medium-impedance group has a significantly higher mean impedance level than the low-impedance group.

41. A *double-blind study* is one in which neither doctor nor patient knows who is receiving a drug and who is receiving a placebo. In such a study of 118 patients with mild to moderate heart difficulty, the mean change in total exercise time (in seconds) for the random sample of 60 patients given a placebo was 22 seconds, and the standard deviation was 13 seconds. For the experimental group of 58 patients given NORVASC, the mean change in total exercise time was 62 seconds with a standard deviation of 17 seconds. Test the hypothesis at the .01 level that NORVASC significantly improves the population mean exercise time.

42. Let's look again at the double-blind study discussed in problem 41. Out of the 60 patients who received a placebo, 17 had symptomatic improvement. Of the 58 patients given NORVASC, 32 had significant improvement. Test the hypothesis at the .01 level that the population percentage of patients who used NORVASC and experienced improvement is significantly greater than those who were given a placebo.

43. In a study comparing chronic schizophrenic patients (the random sample, $n = 26$) to normal subjects ($n = 25$), the mean years of education for the chronic schizophrenic patients was 13.0, and the standard deviation was 2.0. For the normal subjects the mean and standard deviation for years of education were 14.6 and 2.1, respectively. Test the hypothesis at the .05 level that the population variances for the two groups are equal.

44. Use the statistics from problem 43 to test the hypothesis that the population mean years of education is greater for normal patients than chronic schizophrenic patients. Use the .05 level of significance.

45. A study in the *Journal of Consumer Research* sought to determine if a change of intention to buy differed from store to store. The change-of-intention measure is a self-reported value on a scale of -9 to $+9$. At an advertised discount of 40 percent, the mean change of intention to buy store-brand items was 1.90 on this scale for a random sample of 10 customers at the May Company. The sample standard deviation was .15. At the same discount level, a random sample of 8 customers at Nordstrom's reported a mean change-of-intention score of 3.25 for store-brand items and a standard deviation of .21. Test the hypothesis at the .05 level that with a 40 percent discount the population variances at the two stores for the change of intention are equal.

46. Use the statistics given in problem 45 to test the hypothesis at the .05 level that the population mean change-of-intention value is different at the May Company than it is at Nordstrom's.

47. The perceived discount (PD) is measured as follows: PD = (perceived regular price − perceived sale price)/perceived regular price. A study to see if the PD for a high-image store was higher than the PD for a store with a lower image was published in the *Journal of Consumer Research.* The following data represent PD values for brand-name items at various actual discount rates at the May Company and at Nordstrom's:

Actual Discount (%)	May Company's Mean PD	Nordstrom's Mean PD
10	10.51	11.97
20	20.30	19.47
30	27.88	30.19
40	40.56	38.43
50	47.57	46.23
60	58.35	57.15
70	66.63	66.27

Use a paired t procedure to test the hypothesis at the .05 level that the population mean PD values are the same at Nordstrom's and the May Company.

48. Adolescents with learning disabilities in math were divided into two groups in a study recently reported in *Exceptional Children.* The random sample of 14 students in group 1 were taught problem-solving skills involving standard word problems. The sample of 15 students in group 2 were taught by a technique that used real-life problems and a videodisc program. After completing the course, both groups were tested. For group 1, the mean score was 21.39 with a standard deviation of 5.84. For group 2, the mean score was 21.87 with a standard deviation of 6.63. Test the hypothesis at the .01 level that the variances for the two population groups are equal.

49. For the learning-disabled students described in problem 48, test the hypothesis at the .05 level that there's no difference in the population mean scores made by students taught with the two methods.

50. In a recent study of HIV risk behaviors reported in the *American Journal of Psychiatry,* the responses of a random sample of 76 hospitalized adolescents requiring psychiatric help were compared with a sample of 802 school-based adolescents in the same city. Forty of the hospitalized adolescents and 195 of the school group reported that they were sexually active. Test the hypothesis at the .05 level that the population of psychiatrically hospitalized adolescents are more sexually active than school-based adolescents.

51. A recent article in *The New York Times* reported on a study that sought to determine if the spermicide *nonoxynol-9* was effective in protecting against the HIV virus. This study involved a randomly selected group of 138 HIV-negative prostitutes who received either a contraceptive sponge containing the spermicide *nonoxynol-9* or a placebo. Of the 60 women who received *nonoxynol-9,* 27 of them became HIV infected. These figures compared with 20 of the 56 in the placebo group. The remaining women didn't complete the study. Test the hypothesis at the .05 level that there's no difference in the population percentage of HIV infection in the two groups.

Topics for Review and Discussion

1. Explain the purpose of two-sample hypothesis tests of variances, means, and percentages.

2. For the two-sample test of variances, what are the necessary assumptions that must be made about the population if the test is to yield usable results?

3. Compare and contrast the four types of two-sample hypothesis tests of means discussed in this chapter. For each, describe the necessary assumptions.

4. "The paired t test actually uses the techniques discussed in Chapter 8." Why is this statement true?

5. What is a sampling distribution of the differences between sample means? How is such a distribution created?

6. When will the mean of the sampling distribution of the differences between sample means be equal to zero?

7. What assumptions are necessary to perform a two-sample test of percentages?

8. What is a sampling distribution of the differences between sample percentages? How is such a distribution created?

9. When will the mean of the sampling distribution of the differences between sample percentages be equal to zero?

10. Why is it necessary to use an estimated standard error of the differences between percentages when testing a hypothesis about population percentage differences?

Projects/Issues to Consider

1. Go to the library and look through journals that relate to a field that interests you. Locate a source in which there's raw data from two samples. First analyze the type of test that could be done to test the hypothesis that the population means are equal. Then perform the test. Discuss your findings.

2. Locate a library source that involves qualitative data and that gives percentage information from two samples. Perform a hypothesis test about the appropriate population percentages.

3. Identify a current issue—one about which people have differing opinions. Now compose a question about this issue and identify two populations (these could be male-female, student-nonstudent, undergraduate-graduate student, over 25-under 25, smoker-nonsmoker, and so on) from which you will select samples and record responses. Perform a test to see if the percentages in the two populations are likely to be different. Discuss your methodology and results.

Computer Exercises

1–4. Using the data in STR 9.1, problem 9:

1. Determine the standard deviations for the control group and treatment group samples.

2. Assuming that both populations are normally distributed, conduct a test to see if the population variances are likely to be equal for the two groups.

3. Use your software to test the hypothesis at the .05 level that there's no difference after 2 years in the population mean BMI between the treatment group and the control group. (*Hint:* If you're using *Minitab*, you can follow the procedure shown in Figure 9.6.)

4. Discuss the procedures used and the results obtained.

5–6. Using the data in STR 9.2, problem 17:

5. Conduct a paired t test at the .01 level to see if there's likely to be a significant decrease in the population cholesterol level after the diet.

6. Discuss the procedures used and the results obtained.

7–9. We've performed several analyses on the billable hours data (see STR 3.2, page 49) that Bill Alott used to prepare his GOTCHA report. You may already have saved the data in that report that came from the week of March 15. Now, let's assume that we want to compare the hours listed for March 15 with the hours posted for the following week of March 22 to see if there has been a change in the mean number of hours reported. The hours for the week of March 22 are:

41	46	32	50	46	43	40	53	48	37
47	49	47	46	49	41	43	34	46	37
44	44	40	46	40	50	57	45	46	45
41	58	41	47	47					

7. Use your software to find the standard deviations for both large samples.

8. Test the hypotheses at the .05 level that the population means are likely to be equal for the two periods.

9. Discuss the procedures used and the results obtained.

10–11. Using the data in STR 9.2, problem 4:

10. Use a two-sample t test software procedure at the .05 level to test the

hypothesis that the population rating in the computer and office equipment industry is lower than in the electronics industry. (*Hint:* If you're using *Minitab,* you can follow a procedure similar to the one shown in Figure 9.6, but you'll want to use an ALTERNATIVE subcommand for this one-tailed test. Note, too, that since the population variances aren't equal, the POOLED subcommand in Figure 9.6 shouldn't be used.)

11. Discuss the procedures used and the results obtained.

Answers to Odd-Numbered Self-Testing Review Questions

Section 9.1

1. To perform a test of equal variances that gives meaningful results, both populations must be normally distributed, and the samples must be independently selected.

3. *Step 1: State the null and alternative hypotheses.* $H_0 : \sigma_1^2 = \sigma_2^2$ and $H_1 : \sigma_1^2 \neq \sigma_2^2$. *Step 2: Select the level of significance.* Use $\alpha = .05$. *Step 3: Determine the test distribution to use.* We use an F distribution. Since this is a two-tailed test, we use the table with half of .05, or .025, in the right tail. The larger variance has a sample size of 16, so the df for the numerator is 16 − 1 or 15. The sample size for the smaller variance is 7, so 7 − 1 or 6 is the df for the denominator. *Step 4: Define the critical or rejection region(s).* You look across the top line of the F table for a df value of 15 and down the left column until you get to the line with df = 6. The critical F value is 5.27. *Step 5: State the decision rule.* Reject H_0 and accept H_1 if the TR is > 5.27. Otherwise, accept H_0. *Step 6: Compute the test ratio.* The TR = 457.21/99.08 = 4.6146. *Step 7: Make the statistical decision.* Since a TR of 4.6146 does not fall in the rejection region, we accept the H_0 that the variances for the two populations are equal.

5. *Step 1.* $H_0 : \sigma_1^2 = \sigma_2^2$ and $H : \sigma_1^2 \neq \sigma_2^2$. *Step 2.* Use $\alpha = .05$. *Step 3.* We use an F distribution. Since this is a two-tailed test, we use the table with .025 in the right tail. The df for the numerator is 16 − 1 or 15. And the df for the denominator is 13 − 1 or 12. *Step 4.* The critical F value is 3.18. *Step 5.* Reject H_0 and accept H_1 if the TR is > 3.18. Otherwise, accept H_0. *Step 6.* The TR = 17.39/12.83 = 1.3554. *Step 7.* Since a TR of 1.3554 does not fall in the rejection region, we accept the H_0 that the variances are equal for the two shifts.

7. *Step 1.* $H_0 : \sigma_1^2 = \sigma_2^2$ and $H_1 : \sigma_1^2 \neq \sigma_2^2$. *Step 2.* Use $\alpha = .05$. *Step 3.* We use an F distribution. Since this is a two-tailed test, we use the table with .025 in the right tail. The df for the numerator is 10 − 1 or 9. And the df for the denominator is 10 − 1 or 9. *Step 4.* The critical F value is 4.03. *Step 5.* Reject H_0 and accept H_1 if the TR is > 4.03. Otherwise, accept H_0. *Step 6.* The variance for sample 1 = 18.9957/(10 − 1) = 2.1106, and for sample 2 the variance = 3.8601/(10 − 1) = .4289. So the TR = 4.921. *Step 7.* Since a TR of 4.921 falls in the rejection region, we reject the H_0 and accept the H_1 that the variances of the two groups are not equal.

9. *Step 1.* $H_0 : \sigma_1^2 = \sigma_2^2$ and $H_1 : \sigma_1^2 \neq \sigma_2^2$. *Step 2.* Use $\alpha = .05$. *Step 3.* We use an F distribution. Since this is a two-tailed test, we use the table with .025 in the right tail. The df for the numerator is 16 − 1 or 15. And the df for the denominator is 16 − 1 or 15. *Step 4.* The critical F value is 2.86. *Step 5.* Reject H_0 and accept H_1 if the TR is > 2.86. Otherwise, accept H_0. *Step 6.* The variances are 5.8943/(16 − 1) = .3930 and 3.1740/(16 − 1) = .2117, so TR = .3930/.2117 = 1.8561. *Step 7.* Since a TR of 1.8561 does not fall in the rejection region, we accept the H_0 that the variances are equal.

11. *Step 1.* $H_0 : \sigma_1^2 = \sigma_2^2$ and $H_1 : \sigma_1^2 \neq \sigma_2^2$. *Step 2.* Use $\alpha = .05$. *Step 3.* We use an F distribution. Since this is a two-tailed test, we use the table with .025 in the right tail. The df for the numerator is 25 − 1 = 24. And the df for the denominator is 25 − 1 = 24. *Step 4.* The critical F value is 2.27. *Step 5.* Reject H_0 and accept H_1 if the TR is > 2.27. Otherwise, accept H_0. *Step 6.* The TR = 1.44/.9409 = 1.530. *Step 7.* Since a TR of 1.530 does not fall in the rejection region, we accept the H_0 that the variances are equal.

13. *Step 1.* $H_0 : \sigma_1^2 = \sigma_2^2$ and $H_1 : \sigma_1^2 \neq \sigma_2^2$. *Step 2.* Use $\alpha = .05$. *Step 3.* We use an F distribution. Since this is a two-tailed test, we use

the table with .025 in the right tail. The df for the numerator is 25 − 1 = 24. And the df for the denominator is 25 − 1 = 24. *Step 4.* The critical F value is 2.27. *Step 5.* Reject H_0 and accept H_1 if the TR is > 2.27. Otherwise, accept H_0. *Step 6.* The TR = 282.24/134.56 = 2.0975. *Step 7.* Since a TR of 2.0975 does not fall in the rejection region, we accept the H_0 that the variances are equal.

15. *Step 1.* $H_0 : \sigma_1^2 = \sigma_2^2$ and $H_1 : \sigma_1^2 \neq \sigma_2^2$. *Step 2.* Use $\alpha = .05$. *Step 3.* We use an F distribution. Since this is a two-tailed test, we use the table with .025 in the right tail. The df for the numerator is 5 − 1 = 4. And the df for the denominator is 5 − 1 = 4. *Step 4.* The critical F value is 9.60. *Step 5.* Reject H_0 and accept H_1 if the TR is > 9.60. Otherwise, accept H_0. *Step 6.* The TR = 49/16 = 3.0625. *Step 7.* Since a TR of 3.0625 does not fall in the rejection region, we accept the H_0 that the variances are equal.

17. *Step 1.* $H_0 : \sigma_1^2 = \sigma_2^2$ and $H_1 : \sigma_1^2 \neq \sigma_2^2$. *Step 2.* Use $\alpha = .05$. *Step 3.* We use an F distribution. Since this is a two-tailed test, we use the table with .025 in the right tail. The df for the numerator is 21 − 1 = 20. And the df for the denominator is 21 − 1 = 20. *Step 4.* The critical F value is 2.46. *Step 5.* Reject H_0 and accept H_1 if the TR is > 2.46. Otherwise, accept H_0. *Step 6.* The TR = 2.56/1.96 = 1.3061. *Step 7.* Since a TR of 1.3061 does not fall in the rejection region, we accept the H_0 that the variances are equal.

Section 9.2

1. Procedure 1 and the paired t test are used when the samples are dependent and the populations are normally distributed. Procedure 2 and the z distribution are used to test the equality of two means when the samples are independent and when both samples are over 30 or when the populations are normally distributed and the standard deviations for both populations are known. If the independent samples are small (30 or less) and the population variances are not equal, the t distribution is used with Procedure 3. Finally, if the independent samples are small and the population variances are equal, Procedure 4 and the t distribution are used to conduct the test. Both Procedure 3 and Procedure 4 also require the assumption of normality.

3. *Step 1: State the null and alternative hypotheses.* $H_0 : \mu_1 = \mu_2$ and $H_1 : \mu_1 > \mu_2$. *Step 2: Select the level of significance.* $\alpha = .01$. *Step 3: Determine the test distribution to use.* Since the samples are independent and both samples are large, we use Procedure 2 with the z distribution. *Step 4: Define the critical or rejection region(s).* The critical z value is 2.33. *Step 5: State the decision rule.* Reject H_0 and accept H_1 if the TR is > 2.33. Otherwise, accept H_0. *Step 6: Compute the test ratio (TR).* The TR is found as follows:

$$\text{TR} = \frac{\bar{x}_1 - \bar{x}_2}{\sqrt{\dfrac{s_1^2}{n_1} + \dfrac{s_2^2}{n_2}}} = \frac{38.08 - 35.60}{\sqrt{\dfrac{.84^2}{44} + \dfrac{.79^2}{47}}}$$

$$= \frac{2.48}{.1712} = 14.49$$

Step 7: Make the statistical decision. Since the TR of 14.49 falls in the rejection region, we reject the null hypothesis. The PD at Nordstrom's is greater than the PD at the May Company.

5. *Step 1.* $H_0 : \mu_d = 0$ and $H_1 : \mu_d \neq 0$. *Step 2.* $\alpha = .01$. *Step 3.* We use a paired t test to see if there is a match between the before and after values. You'll need a t distribution with 15 df. *Step 4.* The critical t value with 15 df is ±2.974. *Step 5.* Reject H_0 and accept H_1 if the TR is < −2.974 or > +2.974. Otherwise, accept H_0. *Step 6.* The TR = $(\bar{d} - \mu_d)/(s_d/\sqrt{n})$ = $(.044 - 0)/(.469/\sqrt{16})$ = .044/.117 = .3761. *Step 7.* Since the TR of .3761 does not fall in the rejection region, we accept the H_0. The work-site program does not appear to be effective in reducing BMI.

7. *Step 1.* $H_0 : \mu_1 = \mu_2$ and $H_1 : \mu_1 \neq \mu_2$. *Step 2.* $\alpha = .05$. *Step 3.* Since the samples are independent and both samples are large, we use Procedure 2 with a z distribution. *Step 4.* The critical z value is ±1.96. *Step 5.* Reject H_0 and accept H_1 if the TR is < −1.96 or is > +1.96. Otherwise, accept H_0. *Step 6.* The TR = $(\bar{x}_1 - \bar{x}_2)/\sqrt{s_1^2/n_1 + s_2^2/n_2}$ = $(42 - 37.2)/\sqrt{(34.7^2/141) + (33.4^2/123)}$ = 1.1439. *Step 7.* Since the TR of 1.1439 does not fall in the rejection region, we accept the H_0. There is no difference in the amount of time spent between the two groups.

9. *Step 1.* $H_0 : \mu_d = 0$ and $H_1 : \mu_d > 0$. *Step 2.* $\alpha = .05$. *Step 3.* We use a paired t test with 7 df. *Step 4.* The critical t value is 1.895. *Step 5.* Reject H_0 and accept H_1 if the TR is > 1.895. Otherwise, accept H_0. *Step 6.* The TR = $(\bar{d} - \mu_d)/(s_d/\sqrt{n})$ = $(139.75 - 0)/(144.0930/\sqrt{8})$ = 139.75/50.9446 = 2.7432. *Step 7.* Since the TR of 2.7432 falls in the rejection region, we reject the H_0. The high-resolution slides take longer to produce.

11. *Step 1.* $H_0 : \mu_1 = \mu_2$ and $H_1 : \mu_1 \neq \mu_2$.

LOOKING AHEAD

Earlier chapters have shown that the statistical tools that you've now learned may be used to help answer questions and solve problems in many fields of study. Now, in this chapter, we'll take a closer look at one such application area to see how statistical methods can help control the quality of goods and services produced by many organizations.

In recent years, concepts assembled under the general heading of "total quality management" have become indispensable in manufacturing plants, service industries, educational institutions, and other organizations across the United States and around the world. *Statistical process control* (or SPC) is an important component of total quality management, and our focus in this chapter is on SPC and on several of the SPC control charts that are commonly used in a variety of workplace settings.

Thus, after studying this chapter, you should be able to:

- Describe the historical development of statistical process control.
- Construct a control chart for individual values or observations.
- Construct $\overline{X}$ and R control charts.
- Construct a p chart for qualitative data.

10.1 Quality Control Concepts: A Brief Historical Perspective

Before we examine some basic quality control tools, let's briefly look at the origins of these tools. A good place to begin is with the work of Dr. Walter Shewhart, a Bell Laboratories scientist who published a paper in 1924 that outlined many of the principles of statistical quality control. In the course of his work in the 1920s and 1930s, Shewhart (and his Bell Laboratories colleagues H. F. Dodge and H. G. Romig) developed statistical control charts and did additional work on the theory of statistical sampling for quality control purposes. During this time, Shewhart and his colleagues developed techniques to make the processes of providing goods and services more predictable and consistent. In doing so, they used the laws of probability and statistics that we examined in Chapters 4, 5, and 6 to describe the way variation affects sample measures for manufactured goods.

You know from earlier chapters that when goods and services are produced, these outputs will be similar but not identical. Variation is natural and normal! No two things are likely to be exactly alike; there will almost always be some variation. But Shewhart saw that such variability could be viewed as being within the limits set by chance or as being outside those limits. Thus, he observed that data did not always conform to a "normal" pattern, and out of that inconsistency he concluded that while every process produces variation, some processes generate *controlled variation* while others display *uncontrolled variation.*

The Deming Funnel Experiment

A funnel, fastened to a stand, is placed above a bull's-eye drawn on a piece of paper. The purpose of the experiment is to drop a marble through the funnel onto the paper as close to the bull's-eye as possible. The point of the actual impact is then marked with a pen. A pattern showing the extent of the variation in impact points is typically established after 30 or 40 drops. In spite of the best efforts of the participant, the marble won't land exactly on the bull's-eye each time. Rather, common-cause variation will occur. If an attempt is made to move the funnel around in reaction to this variation, the result is likely to be even greater variability. The lesson taught in this experiment is that a stable common-cause process (such as the one with the funnel in a fixed position) should probably be left alone for best results. If results must be improved, the funnel apparatus itself (the common-cause system) should be refined.

A **controlled**, or **common-cause, variation** is one that occurs naturally and is inherent and expected in any stable process that provides products and services. Such acceptable and allowable variation can be attributed to "chance" or random causes.

In Figure 10.1*a*, the process is stable and controlled although variation occurs daily about a central value.

An **uncontrolled**, or **special-cause, variation** is one that occurs when an abnormal action enters a process and produces unexpected and unpredictable changes that can no longer be attributed to chance or random causes.

In Figure 10.1*b*, the process is uncontrolled, and the variation is unpredictable. In Figure 10.1*a*, it's likely that the same variation pattern that has occurred earlier will reappear on Friday, but in Figure 10.1*b*, the Friday variation can't be anticipated. Common-cause variation in manufacturing occurs in the quality of incoming materials, training level of machine operators, and design of machines. Special-cause variation may occur because of power failures, machines out of adjustment, or differences in worker training.

Unfortunately, Shewhart's techniques were not widely adopted in the 1920s and 1930s. Many felt then that his ideas were too complicated. (Of course, those folks generally didn't have the considerable background and expertise in statistics that you've gained in the past several weeks.)

Robotic welding lines and statistical process control techniques have improved the quality of the cars people buy. (*George Haling/Photo Researchers*)

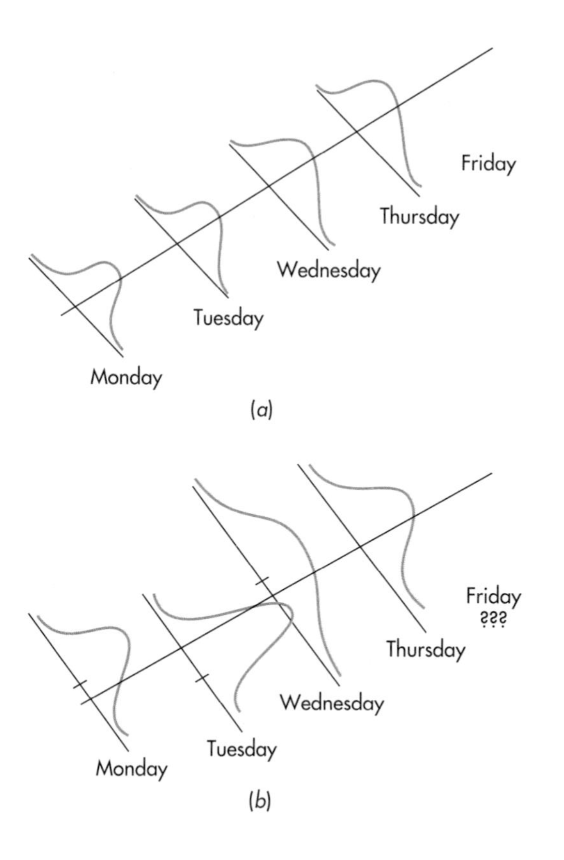

FIGURE 10.1 (*a*) Process is in control. (*b*) Process is out of control.

Opinions changed, though, in the early 1940s when the country entered World War II. The war created a huge new demand for mass-produced products, and the U.S. government and its military branches began to insist that military suppliers use statistical sampling and inspection methods to make sure that their products complied with government quality standards. And it wasn't long before the use of these government-mandated quality control techniques were adopted by suppliers of nonmilitary goods.

W. Edwards Deming, who had worked with Shewhart, was a wartime leader in the introduction of statistical quality control methods in the United States. Deming felt that the struggle to improve quality depended on waste reduction. Suppliers shouldn't accept defective incoming materials that didn't conform to stated specifications since using these materials would result in products that wouldn't meet the output standards that had been specified. The key, Deming felt, was to introduce training systems in which "managers and production workers alike are instructed in statistical methodology."

After the war ended, Deming took his quality control techniques to Japan, where businesses eagerly adopted them to revive a devastated economy. Deming believed that simply meeting specifications wasn't enough. Producers had to do more; they had to study their processes, search out and eliminate sources of variation, and thus constantly improve their products and services. His quality control methodology included the following four-stage cycle that was designed to produce continuous improvement:

1. **PLAN.** Determine the goals and make a plan to achieve them.
2. **DO.** Make the necessary changes.

Strong and Stiff Enough?

Advanced Composite Materials Corporation (ACMC) of Greer, South Carolina, makes materials for aerospace and commercial applications. ACMC must certify the reliability of these materials and guarantee that they are strong and stiff enough to exceed a customer's specifications. A quality control specialist uses statistical process control (SPC) to determine the lower tolerance limits for the strength and stiffness of the materials supplied, and these lower limits, of course, exceed customer specifications.

3. **C**HECK. Evaluate the results of your changes.
4. **A**CT. Does the data confirm the plan? Does the plan need revision?

And when this PDCA cycle was completed, Deming believed, the organization should go back to step 1 and begin another quality improvement cycle:

While the Japanese were zealously adopting Deming's teachings in the years that followed, suppliers in the United States faced a booming economy and saw no reason to change their methods. Businesses in the United States were emphasizing financial controls while the Japanese were stressing process improvement. After several decades of complacency, during which they lost markets to their Japanese competitors, U.S. suppliers were forced to refocus their attention on the quality issue. In 1981, Ford Motor Company adopted some of Deming's methods. Seeing the advantages, other American companies quickly followed. Today, statistical process control (SPC) is an indispensable tool for those providing goods and services around the world.

Self-Testing Review 10.1

1. Who was the original developer of techniques used in SPC?
2. Who was responsible for bringing quality control to Japan?
3. Why didn't the United States use SPC after World War II?
4. Discuss the differences between common-cause variation and special-cause variation.
5. Discuss the four steps of the PDCA cycle.

10.2 An Introduction to Control Charts

Control charts (also called *process charts* or *quality control charts*) are now widely used to detect uncontrolled variation and to thus monitor a process:

> A **control chart** is a graphic display that compares the data produced by a current process to a set of stable control limits that have been established from previous performance data.

Control charts provide a means for communicating information about the performance of a process between producing groups, between suppliers, or between machine operators.

Types of Control Charts

There are many types of control charts, and we can only consider a few in this chapter. Some basic charts are used to show a series of *individual values* or observations, some plot the *means* of subgroups, some graph the *ranges* of subgroups, and others show the *proportion of defects* that occur in subgroups. We'll first consider a chart in this section that may be used to plot individual values. This type of chart is easy and economical to use, and we'll assume that the data under consideration are normally distributed. In the following sections we'll look at the other types of basic charts mentioned above.

Elements of Control Charts

Although there are various types of control charts, each has the following three elements:

- The upper control limit (UCL)
- The center line (CL)
- The lower control limit (LCL)

These elements are shown in Figure 10.2. The center line (CL) in Figure 10.2 corresponds to the expected population mean of the values being observed in a process. We know, of course, that common-cause variation exists in any stable and controlled process, and the upper (UCL) and lower (LCL) control lines show the limits of the variations we expect to see if the process is indeed in control. But if an observation or measurement falls above the UCL or below the LCL lines, then it's likely that this observation represents an uncontrolled or special-cause variation.

From our study of Chapter 5, we know that when a variable is normally distributed, 99.74 percent (or almost all) of the variable measures are located within a distance of 3 standard deviations of the mean. That is, the mean ± 3 standard deviations $=$ 99.74 percent of all values of the variable. Thus, the placement of the UCL line generally represents a value that is 3 standard deviations above the mean, or CL, and the placement of the LCL represents a value that is 3 standard deviations below the CL. The chance of getting an observation that falls *above* the UCL or *below* the LCL in a process

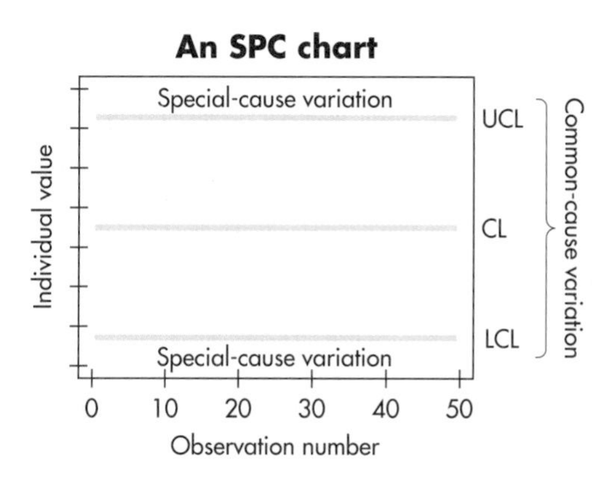

FIGURE 10.2

that *is in control* is only 1 − .9974, or .0026 (that's only 26 times in 10,000 observations).

A Seven-Step Hypothesis-Testing Procedure Using Control Charts

Control charts are used to determine if a process is in *statistical control*:

A process is in **statistical control** if there is no uncontrolled, or special-cause, variation present.

Although other approaches are commonly applied, we'll use control charts and a seven-step hypothesis-testing approach to help determine if a process is in statistical control. Since you've followed similar steps to arrive at statistical decisions in Chapters 8 and 9, the following procedure will seem familiar.

Step 1: State the Null and Alternative Hypotheses The *null hypothesis* in all tests in this chapter is:

H_0: The process is in statistical control.

And the alternative hypothesis is:

H_1: The process in not in statistical control. (This means, of course, that it's likely that uncontrolled, or special-cause, variation exists.)

Step 2: Select the Level of Significance We'll again use the symbol α to represent the risk of erroneously rejecting the H_0. In this chapter, we'll always use the α value of .0026 that was discussed above.

Step 3: Determine the Control Chart(s) and the Test Distribution to Use One or more of the types of control charts listed earlier (individual values, means of subgroups, ranges of subgroups, or proportions of defects) is selected, and an appropriate test distribution is used. Our examples in this chapter will use either normal or binomial probability distributions.

Step 4: Define the Rejection or Critical Regions for the Chart(s) The critical regions are defined by the upper control limit and the lower control limit, and these limits must be determined. Any observation or measurement that is greater than the UCL or less than the LCL falls into a rejection region.

Step 5: State the Decision Rule Our decision rule is to reject H_0 and accept H_1 if at least one data point falls beyond the control limits specified in *Step 4.*

Step 6: Enter the Data on the Control Chart(s) The data points are plotted on the charts(s).

Step 7: Make a Statistical Decision If all the data points fall between the center line and the UCL and LCL lines, then the null hypothesis is accepted. But if one or more data points fall above the UCL or below the LCL, the H_0 is rejected and the H_1 is accepted.

Control Charts for Individual Values

Let's look now at examples of how this hypothesis-testing procedure can be used to see if a series of individual values is likely to have come from a process that is in statistical control.

Example 10.1. Justin Thyme is the chief quality control officer at Highqual Corporation, a firm that makes roller bearings that are used as components in tractors. He has determined from previous batches that the population mean diameter of these bearings is 55 centimeters and the population standard deviation is 1 centimeter. To see if the manufacturing process is in control, Justin performs a hypothesis test using an individual values control chart. He has an employee measure the diameters of the first 30 roller bearings that are produced in a work shift. The results are as follows:

Observation	Diameter	Observation	Diameter
1	55.49	16	54.94
2	54.83	17	55.94
3	53.91	18	56.97
4	54.87	19	55.40
5	54.69	20	53.41
6	53.77	21	54.69
7	55.34	22	54.50
8	55.67	23	56.44
9	53.83	24	53.38
10	54.82	25	55.05
11	53.85	26	55.19
12	53.22	27	55.43
13	54.11	28	55.06
14	54.49	29	55.53
15	52.95	30	54.03

Justin now carries out the following steps:

Step 1: State the Null and Alternative Hypotheses The following hypotheses are established: H_0: The process is in statistical control, and H_1: The process is not in statistical control.

Step 2: Select the Level of Significance The value of $\alpha = .0026$.

Step 3: Determine the Control Chart(s) and the Test Distribution to Use A control chart for individual values based upon normal distribution probabilities is needed.

Step 4: Define the Rejection or Critical Regions for the Chart(s) The critical regions for an individual values control chart are defined as follows:

The upper control limit (UCL) $= \mu + 3\sigma$
The center line (CL) $= \mu$
The lower control limit (LCL) $= \mu - 3\sigma$

On a horizontal line at the bottom of his chart, Justin numbers the observations from 1 to 30, and on a left vertical line he marks a scale for the diameter measurements the employee has supplied. Next, he determines the values for the three essential components of the control chart. The center line is at 55 centimeters since this is the population mean diameter when the process is in statistical control. The upper control limit (UCL) is 3 standard deviations above the center line, so UCL $= 55 + 3(1) = 58$, and the lower control limit (LCL) $= 55 - 3(1) = 52$.

Step 5: State the Decision Rule The decision rule is to reject H_0 and accept H_1 if at least one data point falls beyond the control limits.

Step 6: Enter the Data on the Control Chart(s) Justin enters his data on the control chart as shown in Figure 10.3.

Step 7: Make a Statistical Decision Justin notes that all values lie between the UCL and the LCL, so he accepts the H_0 that the process is in statistical control. Although there is variation from one roller bearing to another, the variation can be explained by common causes. Happy with the results, Justin commends the workers for a job well done.

Example 10.2. Penny Urndt, managing director of the CPA firm of Dollar and Senz, knows that it's important that financial statements be promptly delivered to the firm's clients. Hoping to improve client service, Penny has recently attended a seminar where she has learned statistical process control methods, and she now decides to use her new knowledge to see if her firm is "in control" in the delivery of financial statements. In the past, the mean time for the final delivery of a financial statement was 14 working days, and the standard deviation was 1.5 working days. The following data represent the number of working days it has recently taken to deliver financial statements to 35 clients:

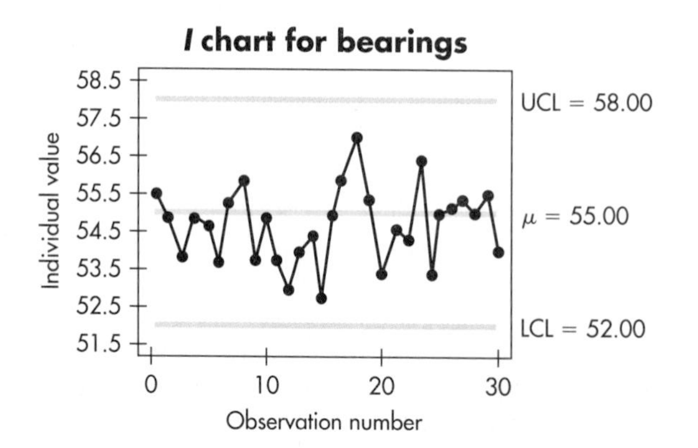

FIGURE 10.3

Client (Observation) Number	Days	Client (Observation) Number	Days
1	13	19	13
2	13	20	11
3	10	21	11
4	10	22	11
5	15	23	7
6	16	24	15
7	12	25	11
8	9	26	18
9	16	27	17
10	13	28	17
11	13	29	5
12	15	30	18
13	14	31	12
14	20	32	14
15	23	33	17
16	9	34	11
17	14	35	12
18	16		

Penny is now ready to perform her hypothesis test.

Step 1 H_0: The process is in statistical control, and H_1: The process is not in statistical control.

Step 2 $\alpha = .0026$.

Step 3 Penny will use a control chart for individual values based on normal distribution probabilities.

Step 4 Computing the essential components for the control chart, Penny knows from past experience that the center line value should be 14 days. The UCL is then drawn at a value of $14 + 3(1.5)$, or 18.5, and the LCL is drawn at $14 - 3(1.5)$, or 9.5.

Step 5 The decision rule is to reject H_0 and accept H_1 if at least one data point falls beyond the control limits.

Step 6 Penny draws the control chart shown in Figure 10.4.

Step 7 Since there are several points beyond the control limits, she'll reject H_0 and accept H_1 that the process is out of control. Observation numbers 8, 16, 23, and 29 fall below the lower control limit, and Penny fears that these statements might have been hastily prepared. And observation numbers 14 and 15 fall above the upper control limit, and she suspects that the lengthy period it has taken to complete these might lead to customer dissatisfaction and perhaps even to the loss of clients. Thus, her next task is to look for any special causes and try to eliminate them.

FIGURE 10.4

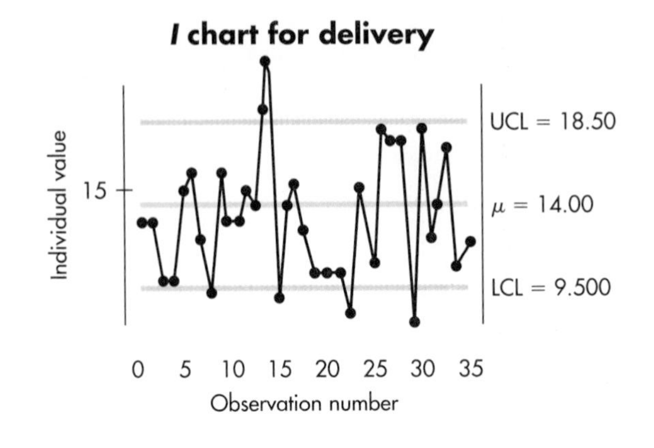

Self Testing Review 10.2

1. The Slurper's Choice Coffee Company buys imported coffee beans in bulk shipments and then uses a secret recipe to freeze-dry these beans and make them into a gourmet instant coffee product. To monitor the first stage of this coffee-making process, Ima Brewer weighs 30 bags of coffee beans from an incoming shipment provided by a new supplier. The results are as follows:

Bag Number	Weight	Bag Number	Weight
1	41	16	43
2	46	17	50
3	45	18	52
4	57	19	47
5	47	20	39
6	29	21	45
7	41	22	46
8	45	23	43
9	48	24	43
10	53	25	41
11	48	26	46
12	40	27	51
13	40	28	39
14	52	29	49
15	52	30	40

From past history, Ima knows the mean weight of such bags is 45 kilograms, and the standard deviation is 5 kilograms. Test the hypothesis at $\alpha = .0026$ that the shipment is in statistical control.

2. Mort Alitee, a district manager for Sooner or Later Insurance Corporation (SLIC), has observed a decrease in new policy sales. Suspecting that worker morale is low and may be out of control, Mort asks 36 employees to respond to a psychological questionnaire. The questionnaire produces morale ratings on a scale from 1 to 10

Statistical data may be entered into a company database from a warehouse location or from a manufacturing facility. (*Courtesy Hewlett Packard*)

for each employee. In the past, the mean morale rating at the company was 6, and the standard deviation was 1.2. Mort's latest survey produces the following results:

Employee Number	Morale Rating	Employee Number	Morale Rating
1	4	19	7
2	6	20	8
3	5	21	4
4	5	22	5
5	6	23	5
6	5	24	6
7	7	25	7
8	1	26	5
9	6	27	5
10	8	28	5
11	6	29	4
12	5	30	5
13	2	31	5
14	5	32	5
15	7	33	6
16	5	34	5
17	8	35	8
18	5	36	5

Test the hypothesis at $\alpha = .0026$ that the morale ratings at SLIC are in control.

10.3 An Introduction to $\overline{X}$ and *R* Charts

Although control charts for individual values are useful in many situations, when the process is not "normal," the interpretation of individual value charts is risky. Charts for individual values are not sensitive to small shifts or gradual trends (slow continuing change). When each item produced is a discrete unit that is independent of other units and the output naturally falls into subgroups, it's more desirable to monitor the variability with two other charts that are usually used together. The first of these charts we'll consider is the $\overline{X}$ *chart.*

An $\overline{\mathbf{X}}$ **chart** is one that monitors the mean value in a process.

And the second chart is the *R chart.*

An ***R*** **chart** is one that monitors the dispersion in a process.

To construct an $\overline{X}$ chart, the first step is to gather the data in subgroups that have n observations each. Groups of four or five are commonly used for subgroup sample sizes since such small samples make data collection easier. In most cases the population process mean and standard deviation aren't known, but they can be approximated. The population process mean is estimated by the *grand mean* ($\overline{\overline{X}}$), which is the *mean of the subgroup means.* This grand mean is calculated by first totaling all the subgroup means and then dividing this total by the number of subgroups. The grand mean thus produced is the *center line* (CL) in an $\overline{X}$ chart. An estimate of 3 standard deviations above the $\overline{\overline{X}}$ produces the upper control limit (UCL), and 3 standard deviations below the $\overline{\overline{X}}$ yields the lower control limit (LCL).

Quality control experts use the following equation to compute the upper control limit for an $\overline{X}$ chart (remember that $\overline{\overline{X}}$ is the center line):

$$\text{UCL} = \overline{\overline{X}} + A_2\overline{R} \tag{10.1}$$

where A_2 = a value that depends on the subgroup sample size
$\overline{R}$ = the mean of all the subgroup ranges; that is, the sum of all the subgroup ranges (remember that a range is the difference between the highest and lowest value) divided by the number of such subgroups
$A_2\overline{R}$ = an estimate of $3\sigma_{\overline{x}}$

Values for A_2 can be found in Appendix 11 at the back of the book. If, for example, the subgroup sample size is 5, then we see in Appendix 11 that the value of A_2 is .577, and if the subgroup sample size is 4, then A_2 is .729.

And the following equation is used to find the lower control limit for an $\overline{X}$ chart:

$$\text{LCL} = \overline{\overline{X}} - A_2\overline{R} \tag{10.2}$$

Thus, an $\overline{X}$ chart will look like the one shown in Figure 10.5.

The *center line* (CL) for an R chart is $\overline{R}$, which we've seen is the mean of all the subgroup ranges. The upper and lower control limits for an R chart are found with these equations:

$$\text{UCL} = D_4\overline{R} \tag{10.3}$$

Down to 2 Percent

Unipar, Inc., a State College, Pennsylvania, rotational molding company, makes hollow plastic products including diesel fuel tanks for Ford vehicles. Although Unipar uses a high-strength material, some of the tanks developed leaks. To reduce defects, Unipar's quality assurance manager began using SPC. Data representing the temperature of the heated plastic and the speed of the rotational process were graphed on $\overline{X}$ and R charts. From these charts, the mean temperature and speed were identified, and control limits were set. To ensure that the manufacturing process was in statistical control, similar charts were produced and monitored by quality control technicians who looked for trends or points outside the control limits that indicated an unsatisfactory process. Unipar was able to reduce defects to a level of 2 percent per month and thus save thousands of dollars.

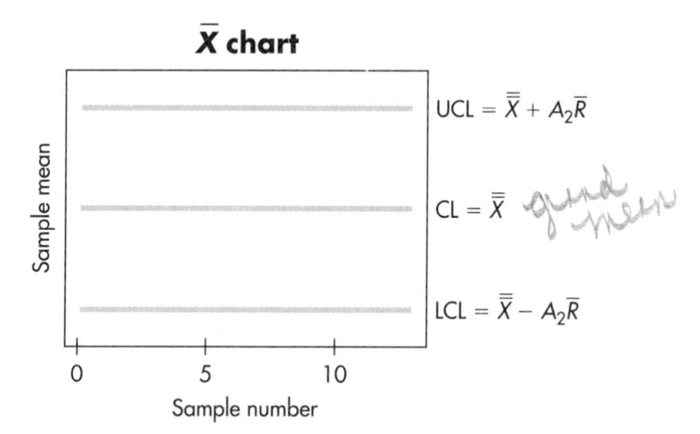

FIGURE 10.5

where D_4 = a value that depends on the subgroup sample size

$$\text{LCL} = D_3\bar{R} \tag{10.4}$$

where D_3 = a value that depends on the subgroup sample size

In equations 10.3 and 10.4, the expressions $D_4\bar{R}$ and $D_3\bar{R}$ are approximations of $\bar{R} + 3\sigma_R$ and $\bar{R} - 3\sigma_R$. Values for D_4 and D_3 can also be found in Appendix 11. If, for example, the subgroup sample size is 7, then D_4 is 1.924 and D_3 is .076, and if the subgroup sample size is 4, then D_4 is 2.282 and D_3 is 0. An R chart will look like the one shown in Figure 10.6.

Let's look now at some applications that make use of $\bar{X}$ and R charts.

Example 10.3 A machine is used to automatically fill 12-ounce cans with Zippy Cola. Philip D. Cannes, a production supervisor, takes subgroups of 4 cans from the filling line each hour and measures the contents of the cans. The results of the first 20 subgroups taken over a 20-hour period are shown below:

Subgroup	Obs. 1	Obs. 2	Obs. 3	Obs. 4
1	11.93	12.06	12.03	12.08
2	11.89	12.15	12.05	11.86
3	12.54	12.63	11.96	11.95
4	12.08	12.08	11.96	11.81
5	12.02	11.89	11.83	11.97
6	11.96	12.00	11.95	11.84
7	12.48	12.07	12.14	11.52
8	12.09	12.18	12.02	11.83
9	12.04	11.96	11.79	12.08
10	11.97	11.95	12.17	12.01
11	12.24	12.09	11.92	11.92
12	11.87	12.00	11.90	12.09
13	12.05	11.91	11.87	12.00
14	11.93	12.00	11.88	12.07
15	12.12	11.89	12.04	12.02
16	12.06	11.94	12.03	12.09
17	12.09	12.06	11.91	11.92
18	12.09	12.00	12.12	11.96
19	11.51	11.63	11.83	11.97
20	12.12	12.09	11.87	11.98

To see if the process is in statistical control, Phil performs the following hypothesis test using an $\bar{X}$ and an R chart.

Step 1 Phil sets up his hypotheses as follows: H_0: The process is in statistical control, and H_1: The process is not in statistical control.

Step 2 $\alpha = .0026$.

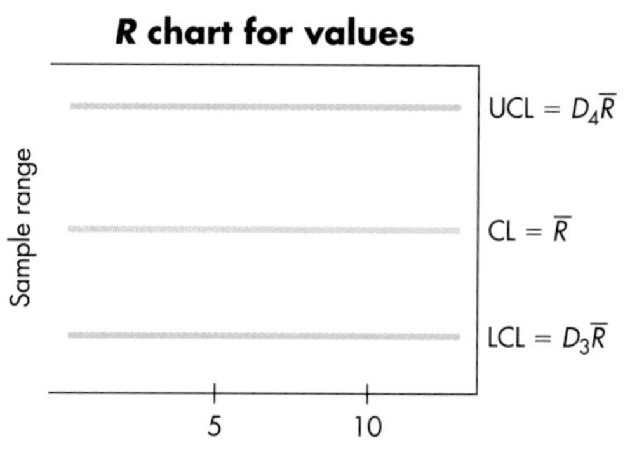

FIGURE 10.6

Step 3 Phil uses an $\overline{X}$ and an R control chart based upon normal distribution probabilities.

Step 4 He computes the mean and the range of each subgroup and obtains the following results:

Subgroup	Obs. 1	Obs. 2	Obs. 3	Obs. 4	X bar	Range
1	11.93	12.06	12.03	12.08	12.0250	.15
2	11.89	12.15	12.05	11.86	11.9875	.29
3	12.54	12.63	11.96	11.95	12.2700	.68
4	12.08	12.08	11.96	11.81	11.9825	.27
5	12.02	11.89	11.83	11.97	11.9275	.19
6	11.96	12.00	11.95	11.84	11.9375	.16
7	12.48	12.07	12.14	11.52	12.0525	.96
8	12.09	12.18	12.02	11.83	12.0300	.35
9	12.04	11.96	11.79	12.08	11.9675	.29
10	11.97	11.95	12.17	12.01	12.0250	.22
11	12.24	12.09	11.92	11.92	12.0425	.32
12	11.87	12.00	11.90	12.09	11.9650	.22
13	12.05	11.91	11.87	12.00	11.9575	.18
14	11.93	12.00	11.88	12.07	11.9700	.19
15	12.12	11.89	12.04	12.02	12.0175	.23
16	12.06	11.94	12.03	12.09	12.0300	.15
17	12.09	12.06	11.91	11.92	11.9950	.18
18	12.09	12.00	12.12	11.96	12.0425	.16
19	11.51	11.63	11.83	11.97	11.7350	.46
20	12.12	12.09	11.87	11.98	12.0150	.25

Next, he computes the mean of the 20 subgroup means and then finds the average range. The mean of the subgroup means—the $\overline{\overline{X}}$—is 239.97/20 = 12.00, and the mean of the ranges—the $\overline{R}$—is 5.90/20 = .2950. Now Phil's ready to compute the key values for the $\overline{X}$ chart as follows:

The UCL = $\overline{\overline{X}} + A_2\overline{R} = 12.00 + .729(.2950) = 12.21$
The CL = the $\overline{\overline{X}}$ value of 12.00
The LCL = $\overline{\overline{X}} - A_2\overline{R} = 12.00 - .729(.2950) = 11.78$

FIGURE 10.7

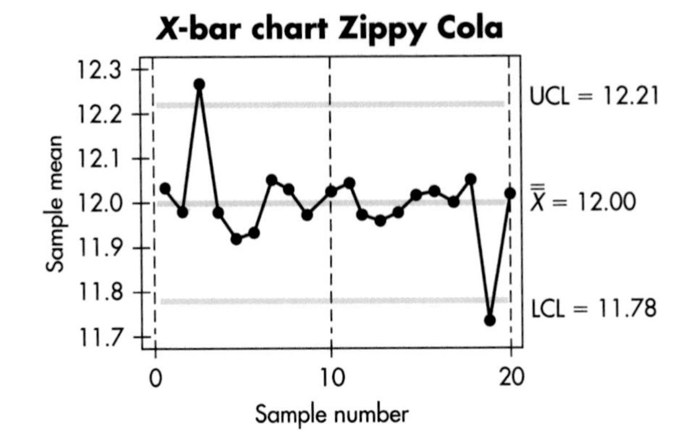

Using the values in the table in Appendix 11, Phil computes the key values for the R chart as follows:

The UCL = $D_4\overline{R}$ = 2.282(.2950) = .6732
The CL = $\overline{R}$ = .2950
The LCL = $D_3\overline{R}$ = 0(.2950) = 0

Step 5 The decision rule now is to reject H_0 and accept H_1 if at least one data point falls beyond the control limits in either control chart.

Step 6 Phil now prepares the two charts. For the $\overline{X}$ chart, he numbers the subgroups on the horizontal axis and locates the point corresponding to the mean of each subgroup above its corresponding subgroup number. This $\overline{X}$ chart is shown in Figure 10.7. And by plotting the range of each subgroup on a chart, the R chart in Figure 10.8 is created.

Step 7 Phil sees that the process is "out of control" in Figure 10.7 since the mean of subgroup 3 falls above the upper control line and the mean of subgroup 19 falls below the lower control line. And he sees in Figure 10.8 that the ranges of subgroups 3 and 7 are above the upper control limit. This indicates that there's too much variability in the process. Phil must now check the groups that are beyond the control limits for either chart and look for special causes. Since subgroup 3 is beyond the control limits in both charts, he'll start there. After a careful check, he decides there is good reason to stop the process and readjust the machine.

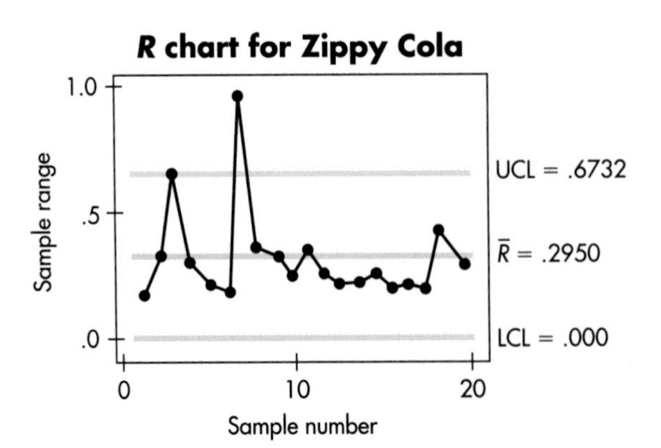

FIGURE 10.8

Example 10.4 Now that Phil has readjusted the filling machine at Zippy Cola, he wants to see if the process of filling the containers has been brought under control. Repeating the data collection step yields the following data:

Subgroup	Obs. 1	Obs. 2	Obs. 3	Obs. 4
1	11.88	11.99	12.06	12.04
2	11.89	12.18	12.08	11.89
3	11.98	12.02	11.99	11.86
4	11.84	12.11	11.99	12.01
5	12.00	11.92	11.86	12.05
6	11.87	12.03	11.98	11.99
7	11.65	12.10	12.07	11.99
8	11.86	12.21	12.05	12.12
9	12.11	11.99	11.82	12.07
10	12.04	11.98	12.20	12.00
11	11.95	12.12	11.95	12.27
12	12.12	12.03	11.93	11.90
13	12.03	11.94	11.90	12.08
14	11.60	12.03	11.91	11.96
15	12.05	11.92	12.07	12.15
16	12.12	11.97	12.06	12.09
17	11.95	12.09	11.94	12.12
18	11.99	12.03	12.15	12.12
19	12.00	11.66	11.86	11.94
20	12.01	12.12	11.90	12.15

Phil now performs another hypothesis test using the new data.

Step 1 He sets up his hypotheses as follows: H_0: The process is in statistical control, and H_1: The process is not in statistical control.

Step 2 $\alpha = .0026$.

Step 3 He'll use $\overline{X}$ and R control charts based upon normal distribution probabilities.

Step 4 He computes the mean and the range of each subgroup and obtains the following results:

Subgroup	Obs. 1	Obs. 2	Obs. 3	Obs. 4	X bar	Range
1	11.88	11.99	12.06	12.04	11.9925	.18
2	11.89	12.18	12.08	11.89	12.0100	.29
3	11.98	12.02	11.99	11.86	11.9625	.16
4	11.84	12.11	11.99	12.01	12.0227	.27
5	12.00	11.92	11.86	12.05	11.9574	.19
6	11.87	12.03	11.98	11.99	11.9675	.16
7	11.65	12.10	12.07	11.99	11.9525	.45
8	11.86	12.21	12.05	12.12	12.0600	.35
9	12.11	11.99	11.82	12.07	11.9975	.29
10	12.04	11.98	12.20	12.00	12.0550	.22
11	11.95	12.12	11.95	12.27	12.0725	.32
12	12.12	12.03	11.93	11.90	11.9950	.22
13	12.03	11.94	11.90	12.08	11.9875	.18
14	11.60	12.03	11.91	11.96	11.8750	.43
15	12.05	11.92	12.07	12.15	12.0475	.23
16	12.12	11.97	12.06	12.09	12.0600	.15
17	11.95	12.09	11.94	12.12	12.0250	.18
18	11.99	12.03	12.15	12.12	12.0725	.16
19	12.00	11.66	11.86	11.94	11.8650	.34
20	12.01	12.12	11.90	12.15	12.0450	.25

Next, he computes the mean of the 20 subgroup means and then finds the average range. The mean of the subgroup means—the $\overline{\overline{X}}$—is 240.02/20 =12.00, and the mean of the ranges—the $\overline{R}$— is 5.02/20 = .2510. Now Phil's ready to compute the key values for the $\overline{X}$ chart as follows:

The UCL $= \overline{\overline{X}} + A_2\overline{R} = 12.00 + .729(.2510) = 12.18$

The CL = the $\overline{\overline{X}}$ value of 12.00

The LCL $= \overline{\overline{X}} - A_2\overline{R} = 12.00 - .729(.2510) = 11.82$

Using the values in the table in Appendix 11, Phil computes the key values for the R chart as follows:

The UCL $= D_4\overline{R} = 2.282(.2510) = .5728$

The CL $= \overline{R} = .2510$

The LCL $= D_3\overline{R} = 0(.2510) = 0$

Step 5 The decision rule now is to reject H_0 and accept H_1 if at least one data point falls beyond the control limits in either control chart.

Step 6 Phil now prepares the two new charts as shown in Figures 10.9 and 10.10.

Step 7 He sees that both charts indicate that the process is "in control" and that only common-cause variation appears to be present.

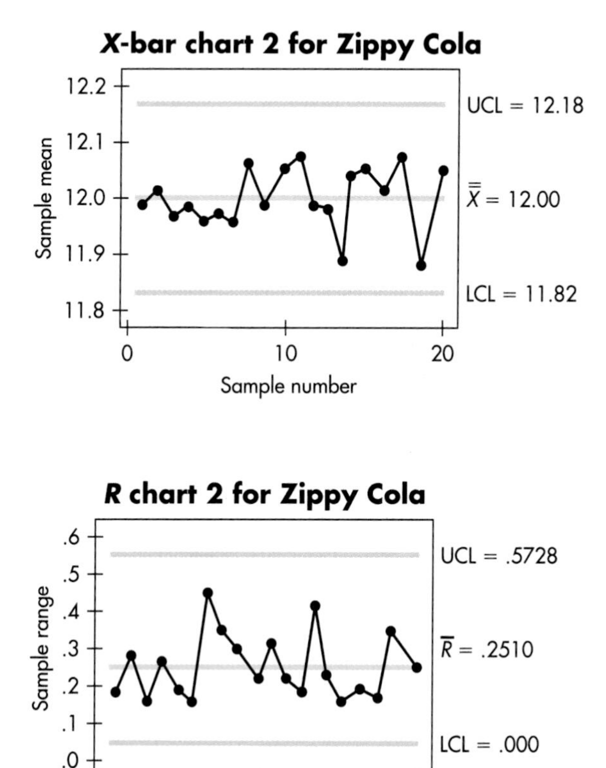

FIGURE 10.9

FIGURE 10.10

Self-Testing Review 10.3

1. Ima Brewer at the Slurper's Choice Coffee Company thinks that it may be economical to use a new supplier of coffee beans. She decides to use this new supplier on a trial basis for 1 month. Form $\overline{X}$ and R charts from the subgroup weights of the coffee bags listed below (in kilograms), and test the hypothesis at $\alpha = .0026$ that the shipping process with the new supplier is in control.

Subgroup	Obs. 1	Obs. 2	Obs. 3	Obs. 4
1	45	44	44	41
2	45	47	51	39
3	49	36	43	45
4	43	51	54	42
5	43	49	43	38
6	46	45	42	43
7	45	43	41	37
8	43	41	39	47
9	46	38	41	42
10	44	48	43	42
11	45	50	46	42
12	34	45	38	47
13	33	41	40	42
14	41	42	44	39
15	46	42	37	54
16	41	46	43	47
17	54	46	48	51
18	48	53	38	41
19	37	53	45	45
20	53	44	47	51

2. Mort Alitee, a district manager at Sooner or Later Insurance Corporation (SLIC), decides to offer bonus incentives to improve employee moral. After the incentive plan is implemented, Mort asks 24 employees to take a battery of 5 psychological tests. A morale rating from 1 to 10 is determined for each test. Using $\overline{X}$ and R control charts, perform a hypothesis test at $\alpha = .0026$ to see if morale ratings are in control at SLIC. The test data are:

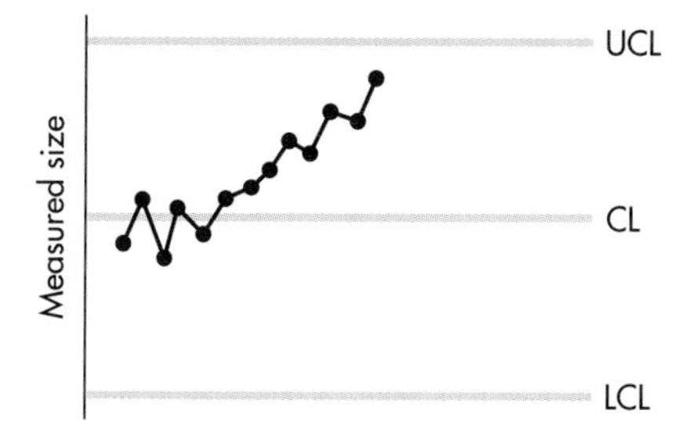

FIGURE 10.12

strong suggestion that the upper limit will soon be breached. A prudent step would thus be to stop the process and correct any problems before unacceptable outputs are produced.

Self-Testing Review 10.4

1. M & D Electronics produces 200 motherboards for the computer industry each hour. An inspection determines if each motherboard conforms or fails to conform to design specifications. The following data represent the number of nonconforming boards produced by the 4 p.m. to midnight shift each hour for a period of 5 days:

0 1 0 2 1 2 0 2 0 1 0 2 1 1 7 0 0 9 3 2 0 8 1 0
2 1 0 1 1 2 1 1 3 0 1 1 4 1 0 2

Use a p chart to test the hypothesis at $\alpha = .0026$ that the data are in statistical control.

2. The E. Z. Money Bank employs clerks who are responsible for entering transactions into the bank's database. At the end of each day a sample of 500 transactions is taken, and a computer audit is completed to find the number of errors made out of the 500 transactions. The following data represent the number of transactions with errors recorded each day for the month of March:

8 4 16 5 3 8 8 6 1 2 10 7 3 16 4 5 8 4 3 3 7 4 3
4 18 5 1 15 19 14 19

Use a p chart to test the hypothesis at $\alpha = .0026$ that the data are in statistical control.

LOOKING BACK

1. Statistical methods for quality control were developed in the 1920s and 1930s by Dr. Walter Shewhart and others. One of Shewhart's colleagues, W. Edwards Deming, was a leader in the introduction of statistical quality control methods in the United States during World War II. After the war, Deming took his concepts to Japan, where they were eagerly adopted. Japanese quality has become legendary in recent

times, and this qualitative advantage has jolted many organizations in the United States out of decades of complacency. But now there are many U.S. organizations that employ the latest statistical process control methods.

2. A distinction must be made between the controlled, or common-cause, variation that is normal and inherent in any stable process and the uncontrolled, or special-cause, variation that occurs when an abnormal action enters a process and produces unexpected changes. Control charts are graphic displays that compare the data produced by a current process to a set of stable control limits. Such charts can thus be used to detect special-cause variation.

3. All control charts have a center line (CL), an upper control limit (UCL), and a lower control limit (LCL). Data are plotted in a time sequence, and any point that falls outside the upper and lower control limits is an indication that the process is out of statistical control. A control chart for individual observations can be constructed if the output of a process is normally distributed. For such a chart, the UCL $= \mu + 3\sigma$, the CL $= \mu$, and the LCL $= \mu - 3\sigma$.

4. Often $\overline{X}$ and R charts are used together to monitor the average of subgroups and the range of subgroups. For the $\overline{X}$ chart, the UCL $= \overline{\overline{X}} + A_2\overline{R}$, the CL $= \overline{\overline{X}}$, and the LCL $= \overline{\overline{X}} - A_2\overline{R}$. Values for A_2 can be found in Appendix 11, and $A_2\overline{R}$ is an estimate of $3\sigma_{\overline{X}}$. For the R chart, the UCL $= D_4\overline{R}$, the CL $= \overline{R}$, and the LCL $= D_3\overline{R}$. Values for D_4 and D_3 are listed in Appendix 11, and the expressions $D_4\overline{R}$ and $D_3\overline{R}$ are approximations of $\overline{R} + 3\sigma_R$ and $\overline{R} - 3\sigma_R$.

5. When data are categorical or qualitative, a p chart is used to monitor the proportion of nonconforming items in a series of samples. An estimate of the population proportion of defective items is given the symbol $\overline{p}$, and:

$$\overline{p} = \frac{\text{number of nonconforming items in all the subgroups}}{\text{total numbers of items in all the subgroups}}$$

For the p chart:

$$\text{UCL} = \overline{p} + 3\sqrt{\frac{\overline{p}(1-\overline{p})}{n}}, \qquad \text{the CL} = \overline{p}, \qquad \text{and the LCL} = \overline{p} - 3\sqrt{\frac{\overline{p}(1-\overline{p})}{n}}$$

Review Exercises

1. An essential step in making a certain type of rod is a plating process in which various composite materials are coated with a thin layer of copper. For this process, the temperature of the plating bath must be controlled so that the plating is uniform. The following data represent hourly temperatures of the plating bath over a 24-hour period:

Hour	Temperature	Hour	Temperature
1	115.6	14	111.4
2	104.5	15	102.2
3	113.0	16	107.0
4	105.1	17	113.0
5	117.3	18	110.4
6	103.9	19	113.4
7	107.1	20	116.7
8	102.3	21	107.9
9	108.8	22	107.3
10	115.5	23	117.5
11	104.1	24	112.1
12	102.6	25	107.8
13	108.0		

In the past the mean temperature has been 110° F, and the standard deviation has been .7° F. Use a control chart for individual values to test the hypothesis at $\alpha = .0026$ that the temperature of the plating bath is in control.

2. The Televac Company makes thermocouple gauge tubes used to measure pressure in a vacuum. An oxidation process protects a tube during production. In the past, the average time needed to oxidize a tube was 3.2 seconds, and the standard deviation time was .6 second. The following represents the oxidation time required for 40 tubes coming consecutively off the production line:

Number	Oxidation Time	Number	Oxidation Time
1	5.1	21	3.5
2	3.5	22	5.1
3	3.2	23	3.6
4	1.7	24	4.0
5	3.2	25	5.2
6	3.6	26	1.9
7	3.4	27	5.7
8	3.9	28	5.0
9	1.5	29	1.8
10	3.5	30	4.1
11	4.1	31	2.0
12	1.6	32	3.9
13	3.0	33	2.8
14	2.4	34	5.0
15	3.8	35	5.7
16	1.7	36	2.6
17	4.0	37	3.5
18	4.6	38	5.4
19	4.1	39	4.0
20	3.0	40	2.3

Use a control chart for individual values to test the hypothesis at $\alpha = .0026$ that the oxidation process time is in statistical control.

3. A production process cuts aluminum rods into specified lengths. An industrial engineer wants to see if the process is in control, so he measures the lengths of the first 30 rods in a batch. The results are as follows:

Number	Length	Number	Length
1	9.00	16	8.92
2	5.32	17	8.08
3	11.90	18	10.78
4	9.21	19	11.54
5	7.42	20	7.88
6	8.27	21	10.80
7	9.87	22	7.59
8	11.87	23	7.93
9	9.69	24	11.19
10	11.46	25	10.61
11	9.70	26	6.63
12	13.32	27	7.54
13	7.07	28	6.43
14	9.69	29	9.38
15	10.01	30	10.49

In the past, the mean length has been 9 centimeters, and the standard deviation has been .7 centimeters. Use a control chart for individual values to test the hypothesis at $\alpha = .0026$ that the process is in statistical control.

4. The manager at a plant where copper rods are plated has made some personnel changes and wants to see if the temperature of the plating bath is now in control. He takes 4 temperature measures each hour for 24 hours and obtains the following data:

Subgroup	Obs. 1	Obs. 2	Obs. 3	Obs. 4
1	16.00	112.04	107.92	113.28
2	104.74	117.99	96.37	111.62
3	117.88	131.01	118.31	116.64
4	105.69	101.30	113.93	113.74
5	111.25	96.71	112.28	101.19
6	95.84	113.57	113.55	104.37
7	116.15	101.55	136.68	117.06
8	106.24	106.09	104.70	108.05
9	115.28	130.14	103.98	112.45
10	99.87	117.56	102.61	98.05
11	100.20	122.07	112.05	95.07
12	105.71	112.04	112.49	108.57
13	122.42	133.03	116.80	97.03
14	114.80	113.30	103.64	115.24
15	126.27	124.33	101.69	102.05
16	100.07	130.49	108.38	111.10
17	122.76	98.80	106.17	112.10
18	111.08	108.21	112.89	118.82
19	107.25	117.01	94.93	103.31
20	105.67	118.00	123.98	123.43
21	132.99	118.96	115.07	109.40
22	114.44	104.12	123.59	112.32
23	111.58	106.05	118.09	105.37
24	120.13	93.77	115.35	110.69

Use an $\overline{X}$ and an R chart to test the hypothesis at $\alpha = .0026$ that the temperature of the plating bath is in statistical control.

5. The Televac Company makes thermocouple gauge tubes used to measure pressure in a vacuum. An oxidation process protects a tube during production. The length of the oxidation process is timed, and batches of 5 tubes are taken off the production line each hour for 30 hours. The following data give the oxidation time for the tubes in each batch:

Subgroup	Obs. 1	Obs. 2	Obs. 3	Obs. 4	Obs. 5
1	3.2	4.9	3.4	1.3	3.5
2	3.7	4.6	2.8	1.9	5.8
3	1.6	2.8	3.7	2.8	4.6
4	1.4	3.4	5.1	5.9	2.7
5	4.6	1.9	4.6	1.3	6.2
6	3.1	3.7	3.7	3.6	2.4
7	5.9	5.1	1.2	4.3	5.3
8	1.7	5.3	4.6	2.8	5.5
9	4.6	4.7	2.6	4.6	1.9
10	3.7	2.8	3.6	2.4	3.3
11	4.6	6.4	1.7	1.9	2.1
12	1.5	1.3	1.9	2.4	1.3
13	2.8	1.9	5.5	3.7	1.1
14	2.6	3.2	2.9	1.9	4.6
15	7.3	5.5	7.5	4.6	5.4
16	7.3	2.6	5.5	2.7	4.6
17	5.3	1.9	5.1	1.9	3.7
18	4.6	6.4	5.5	4.6	3.7
19	2.8	3.7	1.9	4.6	3.6
20	3.4	1.3	5.2	4.6	4.6
21	3.7	3.1	5.9	4.8	2.6
22	5.5	4.6	1.9	2.8	3.4
23	7.1	4.6	4.6	5.5	5.5
24	3.7	1.9	4.6	3.7	5.5
25	2.8	2.8	6.4	3.7	3.7
26	3.7	2.8	3.7	3.7	3.7
27	5.5	4.6	2.8	1.0	2.8
28	5.5	2.8	3.7	4.6	4.6
29	4.6	3.7	4.6	3.7	2.8
30	1.9	3.7	3.7	2.8	4.0

Use an $\overline{X}$ and an R chart to test the hypothesis at the .0026 level that the process is in control.

6. A machine used in a production process is supposed to automatically cut aluminum rods into 9-centimeter lengths. At the last inspection, the process was out of control. An industrial engineer has recalibrated some equipment and wants to see if the process is now in control. The following data represent subgroups of 4 rods sampled each hour for 30 hours:

Subgroup	Obs. 1	Obs. 2	Obs. 3	Obs. 4
1	7.17	12.01	9.74	7.66
2	9.91	12.16	10.91	10.85
3	13.80	14.47	8.30	9.68
4	14.77	5.84	7.98	9.68
5	8.60	7.91	10.35	8.20
6	12.50	12.33	10.96	6.11
7	10.34	5.99	8.48	13.23
8	10.32	5.49	8.18	8.83
9	8.20	12.62	11.01	9.72
10	8.87	3.39	11.58	6.92
11	6.47	11.21	12.23	12.62
12	9.79	11.77	7.42	12.53
13	7.09	11.33	10.61	11.14
14	2.23	11.85	9.93	8.98
15	4.09	8.73	9.79	11.57
16	11.23	15.56	6.70	9.09
17	8.88	6.70	6.47	7.33
18	11.96	9.06	11.06	10.06
19	11.47	6.84	10.23	16.45
20	7.84	10.73	8.31	16.56
21	6.52	7.22	8.16	5.47
22	7.54	8.50	10.77	11.32
23	6.23	15.84	10.38	6.62
24	9.84	11.52	8.76	8.06
25	14.22	12.57	6.44	12.45
26	5.39	8.25	8.46	10.74
27	7.13	9.40	10.62	9.82
28	7.32	9.90	11.60	9.23
29	11.27	10.05	10.06	12.22
30	7.75	11.28	7.05	8.41

Use an $\overline{X}$ and an R chart to test the hypothesis at the .0026 level that the process is in control.

7. The Escargot Z-968 sports car is painted at a plant in Mobile, Alabama. Each hour during production, 300 cars are painted and thoroughly inspected for bubbles and defects. The following data give the number of defective paint jobs found in the samples of 300 taken each hour for a 24-hour period:

5 5 18 4 5 3 4 8 9 15 7 4
4 8 9 4 8 6 8 17 7 19 6 6

Use a p chart to test the hypothesis at the .0026 level that the data are in statistical control.

8. A company produces foam material to be placed inside seat cushions. There are 100 cushion inserts in a batch. Each batch is inspected, and the number of inserts not conforming to specifications is recorded. The following data represent the number of nonconforming inserts in samples of 35 batches:

7 3 2 2 2 2 3 2 2 2 3 3 1
5 1 3 2 2 4 0 5 4 2 1 2 5
4 5 1 3 2 4 3 1 1

Use a p chart to test the hypothesis at the .0026 level that the data are in statistical control.

9. There are 800 teachers employed in a local school district. An administrator keeps records of the number of teachers absent each school day for the first marking period of the year. The data are as follows:

48 65 56 44 95 54 56 35 42 93
45 49 54 52 102 44 51 54 59
114 41 47 53 47 109 47 59 50
49 94 30 58 52 39 110 48 40
59 35 105 61 55 54 50 95

Use a p chart to test the hypothesis at the .0026 level that the data are in statistical control.

10. A nursing quality assessment program is in place in the intensive care unit (ICU) at Getwell Hospital. The program requires that an admission assessment be completed for each patient within 8 hours. Each day during the month of October a sample of 40 ICU patients is selected. The following data represent the number of patients in the daily sample whose admissions assessment was not completed within 8 hours:

2 2 3 4 2 2 4 3 4 2 3 4 2
4 2 4 5 3 1 6 2 15 11 19
16 12 12 9 14 12 15

Use a p chart to test the hypothesis at the .0026 level that the data are in statistical control.

Topics for Review and Discussion

1. Discuss the relationship between economic conditions and the development of statistical process control methods.

2. Why are the control limits for an $\overline{X}$ chart narrower than those of a control chart for individual values that have the same mean and standard deviation?

3. It is possible to construct $\overline{X}$ control charts using different size subgroups. Why would this make the charts more complicated?

Projects/Issues to Consider

1. Go to the library and locate an article in a recent periodical describing how total quality management is being used in a particular setting. Summarize the article, and discuss how statistical ideas are used.

2. Locate a manufacturing plant, hospital, or school in your area, and contact the person in charge of quality control. Interview this person, and get his or her opinion about the effectiveness of quality control. What are some of the problems? The advantages?

Computer Exercises

1. Using the data from problem 2 in the end-of-chapter exercises, use your software to prepare a control chart for individual observations. (If you're using *Minitab*, this can be done with the ICHART command. When using *Minitab for Windows*, individual and other control charts are found under the **Stat** option on the menu bar.)

2. Using the data from problem 4 in the end-of-chapter exercises, use your software to prepare $\overline{X}$ and R charts. (If you're using *Minitab*, this can be done with XBARCHART and RCHART commands.)

3. Using the data from problem 8 in the end-of-chapter exercises, use your software to prepare a p chart. (If you're using *Minitab*, this can be done with the PCHART command.)

Answers to Odd-Numbered Self-Testing-Review Problems

Section 10.1

1. Dr. Walter Shewhart

3. The economy was booming, and few saw the need to focus on SPC concepts.

5. **Plan**: Make a plan to achieve your goals. **Do**: Implement necessary changes. **Check**: Evaluate the changes made. **Act**. Take any necessary actions. Is there a need to revise your plan?

Section 10.2

1. *Step 1.* H_0: The process is in statistical control, and H_1: The process is not in statistical control. *Step 2.* $\alpha = .0026$ *Step 3.* We'll use a chart for individual values based on the normal distribution. *Step 4.* The essential components for the control chart are: CL = μ = 45. The UCL = 45 + 3(5) = 60, and the LCL = 45 − 3(5) = 30. *Step 5.* Reject H_0 and accept H_1 if one or more data points fall beyond the control limits. *Step 6.* The control chart is shown to the right:

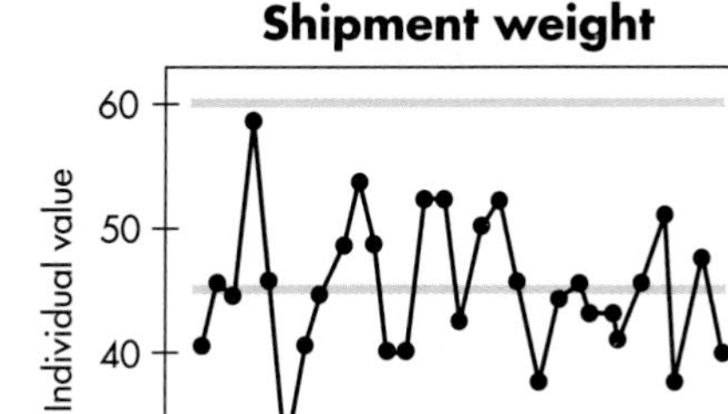
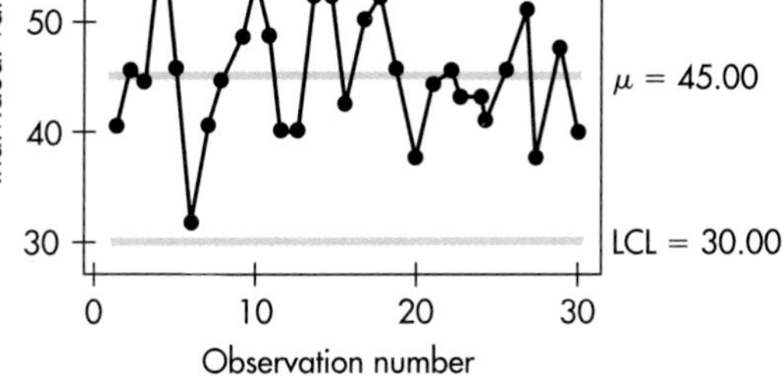

Step 7. Since all points are within the 3 standard deviation control limits, we accept H_0. The process is in statistical control.

Section 10.3

1. *Step 1.* H_0: The process is in statistical control, and H_1: The process is not in statistical control. *Step 2.* $\alpha = .0026$ *Step 3.* We'll use $\bar{X}$ and R charts based on normal probabilities. *Step 4.* The essential components for the control chart are found as follows:

Subgroup	x Bar	Range
1	43.50	4
2	45.50	12
3	43.25	13
4	47.50	12
5	43.25	11
6	44.00	4
7	41.50	8
8	42.50	8
9	41.75	8
10	44.25	6
11	45.75	8
12	41.00	13
13	39.00	9
14	41.50	5
15	44.75	17
16	44.25	6
17	49.75	8
18	45.00	15
19	45.00	16
20	48.75	9

$\Sigma\bar{x} = 881.75$, and since there are 20 subgroups, $\bar{\bar{X}} = 881.75/20 = 44.0875$. The $\Sigma R = 192$, so $\bar{R} = 192/20 = 9.60$. For the $\bar{X}$ chart, the CL = $\bar{\bar{X}}$ = 44.0875. The UCL = 44.0875 + .729(9.60) = 51.08, and the LCL = 44.0875 − .729(9.60) = 37.09. In the R chart, the CL = $\bar{R}$ = 9.60, the UCL = 2.282(9.60) = 21.91, and the LCL = 0(9.60) = 0. *Step 5.* Reject H_0 and accept H_1 if one or more data points fall beyond the control limits in either chart. *Step 6.* The control charts are shown below:

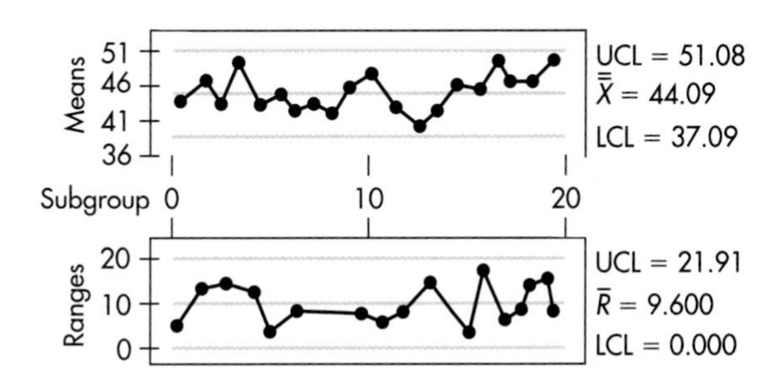

Step 7. We accept the H_0. The process is in statistical control.

Section 10.4

1. *Step 1.* H_0: The process is in statistical control, and H_1: The process is not in statistical control. *Step 2.* $\alpha = .0026$. *Step 3.* We'll use a p chart based on the binomial distribution. *Step 4.* The number of nonconforming items in all the subgroups is 64, and there were 200 (40) = 8000 cushions in all the samples. Thus;

$\overline{p} = 64/8000 = .008$, and

$$s_p = \sqrt{\frac{(.008)(1 - .008)}{200}} = .0063$$

The UCL = .008 + 3(.0063) = .0269, and the LCL = 0 since .008 − 3(.0063) is negative. We next calculate the percent of nonconforming motherboards for each batch by dividing the given numbers by 200. The results for the 40 samples are:

.000	.005	.000	.010	.005	.010	.000
.010	.000	.005	.000	.010	.005	.005
.035	.000	.000	.045	.015	.010	.000
.040	.005	.000	.010	.005	.000	.005
.005	.010	.005	.005	.015	.000	.005
.005	.020	.005	.000	.010		

Step 5. Reject H_0 and accept H_1 if one or more data points fall beyond the control limits. *Step 6.* The control chart is shown below:

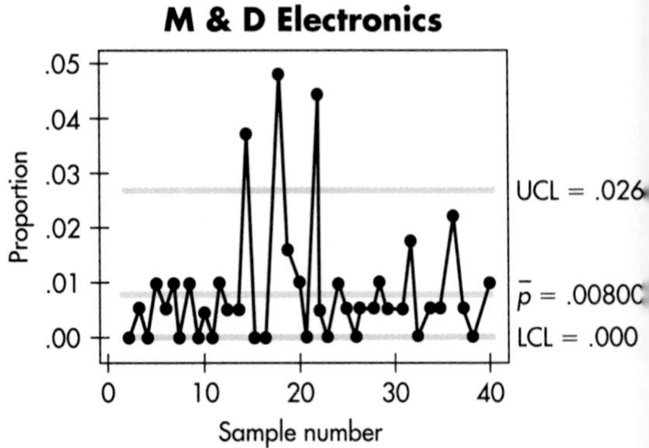

Step 7. There are three points in the control chart that fall above the UCL, so we reject the H_0. The process is out of statistical control.

CHAPTER 11

Analysis of Variance

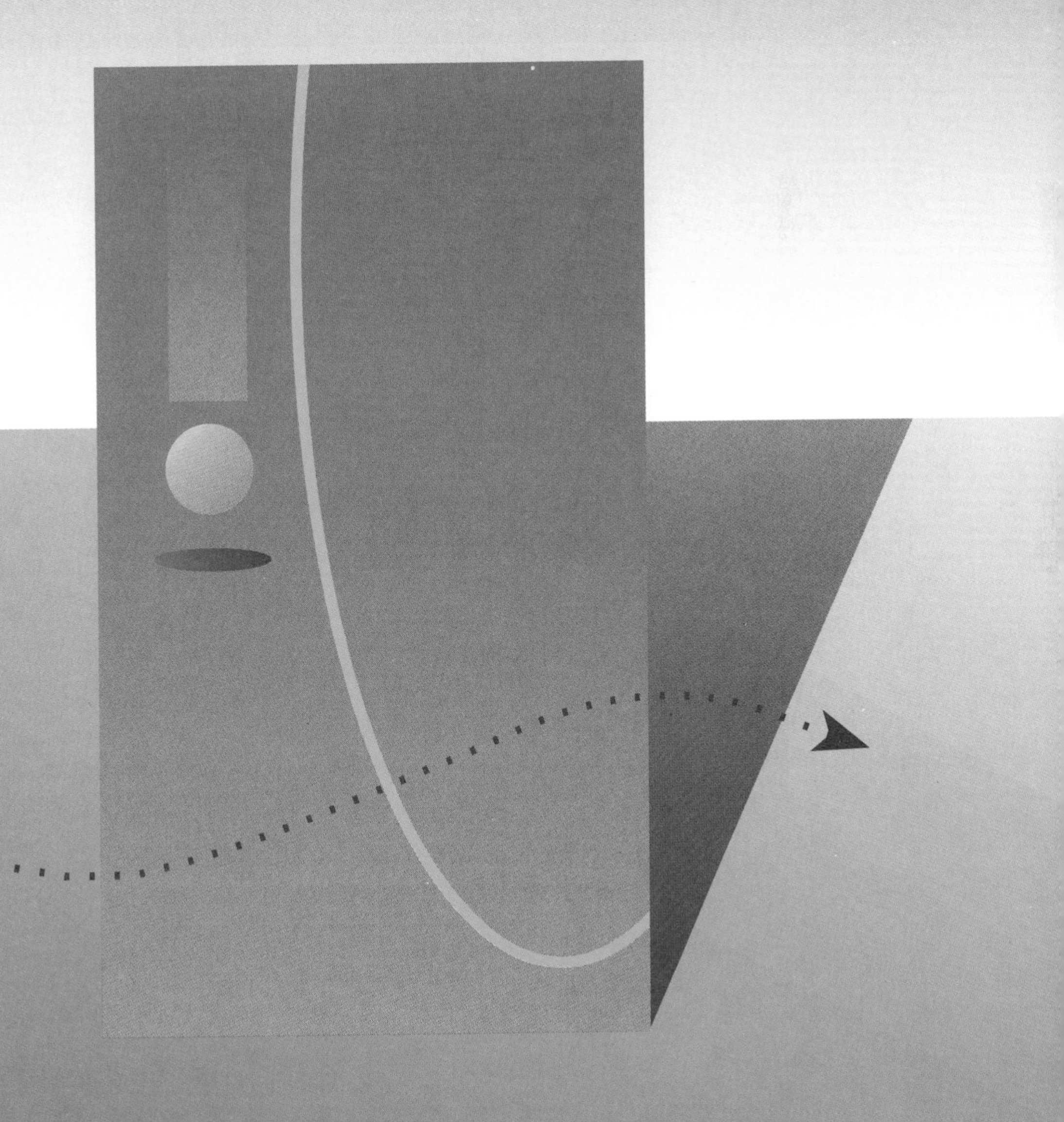

Consequences of Genetics Research

Genetics research makes it possible to conduct tests that predict the risk of acquiring several adult-onset diseases. To study the psychological consequences of this research, 135 Canadian citizens were given a test that predicted their risk of getting Huntington's disease. They were then placed in three groups according to their test results. There were 37 in the increased-risk group, 58 in the decreased-risk group, and 40 in the no-change-in-risk group. Psychosocial distress, depression, and well-being were measured before and at intervals after the participants received the test results. After 12 months, the increased-risk group and the decreased-risk group had lower scores for depression and higher scores for well-being than the no-change group. The study concluded that predictive testing for Huntington's disease has potential benefits for the psychological health of those whose results show either an increase or a decrease in the risk of inheriting the gene for the disease.

LOOKING AHEAD

You saw in Chapter 9 (and in Figure 9.3, page 339) that there are three hypothesis-testing procedures that may be used to see if there's likely to be a significant difference between the means of *two independent* populations. Now in this chapter you'll learn a testing technique called *analysis of variance* (often abbreviated *ANOVA*) that can be used, for example, to help a manager evaluate the performance of three (or more) employees to see if any performance level is different from the others. Using ANOVA techniques, a physical therapist can evaluate the results of four treatments for lower back problems to see if one or more is different from the others. And a marketing executive can see if there's a difference in sales productivity in the five company regions.

We'll first consider the assumptions and procedural steps associated with the ANOVA technique. Next, we'll present an example that uses the ANOVA testing procedure to arrive at a statistical decision. And finally, we'll see how computers can be used to make short work of the ANOVA testing process.

Thus, after studying this chapter, you should be able to:

- Explain the purpose of analysis of variance and identify the assumptions that underlie the ANOVA technique.
- Describe the ANOVA hypothesis-testing procedure.
- Use the ANOVA testing procedure and the *F* distributions tables to arrive at statistical decisions about the means of three or more populations.

11.1 Analysis of Variance: Purpose and Procedure

Purpose and Assumptions

Analysis of variance (ANOVA) is the name given to the approach that allows us to use sample data to see if the values of two or more unknown population means are likely to be equal. If exactly *two means* are compared, the ANOVA procedure discussed in this chapter gives the same results obtained with the two-sample procedure for testing small samples that was outlined in Section 9.2, Procedure 4, in Chapter 9. Our purpose in this chapter, though, isn't to generate more two-sample tests. Rather, our focus now is to look at situations where the need is to compare three or more unknown population means.

We'll limit our attention to cases where the data taken from the populations deal with a *single variable or factor* such as the number of customers handled at three sampled locations or the speed of relief obtained by samples of people using four brands of pain relievers. Our focus, then, is on a **one-way ANOVA test** of means, in which only one classification factor or variable is considered. But you should know that statistical software packages contain programs that allow two or more factors to be studied.

As you've learned in earlier chapters, statistical techniques generally involve assumptions that must be valid if the techniques are to be correctly applied. In the case of analysis of variance, the *following assumptions must be true*:

1. The populations under study have *normal* distributions.
2. The samples are drawn *randomly*, and each sample is *independent* of the other samples.
3. The populations from which the sample values are obtained all have the same unknown population variance (σ^2). That is, this third assumption is:

$$\sigma_1^2 = \sigma_2^2 = \sigma_3^2 = \cdots = \sigma_k^2$$

where k = number of populations

The Procedural Steps for an ANOVA Test

As you certainly know by now, a hypothesis-testing procedure begins with a statement of the hypothesis to be tested, and it concludes when a statistical decision is made. The testing procedure for ANOVA is no different, and it follows seven familiar steps.

Step 1: State the Null and Alternative Hypotheses The *null hypothesis* in analysis of variance is that the independent samples are drawn from different populations with the same mean. In other words, the null hypothesis is always:

H_0: $\mu_1 = \mu_2 = \mu_3 = \cdots = \mu_k$. That is, H_0: All population means are equal.

where k = the number of populations under study

And the *alternative hypothesis* in any analysis of variance is:

H_1: *Not all* population means are equal.

A careful reading of the alternative hypothesis shows that if the H_1 is accepted, you may conclude that *at least one* population mean differs from the other population means. But analysis of variance cannot tell you exactly *how many* population means differ; nor will it give you exact information about *which* means differ. For example, six populations could be under study, and if only one population mean differs from the other five means, which are equal, the H_0 may be rejected and the alternative hypothesis accepted.

But *if* the H_0 is true and *if* the three assumptions listed above are valid, the net effect is conceptually equal to the case where all the samples are picked from the one population shown in Figure 11.1*a*. But *if* the H_0 turns out to be *false* and *if* the three assumptions still remain valid, the population means will not be equal. In this event, the samples in an application might be taken from populations such as those shown in Figure 11.1*b*. Of course, these populations are still normally distributed, and they still have the same variance value.

Step 2: Select the Level of Significance A criterion for rejection of the H_0 is necessary, and tests are typically made where α is specified to be .01 or .05.

FIGURE 11.1

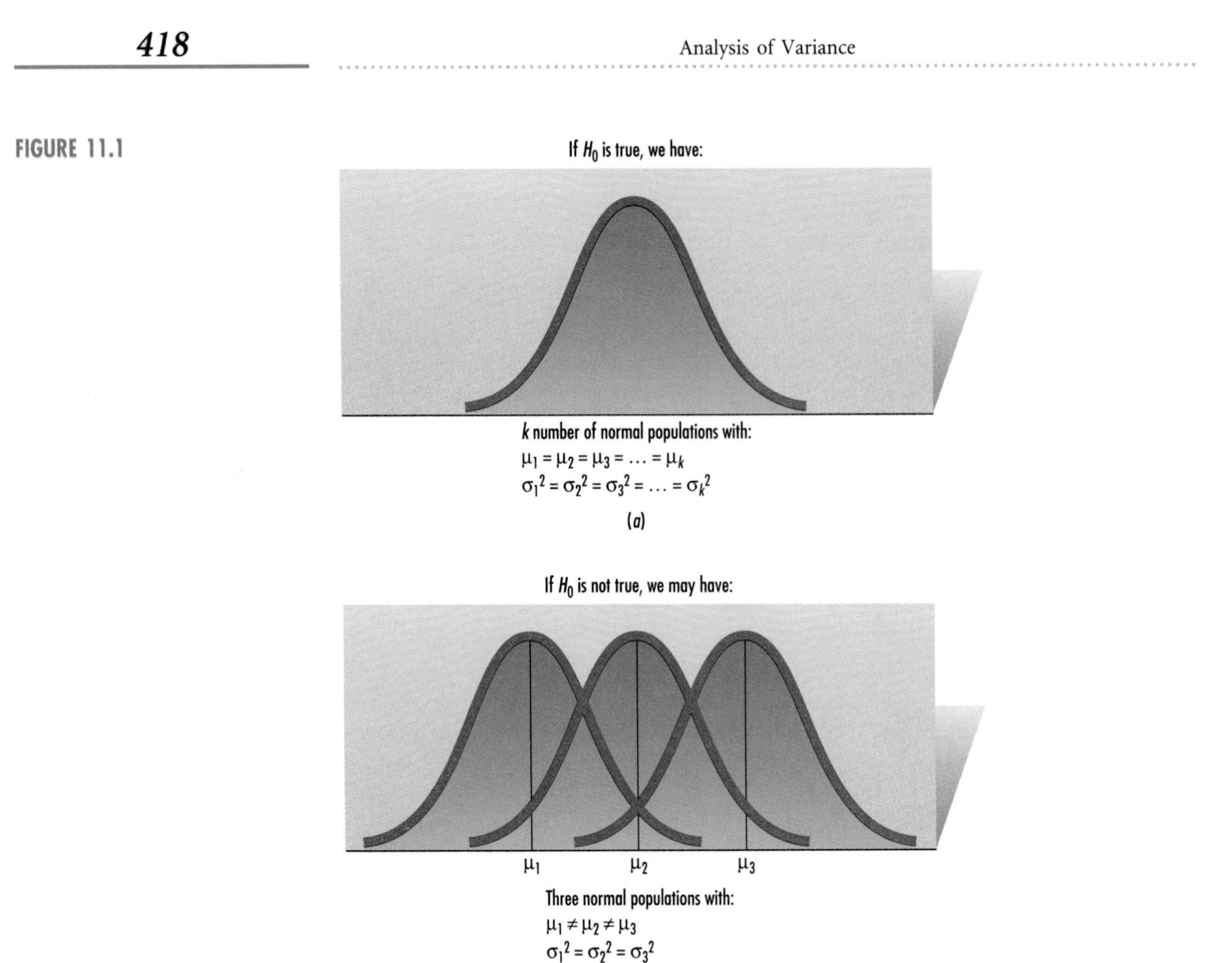
If H_0 is true, we have:
k number of normal populations with:
$\mu_1 = \mu_2 = \mu_3 = \ldots = \mu_k$
$\sigma_1^2 = \sigma_2^2 = \sigma_3^2 = \ldots = \sigma_k^2$
(a)
If H_0 is not true, we may have:
μ_1
μ_2
μ_3
Three normal populations with:
$\mu_1 \neq \mu_2 \neq \mu_3$
$\sigma_1^2 = \sigma_2^2 = \sigma_3^2$
(b)

Step 3: Determine the Test Distribution to Use An *F distribution* is used in an ANOVA test. You'll recall that the properties and shapes of F distributions were introduced in Chapter 9 in the discussion of hypothesis tests of two variances. To summarize here, an F distribution is skewed to the right and has a scale starting at 0 and extending to the right. There are many F distributions, and each one is determined by the number of samples and the number of observations in the samples. The F-distribution tables are found in Appendix 5 at the back of the book.

Step 4: Define Rejection or Critical Regions You may remember from Chapter 9 that to find the F value that separates the area of acceptance from the rejection area, we need to know three things:

1. The *level of significance* (specified in *Step 2* in our procedure).
2. The *degrees of freedom for the numerator* (df_{num}) of the test ratio (TR) that we will compute in a later step. The value of df_{num} is:

$$\text{df}_{\text{num}} = k - 1 \tag{11.1}$$

where k = the number of samples

Thus, if we have taken random and independent samples from three populations:

$df_{num} = 3 - 1$, or 2

3. The *degrees of freedom for the denominator* (df_{den}) of the test ratio (TR) that we will compute later. The value of df_{den} is:

$$df_{den} = T - k \tag{11.2}$$

where T = the total number of items in all samples, or $n_1 + n_2 + n_3 + \cdots + n_k$
k = the number of samples

So if we have three samples with 7 items in the first sample, 9 in the second, and 8 in the third, $df_{den} = (7 + 9 + 8) - 3 = 21$.

If we know that $\alpha = .05$, $df_{num} = 2$, and $df_{den} = 21$, then we can find the critical F value that separates the areas of acceptance and rejection in an ANOVA test. In using the F tables in Appendix 5, we must first locate the table with the relevant α. (With $\alpha = .05$ here, it's the first table in that appendix.) Next, we must locate the critical value of F where the degrees of freedom for the *numerator* (found at the top of the *columns*) and the degrees of freedom for the *denominator* (shown to the left of the *rows*) intersect. In this example, the critical F value is:

$F_{(2,21,\alpha=.05)} = 3.47$

Thus, a test ratio with a value that exceeds 3.47 in this case falls into the rejection region.

Step 5: State the Decision Rule In a situation where $\alpha = .05$, $df_{num} = 2$, and $df_{den} = 21$, the *decision rule* is:

Reject H_0 and accept H_1 if the TR is > 3.47. Otherwise, accept H_0.

Step 6: Compute the Test Ratio The *test ratio* (or F value) is found with this formula:

$$TR_F = \frac{\hat{\sigma}^2_{between}}{\hat{\sigma}^2_{within}} \tag{11.3}$$

As you can see, the test ratio consists of *two estimates of the population variance* ($\hat{\sigma}^2$). Two independent computational methods are used to produce these estimates. (We'll consider the $\hat{\sigma}^2_{within}$ estimate first, but, as you'll see later, it's not necessary for this value to be computed first in the ANOVA procedure.)

The $\hat{\sigma}^2_{within}$ estimate of σ^2 remains appropriate regardless of any differences between population means. In other words, the means of the several populations may differ, but

"Whip Economists Now" —G. Ford

Economic forecasters don't do a very good job of calling the turns in the business cycle. They generally can't tell when the next recession will start, and they usually can't tell us we're in a recession until months after the fact. The minimum time it takes the National Bureau of Economic Research to date a business-cycle turn is about 6 months. In the fall of 1974, the Ford administration launched a campaign to "Whip Inflation Now" but found out a few months later that the economy had actually gone into a severe recession in November 1973.

this estimate of σ^2 *isn't* affected by the possible fact that the H_0 is false. Although any one of the sample variances (s^2) could be used as this estimate of σ^2, the variances of *all the samples* are usually *pooled* and *averaged* to estimate σ^2 because of the greater amount of data thus considered. Therefore, in the ANOVA procedure, a variance from each sample is computed, and the sample variances are then pooled to produce a value for $\hat{\sigma}^2_{\text{within}}$. Any deviation of this estimate from the population variance is due to random error present in any sampling situation. (Keep that word "error" in mind—we'll refer to it later.) The formula for $\hat{\sigma}^2_{\text{within}}$ (regardless of whether the samples are of equal size or not) is:

$$\hat{\sigma}^2_{\text{within}} = \frac{\Sigma\, d_1^2 + \Sigma\, d_2^2 + \Sigma\, d_3^2 + \cdots + \Sigma\, d_k^2}{T - k} \tag{11.4}$$

where $\Sigma\, d_1^2$ = the sum of the squared differences—that is, $\Sigma\,(x_1 - \bar{x}_1)^2$—for the first sample

$\Sigma\, d_2^2$ = the sum of the squared differences—that is, $\Sigma\,(x_2 - \bar{x}_2)^2$—for the second sample, and so on

T = total number of all items in all samples $(n_1 + n_2 + n_3 \cdots + n_k)$

k = number of samples

Since, as noted above, the $\hat{\sigma}^2_{\text{within}}$ estimate isn't affected by the possible fact that the H_0 is false, this computed value cannot, by itself, be used to test the validity of the H_0. As you've seen in the test ratio formula, a second element—$\hat{\sigma}^2_{\text{between}}$—is also needed.

The second method used to estimate σ^2—the $\hat{\sigma}^2_{\text{between}}$ procedure—results in an appropriate estimate of σ^2 *if, and only if,* the population means are equal. This approach produces an estimate that contains the effects of any differences between the population means. If there are *no differences* between means, this computed value of σ^2 doesn't differ too much from $\hat{\sigma}^2_{\text{within}}$ (which is now used as a standard against which this second estimate is evaluated). This $\hat{\sigma}^2_{\text{between}}$ method of estimating σ^2 is based on the variation between the sample means and is founded on the Central Limit Theorem.

If the null hypothesis is true, then, as we saw in Figure 11.1*a*, it's as though all the samples are selected from the same normal population distribution with the same μ. And as we saw in Chapter 6, and as the Central Limit Theorem tells us, if the population is normally distributed, the distribution of the sample means is also normal. Furthermore, you'll remember that the standard deviation of this sampling distribution—the standard error of the sample means—is found by this basic formula:

$$\sigma_{\bar{x}} = \frac{\sigma}{\sqrt{n}}$$

Now, if we square both sides of this basic equation, we get:

$$\sigma_{\bar{x}}^2 = \frac{\sigma^2}{n}$$

which can be manipulated by multiplying both sides by n to yield the population variance

$$n\sigma_{\bar{x}}^2 = \sigma^2$$

Statistical data are used by air-traffic control systems to help satisfy customers' demand for service in an efficient way. (*Comstock*)

Thus, if we knew the square of the standard error ($\sigma_{\bar{x}}^2$), we could compute the precise value of σ^2 merely by multiplying $\sigma_{\bar{x}}^2$ by the sample size.

Without going into further details, we can simply summarize here and supply the formulas needed to effectively (1) compute an estimate of the square of the standard error ($\hat{\sigma}_{\bar{x}}^2$) and (2) multiply this estimate by the sample size to effect an estimate of σ^2. This second way to estimate σ^2 produces a value for $\hat{\sigma}^2_{\text{between}}$. This estimate of σ^2 has merit if (and only if) the H_0 is true. Any deviation of this estimate from the σ^2 is due to factor differences among the samples (keep that word "factor" in mind, too, because we'll also get back to it later).

There are two formulas needed to compute a value for $\hat{\sigma}^2_{\text{between}}$. First, a **grand mean** ($\bar{\bar{X}}$)—the mean of *all* the values in *all* the samples—is required:

$$\bar{\bar{X}} = \frac{\text{total of all sample items}}{\text{number of items from all samples}} = \frac{n_1\bar{x}_1 + n_2\bar{x}_2 + \cdots + n_k\bar{x}_k}{n_1 + n_2 + \cdots + n_k} \tag{11.5}$$

where n_1 = number of items in sample 1
n_2 = number of items in sample 2
n_k = number of items in sample k
$\bar{x}_1$ = mean of sample 1
$\bar{x}_2$ = mean of sample 2
$\bar{x}_k$ = mean of sample k

And then we can produce $\hat{\sigma}^2_{\text{between}}$ with this formula:

$$\hat{\sigma}^2_{\text{between}} = \frac{n_1(\bar{x}_1 - \bar{\bar{X}})^2 + n_2(\bar{x}_2 - \bar{\bar{X}})^2 + \cdots + n_k(\bar{x}_k - \bar{\bar{X}})^2}{k - 1} \qquad (11.6)$$

where k = number of samples

In summary, then, if the H_0 is true, the $\hat{\sigma}^2_{\text{between}}$ value should be a good estimate of the population variance, and it should be approximately the same as the $\hat{\sigma}^2_{\text{within}}$ value. Ideally, of course, if the H_0 is true, then:

$$\text{TR}_F = \frac{\hat{\sigma}^2_{\text{between}}}{\hat{\sigma}^2_{\text{within}}} = 1.00$$

But we know that sampling variation makes this an unlikely result, even when the H_0 is true. Should there be a significant difference between $\hat{\sigma}^2_{\text{within}}$ and $\hat{\sigma}^2_{\text{between}}$, however, it may be concluded that this difference is the result of differences between population means.

Step 7: Make the Statistical Decision If, as noted above in our decision-rule step, we have a critical F value of 3.47, and if our two computed estimates of σ^2 yield similar values that produce a test ratio that is ≤ 3.47, then we will accept the null hypothesis that the means of the three populations sampled are equal.

Self-Testing Review 11.1

1. What necessary assumptions must be met for an analysis of variance test to be valid?
2. What are the null and alternative hypotheses in any ANOVA test?
3. Discuss the concepts of $\hat{\sigma}^2_{\text{within}}$ and $\hat{\sigma}^2_{\text{between}}$. How is each computed? Which is a pooled variance?
4. What is the grand mean ($\bar{\bar{X}}$)? How can it be computed?
5. Which measure—$\hat{\sigma}^2_{\text{within}}$ or $\hat{\sigma}^2_{\text{between}}$—is an estimate of σ^2 if (and only if) the null hypothesis is true?
6. Why is an ANOVA test always a right-tailed test?
7. How is the critical region determined in an ANOVA test? What do you need to know before finding the critical F value?
8. Find the critical F value for an ANOVA test if $\alpha = .01$ and if there are 6 samples with a total of 35 items in all of the samples.
9. Find the critical F value for an ANOVA test if $\alpha = .05$ and if there are 4 samples with a total of 44 items in all of the samples.
10. Find the critical F value for an ANOVA test if $\alpha = .01$ and if there are 3 samples with a total of 29 items in all of the samples.
11. What is the TR_F if $\hat{\sigma}^2_{\text{between}} = 35.7$ and $\hat{\sigma}^2_{\text{within}} = 14.6$?
12. What is the TR_F if $\hat{\sigma}^2_{\text{between}} = 215.23$ and $\hat{\sigma}^2_{\text{within}} = 73.81$?

13–15. The following are true-false questions:

13. When the null hypothesis is rejected, it may be concluded that no two population means are equal.

14. If the null hypothesis is true, $\hat{\sigma}^2_{\text{between}}$ must equal $\hat{\sigma}^2_{\text{within}}$.

15. The $\hat{\sigma}^2_{\text{between}}$ value is used as a pooled variance.

11.2 An ANOVA Example

Callie Fisk, vice president of the Nickel and Dime Savings Bank, is reviewing employee performance for possible salary increases. In evaluating tellers, Callie decides that an important performance criterion is the number of customers served each day. She expects that each teller should handle approximately the same number of customers daily. Otherwise, each teller should be rewarded or penalized accordingly.

Callie randomly selects 6 business days, and customer traffic for each teller during these days is recorded. The factor or variable of interest, then, is the number of customers served. The sample (or teller) data are shown below:

	Customer Traffic Data		
Day	Teller 1 Ms. Munny	Teller 2 Mr. Coyne	Teller 3 Mr. Sentz
1	45	55	54
2	56	50	61
3	47	53	54
4	51	59	58
5	50	58	52
6	45	49	51

The *Null Hypothesis (Step 1)* is that the 3 tellers each serve the same average number of customers per day. That is, Ms. Munny, Mr. Coyne, and Mr. Sentz are assumed to have the same workload. Thus:

H_0: $\mu_1 = \mu_2 = \mu_3$ or H_0: All population means are equal.

The *Alternative Hypothesis* is that not all the tellers are handling the same average number of customers per day. That is, at least 1 of the tellers is performing much better than the others, or, perhaps, at least 1 of the tellers is not performing up to the standards of the others. Thus:

H_1: Not all the population means are equal.

If Callie is to make a judgment about the tellers' performance, she should have some idea of the degree of error possible in her decision. Let's assume that she believes that there should be at most only a 5 percent chance of erroneously rejecting a true H_0. Thus, she specifies a *level of significance* of .05 (*Step 2*).

Callie knows, of course, to use an *F distribution* in her test (*Step 3*), and her next step is to *define the rejection or critical region.* The df_{num} value is $k - 1$, or $3 - 1$, or 2, and the df_{den} value is $T - k$, or $18 - 3$, or 15. So, with $\alpha = .05$, $df_{num} = 2$, and $df_{den} = 15$, the critical F value that separates the areas of acceptance and rejection in this ANOVA test is $F_{(2,15,\alpha=.05)} = 3.68$ (*Step 4*).

Decision Rule (Step 5):

Reject H_0 and accept H_1 if the TR is > 3.68. Otherwise, accept H_0.

Test Ratio (Step 6): Since the formula for the test ratio (F value) is:

$$TR_F = \frac{\hat{\sigma}^2_{\text{between}}}{\hat{\sigma}^2_{\text{within}}}$$

we (and Callie) must first calculate the numerator and denominator values for this formula.

The following procedure is carried out to *compute* $\hat{\sigma}^2_{\text{between}}$ (the numerator of our TR):

1. A mean is computed from the sample data for each teller. The sample means are shown in Table 11.1 and have been designated $\bar{x}_1$, $\bar{x}_2$, and $\bar{x}_3$ for tellers 1, 2, and 3, respectively.

TABLE 11.1 DATA USED IN COMPUTING $\hat{\sigma}^2_{\text{BETWEEN}}$

	Customer Traffic Data		
Day	Teller 1 Ms. Munny	Teller 2 Mr. Coyne	Teller 3 Mr. Sentz
1	45	55	54
2	56	50	61
3	47	53	54
4	51	59	58
5	50	58	52
6	45	49	51
Totals	294	324	330
	$\bar{x}_1 = 49$	$\bar{x}_2 = 54$	$\bar{x}_3 = 55$

$$\bar{\bar{X}} = \frac{n_1\bar{x}_1 + n_2\bar{x}_2 + \cdots + n_k\bar{x}_k}{n_1 + n_2 + \cdots + n_k} = \frac{6(49) + 6(54) + 6(55)}{6 + 6 + 6} = \frac{948}{18} = 52.67$$

$$\hat{\sigma}^2_{\text{between}} = \frac{n_1(\bar{x}_1 - \bar{\bar{X}})^2 + n_2(\bar{x}_2 - \bar{\bar{X}})^2 + n_3(\bar{x}_3 - \bar{\bar{X}})^2}{k - 1}$$

$$= \frac{6(49 - 52.67)^2 + 6(54 - 52.67)^2 + 6(55 - 52.67)^2}{3 - 1}$$

$$= \frac{6(13.469) + 6(1.769) + 6(5.429)}{2} = \frac{124.00}{2} = 62.0$$

2. The grand mean is computed next:

$$\bar{\bar{X}} = \frac{n_1\bar{x}_1 + n_2\bar{x}_2 + \cdots + n_k\bar{x}_k}{n_1 + n_2 + \cdots + n_k} = \frac{6(49) + 6(54) + 6(55)}{6 + 6 + 6} = \frac{948}{18} = 52.67$$

3. After these easy steps, we are now ready to determine the value of $\hat{\sigma}^2_{\text{between}}$:

$$\hat{\sigma}^2_{\text{between}} = \frac{n_1(\bar{x}_1 - \bar{\bar{X}})^2 + n_2(\bar{x}_2 - \bar{\bar{X}})^2 + n_3(\bar{x}_3 - \bar{\bar{X}})^2}{k - 1}$$

$$= \frac{6(49 - 52.67)^2 + 6(54 - 52.67)^2 + 6(55 - 52.67)^2}{3 - 1}$$

$$= \frac{6(13.469) + 6(1.769) + 6(5.429)}{2} = \frac{124.00}{2} = 62.0$$

We can see that the total computed by using the squared differences of the sample means about the $\bar{\bar{X}}$ is 124.00. This total of 124 is often called a *sum of squares between* value, as we'll see later. The computed value for $\hat{\sigma}^2_{\text{between}}$ is 62.0, and this figure is often called a *mean of squares between* that's based on the factor or variable being considered. This 62.0 value is an appropriate estimate of σ^2 *if, and only if,* the null hypothesis is true.

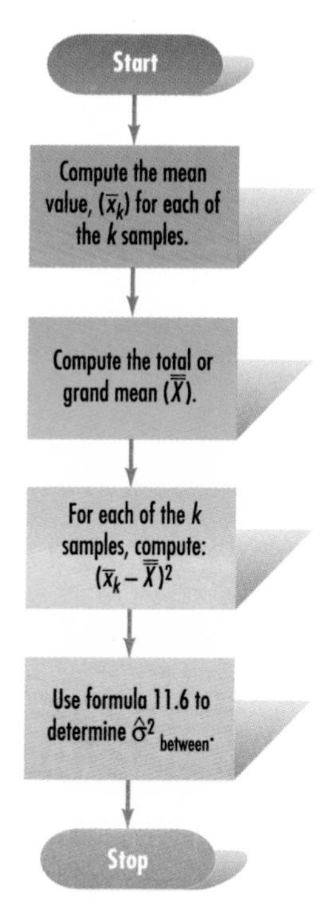

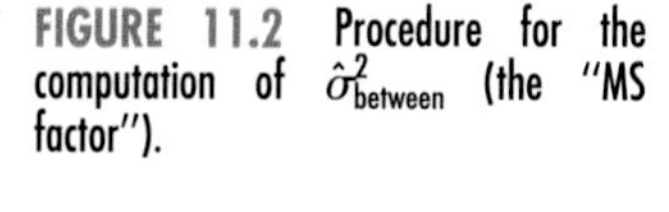
FIGURE 11.2 Procedure for the computation of $\hat{\sigma}^2_{\text{between}}$ (the "MS factor").

The procedure for computing $\hat{\sigma}^2_{\text{between}}$ by focusing on the variance between sample means is summarized in Figure 11.2.

We now have a value for $\hat{\sigma}^2_{\text{between}}$, the numerator of our test ratio, but we still need to find the denominator, $\hat{\sigma}^2_{\text{within}}$. The following procedure is carried out to *compute* $\hat{\sigma}^2_{\text{within}}$:

1. For *each sample*, we compute the *deviation* between each value within the sample and the mean of that sample—that is, compute $x - \bar{x}$ for each sample. Table 11.2 shows the deviation computations for each teller. For teller 1, Ms. Munny, the average customer traffic ($\bar{x}_1$) is 49. Thus, the deviation for day 1 is 45 − 49, or −4; the deviation for day 2 is 56 − 49 = 7; and so forth.

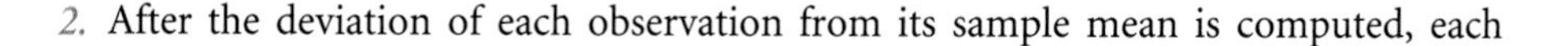
2. After the deviation of each observation from its sample mean is computed, each

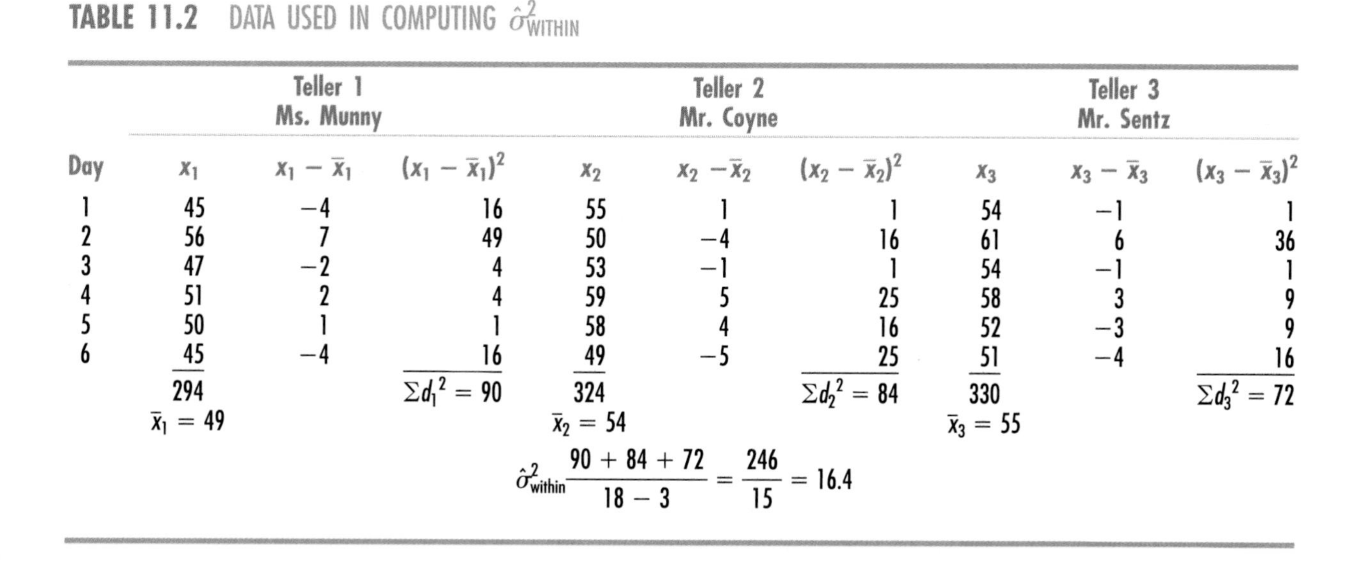
TABLE 11.2 DATA USED IN COMPUTING $\hat{\sigma}^2_{\text{WITHIN}}$

	Teller 1 Ms. Munny			Teller 2 Mr. Coyne			Teller 3 Mr. Sentz		
Day	x_1	$x_1 - \bar{x}_1$	$(x_1 - \bar{x}_1)^2$	x_2	$x_2 - \bar{x}_2$	$(x_2 - \bar{x}_2)^2$	x_3	$x_3 - \bar{x}_3$	$(x_3 - \bar{x}_3)^2$
1	45	−4	16	55	1	1	54	−1	1
2	56	7	49	50	−4	16	61	6	36
3	47	−2	4	53	−1	1	54	−1	1
4	51	2	4	59	5	25	58	3	9
5	50	1	1	58	4	16	52	−3	9
6	45	−4	16	49	−5	25	51	−4	16
	294		$\Sigma d_1^2 = 90$	324		$\Sigma d_2^2 = 84$	330		$\Sigma d_3^2 = 72$
	$\bar{x}_1 = 49$			$\bar{x}_2 = 54$			$\bar{x}_3 = 55$		

$$\hat{\sigma}^2_{\text{within}} = \frac{90 + 84 + 72}{18 - 3} = \frac{246}{15} = 16.4$$

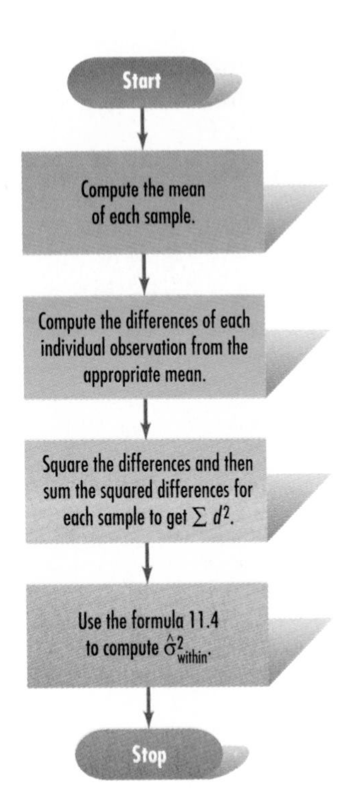

FIGURE 11.3 Procedure for the computation of $\hat{\sigma}^2_{within}$ (the "MS error").

deviation is *squared*—that is, $(x - \overline{x})^2$. These squared deviations are summed—$\Sigma(x - \overline{x})^2$—and the sums are labeled $\Sigma\, d^2$. That is, $\Sigma(x - \overline{x})^2 = \Sigma\, d^2$. For tellers 1, 2, and 3, the sums of the squared deviations are 90, 84, and 72, respectively. This total of 246 is also often called a *sum of squares within* value, as we'll see later.

3. The sum of squares figure of 246 is then divided by the quantity $T - k$, where T is the total number of sample items, or 18 in this example, and k is the number of samples (3). Thus, $T - k = 15$.

Finally, as you can see in Table 11.2, $\hat{\sigma}^2_{within}$ is conveniently computed as follows, using formula 11.4:

$$\hat{\sigma}^2_{within} = \frac{\Sigma\, d_1^2 + \Sigma\, d_2^2 + \Sigma\, d_3^2}{T - k} = \frac{90 + 84 + 72}{18 - 3} = \frac{246}{15} = 16.4$$

This estimate of σ^2 is also referred to as a *mean of squares within*, and any deviation of this value from the true σ^2 is due to random error. Figure 11.3 summarizes the procedure used to compute $\hat{\sigma}^2_{within}$.

Now that we've computed $\hat{\sigma}^2_{between}$ and $\hat{\sigma}^2_{within}$, we're ready to form a ratio of two population variance estimates and calculate the test ratio as follows:

$$\text{TR}_F = \frac{\hat{\sigma}^2_{between}}{\hat{\sigma}^2_{within}} = \frac{62.0}{16.4} = 3.78$$

Conclusion (Step 7): The final step now is to make the *statistical decision.* Since the $\text{TR}_F = 3.78$ and since the critical table value $= 3.68$, Callie should conclude that the H_0 is unlikely and that the alternative hypothesis should be accepted. At least one of the tellers among Munny, Coyne, and Sentz is likely to be handling more or fewer customers than the others. Additional research is needed to precisely define the nature of the difference in work performance.

Self-Testing Review 11.2

1. It has been estimated that 80 percent of all lower back pain cases are caused by weak trunk muscles. Studies have suggested that improved strength of the trunk muscles will aid in preventing lower back pain and in treating cases when they do arise. In a study in *Physical Therapy* to compare the effect of various training frequencies on the development of trunk-muscle strength, random samples of participants were assigned to 1 of 4 groups. These groups and their training frequencies were: group 1 = once every other week, group 2 = once a week, group 3 = twice a week, and group 4 = three times a week. The following data represent the weight in kilograms for each of the participants at the beginning of the program:

Group 1	Group 2	Group 3	Group 4
84	56	70	47
93	78	59	73
83	56	78	104
61	61	53	71
121		104	69
67		110	99
		40	

Use an ANOVA test at the .05 level to test the hypothesis that the population mean weights of the 4 groups are equal.

2. Language comprehension appears to decline with advancing age. A study in *Applied Psycholinguistics* sought to determine which factors contribute significantly to comprehension difficulty. A randomly selected group of paid volunteers was recruited through advertising in local newspapers, and these people agreed to have their hearing tested. The following data represent the pure-tone average (in dB HTL) for the subjects who were separated into different age groups (30s, 50s, 60s, and 70s). Test the hypothesis at the .05 level that the population mean hearing level is the same for all age groups:

Thirties	Fifties	Sixties	Seventies
9	9	19	18
13	5	8	22
5	8	14	24
5	3	26	
10	15		

3. Do opinions about testing differ among sophomore education majors, senior education majors, and teachers who are on the job? In a *Journal of Educational Research* study, a survey form with questions about testing opinions was mailed to random samples from each of the 3 groups. The following data represent samples of responses from each of the 3 groups when asked how many hours should be spent in testing per week. Test the hypothesis at the .05 level that the population mean responses for the 3 groups are equal:

Sophomore	Senior	Teacher
11	10	3
9	4	6
19	0	2
18	13	9
8	18	1
	14	7
		2
		5

4. A study in *The American Journal of Psychology* involved random samples of normal subjects and alcoholics. Some of the normal subjects and some of the alcoholics were given *Lorazepam*, and some from each group were given a placebo. The metabolic value at the cerebellum (a portion of the brain) was then determined (in μ mol/100 g/min) for each subject. Test the hypothesis at the .05 level that the population mean metabolic values are equal for all four groups:

STATISTICS IN ACTION

A Walnut a Day . . . ?

A study reported in *The New England Journal of Medicine* suggests that eating nuts, and in particular walnuts, may be good for the heart. The study subjects were 31,208 members of Seventh-Day Adventist families—people who generally avoid smoking and drinking alcoholic beverages. Those who ate nuts at least five times a week had only half the risk of fatal heart attacks as those who ate nuts once a week. On a diet rich in walnuts, total cholesterol levels dropped from 182 to 160.

Normal Placebo	Normal *Lorazepam*	Alcoholic Placebo	Alcoholic *Lorazepam*
40	38	41	32
49	37	40	39
46	33	33	34
45	34	36	35
47	40		

5. A career counselor claims that there is no difference in career decision-making skills among various groups. A test to determine career decision-making skills is given to random samples of people belonging to 3 groups. The results of this test are shown below. Test the counselor's claim at the .01 level:

African American	Hispanic	Caucasian
17	12	13
9	10	14
13	15	14
16	13	15
12		15
		14

6. A large accounting firm wants to see if the accuracy of its employees' work is related to the school from which the employees graduated. Accountants representing 4 schools were randomly selected, and the number of errors committed by each accountant over a 2-week period was recorded as shown below:

School A	School B	School C	School D
14	17	19	23
16	16	20	12
17	18	22	21
13	15	21	10
22	16	18	9
9	12	19	15
10	14	15	16

Conduct an ANOVA test at the .01 level. Is there a significant difference in accuracy?

7. A track coach has learned of 2 new training techniques that are designed to reduce the time required to run a mile. Three random samples of runners have been selected for the experiment. Group A trained under the old approach, group B under one of the new techniques, and group C under the other new technique.

After a month of training, each runner was timed in a mile run. The data are as follows (times in minutes):

Group A	Group B	Group C
4.81	4.43	4.38
4.62	4.50	4.29
5.02	4.32	4.33
4.65	4.37	4.36
	4.41	

Conduct an ANOVA test at the .05 level to see if the population mean times for all those using the 3 methods are equal.

8. Ms. Anne Tenna, a first-year middle school teacher, claims that television-viewing habits are the same for all students in the different middle school grades. She questioned a random sample of middle school students about how many minutes they watch TV each day after school and until bedtime, and produced the following data:

Sixth Grade	Seventh Grade	Eighth Grade
459	115	272
311	153	88
152	201	374
293	30	178

Test Ms. Tenna's claim at the .05 level.

9. The Tackey Toy Company plans to install special battery packs in its new line of Tackey robots. Three suppliers can produce packs with equal prices that meet Tackey's needs. But Tackey managers want to examine the data on the life expectancy of each brand before selecting a supplier. A random sample of 6 battery packs is selected from each vendor, and the useful life of each battery pack (in hours) is determined. Use the following data to make a statistical decision at the .05 level. Is there a significant difference in the population life expectancy of the different brands of battery packs?

Lifelong	Neverstop	Everrun
144	168	184
136	150	172
146	142	168
134	166	187
150	136	176

10. A study in *The American Journal of Psychiatry* dealt with scores made by individuals on a Wide Range Achievement Test (WRAT). These scores are given below for random samples of first-episode schizophrenic patients, chronic schizophrenic patients, and normal comparison subjects. Test the hypothesis at the .01 level that there's no difference in the WRAT scores of the 3 populations:

First Episode	Chronic Schizophrenics	Normal Comparison
93	87	118
84	91	107
74	116	110
109	86	83
116	116	

11. Transcutaneous electrical nerve stimulation (TENS) devices are frequently used in the management of acute and chronic pain conditions. An important component of the TENS system is the skin electrode. A study in *Physical Therapy* was carried out to determine the conductive differences among electrodes used with TENS devices. The impedance measures (in ohms) for random samples of TENS devices in the low-impedance, medium-impedance, and high-impedance groups are given below:

Low Impedance	Medium Impedance	High Impedance
1200	3100	7500
1200	3100	6400
1000	2900	
1600	2700	
1400	2500	
1400	3600	
1200	3100	
1700	3100	
1600	2100	
1300	4300	
1600	3000	
	2600	

Test the hypothesis at the .01 level that the population mean impedance values for all 3 groups are equal.

12. A large retailer must make a choice between 3 sales locations in a shopping mall. The retailer wants to see if the mean daily traffic count is the same for all locations. The following data are traffic counts for a random 7-day period:

Location 1	Location 2	Location 3
643	249	458
542	404	513
537	564	475
484	745	536
464	353	364
369	647	738
478	351	594

At the .05 level, is there likely to be a significant difference in the population mean traffic count at the 3 locations?

13. A major distributor of cameras suspects that consumers are insensitive to price changes for the highest-quality camera. To test this suspicion, 4 retail outlets are selected at random and asked to sell the camera at 1 of 4 predetermined prices. After 4 weeks, the number of cameras sold each week is as follows:

Price 1	Price 2	Price 3	Price 4
3	5	10	8
5	4	9	4
7	6	4	5
4	7	5	7

What conclusions can be made about the population mean selling price at the .05 level of significance?

11.3 The One-Way ANOVA Table and Computers to the Rescue

The One-Way ANOVA Table

It's often desirable to place the computations of the last several pages in a **one-way ANOVA table**—a summary listing of the values needed to produce an ANOVA test. Table 11.3 shows the general format of this type of ANOVA table. The first column lists the *source*, or type, of variation. As we've seen, this variation is measured *between samples*—the **factor variation**—and *within samples*—the **error variation**. (The word "treatment" is often substituted for the word "factor" because many ANOVA techniques originated in agricultural research where the concern was with the different treatments—e.g., different fertilizers or seeds—that could be applied to the soil.) Column 2 in Table 11.3 shows the *degrees of freedom* associated with each source of variation (the df values we've computed earlier).

Hair Today, Gone Tomorrow

A study of 872 male electronics factory workers in Italy was conducted by Dr. Maurizio Trevisan of the University of Naples. Of this group, 278 had "male-pattern baldness" (a receding hairline and a bald spot on the crown of the head), 273 had receding hairlines but not a balding crown, and 321 had full heads of hair. The blood cholesterol and blood pressure levels rose with age among those with male-pattern baldness but not with those in the other two categories. Trevisan and his colleagues plan additonal research to see if those with male-pattern baldness are more prone to heart attack and stroke.

TABLE 11.3 THE FORMAT OF A GENERAL ONE-WAY ANOVA TABLE

Source of Variation	Degrees of Freedom (df)	Sum of Squares (SS)	Mean of Squares (MS)	Computed F Value (TR_F)
Between samples (factor variation)	$k - 1$	SS between (SSB) or SS factor	$\hat{\sigma}^2_{between}$ (MS factor)	$\dfrac{\hat{\sigma}^2_{between}}{\hat{\sigma}^2_{within}}$
Within samples (error variation)	$T - k$	SS within (SSW) or SS error	$\hat{\sigma}^2_{within}$ (MS error)	
Total	$T - 1$	SSB + SSW		

Column 3 in Table 11.3 records the *sum of squares* (SS) figures computed during our $\hat{\sigma}^2_{between}$ and $\hat{\sigma}^2_{within}$ calculations. (You'll remember that these terms were mentioned earlier.) The fourth column is labeled **mean of squares (MS)**—the designation in an ANOVA table for our computed values of $\hat{\sigma}^2_{between}$ and $\hat{\sigma}^2_{within}$. The $\hat{\sigma}^2_{between}$ value is often called the **MS factor**, and the $\hat{\sigma}^2_{within}$ value is often referred to as the **MS error.** Finally, column 5 presents the computed F value (TR_F), which, we've seen, is $\hat{\sigma}^2_{between}/\hat{\sigma}^2_{within}$ (or MS factor/MS error).

Computers to the Rescue

The *Minitab* statistical package makes short work of the procedures we've considered in this chapter. The top lines in Figure 11.4 show the number of customers served by the 3 bank tellers in our example problem. Once the sample data are entered, the entire ANOVA procedure is carried out by the package in response to the single AOVONEWAY C1-C3 command. As you can see in Figure 11.4, a one-way ANOVA table is provided. The same values we've painstakingly calculated are reproduced in the computer output, along with some additional useful information.

A *p value* of .047 is given at the top right corner of the ANOVA table. You know from our discussion in Chapter 8 that the p-value approach is often used in hypothesis testing. When that's the case, the *decision rule* is to reject H_0 and accept H_1 if the p value is $< \alpha$. Otherwise, the H_0 is accepted. Since the p value of .047 in Figure 11.4 is less than .05, we know to reject the null hypothesis at the 5 percent level of significance.

The 95 percent confidence intervals produced below the ANOVA table give us an idea of the intervals that are likely to include the three population means. Each interval is calculated with the formula:

$$\bar{x} \pm t\left(\frac{s_{p_0}}{\sqrt{n}}\right)$$

where s_{p_0} is the estimated population standard deviation found by pooling and averaging the sample variances. In this case, the value of s_{p_0} is 4.05, which is the square root of the $\hat{\sigma}^2_{within}$ value of 16.4. Thus, the 95 percent confidence interval that's likely to include μ_1 is found as follows:

$$49.00 \pm t\left(\frac{4.05}{\sqrt{6}}\right) = 49.00 \pm 2.131(1.653) = 45.48 \text{ to } 52.52$$

(The t value from the t table in Appendix 4 is at the 95 percent level, and the df is that of the MS error value, or 15 in this case, so t is 2.131 in this example.) The other intervals likely to include μ_2 and μ_3 are found in the same way. Since there's a slight overlap of all three intervals, there's a slim chance that the H_0 is true, but as we've seen, that chance is less than .05. The greater the interval overlap, of course, the more likely it is that H_0 is true.

Congratulations! You've made it through a rather detailed chapter.

STATISTICS IN ACTION

Charleston's Numbers

Albert Parish, Jr., chief economist for the Charleston, South Carolina, Chamber of Commerce, uses statistical software to monitor the local economy. Monthly data are kept on such variables as retail sales, unemployment levels, port tonnage (Charleston has a large seaport), automobile registrations, airport boardings and deboardings, treasury yields, FHA mortgage rates, prime lending rates, building starts, the Consumer Price Index, lodging occupancy, and tourist attraction attendance. Every quarter, Parish merges the new data with monthly data from the last 6 years and then uses his software to obtain descriptive statistics and to forecast levels of economic activity for the next 3 months.

Self-Testing Review 11.3

1–13. You'll recall that the data in problem 1, STR 11.2, are as follows:

Group 1	Group 2	Group 3	Group 4
84	56	70	47
93	78	59	73
83	56	78	104
61	61	53	71
121		104	69
67		110	99
		40	

```
MTB > Set C1
DATA> 45  56  47  51  50  45
DATA> Set C2
DATA> 55  50  53  59  58  49
DATA> Set C3
DATA> 54  61  54  58  52  51
DATA> End
MTB > Name C1 'Munny', C2 'Coyne', C3 'Sentz'
MTB > AOVONEWAY C1-C3

ANALYSIS OF VARIANCE
SOURCE     DF        SS        MS        F         p
FACTOR      2     124.0      62.0     3.78     0.047
ERROR      15     246.0      16.4
TOTAL      17     370.0
                                    INDIVIDUAL 95 PCT CI'S FOR MEAN
                                    BASED ON POOLED STDEV
 LEVEL      N      MEAN     STDEV   -------+---------+---------+--------
Munny       6    49.000     4.243    (-------*--------)
Coyne       6    54.000     4.099              (--------*--------)
Sentz       6    55.000     3.795                 (-------*--------)
                                    -------+---------+---------+--------
POOLED STDEV =    4.050                48 .0      52.0      56.0
```

FIGURE 11.4 The *Minitab* output for our chapter example problem.

When supplied with this data set, *Minitab* produces the following output:

```
ANALYSIS OF VARIANCE
SOURCE      DF        SS       MS         F         p
FACTOR       3      1220      407      0.87     0.473
ERROR       19      8868      467
TOTAL       22     10088
                                   INDIVIDUAL 95 PCT CI'S FOR MEAN
                                   BASED ON POOLED STDEV
 LEVEL       N      MEAN    STDEV  ----------+---------+---------+------
Group 1      6     84.83    21.28               (--------*---------)
Group 2      4     62.75    10.44  (----------*-----------)
Group 3      7     73.43    25.97             (--------*-------)
Group 4      6     77.17    21.11               (---------*--------)
                                   ----------+---------+---------+------
POOLED STDEV =     21.60                     60        80       100
```

1. What does the term "factor" mean in the *Minitab* output? What other term has been used to designate the same meaning?
2. What does the term "error" mean?
3. Interpret the degrees of freedom (df) entries in the *Minitab* output, and explain how they are computed.
4. How is the SS error computed?
5. How is the SS factor computed?
6. What is another name for the MS error?
7. What is another name for the MS factor?
8. What is the test ratio, and how is it computed from the ANOVA table entries?
9. Interpret the p value in the *Minitab* output, and make the statistical decision at the .05 level.
10. A LEVEL heading is used in the output. What's the meaning of this label?
11. What information is easily available from the bottom half of the *Minitab* output?
12. What is the purpose of the confidence intervals chart shown below the ANOVA table?
13. What is the meaning of the pooled standard deviation shown in the *Minitab* output?

14–17. The following gives market share data in four locations:

Loc. 1	Loc. 2	Loc. 3	Loc. 4
28.42	10.81	17.78	32.25
23.00	12.83	13.58	22.75
29.00	10.66	18.50	27.45
28.79	10.11	18.37	11.79

When supplied with this data set, *Minitab* produces the following output:

```
ANALYSIS OF VARIANCE
SOURCE      DF        SS        MS        F         p
FACTOR       3     614.3     204.8     8.92     0.002
ERROR       12     275.4      22.9
TOTAL       15     889.7
                                    INDIVIDUAL 95 PCT CI'S FOR MEAN
                                    BASED ON POOLED STDEV
 LEVEL       N      MEAN     STDEV  ---+---------+---------+---------+---
Loc 1        4    27.302     2.878                        (-----*------)
Loc 2        4    11.102     1.190  (------*-----)
Loc 3        4    17.058     2.339          (-----*------)
Loc 4        4    23.560     8.753                  (-----*------)
                                    ---+---------+---------+---------+---
POOLED STDEV =     4.790              8.0      16.0      24.0      32.0
```

14. Explain the df entries, and tell how they are computed.

15. What is the test ratio, and how is it computed from the table entries?

16. Interpret the *p* value and make the statistical decision at the .05 level.

17. What is the purpose of the confidence intervals chart shown below the ANOVA table?

18–24. Consider the following ANOVA output:

```
ANALYSIS OF VARIANCE
SOURCE      DF        SS        MS        F         p
FACTOR       3     491.6     163.9     6.56     0.006
ERROR       13     324.6      25.0
TOTAL       16     816.2
                                    INDIVIDUAL 95 PCT CI'S FOR MEAN
                                    BASED ON POOLED STDEV
 LEVEL       N      MEAN     STDEV  ------+---------+---------+---------+
Thirties     5     8.400     3.435   (------*------)
Fifties      5     8.000     4.583   (-----*------)
Sixties      4    16.750     7.632             (-------*-------)
Seventies    3    21.333     3.055                  (-------*--------)
                                    ------+---------+---------+---------+
POOLED STDEV =     4.997                 7.0      14.0      21.0      28.0
```

18. How many different factors are considered?

19. How many values are in each factor group, and what is the total number of values?

20. How would you compute the various df entries?

21. How are the MS entries computed?

22. What two values in the table do you need to compute the test ratio?

23. What is the p value for this test? Interpret it, and tell what decision you would make from it.

24. How would you interpret the confidence intervals chart below the ANOVA table?

25–31. The data from problem 4, STR 11.2, produce the following output:

```
ANALYSIS OF VARIANCE
SOURCE      DF        SS        MS         F         p
FACTOR       3     311.1     103.7      9.98     0.001
ERROR       14     145.4      10.4
TOTAL       17     456.5
                                   INDIVIDUAL 95 PCT CI'S FOR MEAN
                                   BASED ON POOLED STDEV
 LEVEL       N      MEAN     STDEV  -------+---------+---------+--------
NormPlac     5    45.400     3.362                      (-----*-----)
NormLora     5    36.400     2.881       (-----*-----)
AlcoPlac     4    37.500     3.697         (------*------)
AlcoLora     4    35.000     2.944  (------*------)
                                    -------+---------+---------+--------
POOLED STDEV =    3.223                  35.0      40.0      45.0
```

25. How many different factors are considered?

26. How many values are in each factor group, and what is the total number of values?

27. How would you compute the various df entries?

28. How are the MS entries computed?

29. What two values in the table do you need to compute the test ratio?

30. What is the p value for this test? What decision would you make from it?

31. How would you interpret the confidence intervals chart below the ANOVA table?

32–38. The data from problem 10, STR 11.2, produce the following output:

```
ANALYSIS OF VARIANCE
SOURCE      DF        SS        MS         F         p
FACTOR       2       192        96      0.37     0.697
ERROR       11      2839       258
TOTAL       13      3031
                                   INDIVIDUAL 95 PCT CI'S FOR MEAN
                                   BASED ON POOLED STDEV
 LEVEL       N      MEAN     STDEV  ----+---------+---------+---------+--
FirstEpi     5     95.20     17.34  (------------*-------------)
Chronic      5     99.20     15.45     (-------------*-------------)
Normal       4    104.50     15.07        (--------------*--------------)
                                    ----+---------+---------+---------+--
POOLED STDEV =     16.06                84        96       108       120
```

32. How many different factors are considered?

33. How many values are in each factor group, and what is the total number of values?

34. How would you compute the various df entries?

35. How are the MS entries computed?

36. What two values in the table do you need to compute the test ratio?

37. What is the p value for this test? Interpret it, and tell what decision you would make from it.

38. How would you interpret the confidence intervals chart below the ANOVA table?

39–44. Fill in the missing entries for the following ANOVA table:

ANALYSIS OF VARIANCE

SOURCE	DF	SS	MS	F
FACTOR	3	73.6	C?	E?
ERROR	A?	816.1	D?	
TOTAL	15	B?		

39. A =

40. B =

41. C =

42. D =

43. E =

44. What decision should be made about the equality of the population means at the .01 level?

45–50. Fill in the missing entries for the following ANOVA table:

ANALYSIS OF VARIANCE

SOURCE	DF	SS	MS	F
FACTOR	A?	363.6	C?	E?
ERROR	12	727.3	D?	
TOTAL	14	B?		

45. A =

46. B =

47. C =

48. D =

49. E =

50. What decision should be made about the equality of the population means at the .01 level?

51. How would you interpret the following confidence intervals chart?

```
                                      INDIVIDUAL 95 PCT CI'S FOR MEAN
                                      BASED ON POOLED STDEV
LEVEL       N     MEAN     STDEV     -------+---------+---------+--------
Summer      7     4.571    1.512      (--------*--------)
Fall        9     8.111    3.180                        (-------*-------)
Spring      8     5.000    1.512         (-------*-------)
                                     -------+---------+---------+--------
POOLED STDEV =    2.295                    4.0       6.0       8.0
```

LOOKING BACK

1. The purpose of the one-way ANOVA technique discussed in this chapter is to enable a decision maker to compare three or more independent sample means to see if there are statistically significant differences between the means of the populations from which the samples are taken. The following three assumptions must be true before the ANOVA technique can be applied to a decision-making situation: (1) The population distributions approximate the normal distribution, (2) the samples are random and independent of each other, and (3) the variances of all populations are equal.

2. As is the case with other hypothesis-testing procedures, we follow a seven-step procedure to complete an ANOVA test. This test begins with a statement of the null and alternative hypotheses (*Step 1*); it then requires that the test be made at a suitable level of confidence (*Step 2*). An F distribution is used in an ANOVA test (*Step 3*). Three things must be known to find the F value in Appendix 5 that separates the area of acceptance from the rejection area (*Step 4*). These three items are (*a*) the level of significance, (*b*) the degrees of freedom for the numerator of the test ratio, and (*c*) the degrees of freedom for the denominator of the test ratio. When these three items are determined, the critical F value for the test can be found, and the decision rule can be formulated (*Step 5*).

3. Finding the test ratio (*Step 6*) requires that two estimates of the population variance be using two independent computational methods. In one approach—the one that produces $\hat{\sigma}^2_{within}$—the population variance is estimated by computing the variances found in each sample. These sample variances are pooled and averaged to get an estimate of σ^2. Any deviation of this estimate from the true population variance is due to random sampling error. In the other approach, $\hat{\sigma}^2_{between}$, an estimate of σ^2 is found by measuring the variation between the sample means. This estimate of σ^2 has merit if, and only if, the H_0 is true. Any deviation of this estimate from the σ^2 is due to factor differences among the samples. The two estimates of σ^2 are then used to compute the test ratio (or F value). Ideally, if the H_0 is true, this ratio will

have a value of 1.00 since the two estimates of σ^2 will yield the same results. Realistically, however, sampling variation will normally cause the test ratio to exceed 1.00 even when the H_0 is true. Once the test ratio (TR_F) is computed, a statistical decision (*Step 7*) can be made. If a computed (TR_F) value is found to exceed the appropriate *F*-table value at a specified level of confidence, the ANOVA test concludes with the rejection of the null hypothesis. But if the TR_F value $\leq$ the table value, the null hypothesis is accepted.

4. It's often desirable to prepare a one-way ANOVA table—a summary listing of the values needed to produce an ANOVA test. The first column of this table lists the sources of variation, and subsequent columns show degrees of freedom, sum of squares (SS) figures, mean of squares (MS) values (the $\hat{\sigma}^2_{between}$ and $\hat{\sigma}^2_{within}$ calculations), and the computed test (*F*) ratio. Statistical software packages typically display ANOVA output in this table format.

Review Exercises

1. Find the critical *F* value for each of the following situations:

a. $\alpha = .05$, $df_{num} = 5$, and $df_{den} = 8$.

b. $\alpha = .05$, $df_{num} = 5$, and $df_{den} = 20$.

c. $\alpha = .05$, $df_{num} = 5$, and $df_{den} = 30$.

2. Examine your answers for problem 1. What can you conclude about the critical *F* value when all other conditions remain the same and the df_{den} increases?

3. Find the critical *F* value for each of the following situations:

a. $\alpha = .01$, $df_{num} = 5$, and $df_{den} = 25$.

b. $\alpha = .01$, $df_{num} = 7$, and $df_{den} = 25$.

c. $\alpha = .01$, $df_{num} = 10$, and $df_{den} = 25$.

4. Examine your answers for problem 3. What can you conclude about the critical *F* value when all other conditions remain the same and the df_{num} increases?

5. Find the critical *F* value for each of these situations:

a. $\alpha = .05$, $df_{num} = 8$, and $df_{den} = 15$.

b. $\alpha = .01$, $df_{num} = 5$, and $df_{den} = 20$.

6. In a test to see if mental skills vary for different age groups, a random sample of subjects in their 30s, 50s, 60s, and 70s were given a "digits backward" test. (A psychologist reads a sequence of digits, and each subject is asked to recite the digits in reverse order.) The following data represent the number of digits correctly recited by each subject:

Thirties	Fifties	Sixties	Seventies
4	3	4	5
5	5	7	3
6	3	7	7
8	7	4	
6	7		

Test the hypothesis at the .01 level that the population means for all 4 age groups are equal.

7. A study in *Physical Therapy* compared the effect of various training frequencies on the development of strength. Participants were

randomly selected from 4 training groups. Those in group 1 worked out once every other week, those in group 2 had a weekly workout, group 3 participants met twice a week, and group 4 members worked out three times a week. The following data represent the age for each of the participants:

Group 1	Group 2	Group 3	Group 4
27	39	37	24
50	36	28	53
43	47	44	51
31	51	36	51
37		30	45
37		27	65
		44	

Test the hypothesis at the .05 level that the population mean ages for the participants of the 4 groups are equal.

8. The marketing director for a pasta manufacturer claims that the height of the shelves on which her product is displayed will affect the sales volume. With the cooperation of 3 retail stores, an experiment is carried out. In one store, the product is placed at eye level on the shelves. In the second store, the product is displayed at waist level, and in the third store, the product is placed at knee level. The number of pasta packages sold for each of 5 randomly selected days is recorded below:

Eye Level	Waist Level	Knee Level
98	106	103
106	105	95
111	98	87
85	93	94
108	96	92

At the .01 level, is there a significant difference in the population mean sales based on the shelf location of the product?

9. A career counselor claims in *Career Development Quarterly* that there is no difference in career decision-making attitudes among the population of students from various socioeconomic classes. The results of scores from an attitudes test given to random samples of students are as follows:

Lower	Middle	Upper
32	45	38
36	42	38
40	34	31
32	42	41
33	29	
37	33	
	34	

Test the counselor's claim at the .01 level.

10. A psychology professor believes that the amount of time his students spend studying is affected by the school term. In a yearlong experiment, students were randomly selected during the 3 different terms and asked to estimate the number of hours they spent studying each week. The hourly estimates were:

Summer	Fall	Spring
4	7	7
3	11	5
6	12	6
7	8	4
5	13	4
3	6	3
4	5	4
	4	7
	7	

CHAPTER 12

Chi-Square Tests: Goodness-of-Fit and Contingency Table Methods

LOOKING AHEAD

In this chapter you'll learn how a management consultant can evaluate the market response to his client's new product and how a political analyst can improve her candidate's campaign strategy. In the early pages, we'll first present an overview of the purpose of two types of hypothesis tests that use the chi-square probability distributions. This orientation leads us to a discussion of the procedural steps needed to conduct such tests. Then, we'll follow these procedural steps to carry out goodness-of-fit and contingency table hypothesis tests.

Thus, after studying this chapter, you should be able to:

- Explain the purpose of the (*a*) goodness-of-fit test and (*b*) contingency table test.
- Describe the steps needed to carry out such hypothesis tests.
- Use these procedural steps to arrive at statistical decisions when goodness-of-fit and contingency table tests are made.

12.1 Chi-Square Testing: Purpose and Procedure

Purpose and Assumptions

You saw in Chapter 3 that a variable of interest is typically obtained by counting or by using some measuring device. In this chapter, we'll concentrate on techniques that can be used for qualitative data and that involve *counting* the data items that fall into selected testing categories. And we'll be dealing with two types of hypothesis tests—the goodness-of-fit and contingency table tests—that focus on count data.

The Goodness-of-Fit Test. In Chapter 5, we saw that in a *binomial experiment* the same action (trial) is repeated a fixed number of times under identical conditions and each action is independent of the others. For each trial there are only *two* possible outcomes (success or failure). Now, though, our interest in conducting a goodness-of-fit test is to evaluate a *multinomial experiment* in which there can be more than two possible outcomes for each trial:

In a **multinomial experiment**, the number of identical and independent trials is fixed, the outcome for each trial falls into only one of the several (or k) possible classes or cells, and the probability that a single trial will result in a specified outcome remains the same throughout the experiment.

Thus,

A **goodness-of-fit test** is one that's used to see if the distribution of the observed outcomes of the sample trials supports a hypothesized population distribution.

For example, a researcher can use a goodness-of-fit test to see if a new product is likely to have altered the market shares commanded by competing products. Or a geneticist can crossbreed flowers of a certain species and then check to see if the four color patterns produced follow a theoretical Mendelian ratio of 9:3:3:1.

In a goodness-of-fit test, then, our purpose is to establish a hypothesis about how we think the data items will be distributed into the different testing categories of interest. Then, we'll conduct a test to see if the observed results conform to the results that would be *expected* if our hypothesis were true. The *assumptions in a goodness-of-fit test are*:

1. The sample is a random sample, and frequency counts are obtained for the possible k classes or cells.
2. The expected frequency for each of the k categories is at least 5.

The Contingency Table Test. You saw in Chapter 4 that a contingency table is one that shows all the classifications of the variables being studied—that is, it accounts for all contingencies in a particular situation. In this chapter, we'll use a contingency table to see if two classification variables are likely to be independent. Each cell in a contingency table is at the intersection of a row and a column. The rows represent one classification category, and the columns represent another such category.

Thus, for our present needs, the table contains values that are grouped in row-variable and column-variable ways, and the purpose of the contingency table test is to see if the data are being classified in *independent* ways.

A **contingency table test** (or **test of independence**) is one that tests the hypothesis that the data are cross-classified in independent ways.

For example, a medical school researcher may use a contingency table test to evaluate the condition of patients over a period of time (one variable) according to the drug treatments they have received (the other variable). Or a political scientist may classify a sample of voters by location of residence (one variable) and preference for candidates seeking office (the other variable).

The *assumptions in a contingency table test are*:

1. The sample data items are obtained through random selection.
2. The expected frequency for each cell in the table is at least 5.

Testing, Testing, 1 . . . 2 . . . 3 . . .

"Our world is inundated with statistics. Every medical fear or triumph is charted by a complex analysis of chances. Think of cancer, heart disease, AIDS: the less we know, the more we hear of probabilities. This daily barrage is not a matter of mere counting but of inference and decision in the face of uncertainty. . . . Money markets, drunken driving, family life, high energy physics, and deviant human cells are all subject to tests of significance and data analysis. This all began in 1900 when Karl Pearson published his chi-square test of goodness of fit, a formula for measuring how well a theoretical hypothesis fits some observations. . . . The chi-square test was a tiny event in itself, but it was the signal for a sweeping transformation in the ways we interpret our numerical world."

From Ian Hacking, "Trial By Number," *Science 84*, November 1984, pp. 69–70. Copyright 1984 by the AAAS. Reprinted with permission.

The Procedural Steps for These Chi-Square Tests

Once again, our topic is a hypothesis-testing procedure, and once again we follow a familiar seven-step process that begins with a statement of the hypotheses to be tested and concludes when a statistical decision is made. We'll outline the general testing procedure that applies to both goodness-of-fit and contingency table tests here and leave the differing details to Sections 12.2 and 12.3.

Step 1: State the Null and Alternative Hypotheses The *null hypothesis* for a *goodness-of-fit test* is essentially that the population being analyzed fits some specified probability distribution pattern. This pattern can vary, of course, according to the logic

The Economic Facts

A Gallup poll randomly selected 1,005 heads of households to represent the "general public" and then randomly selected an additional 300 high school seniors and 300 college seniors. Questions were presented to the three groups to test their understanding of basic economics. The general public answered 39 percent of the questions correctly, the high school seniors answered 35 percent of the questions correctly, and the college seniors answered 51 percent of the questions correctly. The poll showed that most Americans pay little attention to the economic facts of life. The margin of error was ±3 percentage points for heads of households and ±6 percentage points for students.

of the experiment. The *alternative hypothesis*, then, is simply that the population doesn't fit the specified distribution. For example, in one test, the H_0 might be that the population distribution is uniform. A **uniform distribution** is one in which all outcomes considered have an equal or uniform probability. Thus, if an experiment with a hypothesized uniform distribution has k outcomes, each outcome should contain $100/k$ percent of all the sample outcomes. The alternative hypothesis in this case is that the population distribution is *not* uniform. Or the null hypothesis in a goodness-of-fit test could be that the population percentage breakdown of four possible blood types A, B, O, and AB is 42 percent, 10 percent, 45 percent, and 3 percent. The alternative hypothesis in such a test is then that at least one of the stated percentages isn't supported by the test results.

The *null hypothesis for a contingency table test* is that the two variables under study are *independent*. The *alternative hypothesis* in this test is that the two variables are *not* independent—that is, one variable is dependent on (or related to) the other.

Step 2: Select the Level of Significance This step doesn't differ from similar steps in other tests. A criterion for rejection of the H_0 is necessary, and an α value of .01 or .05 is typically used.

Step 3: Determine the Test Distribution to Use Both goodness-of-fit and contingency table tests are based on the chi-square probability distributions introduced in Chapter 7. To summarize here, a χ^2 distribution has the following properties:

- The scale of possible χ^2 values extends from zero indefinitely to the right.
- A χ^2 distribution is not symmetrical, and a different one is used for each change in the number of degrees of freedom that can exist in a sampling situation (see Figure 12.1).

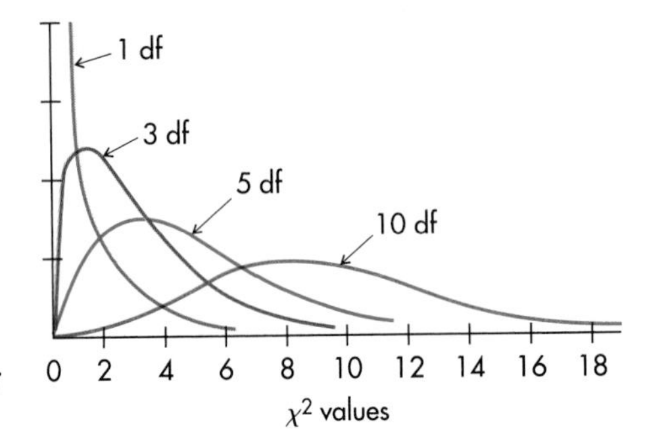

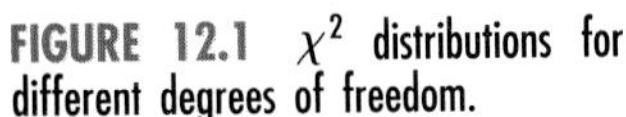
FIGURE 12.1 χ^2 distributions for different degrees of freedom.

The chi-square table we'll use in this chapter is found in Appendix 6 at the back of the book.

Step 4: Define the Rejection or Critical Region We must know the chi-square value that separates the area of acceptance from the area of rejection in a goodness-of-fit test or in a contingency table test. To find this critical χ^2 value, we must know two things:

1. *The level of significance* (specified in ***Step 2*** in our procedure).
2. *The degrees of freedom*. In a *goodness-of-fit test*, the value of the degrees of freedom is:

$$\text{df (goodness of fit)} = k - 1 \tag{12.1}$$

where k = the number of possible outcomes in the experiment

And in a *contingency table test*, the value of the degrees of freedom is:

$$\text{df (contingency table)} = (r - 1)\ (c - 1) \tag{12.2}$$

where r = number of rows in the table
c = number of columns in the table

Let's suppose first that we are making a *goodness-of-fit test* at the .05 level, and there are four possible outcomes in our experiment. In that case, the critical χ^2 value that separates the areas of acceptance and rejection can be found in Appendix 6 under the column labeled .05. (You'll be pleased to know that the area of rejection in all goodness-of-fit and contingency table tests is found only in the *right tail* of the χ^2 distribution.) To find the cell in the .05 column that we need, we must also know the proper

Naval architecture students can use statistical techniques to analyze the hull designs they create. (*Hank Morgan/ Science Source/Photo Researchers*)

degrees of freedom row to use. With four possible outcomes, df = $k - 1$, or $4 - 1$, or 3. Thus, the critical χ^2 value for our test is 7.81 (.05 column, df-3 row).

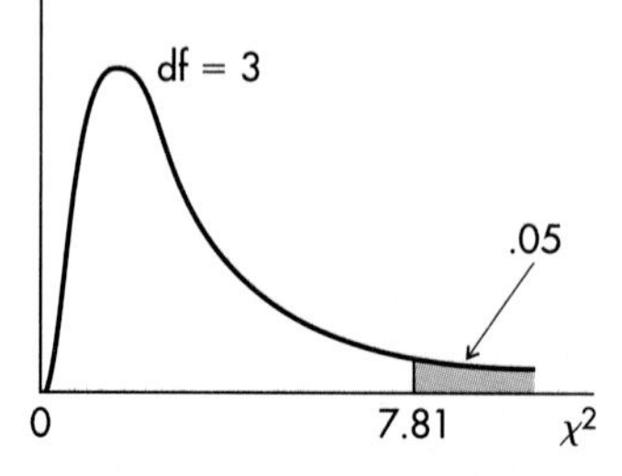

Now, let's assume we are making a *test of independence* at the .01 level, and our contingency table has 2 rows and 3 columns—that is, we have a 2×3 table. The degrees of freedom value we need is $(r - 1)(c - 1)$, or $(2 - 1)(3 - 1)$, or 2. And the critical χ^2 value at the intersection of the 2-df row and the .01 column in Appendix 6 is 9.21:

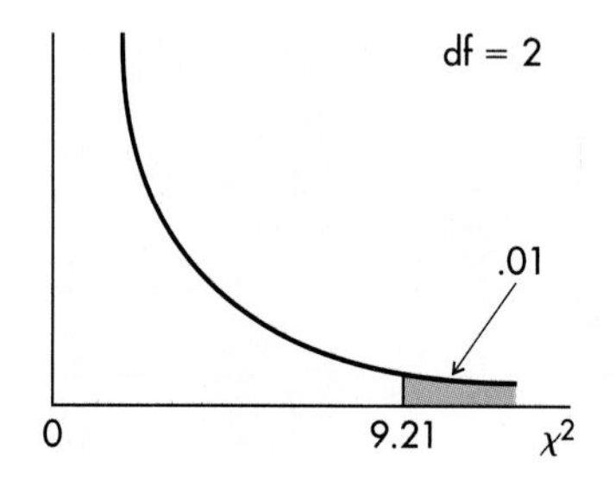

Step 5: State the Decision Rule The general form of the *decision rule* for both goodness-of-fit and contingency table tests is:

Reject H_0 and accept H_1 if the test ratio (TR) is $> \chi^2$ value found in ***Step 4*** above. Otherwise, accept H_0.

Thus, for our goodness-of-fit test at the .05 level with four possible outcomes, the decision rule is:

Reject H_0 and accept H_1 if the TR is > 7.81. Otherwise, accept H_0.

And for our 2×3 contingency table test at the .01 level, the decision rule is:

Reject H_0 and accept H_1 if the TR is > 9.21. Otherwise, accept H_0.

Step 6: Compute the Test Ratio The test ratio is a chi-square test statistic. Computational differences arise in finding this value in goodness-of-fit and contingency table experiments, and so we'll defer any actual computations to later pages. But the following formula for the test ratio is the same for both types of tests:

$$\text{TR}_{\chi^2} = \Sigma\left[\frac{(O - E)^2}{E}\right] \tag{12.3}$$

where O = an *observed* (sample) frequency
E = an *expected* (hypothetical) frequency if the H_0 is true

The O, or observed, values are found in the problem (sample) data set, and the E, or expected, values are computed to conform to a null hypothesis that is assumed to be true. If the observed frequencies, or counts, and the expected frequencies are *identical*, formula 12.3 shows us that the test ratio is *zero*. From the standpoint of verifying the null hypothesis, a computed test ratio of zero is ideal since our sample data will be exactly what we had expected. But you're sophisticated enough by now in matters statistical to know that it's very unlikely that the O and E values in an application will be identical even when the H_0 is true. Sampling variation will, after all, usually cause some discrepancy between the O and E values.

Step 7: Make the Statistical Decision If the test ratio computed in ***Step 6*** is less than or equal to the chi-square critical value specified in the decision rule (***Step 5***), then it falls into the area of acceptance (see Figure 12.2), and the H_0 is accepted. If, however, the TR value exceeds this χ^2 value, then it falls into the area of rejection, and the null hypothesis cannot be accepted.

Self-Testing Review 12.1

1. What is a multinomial experiment?
2. What is the objective of a goodness-of-fit test?
3. List the assumptions in a goodness-of-fit test.
4. What is the objective of a contingency table test or test of independence?
5. List the assumptions in a contingency table test.
6. What information is needed to find the critical χ^2 value?
7. Find the critical χ^2value for a goodness-of-fit test where α = .05 and there are 6 possible outcomes.
8. Find the critical χ^2 value for a goodness-of-fit test where α = .01 and there are 4 possible outcomes.

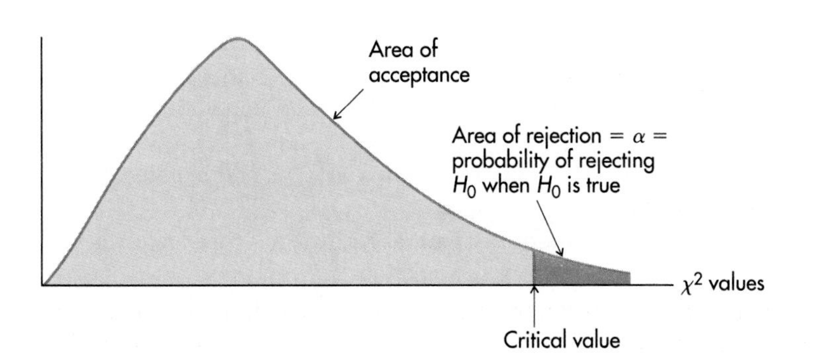

FIGURE 12.2 Areas of acceptance and rejection in a χ^2 distribution.

A Declining Trend

The National Health Interview Survey of adults reports that tobacco smoking is on the decline. In 1965, 52 percent of adult males and 34 percent of adult females were listed as smokers. Overall, 42 percent of adults smoked in 1965. But 25 years later, these percentages had dropped to 28.4 percent for males and 22.8 percent for females, and the overall percentage had declined to 25.5.

9. Find the critical χ^2 value for a test of independence where $\alpha = .01$ and the contingency table has 4 rows and 5 columns.

10. Find the critical χ^2 value for a test of independence where $\alpha = .05$ and the contingency table has 2 rows and 3 columns.

12.2 The Goodness-of-Fit Test

We've seen that the goodness-of-fit test is used to find out if a population under study "fits" or follows one with given probability distribution values. For example, such a test may be used to see how well a study group follows a normal, binomial, Poisson, or uniform probability distribution.

The following example shows how χ^2 concepts can be used to test the goodness of fit between sample data and a uniform distribution. The test for goodness of fit between sample data and a uniform distribution is the simplest, and it is quite useful.

Example 12.1. The Bitter Bottling Company has developed "Featherweight," the cola with fewer calories and less taste. To evaluate this new product, the marketing manager gives a taste test to a random sample of 300 people. Each person in the sample tastes Featherweight and 4 other diet cola brands. To avoid bias, the actual brand labels are replaced by the letters A, B, C, D, and E. The results of the sample are shown in the following table:

RESULTS OF BITTER BOTTLING'S TASTE TEST

Brand	Number Preferring Brand (O)
A	50
B	65
C	45
D	70
E	70
	300

Let's now follow our familiar seven-step testing procedure to evaluate Featherweight.

Step 1: State the Null and Alternative Hypotheses The hypotheses are:

H_0: The population distribution is uniform—that is, each of the cola brands is preferred by an equal percentage of the population.

H_1: The population distribution is *not* uniform—that is, the taste preference frequencies are not equal.

Step 2: Select the Level of Significance Let's assume that Bitter's marketing manager wants to conduct the test at the .05 level of significance.

Step 3: Determine the Test Distribution to Use We know to use a χ^2 distribution.

Step 4: Define the Rejection or Critical Region The level of significance is .05, and the degrees of freedom value is $k - 1$, or $5 - 1$, or 4. Thus, the χ^2 table value that separates the areas of acceptance and rejection is found in the table in Appendix 6 at the intersection of the .05 column and the 4-df row. This table value is 9.49.

Step 5: State the Decision Rule The decision rule is:

Reject H_0 and accept H_1 if the TR is >9.49. Otherwise, accept H_0.

Step 6: Compute the Test Ratio The observed frequencies (O) are given in our sample data set. And 4 additional columns are added to this data set in Table 12.1. If the null hypothesis is true, we would expect an equal, or uniform, number of people to prefer each of the 5 cola brands. That is, there would be no significant difference in taste preference, and one-fifth, or 20 percent, of the tasters should prefer brand A, 20 percent should prefer brand B, and so on. Thus, the E value for each brand should be 20 percent of 300, or 60.

TABLE 12.1 COMPUTATION OF THE TEST RATIO FOR EXAMPLE 12.1

Brand	Number Preferring (O)	E	$O - E$	$(O - E)^2$	$\frac{(O - E)^2}{E}$
A	50	60	−10	100	1.667
B	65	60	5	25	.417
C	45	60	−15	225	3.750
D	70	60	10	100	1.667
E	70	60	10	100	1.667
	300	300	0		9.168

$$TR_{\chi^2} = \Sigma\left[\frac{(O - E)^2}{E}\right] = 9.168$$

As we've seen, the appropriate formula for our test ratio is:

$$TR_{\chi^2} = \Sigma\left[\frac{(O - E)^2}{E}\right]$$

And Table 12.1 shows the computation of the test ratio for this goodness-of-fit example problem. The steps to compute the test ratio are:

1. Subtract the E value from the O value—that is, $O - E$—and record the difference (as shown in column 4 of Table 12.1). As a check of your math, note that $\Sigma (O - E)$ must equal zero, and note as well that $\Sigma O = \Sigma E$.
2. To eliminate negative values, square the $O - E$ differences to get $(O - E)^2$ (column 5, Table 12.1).
3. Divide each of these squared differences—$(O - E)^2$—by the E value to get $(O - E)^2/E$ (column 6, Table 12.1).
4. Add the $(O - E)^2/E$ values to get the computed test ratio value (column 6, Table 12.1). As you can see, the test ratio is 9.168.

(Figure 12.3 shows how a *Minitab* software user can supply observed and expected frequencies and a test ratio formula line to arrive at the same computed value.)

Step 7: Make the Statistical Decision We accept our H_0 because the test ratio of 9.168 is less than the table value of 9.49. At the .05 level, we cannot reject the hypothesis that all the cola brands are preferred by an equal percentage of the population. We must conclude that at the .05 level Featherweight does not taste significantly better (or worse) than the other brands.

Example 12.2. Suppose that Bill Alott, management consultant with Global Technologies (and author of the famous GOTCHA report in Chapter 3), is hired to evaluate the market response to a new and improved irrigation pump developed by Delta Corporation. Two other companies—Alpha Products and Beta Industries—supply competitive pumps. For some time, Alpha has controlled 50 percent of this market, Beta has had 30 percent of the market, and Delta has had a 20 percent market share. Bill's assignment now is to see if Delta's new pump is likely to cause a shift in these market percentages.

Bill assembles a random sample of 200 pump users who are familiar with Delta's new product and with the competing equipment. The purchase preferences of these 200 users is given on the next page:

```
MTB > PRINT C1

C1
   50     65     45     70     70

MTB > PRINT C2

C2
   60     60     60     60     60

MTB > LET K1 = SUM ((C1 - C2) **2/C2)
MTB > PRINT K1
K1         9.16667
```

FIGURE 12.3 ***Minitab*** **help with the goodness-of-fit test example.**

Product Evaluated	Number Preferring (*O*)
Alpha's pump	74
Beta's pump	62
Delta's new pump	64
	200

What conclusions can Bill draw from this survey response?

Step 1: State the Null and Alternative Hypotheses The null hypothesis is that the new pump has no effect on the market shares controlled by the 3 companies. That is:

H_0: The population market share percentages are: Alpha = 50, Beta = 30, and Delta = 20.

The alternative hypothesis is:

H_1: The population market share percentages are no longer as specified in the H_0.

Step 2: Select the Level of Significance Let's assume that Bill wants to make this test at the .01 level of significance.

Step 3: Determine the Test Distribution to Use Bill employs the chi-square distribution.

Step 4: Define the Rejection or Critical Region The level of significance is .01, and the degrees of freedom value is $k - 1$, or $3 - 1$, or 2. Thus, the χ^2 table value that separates the .99 area of acceptance from the .01 area of rejection is found in Appendix 6 at the intersection of the .01 column and the 2-df row. This table value is 9.21.

Step 5: State the Decision Rule The decision rule is:

Reject H_0 and accept H_1 if the TR is > 9.21. Otherwise, accept H_0.

Step 6: Compute the Test Ratio The observed frequencies (O) are given in Bill's sample data set. And 4 additional columns are added to this data set in Table 12.2. *If the null hypothesis is true*, the expected number (E) preferring each pump is shown in column 3 of Table 12.2. Thus, Bill would expect that 50 percent of the 200 pump users, or 100 users, would prefer the Alpha product, 60 users would be expected to opt for the Beta pump, and the remaining 40 users would be expected to favor Delta's new offering. With the observed and expected values now available, Bill can compute the test ratio as shown in Table 12.2.

TABLE 12.2 BILL ALOTT'S WORKSHEET TO COMPUTE THE TEST RATIO FOR THE PUMP PREFERENCE DATA IN EXAMPLE 12.2

Product Evaluated	Number Preferring (O)	E	O − E	$(O-E)^2$	$\frac{(O-E)^2}{E}$
Alpha's pump	74	100	−26	676	6.76
Beta's pump	62	60	2	4	.07
Delta's new pump	64	40	24	576	14.40
	200	200	0		21.23

$$TR_{\chi^2} = \Sigma\left[\frac{(O-E)^2}{E}\right] = 21.23$$

Step 7: Make the Statistical Decision Since the test ratio of 21.23 is greater than the chi-square table value of 9.21, Bill must reject the H_0 that the population market share percentages controlled by the 3 companies remain unchanged. Bill can happily report to his client that the introduction of Delta's new pump has apparently altered the market share structure. Although the χ^2 test itself doesn't support any further conclusions, an examination of Bill's data suggests that Delta's new product will increase its market share at Alpha's expense (at least until Alpha makes its own χ^2 test and rushes to bring out its new version).

Self-Testing Review 12.2

1. The personnel director for a large insurance company claims there has been a change in the level of education of management personnel in the last 5 years. Research shows that 5 years ago the highest level of education for 15 percent of the managers was a high school diploma, for 64 percent it was a college degree, for 8 percent it was some graduate school, and for 13 percent it was a master's degree or higher. A recent random sample of 216 of today's managers yielded the following results:

Highest Level of Education	Number of Managers
High school diploma	19
College degree	132
Graduate school	30
Master's degree or higher	35

At the .01 level, does a goodness-of-fit test indicate that there has been a change in the level of education for the management population in the last 5 years?

2. Last year 24 percent of the sales and marketing managers at a computer software corporation earned less than $30,000, 15 percent earned at least $30,000 but less than $40,000, 19 percent earned at least $40,000 but less than $50,000, 36 percent earned at least $50,000 but less than $75,000, and 6 percent earned over $75,000. The personnel manager wants to see if there has been a recent change in the distribution of salaries. The following data represent this year's salary distribution for a random sample of sales and marketing executives at this company:

Salary	Current Number
Less than $30,000	27
$30,000 and <$40,000	30
40,000 and <50,000	30
50,000 and <75,000	72
75,000 and over	57

At the .01 level, does a goodness-of-fit test indicate that there has been a change in the population distribution of salaries at this company?

3. The *American Journal of Public Health* reported that in 1987, 10 percent of nursing home residents were younger than 65, 13 percent of these residents were at least 65 but less than 75, 32 percent were at least 75 but less than 84, and 45 percent were 84 and older. The following data are from a recent random sample of 6,629 nursing home residents:

Age	Current Number
Younger than 65	398
65 and <75	663
75 and <84	2,254
84 and over	3,314

At the .05 level, has the age distribution of all nursing home residents changed since 1987?

4. The production manager at a food products plant claims that absenteeism among workers is more common on some weekdays than on others. The following sample data were recorded last week:

Day	Number Absent
Monday	24
Tuesday	17
Wednesday	15
Thursday	19
Friday	25

Test the hypothesis at the .05 level that absences occur uniformly over the 5 days.

5. In December 1992, about 15 percent of those surveyed thought the U.S. economy was excellent or good, about 41 percent thought the economy was only fair, and about 43 percent thought it was poor. To see if people's views of the economy had changed, a nationwide *USA Today/CNN/Gallup* poll was conducted in August 1993. The responses from the 1,003 adults who participated were:

View of Economy	Number
Excellent/good	100
Only fair	492
Poor	411

Test the hypothesis at the .01 level that there was a change in public opinion about the economy between the 2 periods.

6. A botanist checking for dominant and recessive traits interbreeds 2 types of plants and expects 4 classes of hybrid results to appear in the Mendelian ratio of 9:3:3:1 (56.25 percent, 18.75 percent, 18.75 percent, and 6.25 percent). The results of her experiment show that there are 860 plants of one class, 350 of another, 300 of a third, and 90 of a fourth class. Test the hypothesis at the .05 level that her results are in accordance with the Mendelian ratio.

7. There are 4 branches of a local bank. Last year, branch A handled 28.42 percent of all transactions, branch B had 21.81 percent of these transactions, branch C had 17.81 percent, and branch D had 31.96 percent. A new shopping mall has recently opened, and the vice president of the bank wants to see if this opening has produced any changes in the distribution of transactions. The transactions for a week selected at random after the mall opening were:

Branch	Transactions
A	583
B	749
C	427
D	732

Test the hypothesis at the .01 level that the distribution of transactions has changed after the construction of the new mall.

8. A random sampling of police records shows the following number of crimes were committed in a midsized California city for each day of the week:

Day	Number of Crimes
Sunday	74
Monday	63
Tuesday	57
Wednesday	68
Thursday	79
Friday	98
Saturday	119

Test the hypothesis at the .05 level that there is a uniform distribution in the number of crimes committed each day of the week.

9. The QRS Tee-Shirt Company has traditionally made 20 percent of its shirts in a small size, 40 percent in a medium size, 30 percent in the large size, and 10 percent in the extra-large size. The marketing vice president claims that this distribution has now changed. Her research shows the following demand in a random sample of shirt orders:

Size	Number Ordered (In Thousands)
Small	7.8
Medium	10.7
Large	23.8
Extra large	18.5

Test the hypothesis at the .01 level that the vice president's claim is correct.

10. The *Chronicle of Higher Education Almanac* reported a few years ago that 37.6 percent of college freshmen applied to no college other than the one they were attending, 14.7 percent applied to 1 other school, 15.8 percent applied to 2 others, 13.7 percent to 3 others, and 18.2 percent applied to 4 or more other schools. A guidance counselor is now investigating the attitudes and characteristics of freshmen. A random sample of 308 freshmen were asked for the number of colleges they applied to other than the one they were attending, and the responses were as follows:

Other Colleges Applied to	Number of Students
0	85
1	47
2	58
3	37
4 or more	81

Test the hypothesis at the .05 level that the sample distribution fits the one spelled out in the publication.

11. During the first 6 months of a recent year, the *Grateful Dead* played 115 different songs at 1 or more concerts. A statistical analyst claimed that 17.4 percent of these songs were played only once, 33.9 percent were played twice, 23.5 percent were played 3 times, 19.1 percent were played 4 times, and 6.1 percent were played 5 or more times. A fan (who was also taking a statistics course) decided to see if the distribution for a later period was similar. He obtained the following data for a randomly sampled period:

Number of Times	Number of Songs
1	18
2	39
3	35
4	19
5 or more	7

Using a goodness-of-fit test at the .05 level, is the song frequency distribution still the same as the analyst claimed?

12. A recent study in *The American Journal of Psychiatry* reported on the clinical characteristics of autistic children. In the study, the children were ranked on a developmental scale of 1 to 5, where 1 represented normal development and 5 equaled severe retardation. It was claimed that 19 percent of these children had a developmental score of 1, 5 percent had a score of 2, 14 percent had a score of 3, 48 percent had a score of 4, and 14 percent had a score of 5. A clinical psychologist decided to verify these figures and tested a random sample of 115 autistic children with the following results:

Developmental Score	Number of Children
1	17
2	10
3	19
4	41
5	15

Does this distribution fit the one in *The American Journal of Psychiatry* at the .05 level?

12.3 The Contingency Table Test

The concept of independent variables was discussed in Chapter 4. You'll recall that two variables are independent if the occurrence (or nonoccurrence) of one doesn't affect the chances of the occurrence of the other. How, you may have wondered then, is it possible to tell if two variables are independent? Well, the following example shows how χ^2 concepts can be used in a contingency table test to see if the variables under study are likely to be cross-classified independently.

Friendly Predators?

A recent study in *Ecology* dealt with the introduction of predators in selected artificial ponds. The study found that later larvae levels were comparable in all of the ponds studied. That is, the larvae levels were about the same in ponds that had received predators and in those that had not been so stocked. Thus, the larvae levels were found to be independent of predator treatment.

Example 12.3. Three candidates are running for sheriff of Lawless County. These aspirants to the public trough are Larson E. Bound, Graff D. Lux, and Emma Nocruk. Bound's campaign manager has conducted candidate preference polls in the county's 3 towns. The results of these random samples of county voters are shown in a contingency table (Table 12.3). As you can see, the samples of voters are cross-classified by town of residence and by candidate preference. In planning future campaign strategies, Bound's manager would like to know if the candidate preference of voters is independent of their town of residence.

TABLE 12.3 SURVEY OF VOTERS CLASSIFIED BY TOWN OF RESIDENCE AND CANDIDATE PREFERENCE

	Towns			
	White Lightning	Casino City	Smugglersville	
Bound	50	40	35	125 (row totals)
Lux	30	45	25	100 (row totals)
Nocruk	20	45	20	85 (row totals)
column totals	100	130	80	310 (grand total)

Step 1: State the Null and Alternative Hypotheses The null hypothesis is that the population percentages favoring each of the 3 candidates are unchanged from town to town. That is, the voter preference and town of residence are independent. This does not mean that our H_0 is that each candidate has an equal population percentage of 33.33. Rather, our H_0 is that the population percentage (π) of voters favoring Bound is the same in all 3 towns, the π favoring Lux is the same regardless of location of residence, and the π favoring Nocruk is equal in the 3 locations. Of course, the population percentage favoring Lux can be different from the π favoring Bound in the context of our null hypothesis. Thus, our H_0 may be stated as follows:

H_0: The population percentage favoring each candidate is the same from town to town.

And the alternative hypothesis is:

H_1: The population percentage favoring each candidate is *not* the same from town to town.

The null hypothesis could also be expressed like this:

H_0: The candidate preference of voters is *independent* of their town of residence.

As noted earlier, a contingency table test is often referred to as a *test of independence* because the H_0 tested is essentially that sample data are being classified in independent ways. And the alternative hypothesis is then:

H_1: The candidate preference of voters is dependent on (or related to) their place of residence.

Step 2: Select the Level of Significance We'll assume here that Bound's campaign manager has specified that the test be made at the .05 level of significance.

Step 3: Determine the Test Distribution to Use It's χ^2 time again.

Step 4: Define the Rejection or Critical Region We know that $\alpha = .05$, and we know that our contingency table has 3 rows and 3 columns. Thus, we can use formula 12.2 to find the degrees of freedom (df) value we need:

$$\text{df} = (r - 1)(c - 1) = (3 - 1)(3 - 1) = 4$$

With $\alpha = .05$ and with df = 4, the critical χ^2 value in Appendix 6 that separates the .95 area of acceptance from the .05 area of rejection is 9.49.

Step 5: State the Decision Rule The decision rule is:

Reject H_0 and accept H_1 if the TR is > 9.49. Otherwise, accept H_0.

Step 6: Compute the Test Ratio The actual data from the random samples of voters taken in the 3 towns—the observed, or *O*, values—are shown in Table 12.3. But we also need the expected (*E*) values to compute the test ratio.

You'll recall from Chapter 4 that two variables, say, *B* and *W*, are independent if and only if their joint probability [$P(B \text{ and } W)$] is equal to the product $P(B) \cdot P(W)$. If the H_0 is true that the candidate preference of voters is independent of their town of residence, then we should be able to compute the probability that a sample voter favors Bound and lives in White Lightning by multiplying the individual probabilities of P(favors Bound) $\cdot$ P(lives in White Lightning). This $P(B) \cdot P(W) = (125/310) \cdot (100/310) = (.4032) \cdot (.3226) = .13007$. Thus, if the variables are independent, the probability is .13007 that a selected voter favors Bound and lives in White Lightning. Since there are a total of 310 voters in the survey, the expected number favoring Bound and living in White Lightning would be 310(.13007), or 40.32, if the H_0 is true. And if our H_0 is true and if the population percentage of voters favoring Bound is the same in each of the 3 towns, we could follow the above procedure to compute the number of sample responses favoring Bound that would be expected in the other 2 locations.

Another way to view the same procedure is to consider that a total of 310 voters were polled and 125 of these voters expressed a preference for Bound. Since the 125 "votes" received by Bound is 40.32 percent of the total cast in the 3 towns—$(125/310) \times 100 = 40.32$ percent—Bound should be favored by 40.32 percent of those interviewed in each of the 3 towns if the H_0 is true. Thus, of the 100 people interviewed in White Lightning, we would expect 40.32 percent (or 40.32) of them to favor Bound. Similarly, of the 130 persons polled in Casino City, we would anticipate that 40.32 percent (or 52.42) of them would prefer Bound. And we would expect 32.26 people in Smugglersville (40.32 percent of the 80 persons interviewed) to be in Bound's corner.

The same analysis can also be used to compute the E values for the other candidates in each of the towns. (Since 100/310, or 32.26 percent, of all those sampled showed a preference for Lux, for example, we would expect that Lux would be favored by about 32.26 of the 100 persons interviewed in the White Lightning sample.) But the *computations of the E values are even easier to follow* if we refer to the contingency table in Table 12.3 and compute the hypothetical or expected frequencies for each cell in the table. A cell is formed by the intersection of a column and a row. Since there are 3 rows and 3 columns in Table 12.3, there are 3 × 3, or 9, cells. (Tables with r rows and c columns are often referred to as *$r \times c$ tables*; in our example, we have a 3 × 3 table.)

The expected values may be computed for each cell of a contingency table by the following formula:

$$E = \frac{(\text{row total})(\text{column total})}{\text{grand total}} \tag{12.4}$$

The use of this formula may be illustrated by computing the number of persons who would be *expected* to favor Bound in the town of White Lightning if the H_0 is true. From Table 12.3, you can see that the total for the row in which this cell is located is 125, the total for the column for this cell is 100, and the grand total is 310. Thus, the E value for the cell is computed as follows:

$$E = \frac{(125)(100)}{310} = 40.32$$

The same procedure can be used to get the E values for all the other cells in the table. Table 12.4 duplicates all the observed frequencies from Table 12.3 and shows the E values for each cell in parentheses. Note that the total of the expected values in *each row*

TABLE 12.4 OBSERVED AND EXPECTED FREQUENCIES

	Towns			
	White Lightning	Casino City	Smugglersville	Total
Bound	50 (40.32)	40 (52.42)	35 (32.26)	125
Lux	30 (32.26)	45 (41.94)	25 (25.81)	100
Nocruk	20 (27.42)	45 (35.65)	20 (21.94)	85
Total	100	130	80	310

Row totals: 125, 100, 85; column totals: 100, 130, 80; grand total: 310.

must equal the total of the observed values in *the row*. And the total of the expected values in *each column* must equal the total of the observed values in *the column*. That is, the expected row values for Bound in the 3 towns = 40.32 + 52.42 + 32.26 = 125, and the expected column values for each candidate in White Lightning = 40.32 + 32.26 + 27.42 = 100. Thus, if you know the row total for Bound in Table 12.4 equals 125 and if you have two of the three expected values in the row (say, 40.32 and 52.42), you can easily find the third expected value as follows: 125 − (40.32 + 52.42) = 32.26.

Now that we have all the E values for our contingency table, we are ready to calculate the test ratio. Table 12.5 shows this computation. The columns numbered 1 and 2 in Table 12.5 reproduce the data from Table 12.4 in a more convenient format. As a check of your arithmetic, note that $\Sigma O = \Sigma E$. The steps to compute the test ratio (shown in Table 12.5) are:

1. Subtract each E value from the corresponding O value—that is, $O - E$—for each cell in the contingency table, and record the difference (as shown in column 3 of Table 12.5). As another check of your math, note also that $\Sigma(O - E)$ must equal zero.
2. Square the $O - E$ differences to get $(O - E)^2$ (column 4, Table 12.5).
3. Divide each of these squared differences—$(O - E)^2$—by the E value for each cell to get $(O - E)^2/E$ (column 5, Table 12.5).
4. Add the $(O - E)^2/E$ values to get the computed test ratio value (column 5, Table 12.5). As you can see, the test ratio is 10.539.

TABLE 12.5 COMPUTATION OF THE TEST RATIO

Row-Column (Cell)	O (1)	E (2)	$O - E$ (3)	$(O - E)^2$ (4)	$\frac{(O - E)^2}{E}$ (5)
1-1	50	40.32	9.68	93.702	2.323
1-2	40	52.42	−12.42	154.256	2.942
1-3	35	32.26	2.74	7.508	.233
2-1	30	32.26	−2.26	5.108	.158
2-2	45	41.94	3.06	9.364	.224
2-3	25	25.81	−.81	.656	.025
3-1	20	27.42	−7.42	55.056	2.008
3-2	45	35.65	9.35	87.423	2.455
3-3	20	21.94	−1.94	3.764	.171
	310	310.0	0		10.539

$$TR_{\chi^2} = \Sigma\left[\frac{(O - E)^2}{E}\right] = 10.539$$

Step 7: Make the Statistical Decision Since our test ratio of 10.539 is greater than 9.49, we reject the H_0 and conclude that the population percentage favoring each candidate for sheriff of Lawless County is not the same from town to town. That is, we reject the H_0 that the candidate preference of voters is independent of their town of residence and accept the alternative hypothesis that the candidate preference of voters is dependent or related to their place of residence. Bound's campaign manager may decide, as a result of this conclusion, to conduct campaign activities in a more selective fashion.

Computers Make It Easy

In virtually every case in this book where we've encountered a *standardized* test or procedure that requires numerous calculations, we've also located a program in a statistical software package that takes over the chore of carrying out those calculations. Thus, you'll not be surprised now to learn that programs are readily available to process contingency table tests.

Figure 12.4 shows how the *Minitab* package processes the example problem we've just considered. The voters in White Lightning preferring Bound, Lux, and Nocruk are

Right on the Mark

Technology-market forecasters produce studies that are more often wrong than right. But one of the success stories of such market research was Dataquest's prediction in 1981 that low-cost laser printers would be marketed by 1985 or 1986. This prediction was based on an analyst's understanding of the inexpensive and compact semiconductor laser engine that Canon had just developed for use in its photocopy machines. Dataquest correctly observed that the same laser engine could be used in printers. Accurate long-term predictions such as this are rare, but such forecasts are reported regularly in the media.

```
MTB > PRINT C1

WHTLTNG
   50    30    20

MTB > PRINT C2

CASCITY
   40    45    45

MTB > PRINT C3

SMGVILLE
   35    25    20

MTB > CHISQUARE C1-C3

Expected counts are printed below observed counts

       WHTLTNG  CASCITY SMGVILLE    Total
    1       50       40       35      125
         40.32    52.42    32.26

    2       30       45       25      100
         32.26    41.94    25.81

    3       20       45       20       85
         27.42    35.65    21.94

Total      100      130       80      310

ChiSq =  2.323 +  2.942 +  0.233 +
         0.158 +  0.224 +  0.025 +
         2.008 +  2.455 +  0.171 = 10.539
df = 4
```

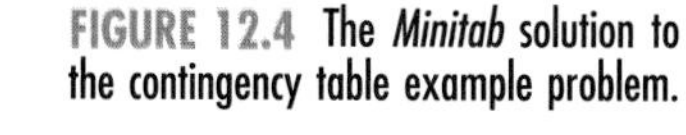
FIGURE 12.4 The *Minitab* solution to the contingency table example problem.

stored in that order in column 1 (C1) of a *Minitab* worksheet, just as they were shown in Table 12.3. The data for the other 2 towns are similarly stored in columns 2 and 3, as you can see in the top lines of Figure 12.4. Once these facts are entered and available to the program, the software produces (1) the observed and expected frequencies found in Table 12.4, (2) the results in column 5 of Table 12.5 and the computed test ratio, which is the total of the column 5 results, and (3) the degrees of freedom needed (4 in this case) to use the χ^2 table in Appendix 6. The user then finds the appropriate χ^2 table value at the intersection of the selected α column and the 4-df row. Finally, the last step (which we've already done) is to compare the computed χ^2 value (the test ratio) with the table value and make the statistical decision.

A summary of the steps in a contingency table test is outlined in Figure 12.5.

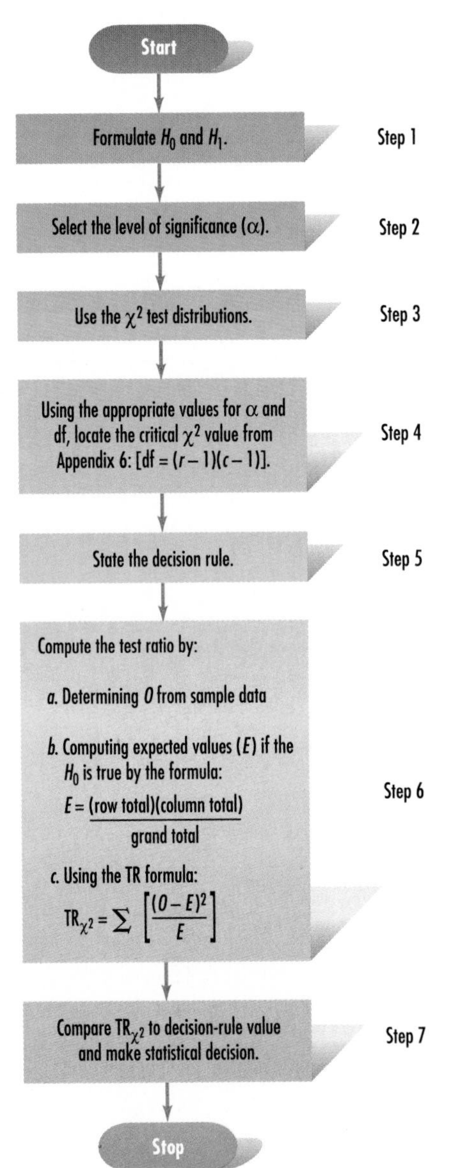
Start
Formulate H_0 and H_1.
Step 1
Select the level of significance (α).
Step 2
Use the χ^2 test distributions.
Step 3
Using the appropriate values for α and df, locate the critical χ^2 value from Appendix 6: [df = $(r-1)(c-1)$].
Step 4
State the decision rule.
Step 5
Compute the test ratio by:
a. Determining O from sample data
b. Computing expected values (E) if the H_0 is true by the formula:
$E = \frac{\text{(row total)(column total)}}{\text{grand total}}$
c. Using the TR formula:
$TR_{\chi^2} = \sum \left[\frac{(O-E)^2}{E}\right]$
Step 6
Compare TR_{χ^2} to decision-rule value and make statistical decision.
Step 7
Stop

FIGURE 12.5 Steps in a contingency table test.

Self-Testing Review 12.3

1. To meet the needs of their students, schools in southern Arizona have students take the Language Assessment Scales (LAS) test to determine their dominant or strongest language. The results of a random sample of kindergarten and first-grade students in this region are as follows:

Dominant Language	Kindergarten	First Grade	Total
English	160	197	357
Spanish	75	92	167
Mixed	284	348	632
Bilingual	16	20	36
Total	535	657	1,192

At the .01 level, is language dominance in the region independent of school grade?

2. The contingency table that follows describes the educational background of a random sample of male Vietnam veterans who suffer from (and are free of) the effects of posttraumatic stress disorder (PTSD). (The information comes from a recent issue of *The Journal of Clinical Psychology*.)

Educational Background	No PTSD	PTSD	Total
Less than high school	45	35	80
High school graduate	299	110	409
Some college	363	154	517
College graduate	71	9	80
Graduate school	93	11	104
Total	871	319	1,190

At the .05 level, test the hypothesis that the presence or absence of PTSD is independent of educational background.

3. To study the traits of heredity, randomly selected pairs of identical and fraternal twins were assessed for mental retardation. Records were kept as to whether the pairs were concordant (both retarded or both not retarded) or not concordant (one retarded and the other not retarded). The data were:

Twin Type	Both the Same	Different	Total
Identical	129	6	135
Fraternal	195	332	527
Total	324	338	662

At the .01 level, test the hypothesis that the percentage of concordance for the mental retardation trait is the same for both types of twins.

4. A large random sample of public school dropouts in grades 7 through 12 in 5 Pennsylvania counties was gathered by the state Department of Education. The sample data are:

County	Male	Female	Total
Bucks	379	247	626
Chester	283	151	434
Delaware	399	263	662
Montgomery	322	222	544
Philadelphia	5,477	4,259	9,736
Total	6,860	5,142	12,002

Test the hypothesis at the .05 level that the gender of a high school dropout in this region is independent of the county he or she comes from.

5. In a study reported in the *Journal of Educational Issues*, a random sample of inner-city middle school students were given a questionnaire to learn their perceptions of school. When males and females were asked if they planned to stay in school, the responses were as follows:

Stay in School	Male	Female	Total
Yes	76	118	194
No	37	157	194
Total	113	275	388

Test the hypothesis at the .05 level that the decision to remain in school is independent of gender for inner-city middle school students.

6. A random sample of apartment units available to students at the University of Transylvania produced the following results:

Type of Apartment	West	South	Center City	Total
One bedroom	97	277	175	549
Two bedrooms	156	315	261	732
Three bedrooms	99	229	97	425
Total	352	821	533	1,706

Test the hypothesis at the .01 level that the type of available apartment is independent of location.

7. Title VII of the Health Professions Educational Assistance Act of 1976 was created to encourage the production of primary care physicians. The following career choices were made by physicians during 2 periods after the act was passed:

Career Choice	Graduates 1976–1980	Graduates 1981–1985	Total
Family medicine	8,597	9,605	18,202
Internal medicine	8,591	10,672	19,263
Pediatrics	3,962	4,755	8,717
Total	21,150	25,032	46,182

Test the hypothesis at the .01 level that the type of career choice made by graduating physicians is independent of the date of graduation.

8. A survey is conducted at random to determine eye and hair color in a population. The results are given below:

Hair Color	Blue Eyes	Brown Eyes	Total
Blond	10	30	40
Black	40	100	140
Red	5	25	30
Total	55	155	210

Test the hypothesis at the .01 level that hair color is independent of eye color for this population.

9. After President Clinton's 1993 budget was passed by Congress, a survey was taken to see if people thought they would be paying a little more in taxes, a lot more, or not any more to the government. The following responses, classified by income, were obtained from this random sample:

Income	A Little More	A Lot More	Not Any More	Didn't Know	Total
Under \$20,000	65	30	50	5	150
20,000 < 30,000	154	77	90	14	335
30,000 < 50,000	180	111	38	18	347
Over 50,000	72	63	33	3	171
Total	471	281	211	40	1,003

Test the hypothesis at the .05 level that income level and the population's opinion about the tax consequences of this budget are independent.

10. In a recent study in the *American Journal of Public Health*, a random sample of 110 disabled elderly subjects were assigned to an experimental primary home care team that provided 24-hour service and was led by a physician. A control group of 73 randomly selected elderly and disabled patients were offered ordinary care. The following is a record of the types of complications each group experienced:

Type of Complication	Team Patients	Control Patients	Total
Accidents	25	24	49
Drug side effects	7	6	13
Mistreatment	4	7	11
Total	36	37	73

At the .05 level, is the population percentage of various types of complications likely to be the same for the 2 groups?

11. The following contingency table shows the number of good and defective parts on each of the 3 work shifts at a manufacturing plant during randomly sampled periods. Using the .05 level, test the hypothesis that there's no significant difference among the population percentages of defective parts on the 3 shifts:

Shift	Good	Defective	Total
First	90	10	100
Second	70	20	90
Third	60	20	80
Total	220	50	270

12. A study to determine optimal time after injury for ligament reconstruction was recently reported in *The American Journal of Sports Medicine.* Time from injury to surgery was established as acute, subacute, and chronic. Subjects in these randomly selected groups were cross-categorized as to whether their knee cartilage was normal, needed replacement, or needed repair. Results were recorded as follows:

Time Group	Normal Cartilage	Cartilage Replacement	Cartilage Repair	Total
Acute	12	21	14	47
Subacute	4	5	6	15
Chronic	2	16	11	29
Total	18	42	21	81

Test the hypothesis at the .05 level that for the population of ligament replacement patients, time after injury and before surgery and cartilage condition are independent.

13. The director of programming of a local cable company wants to provide appropriate television shows for different areas of his viewing market. A random sample is taken to determine programming preferences in rural and urban areas, and the following results are obtained:

Type of Program Preferred	Urban	Rural	Total
Western	82	62	144
Comedy	98	72	170
Variety	63	37	100
Mystery	97	53	150
Total	340	224	564

At the .05 level, is the type of program preferred independent of the area of residence?

14. Integrated circuits are supplied to a computer manufacturer by 2 vendors. Each circuit is tested by the manufacturer for 4 common defects before it is accepted. The following data, representing random test results over a 2-week period, have been supplied to the purchasing department by the quality control manager:

Vendor	Defect W	Defect X	Defect Y	Defect Z	Total
A	63	77	39	31	210
B	27	35	26	19	107
Total	90	112	65	50	317

At the .01 level, are the population percentages of common defects the same for both vendors?

15. A study in *The American Journal of Psychiatry* dealt with the characteristics of those who had ever smoked. Subjects in a random sample were classified as nicotine dependent and nondependent. The following data were reported:

Gender	Nicotine-Dependent Smoker	Nondependent Smoker	Total
Male	75	69	144
Female	127	123	250
Total	202	192	394

At the .01 level, test the hypothesis that the type of smoker is independent of gender in the population.

16. A *Teaching of Psychology* article dealt with the relationship between a college professor's school and whether or not the professor discussed peace, war, nuclear issues, and international relations in a psychology class. A random sample of 127 professors responded to the question of whether they discussed these topics in their classes, and the following results were obtained:

School Type	Discuss	Do Not Discuss	Total
University	30	17	47
4-year college	32	13	45
2-year college	16	19	35
Total	78	49	127

Test the hypothesis at the .05 level that the type of school is independent of whether or not professors discuss the issues mentioned above.

17. A 5-year follow-up study in *The American Journal of Sports Medicine* showed the results of an experiment that evaluated the use of a *Dacron* prosthetic ligament in reconstructive knee surgery. Patients in a random sample were divided into 2 groups. Group 1 consisted of 50 patients with no prior knee surgery, and group 2 had 34 patients who were operated on for a second time after a failed previous surgery. The results of the surgery after 5 years were as follows:

Group	Failure	Success	Total
1	19	31	50
2	11	23	34
Total	30	54	84

Test the hypothesis at the .01 level that the population outcome of success or failure is independent of prior surgical history.

18. A study in *The American Journal of Psychiatry* dealt with the characteristics of those who had ever smoked. Subjects in a random sample were classified as nicotine-dependent and nondependent. The following data were reported:

Race	Nicotine-Dependent Smoker	Nondependent Smoker	Total
Black	16	40	56
White	186	152	338
Total	202	192	394

At the .01 level, is the type of smoker in the population independent of race?

19. The sales manager at the Tackey Toy Company has hired a researcher to gather and evaluate opinion data about Tackey products from people in different parts of the country. The researcher randomly sampled toy buyers in each of 4 regions. Each respondent was first asked if he or she had ever heard of Tackey toys. Those who were familiar with Tackey products were then asked if Tackey toys were competitively priced or if they were overpriced. The following sample results were obtained:

Geographic Region	Unfamiliar with Tackey Toys	Toys Priced Competitively	Toys Overpriced	Total
New England	64	28	106	198
West Coast	84	42	76	202
Gulf Coast	56	14	130	200
Midwest	60	20	120	200
Total	264	104	432	800

At the .01 level, would you accept the hypothesis that there were no regional differences in the way the population of toy buyers responded to the Tackey survey?

20. A random sample of Titan Corporation employees is selected, and data are gathered about the presence or absence of dependent children and the type of medical insurance each employee has. These health insurance categories are a health maintenance organization (or HMO) plan, a Blue Cross/Blue Shield (or BC/BS) plan, or no plan at all. The information gathered is described in the following contingency table:

Dependent Status	HMO Plan	BC/BS Plan	No Plan	Total
Dependent children	145	85	23	253
No dependent children	39	42	52	133
Total	184	127	75	386

Test the hypothesis at the .01 level that whether or not an employee has dependent children is independent of the type of health insurance coverage he or she carries.

LOOKING BACK

1. A goodness-of-fit test is conducted to evaluate a multinomial experiment in which there can be more than two possible outcomes for each trial. Such a test is used to see if the distribution of the observed outcomes of the sample trials supports a

hypothesized population distribution. The researcher establishes a hypothesis about how he or she thinks the data items will be distributed into the different testing categories of interest. Then the test is carried out to see if the observed results conform to the results that would be expected if the hypothesis were true. The assumptions in a goodness-of-fit test are that the sample is a random sample, frequency counts are obtained for the possible k cells, and the expected frequency is at least 5 for each of the k categories.

2. The contingency table tests considered in this chapter deal with two classification variables. Each cell in a contingency table is at the intersection of a row and a column. The rows represent one classification category, and the columns represent another such category. Thus, a contingency table test (or test of independence) is one that tests the hypothesis that the data are cross-classified in independent ways. The assumptions in a contingency table test are that the sample data items are obtained through random selection and the expected frequency for each cell in the table is at least 5.

3. A familiar seven-step hypothesis-testing procedure is used to carry out both the goodness-of-fit and contingency table tests. In this procedure, the null and alternative hypotheses are established (*Step 1*), the level of significance is specified (*Step 2*), and an appropriate chi-square distribution is used (*Step 3*) to define the rejection or critical region (*Step 4*) that allows the analyst to formulate a decision rule (*Step 5*). A test ratio is then computed (*Step 6*) for the goodness-of-fit or contingency table test, and the statistical decision (*Step 7*) is then possible.

Review Exercises

1. A study in the *American Journal of Sports Medicine* discussed two different operative techniques that were used in clinical trials involving a random sample of patients who received reconstructive knee surgery. Technique A used a hole drilled in the femur, and technique B used an "over-the-top" reconstruction. The 5-year results of these techniques were as follows:

Technique	Failure	Success	Total
A	8	22	30
B	22	32	54
Total	30	54	84

Test the hypothesis at the .01 level that the chance of failure is independent of the technique used.

2. A study in *The American Journal of Psychiatry* dealt with the characteristics of those who had ever been tobacco smokers. Subjects in a random sample were classified as nicotine dependent and nondependent. The following data were reported:

Current Smoker	Nicotine-Dependent Smoker	Nondependent Smoker	Total
Yes	169	123	292
No	33	69	102
Total	202	192	394

At the .01 level, is there a significant difference between the proportion of current smokers (defined as those reporting they had smoked during the year preceding the interview) in the 2 groups?

3. Participants in a consumer product research study were given 4 types of orange juice to sample and were asked to rate the taste quality for each type of juice. A random sample of 30 subjects received 1 brand to drink at a time. Five minutes elapsed between consecutive juice samples, during which time the subjects were asked to eat a cracker. The following table shows the preferences:

Brand	Number Preferring
A	10
B	8
C	5
D	7

At the .01 level, test the hypothesis that there was an equal preference for each brand (a uniform distribution).

4. To test the fairness of a die, it is rolled 100 times with the following results:

Face	Frequency of Appearance
1	15
2	20
3	22
4	13
5	19
6	11

At the .05 level, is the die fair?

5. Are marital status and race independent variables for those who have legal abortions? Use the following data (numbers in thousands) from the Center for Disease Control to test the independence of these variables at the .01 level:

Marital Status	White	Not White	Total
Married	68	31	99
Unmarried	279	141	420
Total	347	172	519

6. A recent study in *The American Journal of Public Health* examined specific smoking habits of a random sample of physicians in Minnesota. Use the following data to test the independence of age and smoking status at the .05 level of significance.

Age	Current Smokers	Former Smokers	Never Smokers	Total
Less than 40	5	33	94	132
40 and <50	13	48	62	123
50 and over	17	64	53	134
Total	35	145	209	389

7. A politician running for county supervisor is planning his campaign strategy on the issue of gun control and wants to know if there is a significant difference in the percentage of voters who favor the issue in districts A, B, and C. A random sample of 355 voters yields the following data:

District	In Favor	Opposed	Total
A	89	46	135
B	65	45	110
C	60	50	110
Total	214	141	355

Test the hypothesis at the .01 level that there is no difference in the percentage favoring gun control in the 3 districts.

8. To gauge the opinion of workers about a proposed change in the union constitution, the union's executive committee sent questionnaires to a random sample of 100 members in 3 locals. The survey results are as follows:

Opinion	Local X	Local Y	Local Z	Total
Favor change	17	23	10	50
Oppose change	9	13	8	30
No response	4	4	12	100
Total	30	40	30	100

At the .05 level, is there a significant difference in the reactions of the workers in the 3 locals to the proposed change?

9. The production manager of a company that does steel plating has collected the following data during a randomly selected 24-hour period:

Shift	Number of Defects
8 a.m. to 4 p.m.	27
4 p.m. to midnight	35
Midnight to 8 a.m.	49

At the .01 level, are the defects uniformly distributed among the shifts?

10. A survey conducted by *TV Guide* found that men controlled the TV remote control device in 41 percent of the homes, women had control of the device in 19 percent of the homes, men and women shared control in 27 percent of the homes, and the remaining respondents said they didn't know who had control or they didn't have a remote control device. A random sample later was selected, and the respondents were asked who controlled the device in their households. The following data were collected from this sample:

Who Controls?	Number
Men	87
Women	41
Shared	52
Don't know or don't have device	18
Total	198

At the .05 level, do the data from this sample fit the *TV Guide* survey results?

11. In studies of traits of heredity, a random sample of pairs of identical and fraternal twins were assessed for left- or right-handedness. Records were kept as to whether a pair of twins were concordant (both left- or both right-handed) or not concordant (one left- and the other right-handed). The data were:

Twin Type	Both the Same	Different	Total
Identical	107	28	135
Fraternal	406	121	527
Total	513	149	662

At the .01 level, is the proportion of concordance for the handedness trait the same for both types of twins?

12. The American Journal of Economics and Sociology reports that 1.1 percent of single mothers are under 18 years of age, 16.5 percent are at least 18 but less than 25, 40.2 percent are at least 25 but less than 35, 31.6 percent are at least 35 but less than 45, and 10.6 percent are 45 or over. A random sample of single mothers yielded the following results:

Age	Number
Under 18	15
18 and <25	35
25 and <35	47
35 and <45	27
45 or over	11
Total	135

At the .01 level, does the distribution in this sample fit the one in the report?

13. In clinical trials, 103 of the randomly selected patients discontinued using NORVASC or a placebo because of adverse effects. Test the hypothesis at the .01 level that the type of adverse effect and the dosage level are independent variables.

Adverse Event	5-mg Dose	10-mg Dose	Placebo	Total
Edema	9	29	5	43
Dizziness	10	9	9	28
Flushing or palpitation	8	19	5	32
Total	27	57	19	103

14. In a nationwide poll conducted in December 1992, 34 percent of the respondents thought that the U.S. economy was improving, 16 percent thought it was getting worse, and 48 percent thought it was staying the same. In a later nationwide *USA Today/CNN/Gallup* poll, the following responses were reported:

Opinion	Number
Economy improving	130
Getting worse	261
Staying the same	612
Total	1,003

Test the hypothesis at the .05 level that there was no change in opinion between December 1992 and the date of the later poll.

15. The following contingency table describes the marital status of male Vietnam veterans with and without posttraumatic stress disorder (PTSD). (The data are from *The Journal of Clinical Psychology.*)

Marital Status	No PTSD	PTSD	Total
Married	678	200	878
Living as married	45	44	89
Separated	9	27	36
Divorced/widowed	97	17	114
Never married	42	31	73
Total	871	319	1,190

Test the hypothesis at the .05 level that the presence or absence of PTSD is independent of marital status.

16. To meet the needs of their students, schools in southern Arizona have students take the Language Assessment Scales (LAS) test to determine their dominant reading language. The results of a random sample of first- and second-grade students in this region are as follows:

Grade	Spanish Reading	English Reading	Total
First	242	293	535
Second	295	362	657
Total	537	655	1,192

At the .01 level, is the primary reading language independent of grade?

17. An automobile sales manager in Denver claims that on 20 percent of the selling days, no cars are sold, on 30 percent of the selling days, 1 car is sold, on 30 percent of the days, 2 cars are sold, on 15 percent of the days, 3 cars are sold, and on 5 percent of the days, 4 or more cars are sold. In a recently sampled period, the following sales results are achieved:

Number Sold per Day	Number of Days
0	8
1	11
2	10
3	6
4 or more	6

Test the hypothesis at the .05 level that the population distribution for the number of cars sold each day fits the one claimed by the sales manager.

18. A children's hospital in Wisconsin uses certain criteria to admit as inpatients or outpatients those who have contracted measles. These criteria and a random sample of patients admitted in the in-out patient categories are shown below:

Criteria for Admission	Inpatients	Outpatients	Total
Physiologic instability	128	88	216
Dehydration	103	26	129
Severe complications	100	7	107
Meets more than one criterion	196	113	309
Total	527	234	761

At the .05 level, is inpatient-outpatient status independent of the criteria for admission?

19. According to geneticists, the distribution of blood type in the general population is 42 percent type A, 10 percent type B, 3 percent type AB, and 45 percent type O. In 1919, a study of the 4 blood groups was conducted for various populations. In a sample of 116 residents in 18 villages on Bougainville Island, it was observed that 74 were type A, 12 were type B, 11 were type AB, and 19 were type O. At the .01 level, did the distribution on Bougainville Island fit that of the general population?

20–25. The following is a *Minitab* printout for a test of independence:

```
Expected counts are printed below observed counts

         current   former     never    Total
    1          5       33        94      132
           11.88    49.20     70.92

    2         13       48        62      123
           11.07    45.85     66.08

    3         17       64        53      134
           12.06    49.95     71.99

Total         35      145       209      389

ChiSq =  3.982 +  5.336 +  7.511 +
         0.338 +  0.101 +  0.252 +
         2.027 +  3.953 +  5.012 = 28.511
df = 4
```

20. What is the total number in the sample?

21. How many categories are there for each variable?

22. What is the number of degrees of freedom for this test, and how could you compute it without this printout?

23. What is the value of the test ratio?

24. Find the cell (row and column intersection) that contributed the most to this test ratio value.

25. What additional information is needed to perform this test?

26–30. The following is a *Minitab* printout for a test of independence:

```
            male    female     Total
    1        379       247       626
          357.80    268.20

    2        283       151       434
          248.06    185.94

    3        399       263       662
          378.38    283.62

    4        322       222       544
          310.93    233.07

    5       5477      4259      9736
         5564.82   4171.18

Total       6860      5142     12002

ChiSq =  1.256 +  1.675 +
         4.921 +  6.565 +
         1.124 +  1.499 +
         0.394 +  0.525 +
         1.386 +  1.849 = 21.193
df = 4
```

26. What is the total number in the sample?

27. How many categories are there for each variable?

28. What is the number of degrees of freedom for this test, and how could you compute it without this printout?

29. What is the value of the test ratio?

30. What additional information is needed to perform this test?

31–35. Use the following *Minitab* printout to answer the questions below:

```
Expected counts are printed below observed counts

         failure  success    Total
    1        19       31        50
          17.86    32.14

    2        11       23        34
          12.14    21.86

Total        30       54        84

ChiSq =  0.073 +  0.041 +
         0.108 +  0.060 = 0.281
df = 1
```

31. What is the total number in the sample?

32. How many categories are there for each variable?

33. What is the number of degrees of freedom for this test, and how could you compute it without this printout?

34. What is the value of the test ratio?

35. What additional information is needed to perform this test?

Topics for Review and Discussion

1. What are the similarities and differences between the test for goodness of fit and the contingency table procedure that tests to see if the data are cross-classified in independent ways?

2. Examine the formula used to compute the test ratio, and discuss the significance of a TR_{χ^2} value of zero. Why is the null hypothesis for both tests examined in this chapter equivalent to having a TR_{χ^2} value of zero?

3. Even if the null hypothesis is true, why is it unlikely that the computed test ratio will be zero?

4. Discuss the purpose of a goodness-of-fit test. List three possible situations in which this test would be of use.

5. Discuss the purpose of a contingency table test for independence of variables. List three practical applications for this test.

6. How do the tests examined in this chapter differ from those in Chapters 8, 9, and 11? (*Hint*: Compare the types of data used in each chapter.)

Projects/Issues to Consider

1. You may wish to identify a variable that can be subdivided into 3 or more categories, or you may choose to focus on a controversial issue with 3 or more possible opinions. Obtain measurements, counts, or responses on your campus or elsewhere for your variable or issue. Establish hypotheses about the probability distribution of the possible outcomes, and then perform a goodness-of-fit test. Discuss your results in a class report.

2. (*a*) You'll need to repeat a multinomial experiment 100 times. For example, you can roll a die and note the top surface after each roll; draw a card from a standard deck, note the suit, and replace it; or even write A, B, C, and D on an equal number of equal-sized slips of paper and then draw one slip at a time, note the letter, and replace it. Before carrying out your experiment, though, you should establish hypotheses about the probability distribution of the outcomes. Finally, use the goodness-of-fit test and discuss the results obtained.

(*b*) If you worked with a die, try "loading" the die (use tape or glue on one face). Then, redo the goodness-of-fit test with the loaded die. If you worked with a deck of cards, try altering the deck by, for example, removing 7 of the cards in the heart suit and redoing the goodness-of-fit test. And if you worked with slips of paper, change the distribution of slips, and repeat your test procedure.

(*c*) Compare the results of parts a and b in a class report.

3. First, select a current and controversial issue, and then establish response categories. For example, these categories could be "agree," "disagree," or "don't know." Next, split those you will question into 3 or more groups (male-female, freshman-sophomore-junior-senior, Republican-Democrat-Independent, and so on). Then create a questionnaire that will elicit a response and that will also place the person into his or her proper group. Organize your responses into a contingency table, and perform a test of the independence of the variables. Discuss your results in a class report.

4. Locate a journal article in which a chi-square test either has been used or could be used. In either case, describe the data categories, and discuss how the test was (or could be) set up to either test for a goodness of fit or for the independence of variables.

Computer Exercises

1–4. The *Cancer Journal for Clinicians* of the American Cancer Society predicted that there would be 832,000 new cases of cancer nationwide in a recent year. The types of cancer and the expected percentage for each category are given below:

Type of Cancer	Percent of New Cases
Female breast	21.9
Colon/rectum	18.3
Lung	20.4
Oral	3.6
Uterus	5.3
Prostate	19.8
Skin melanoma	3.8
Pancreas	3.3
Leukemia	3.6

An environmentalist in California believes the distribution of new cancer cases is different in that state. He collects the following data (in thousands of cases) for his state:

Type of Cancer	New Cases (Thousands)
Female breast	18.5
Colon/rectum	14.5
Lung	17.0
Oral	3.2
Uterus	4.6
Prostate	16.0
Skin melanoma	3.8
Pancreas	3.0
Leukemia	3.1

1. What are the null and alternative hypotheses for this test?

2. Assuming that the test is conducted at the .05 level, what is the appropriate decision rule for this test?

3. Use your software to find the value of the test ratio for this problem. (Hint: If you're using *Minitab*, a procedure similar to the one shown in Figure 12.3 may be used.)

4. What is the appropriate statistical decision?

5. Use your software to conduct the test called for in end-of-chapter problem 12. Explain your results.

6. Use your software to conduct the test called for in end-of-chapter problem 2. (If you're using *Minitab*, the procedure shown in Figure 12.9 may be helpful.) Explain your results.

7. Use your software to conduct the test called for in end-of-chapter problem 6. Explain your results.

8. Use your software to conduct the test called for in end-of-chapter problem 8. Explain your results.

Answers to Odd-Numbered Self-Testing Review Questions

Section 12.1

1. A multinomial experiment has a fixed number of trials, and the outcome of each trial is independent of the outcomes of the other trials. A single trial results in one of k possible outcomes.

3. The sample is a random sample, and the expected frequency for each of the categories is 5 or more.

5. Sample data are obtained randomly, and the expected frequency in each cell is 5 or more.

7. There are 6 − 1, or 5, df, and the critical χ^2 value at the .05 level is 11.07.

9. The degrees of freedom = (4 − 1)(5 − 1) = (3)(4) = 12, and the critical χ^2 value at the .01 level is 26.2.

Section 12.2

1. *Step 1: State the null and alternative hypotheses.* H_0: There has been no change in the last 5 years in the level of education of management personnel. That is, we expect the 4 education levels, in the order given, to be represented by 15, 64, 8, and 13 percent of the sample. H_1: There has been a change in education levels. *Step 2: Select the level of significance.* $\alpha = .01$. *Step 3: Determine the test distribution to use.* We'll use a χ^2 distribution for all problems in this chapter. *Step 4: Define the rejection or critical region.* The critical χ^2 value with 4 − 1, or 3, df is 11.34. *Step 5: State the decision rule.* Reject H_0 and accept H_1 if the test ratio (TR) is > 11.34. Otherwise, accept H_0. *Step 6: Compute the test ratio.* The total number of observed values is 216, and the expected values for each educational level are found by multiplying the observed value by the appropriate percentage in the null hypothesis (use .15 for 15 percent, .64 for 64 percent, and so on). The following columns are thus produced:

O	E	O − E	(O − E)²	(O − E)²/E
19	32.40	−13.40	179.560	5.54198
132	138.24	−6.24	38.938	0.28167
30	17.28	12.72	161.798	9.36333
35	28.08	6.92	47.886	1.70536
216	216.00	0		16.89234

The $TR_{\chi^2} = 16.89$. *Step 7: Make the statistical decision.* Since a TR value of 16.89 falls in the rejection region, we reject the H_0. The personnel director is likely to be correct about the change in the level of education. It appears that current management personnel have higher educational levels.

3. *Step 1.* H_0: There has been no change in the distribution of age groups. That is, there are still 10, 13, 32, and 45 percent of the residents in the 4 respective age categories. H_1: The age distribution has changed since 1987. *Step 2.* $\alpha = .05$. *Step 3.* χ^2. *Step 4.* The critical χ^2 value with $4 - 1$, or 3, df is 7.81. *Step 5.* Reject H_0 and accept H_1 if the test ratio (TR) is > 7.81. Otherwise, accept H_0. *Step 6.* The $\Sigma (O - E)^2/E = 105.856 + 45.847 + 8.304 + 36.717 = 196.724$. Thus, the $TR_{\chi^2} = 196.72$. *Step 7.* Since a TR value of 196.72 falls in the rejection region, we reject the H_0. There has been a significant change in the age distribution.

5. *Step 1.* H_0: The distribution of public opinion remains the same as it was in December 1992. H_1: The distribution has changed by August 1993. *Step 2.* $\alpha = .01$. *Step 3.* χ^2. *Step 4.* The critical χ^2 value with $3 - 1$, or 2, df is 9.21. *Step 5.* Reject H_0 and accept H_1 if the test ratio (TR) is > 9.21. Otherwise, accept H_0. *Step 6.* The $\Sigma (O - E)^2/E = 16.9173 + 15.8641 + .9545 = 33.74$. Thus, the $TR_{\chi^2} = 33.74$. *Step 7.* Since a TR value of 33.74 falls in the rejection region, we reject the H_0. There was a significant change in public opinion between the time periods.

7. *Step 1.* H_0: There has been no change in the distribution of transactions. H_1: The distribution has changed. *Step 2.* $\alpha = .01$. *Step 3.* χ^2. *Step 4.* The critical χ^2 value with $4 - 1$, or 3, df is 11.34. *Step 5.* Reject H_0 and accept H_1 if the test ratio (TR) is > 11.34. Otherwise, accept H_0. *Step 6.* The $\Sigma (O - E)^2/E = 22.0506 + 77.8921 + .6247 + 5.1648 = 105.73$. Thus, the $TR_{\chi^2} = 105.73$. *Step 7.* Since a TR value of 105.73 falls in the rejection region, we reject the H_0. The distribution has changed significantly since the new mall was opened.

9. *Step 1.* H_0: The distribution of shirt sizes is still 20, 40, 30, and 10 percent for the sizes listed. H_1: The distribution of sizes has changed. *Step 2.* $\alpha = .01$. *Step 3.* χ^2. *Step 4.* The critical χ^2 value with $4 - 1$, or 3, df is 11.34. *Step 5.* Reject H_0 and accept H_1 if the test ratio (TR) is > 11.34. Otherwise, accept H_0. *Step 6.* The $\Sigma (O - E)^2/E = 1.5633 + 7.6276 + 1.6948 + 25.3711 = 36.26$. Thus, the $TR_{\chi^2} = 36.26$. *Step 7.* Since a TR value of 36.26 falls in the rejection region, we reject the H_0. The demand for shirt sizes has changed.

11. *Step 1.* H_0: The distribution of the number of times a song is played agrees with the analyst's claim. H_1: The distribution doesn't agree with the claim. *Step 2.* $\alpha = .05$. *Step 3.* χ^2. *Step 4.* The critical χ^2 value with $5 - 1$, or 4, df is 9.49. *Step 5.* Reject H_0 and accept H_1 if the test ratio (TR) is > 9.49. Otherwise, accept H_0. *Step 6.* The $\Sigma (O - E)^2/E = .3123 + .0251 + 1.9060 + .5554 + .0055 = 2.80$. Thus, the $TR_{\chi^2} = 2.80$. *Step 7.* Since a TR value of 2.80 does not fall in the rejection region, we accept the H_0 and the analyst's claim.

Section 12.3

1. *Step 1: State the null and alternative hypotheses.* H_0: The strongest language for kindergarten and first-grade students is independent of his or her grade. H_1: The variables are not independent. *Step 2: Select the level of significance.* $\alpha = .01$. *Step 3: Determine the test distribution to use.* χ^2. *Step 4: Define the rejection or critical region.* The critical χ^2 value with $(4 - 1)(2 - 1)$, or 3, df is 11.34. *Step 5: State the decision rule.* Reject H_0 and accept H_1 if the test ratio (TR) is > 11.34. Otherwise, accept H_0. *Step 6: Compute the test ratio.* The following columns are produced:

O	E	$O - E$	$(O - E)^2$	$\frac{(O - E)^2}{E}$
160	160.23	−.23	.05	.0003
197	196.77	.23	.05	.0003
75	74.95	.05	.00	.0000
92	92.05	−.05	.00	.0000
284	283.66	.34	.12	.0004
348	348.34	−.34	.12	.0003
16	16.16	−.16	.03	.0019
20	19.84	.16	.03	.0015
1,192	1,192.00	0		.0047

The $TR_{\chi^2} = .0047$. *Step 7: Make the statistical decision.* Since a TR value of .0047 does not fall in the rejection region, we accept the H_0. The percent of students having the given language dominance is the same for both grades; therefore, the variables are independent.

3. *Step 1.* H_0: The percentage of concordance for the mental retardation trait is the same for identical and fraternal twins. H_1: The percentages are different for these 2 types of twins. *Step 2.* $\alpha = .01$. *Step 3.* χ^2. *Step 4.* The critical χ^2 value with $(2 - 1)(2 - 1)$, or 1, df is 6.63. *Step 5.* Reject H_0 and accept H_1 if the test ratio (TR) is > 6.63. Otherwise, accept H_0. *Step 6.* The $\Sigma\ (O - E)^2/E = 59.9392 + 57.4522 + 15.3537 + 14.7180 = 147.46$. Thus, $TR_{\chi^2} = 147.46$. *Step 7.* Since a TR value of 147.46 falls in the rejection region, we reject the H_0. The variables are not independent; there's a difference in the percentage of concordance for the 2 types of twins.

5. *Step 1.* H_0: There's an equal percentage of male and female inner-city middle school students who want to stay in school. H_1: The variables are not independent. *Step 2.* $\alpha = .05$. *Step 3.* χ^2. *Step 4.* The critical χ^2 value with $(2 - 1)(2 - 1)$, or 1, df is 3.84. *Step 5.* Reject H_0 and accept H_1 if the test ratio (TR) is > 3.84. Otherwise, accept H_0. *Step 6.* The $\Sigma\ (O - E)^2/E = 6.7301 + 2.7655 + 6.7301 + 2.7655 = 18.99$. Thus, $TR_{\chi^2} = 18.99$. *Step 7.* Since a TR value of 18.99 falls in the rejection region, we reject the H_0. The variables are not independent.

7. *Step 1.* H_0: The career choice for primary care physicians is independent of the year they graduated. H_1: The choice is not independent of the year of graduation. *Step 2.* $\alpha = .01$. *Step 3.* χ^2. *Step 4.* The critical χ^2 value with $(3 - 1)(2 - 1)$, or 2, df is 9.21. *Step 5.* Reject H_0 and accept H_1 if the test ratio (TR) is > 9.21. Otherwise, accept H_0. *Step 6.* The $\Sigma\ (O - E)^2/E = 8.1732 + 6.9057 + 6.0429 + 5.1058 + .2274 + .1921 = 26.65$. Thus, $TR_{\chi^2} = 26.65$. *Step 7.* Since a TR value of 26.65 falls in the rejection region, we reject the H_0. The type of career choice is not independent of the graduation date.

9. *Step 1.* H_0: The opinion about Clinton's budget is independent of income level. H_1: The opinion is not independent of income. *Step 2.* $\alpha = .05$. *Step 3.* χ^2. *Step 4.* The critical χ^2 value with $(4 - 1)(4 - 1)$, or 9, df is 16.92. *Step 5.* Reject H_0 and accept H_1 if the test ratio (TR) is > 16.92. Otherwise, accept H_0. *Step 6.* The $\Sigma\ (O - E)^2/E = .4201 + 3.4384 + 10.7741 + .1605 + .0697 + 3.0253 + 5.4125 + .0307 + 1.7840 + 1.9532 + 16.7808 + 1.2507 + .8579 + 4.7529 + .2452 + 2.1393 = 53.10$. Thus, $TR_{\chi^2} = 53.10$. *Step 7.* Since a TR value of 53.10 falls in the rejection region, we reject the H_0. Opinion about the budget and income level are not independent variables.

11. *Step 1.* H_0: There's no significant difference between the population percentages of defective parts on the 3 shifts. H_1: The percentages are not the same. *Step 2.* $\alpha = .05$. *Step 3.* χ^2. *Step 4.* The critical χ^2 value with $(3 - 1)(2 - 1)$, or 2, df is 5.99. *Step 5.* Reject H_0 and accept H_1 if the test ratio (TR) is > 5.99. Otherwise, accept H_0. *Step 6.* The $\Sigma\ (O - E)^2/E = .8906 + 3.9185 + .1515 + .6667 + .4125 + 1.8148 = 7.855$. Thus, $TR_{\chi^2} = 7.855$. *Step 7.* Since a TR value of 7.855 falls in the rejection region, we reject the H_0. The percentage of defective parts isn't the same on the 3 shifts.

13. *Step 1.* H_0: There's no significant difference in programming preference in rural and

urban areas. H_1: There is such a difference in preference. *Step 2.* $\alpha = .05$. *Step 3.* χ^2. *Step 4.* The critical χ^2 value with $(4 - 1)(2 - 1)$, or 3, df is 7.81. *Step 5.* Reject H_0 and accept H_1 if the test ratio (TR) is > 7.81. Otherwise, accept H_0. *Step 6.* The $\Sigma\ (O - E)^2/E = .266 + .404 + .196 + .298 + .122 + .186 + .478 + .726 = 2.68$. Thus, $TR_{\chi^2} = 2.68$. *Step 7.* Since a TR value of 2.68 does not fall in the rejection region, we accept the H_0. Viewing preference is the same in urban and rural areas.

15. *Step 1.* H_0: The type of smoker is independent of gender. H_0: The variables are not independent. *Step 2.* $\alpha = .01$. *Step 3.* χ^2. *Step 4.* The critical χ^2 value with $(2 - 1)(2 - 1)$, or 1, df is 6.63. *Step 5.* Reject H_0 and accept H_1 if the test ratio (TR) is > 6.63. Otherwise, accept H_0. *Step 6.* The $\Sigma\ (O - E)^2/E = .019 + .020 + .011 + .011 = .06$. Thus, $TR_{\chi^2} = .06$. *Step 7.* Since a TR value of .06 does not fall in the rejection region, we accept the H_0 that the type of smoker is independent of gender.

17. *Step 1.* H_0: The success or failure of the surgery is independent of prior surgical history. H_1: The variables are not independent. *Step 2.* $\alpha = .01$. *Step 3.* χ^2. *Step 4.* The critical χ^2 value with $(2 - 1)(2 - 1)$, or 1, df is 6.63. *Step 5.* Reject H_0 and accept H_1 if the test ratio (TR) is > 6.63. Otherwise, accept H_0. *Step 6.* The $\Sigma\ (O - E)^2/E = .073 + .041 + .108 + .060 = .28$. Thus, $TR_{\chi^2} = .28$. *Step 7.* Since a TR value of .28 does not fall in the rejection region, we accept the H_0. The rate of success was not significantly different for the two groups.

19. Step 1. H_0: The opinions of toy buyers are the same from region to region. H_1: There are regional differences of opinion. *Step 2.* $\alpha = .01$. *Step 3.* χ^2. *Step 4.* The critical χ^2 value with $(4 - 1)(3 - 1)$, or 6, df is 16.81. *Step 5.* Reject H_0 and accept H_1 if the test ratio (TR) is > 16.81. Otherwise, accept H_0. *Step 6.* The $\Sigma\ (O - E)^2/E = .027 + .198 + .008 + 4.511 + 9.434 + 10.032 + 1.515 + 5.538 + 4.481 + 0.545 + 1.385 + 1.333 = 39.01$. Thus, $TR_{\chi^2} = 39.01$. *Step 7.* Since a TR value of 39.01 falls in the rejection region, we reject the H_0. There are regional differences in the way toy buyers have responded to the survey.

CHAPTER 13

Linear Regression and Correlation

LOOKING AHEAD

In this chapter you'll see how statistical techniques may be used to help a manager make a hiring decision, and you'll see how an executive can arrive at sales forecasts for her company. We'll consider equations in this chapter that describe the relationship that exists between two or more variables. Such equations are usually solved with the help of modern computers or calculators. The name given to the study methodology that uses these modern tools is *regression analysis*—an anachronistic term dating to Sir Francis Galton's nineteenth-century study of the relationship between parents' heights and children's heights. Since Galton found that the stature of the offspring of very short or very tall parents tended to "regress" toward the mean population height, he was legitimately engaged in a regression analysis. Subsequent studies, of course, had nothing to do with Galton's regression analysis, but the name has stuck.

First, we'll examine some introductory concepts. Next, we'll focus on simple linear regression analysis and some relationship tests and prediction intervals that may be prepared in such an analysis. The focus then shifts to simple linear correlation analysis. Finally, we'll consider multiple linear regression and correlation topics.

Thus, after studying this chapter, you should be able to:

- Explain the purposes of regression and correlation analysis.
- Compute and interpret the meaning of regression equations and standard errors of estimate for simple linear and multiple linear regression analysis situations, and then use these measures to prepare interval estimates of the dependent variable for forecasting purposes.
- Compute and explain the meaning of the coefficients of determination and correlation when simple linear and multiple linear correlation analysis techniques are used.

13.1 Introductory Concepts

It's often necessary to prepare a *forecast*—an expectation of what the future holds—before a decision can be made. For example, it may be necessary to predict revenues before a budget can be prepared. And a university must predict enrollment before making up class schedules. These and other decisions can be made easier *if a relationship can be established between the variable to be predicted and some other variable* that is either known or is significantly easier to anticipate.

For our purposes, the term "relationship" means that changes in two or more variables are *associated with each other.* For example, we might find a high degree of relationship between the consumption of fuel oil and the number of cold days during the winter, or between a change in the price of a product and consumer demand for the product. That is, we might logically expect a change (increase) in the number of cold days to be accompanied by a change (increase) in the consumption of fuel oil, or a change (increase) in the price of a product to be accompanied by a change (decrease) in the demand for the product.

Regression and Correlation Analysis: A Preview

The tools of regression analysis and correlation analysis have been developed to study and measure the statistical relationship that exists between two or more variables. The terms **simple regression** and **simple correlation** refer to those studies dealing with just two variables. When three or more variables are considered, the study deals with **multiple regression** and **multiple correlation.**

In **regression analysis**, an estimating, or predicting, equation is developed to describe the pattern or functional nature of the relationship that exists between the variables. As the name implies, an analyst prepares an **estimating** (or **regression**) **equation** to make estimates of values of one variable from given values of the other(s).

The **dependent** (or **response**) **variable** is the variable to be estimated; it is customarily plotted on the vertical or y axis of a chart and is therefore identified by the symbol y.

An **independent** (or **explanatory**) **variable** (*one in the simple regression/correlation case, two or more in multiple regression/correlation*) is one that presumably exerts an influence on or explains variations in the dependent variable. In simple regression/correlation, the independent variable is customarily plotted on the horizontal or x axis and is therefore identified by the symbol x.

Too Many Sails in the Sunset

An Australian sailor uses a statistical software program to collect and analyze volumes of data on temperature, wind velocity, wind direction, barometric pressure, and a host of other variables. His goal is to predict the weather on the day of a sailing race more accurately than his competitors, and his analysis helps him decide which sails to bring aboard the day of the race (too many sails and the boat is too heavy; too few sails and the right sail could be left behind).

Let's assume that Hiram N. Hess, personnel manager of the Tackey Toy Manufacturing Company, finds that there's a close and logical relationship between the productive output of employees in a certain department of the company and their earlier performance on an aptitude test. Thus, *if* Hiram computes an estimating or regression equation (as we'll do in a few pages), and *if* he has the aptitude test score of a job applicant, *then* he may use his estimating equation to predict the future output (the dependent variable) of the applicant based on test results (the independent variable). Included in the techniques used to make estimates of the dependent variable are regression analysis procedures that measure the dependability of these measures.

In **correlation analysis**, the purpose is to measure the strength or closeness of the relationship between the variables. In other words, regression analysis asks, "What is the pattern of the existing relationship?" and correlation analysis asks, "How strong is the relationship described in the regression equation?" Although it's possible to be concerned with only regression analysis or with only an analysis of correlation, the two are typically considered together.

To summarize, in the following pages we'll concentrate on three main tasks (the first two of which are a part of regression analysis):

1. Computing the regression, or estimating, equation and then using it to provide an estimate of the value of the dependent, or response, variable y when given one or more values of the independent or explanatory variable(s) x.
2. Computing measures that show the possible errors that may be involved in using the estimating equation as a basis for forecasting.
3. Preparing measures that show the closeness of the association or correlation that exists between the variables.

I Grieve

Bill Illing, director of employee services at Kansas City Power and Light, has used statistical software to investigate the handling of employee grievances. His study involved dependent variables such as the number of grievances filed per month against middle and top managers. His independent variables included the managers involved in the grievance, the presence or absence of a collective-bargaining agreement when the grievance was filed, and the presence or absence of an "employee involvement" program when the grievance was filed. The results of the study helped managers improve their negotiating skills, and it helped them do a better job of resolving employee grievances.

A Logical Relationship: The First Step

Your sales last year just paralleled the sales of rum cokes in Rio de Janeiro, as modified by the sum of the last digits of all new telephone numbers in Toronto. So, why bother with surveys of your own market? Just send away for the data from Canada and Brazil.—*Lydia Strong*

Two variables or series may move together for several reasons. And analysts may correctly assume a causal relationship in their interpretation of correlation measures. But just because two variables are associated in a statistical sense does not guarantee the existence of a causal relationship. In other words, the existence of a causal relationship usually does imply correlation, but, as the above quote adequately shows, statistical correlation alone does not in any way prove causality. (If you believe that it does, you should indeed send away for the data from Canada and Brazil!)

Causal relationships may fit into *cause-and-effect* or *common-cause* categories. A **cause-and-effect relationship** exists if a change in one variable produces a change in another variable. For example, an increase in the temperature of a chemical process may cause a decrease in the yield of the process, and an increase in the level of production output may cause an increase in total production cost. Alternatively, two series may vary together because of a **common-cause factor** that impacts both series in the same way. One could probably find a close relationship between jewelry sales and compact disk sales, but one of these series is not the cause and the other the effect. Rather, changes in both series are probably a result of changes in consumer income.

Of course, as we saw in the quote at the beginning of this section, some relationships are purely accidental. So if a relationship were found between furniture sales in the United States and the average temperature in Tanzania, for example, it would be a meaningless exercise to analyze the data. Relationships such as this one are known as *spurious correlations.*

Probably the first step in the study of regression and correlation, therefore, is to see if *a logical relationship* may exist between the variables to support further analysis. Unfortunately, presenting a summary of many types of variables, and the forces at work on those variables, is beyond the scope of this book. Fortunately, though, such a summary is to be found in the psychology, economics, biology, business, and other courses that you've taken or will take. It's only through the use of reason and judgment (along with the application of knowledge about the variables and the forces at work) that an analysis may assume causality. Yet without this assumption, there's not much point in proceeding with regression and correlation analyses.

Other students (not you) sometimes go to one of two extremes at about this point in their introduction to regression and correlation: (1) They fail to realize the importance of seeing if a logical relationship exists, and they mechanically apply the statistical procedures in the following pages and arrive at possibly spurious correlations to which they erroneously assign interpretations of cause and effect. Or (2) they think that since one cannot prove causality from correlation, it's necessary to conclude that there's no connection at all between correlation and causality. You, of course, will avoid either of these extremes.

The Scatter Diagram

Let's assume that a logical relationship may exist between *two variables.* To support further analysis, the next step, then, is to use a graph to plot the available data. This graph—called a **scatter diagram**—shows plotted points. Each point represents a study member for which we have an x (independent or explanatory) value and a y (dependent or response) value. The *scatter diagram serves two purposes*: (1) It helps us see if there's a useful relationship between the two variables, and (2) it helps us determine the type of equation to use to describe the relationship.

We can illustrate the purposes of a scatter diagram by using the data in Table 13-1.

TABLE 13.1 OUTPUT AND APTITUDE TEST RESULTS OF 8 EMPLOYEES OF TACKEY TOY MANUFACTURING COMPANY

Employee	Aptitude Test Results (x)	Output (Dozens of Units) (y)
A	6	30
B	9	49
C	3	18
D	8	42
E	7	39
F	5	25
G	8	41
H	10	52

This table gives us the output for a time period in dozens of units (the dependent or response variable) and the aptitude test results (the independent or explanatory variable) for eight employees in a department of the Tackey Toy Manufacturing Company. If the aptitude test does what it's supposed to do, it's reasonable to assume that employees with higher aptitude scores will be among the higher producers. Our group of eight employees represents a *small sample* of workers. We've kept the employee count small and the data simple, though, to minimize the computational effort necessary in later sections.

The data for each employee represent *one* point on the scatter diagram shown in Figure 13.1. The points representing employees C and F are labeled to show you how the pairs of observations for the employees are used to prepare the points on the chart. As you'll notice in Figure 13.1, the eight points form a path that can be approximated by a *straight line*. Thus, there appears to be a **linear relationship** between the variables. And a high degree of relationship is indicated by the fact that the points are all close to this straight-line path. You'll also notice that there's a **positive** (or **direct**) **relationship** between the variables—that is, as aptitude test results *increase*, output also *increases*. Of course, it's quite possible for variables to have a **negative** (or **inverse**) **relationship** (as the x value increases, the y value decreases).

Figure 13.2*a* shows how the *Minitab* statistical software package prepares a scatter diagram with the same data set used to produce Figure 13.1. The eight points must still follow a positive straight-line pattern, but the program has elected to use different scales for the two variables. Another scatter diagram that plots the relationship between per-capita gross domestic product of selected nations (the independent variable) and the per-capita amount spent on health care (the dependent variable) by those nations is shown in Figure 13.2*b*. This diagram also shows a positive linear relationship.

Figure 13.3 summarizes some possible scatter diagram forms. Figures 13.3*a*, *b*, and *c* show the positive and negative linear patterns we've been considering. We'll consider only linear relationships in this chapter's computations. However, relationships need

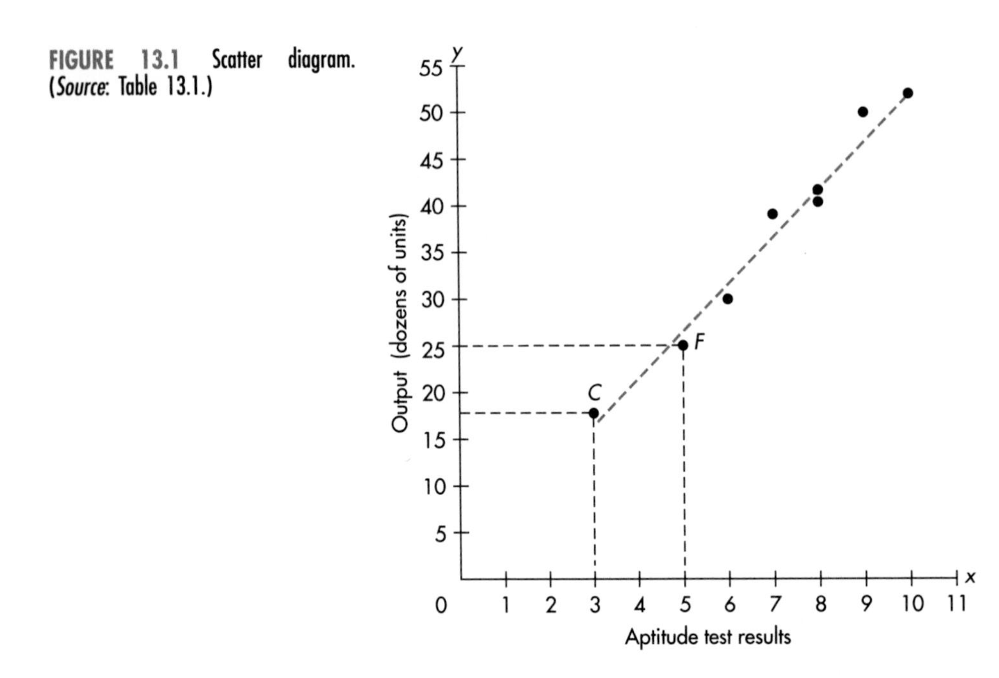

FIGURE 13.1 Scatter diagram. (*Source*: Table 13.1.)

not be linear as shown in Figures 13.3*d* to *f*. In Figure 13.3*f*, the variables might be family income and age of the head of the household. (Income tends to rise for a period and then fall off when retirement age is reached.) Finally, it's possible that a scatter diagram such as the one in Figure 13.3*g* might show *no relationship* at all between the variables.

Self-Testing Review 13.1

1. What is the purpose of linear regression analysis?
2. Discuss the roles of the dependent (response) variable and the independent (explanatory) variable(s).

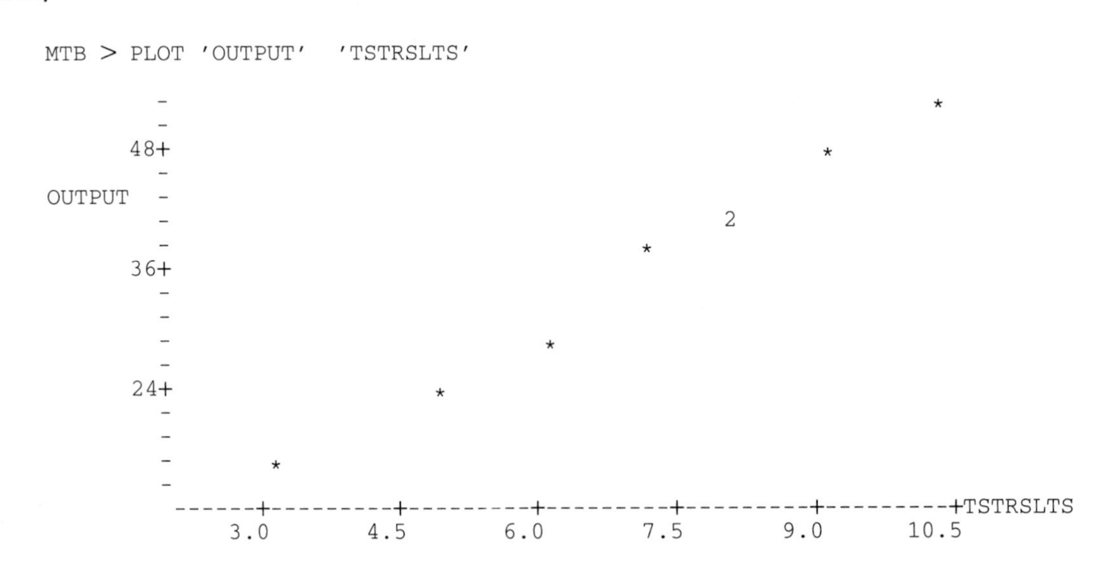

FIGURE 13.2 (*a*) A scatter diagram prepared by the *Minitab* statistical package. (*b*) An example of a positive linear relationship. (*Source: Insight*, August 8, 1988, p. 10. Figure *b* reprinted with permission of *Insight* magazine/M. Vey Martini.)

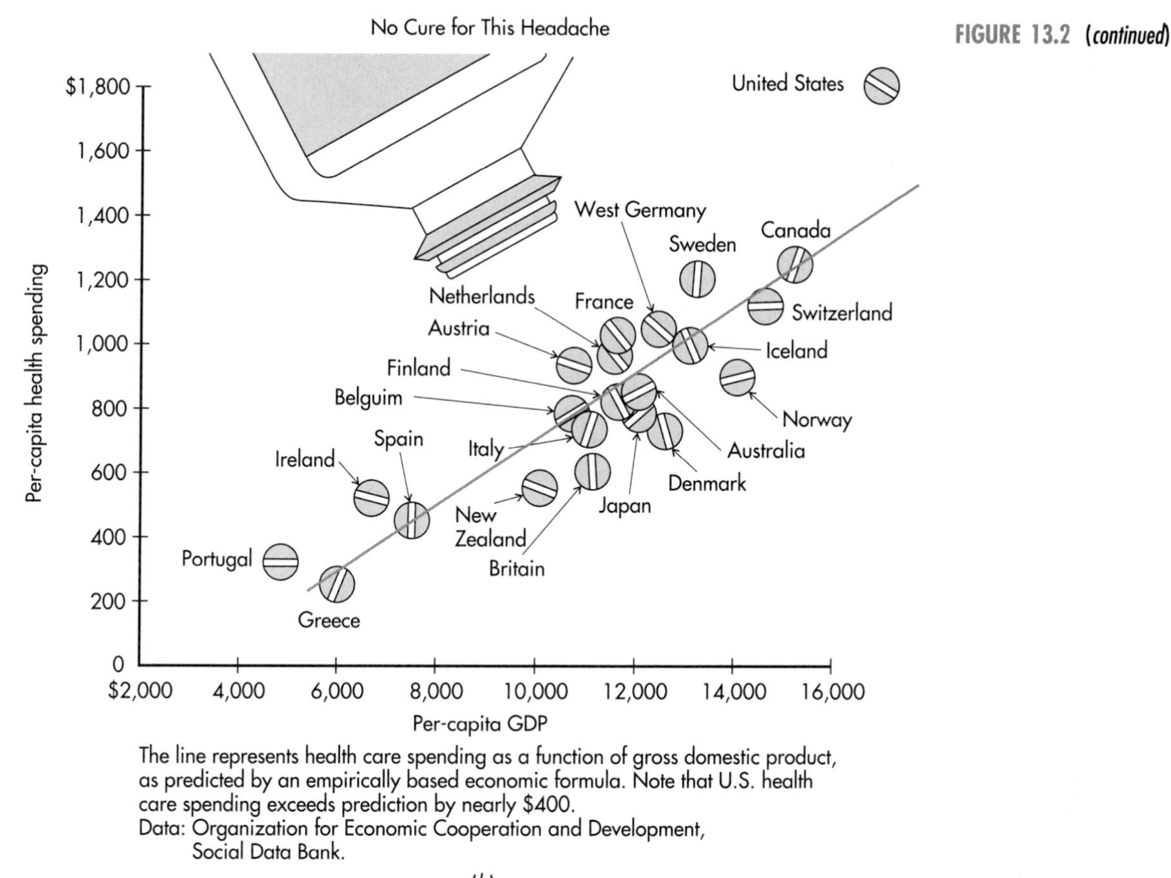

The line represents health care spending as a function of gross domestic product, as predicted by an empirically based economic formula. Note that U.S. health care spending exceeds prediction by nearly $400.
Data: Organization for Economic Cooperation and Development, Social Data Bank.

(*b*)

FIGURE 13.2 (*continued*)

3. What is correlation analysis?

4–8. For each of the following scatter diagrams, determine if there is (*a*) a positive linear relationship, (*b*) a negative linear relationship, or (*c*) no linear relationship:

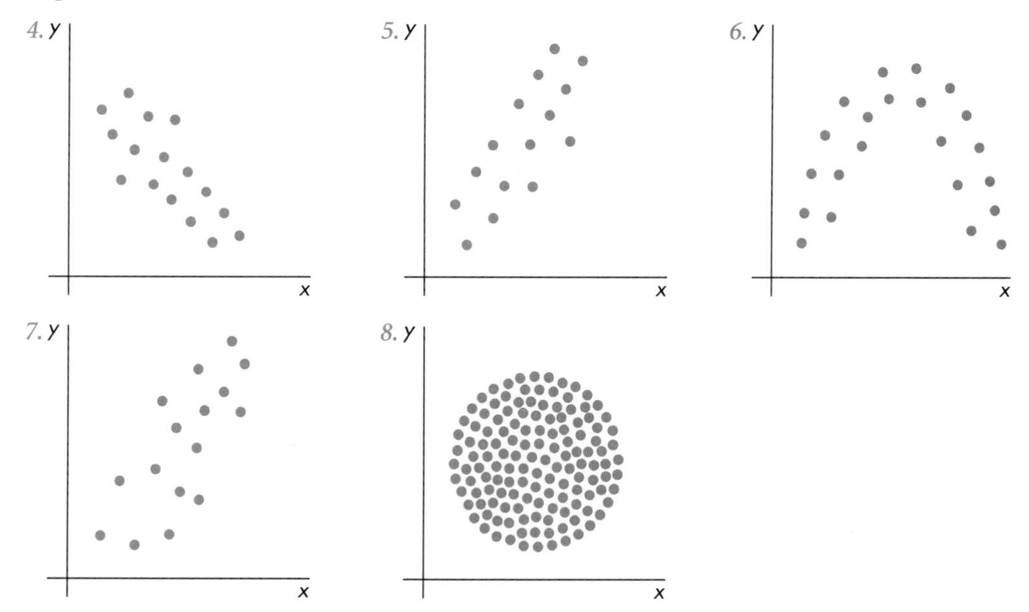

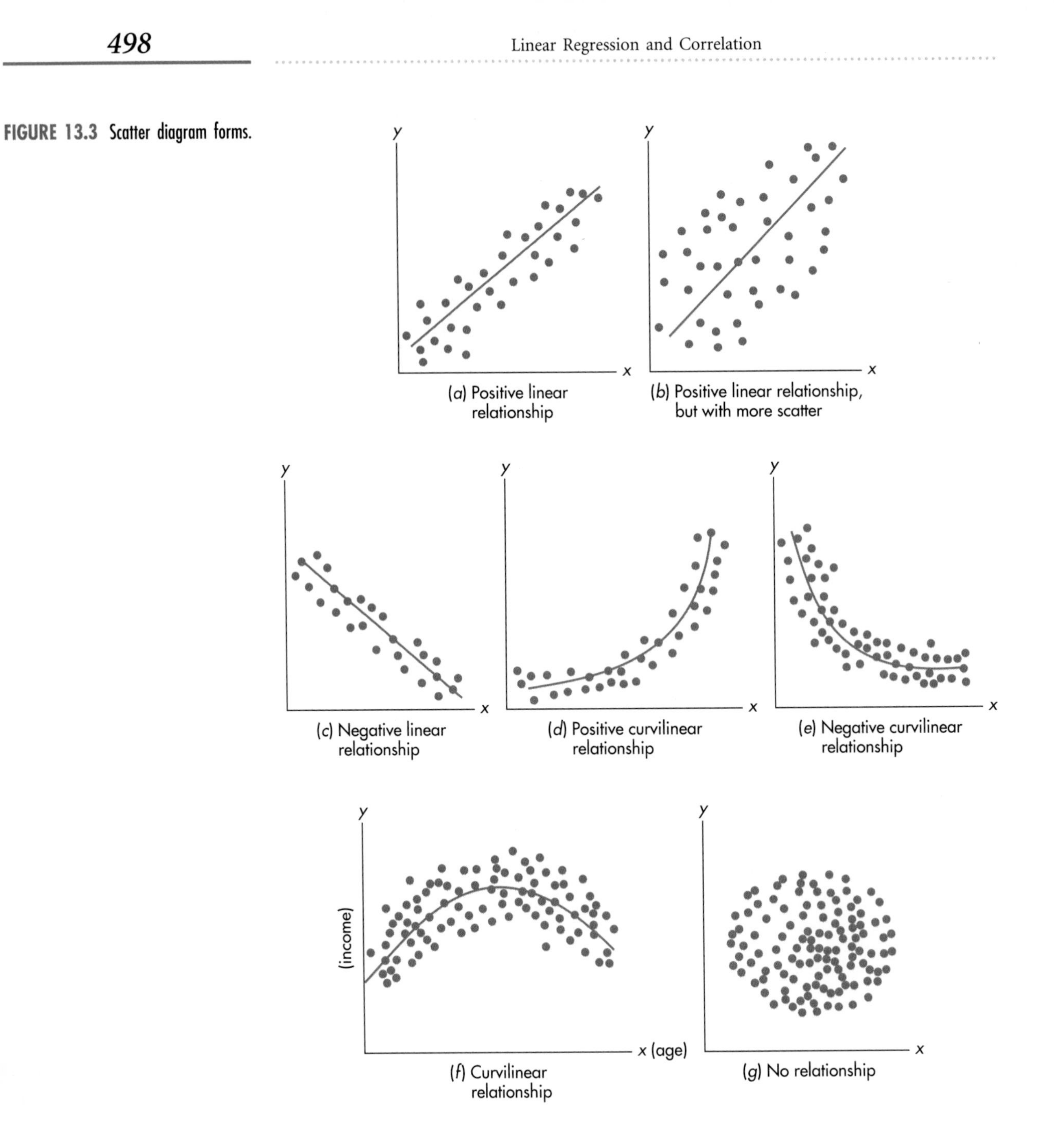

FIGURE 13.3 Scatter diagram forms.

9–16. For each of the following data pairs, do you think there would tend to be a positive linear relationship, a negative linear relationship, or no relationship?

9. $x =$ the height of twin A, and $y =$ the height of twin B.

10. $x =$ the number of hours in safety training, and $y =$ the number of accidents.

11. $x =$ a person's IQ, and $y =$ the number of books the person read last year.

12. $x =$ a person's IQ, and $y =$ the person's height.

13. $x =$ the height of a male, and $y =$ the shoe size of the male.

14. $x =$ the hours a student watches television per week, and $y =$ the student's grade point average.

15. $x =$ the dollar amount of sales to customers, and $y =$ the bonus received.

16. $x =$ a person's income, and $y =$ the dollars the person spent for clothing.

17–20. Draw a scatter diagram for each of the following data pairs. State if you think there's a positive linear relationship, a negative linear relationship, or no relationship.

17. We'll let $x =$ the number of years a person has spent at a company and $y =$ the width in feet of the person's office space:

x	3	16	7	4	15	7	8	5
y	4	40	16	9	38	16	17	10

18. Let $x =$ a sixth grader's IQ score and $y =$ the hours the student spent watching TV each week:

x	125	116	97	114	85	107	105
y	5	14	30	16	41	25	21

19. Let $x =$ a child's age and $y =$ the number of visits the child makes to a doctor during the year:

x	5	6	8	14	15	7	8	12
y	9	2	12	17	9	16	6	15

20. Let $x =$ the number of days a patient is hospitalized and $y =$ the age of the patient.

x	1	6	2	4	5	3
y	15	75	27	3	15	15

13.2 Simple Linear Regression Analysis

The straight line in the scatter diagram in Figure 13.1 (and the straight lines in Figures 13.2 and 13.3) that describe the relationship between the variables is called a **regression** (or **estimating**) **line**. We've seen in the Looking Ahead section that Sir Francis Galton's study showed that the height of the children of tall parents tended to regress (or move back) toward the average height of the population. Galton called the line describing this relationship a "line of regression." The word "regression" has stuck with us, but other words such as "estimating" or "predictive" are probably more apt.

The Linear Regression Equation

We'll compute an estimating, or regression, equation in this section to describe the relationship between the variables. Our interest here is limited to an analysis of **simple linear regression** of y on x—that is, to the case in which the relationship between two variables can be adequately described by a straight line.

The Straight-Line Equation. Figure 13.4a shows a straight line (identified by the symbol $\hat{y}$) of the type that we'll soon be computing. To define this line, we must know *two things* about it. First, we must know the value of the *y intercept*—that is, the value (read on the y axis) of a in Figure 13.4a when x is equal to zero. And second, we need to know the *slope of the line (b)*. This slope as shown in Figure 13.4a is found by (1) taking a segment of the line, (2) measuring a change of one unit of the x variable, (3) measuring the corresponding change in $\hat{y}$ on the y axis, and (4) dividing the change in

Sample data are analyzed at the U.S. Olympic Training Center to help improve athletic performance. (*Andre Perlstein/Jerrican/Photo Researchers*)

$\hat{y}$ by the change in x. In Figure 13.4*a*, the slope of the line has a *positive value*; in Figure 13.4*b*, the slope has a *negative* value. In both cases, however, the formula for $\hat{y}$ (and the formula that we will use to compute the straight-line equation) is:

$$\hat{y} = a + bx \tag{13.1}$$

where $\hat{y}$ = a computed estimate of the dependent variable
a = the y intercept or the value of $\hat{y}$ when x is equal to zero
b = the slope of the regression line, or the increase or decrease in $\hat{y}$ for each change of one unit of x
x = a given value of the independent variable

Thus, if we selected a value of the independent or explanatory variable x_1 as shown in Figure 13.4*c*, drew a vertical line up to the *regression line* ($\hat{y}$), and then drew a horizontal line to the y axis, the value of y_1 would be an estimate of the dependent or response variable y, given the value of the independent variable x_1.

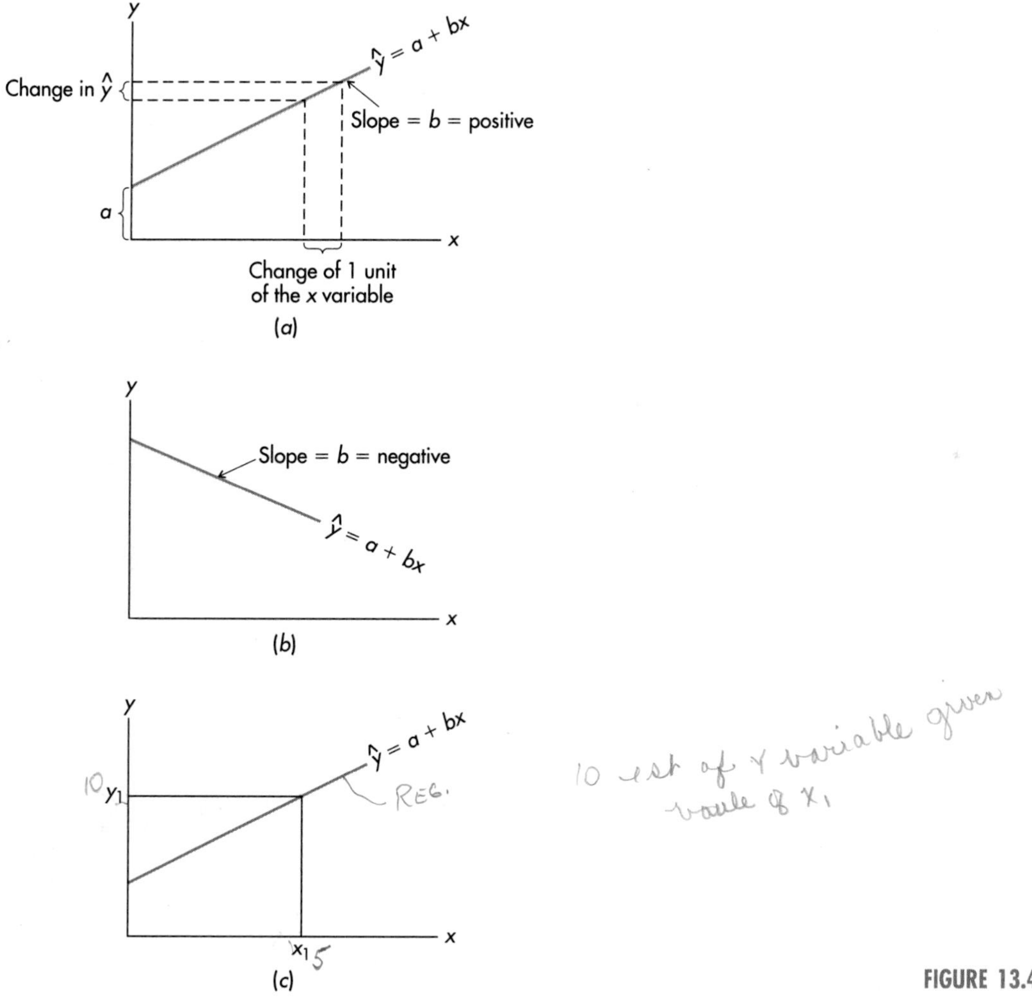

FIGURE 13.4

Properties of the Linear Regression Line. There are *two properties* of the linear regression line. The first of these properties is demonstrated in Figure 13.5, which shows the regression line fitted to the scatter diagram data that are plotted on a graph. Each point on the scatter diagram represents a value for x and y. The regression line ($\hat{y}$) is drawn in a straight line through these points. Sometimes the $\hat{y}$ value for a given value of x is greater than the y value, and sometimes $\hat{y}$ is less than y for a given x quantity. But the fact is that the regression line is fitted to the data in the scatter diagram in such a way that the *positive* deviations of the scatter points *above* the line in the diagram cancel out the *negative* deviations of the scatter points *below* the line, and the resulting sum is zero (see Figure 13.5). Thus, the *first property* of the regression line is:

$$\Sigma (y - \hat{y}) = 0$$

And the sum of the *squares* of the deviations is less than would be the case if any other straight line were substituted for the $\hat{y}$ line and the process of computing and squaring deviations was carried out. In other words, the *second property* is:

$$\Sigma (y - \hat{y})^2 = \text{a minimum or least value}$$

And so the name **method of least squares** is used to describe the approach we'll follow to fit a straight line to a linear regression data set.

The values of a and b in the regression equation are computed with the following formulas:

$$b = \frac{n(\Sigma xy) - (\Sigma x)(\Sigma y)}{n(\Sigma x^2) - (\Sigma x)^2} \qquad (13.2)$$

where n = number of paired observations
(Σx^2) = the sum of all the values in an x^2 column
$(\Sigma x)^2$ = the square of the sum of an x column

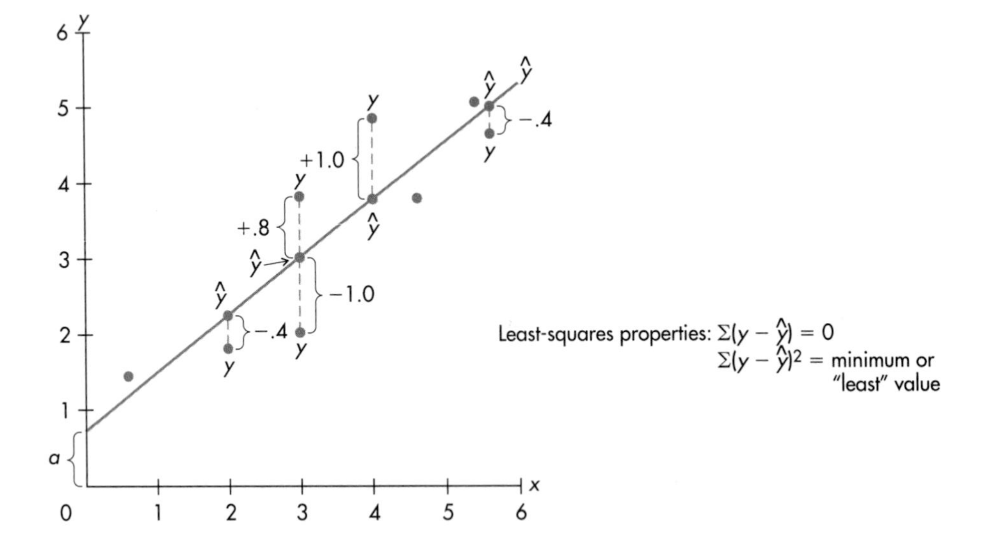

FIGURE 13.5

$$a = \bar{y} - b\bar{x} \qquad (13.3)$$

where $\bar{y}$ = the mean of the y variable
$\bar{x}$ = the mean of the x variable

The linear regression line always passes through the point at the intersection of $\bar{y}$ and $\bar{x}$.

Computing the Linear Regression Equation. We're now ready to find the regression equation for the data presented in Table 13.1. A work sheet for computing the values required to solve for b and a (using formulas 13.2 and 13.3) is given in Table 13.2.

TABLE 13.2

Employee	Aptitude Test Results (x)	Output (Dozens of Units) (y)	xy	x^2	y^2
A	6	30	180	36	900
B	9	49	441	81	2,401
C	3	18	54	9	324
D	8	42	336	64	1,764
E	7	39	273	49	1,521
F	5	25	125	25	625
G	8	41	328	64	1,681
H	10	52	520	100	2,704
Total	56	296	2,257	428	11,920

$$\bar{x} = \frac{\Sigma x}{n} = \frac{56}{8} = 7 \qquad \bar{y} = \frac{\Sigma y}{n} = \frac{296}{8} = 37$$

A y^2 column is also included in Figure 13.2. We don't need that column to calculate the regression equation, but we'll use it later. The values of b and a are found as follows:

$$b = \frac{n(\Sigma xy) - (\Sigma x)(\Sigma y)}{n(\Sigma x^2) - (\Sigma x)^2} = \frac{8(2{,}257) - (56)(296)}{8(428) - (56)^2} = \frac{1{,}480}{288} = 5.1389$$

$$a = \bar{y} - b\bar{x} = 37 - (5.1389)(7) = 1.0277$$

Therefore, the regression equation that describes the relationship between the output of our sample of Tackey Toy employees and their aptitude test results is:

$$\hat{y} = 1.0277 + 5.1389(x)$$

An Alternative

Regression analysis is a commonly applied tool in the social and physical sciences, and it is used to forecast future business and economic results. An alternative forecasting approach would be for you to sit in a trance on a tripod above a chasm that emitted noxious vapors and make oracular utterances for someone to record. This was the approach used by the Greeks in ancient Delphi.

Making Preliminary Predictions with the Regression Equation

There was once a young manager named Hess
Whose forecasts were always a mess.
So his boss did appear,
And in voice loud and clear,
Said, "Hess, son, try regression, or consider another career!"

The primary use of the regression equation is to estimate values of the dependent variable given values of the independent variable. Suppose, for example, that the unfortunate Hiram Hess, personnel manager for Tackey Toys, is considering hiring an applicant who scored a 4 on the aptitude test. The supervisor of the department wants someone hired who can produce an average of 30 dozen units. Of course, it's not possible to tell exactly what the applicant's future production might be, but Hiram can use the equation computed in the preceding section to arrive at a preliminary estimate or forecast of the average amount of output produced by those who score a 4 on the aptitude test. How? By simply substituting 4 for x in the regression equation. The estimate is computed as follows:

$$\hat{y} = 1.0277 + 5.1389(4) = 1.0277 + 20.556 = 21.58 \text{ dozen units of output}$$

This prediction is shown graphically in Figure 13.6.

But some words of caution are needed here. *First*, we do not yet know how dependable our estimate is likely to be (that's the subject of the next section). And *second*, when we make an estimate from a regression equation, it's incorrect to extend this estimate beyond the range of the original observations. There's no way of knowing what the nature of the regression equation will be if we encounter values of the dependent variable larger or smaller than those we've observed. For example, in our Tackey

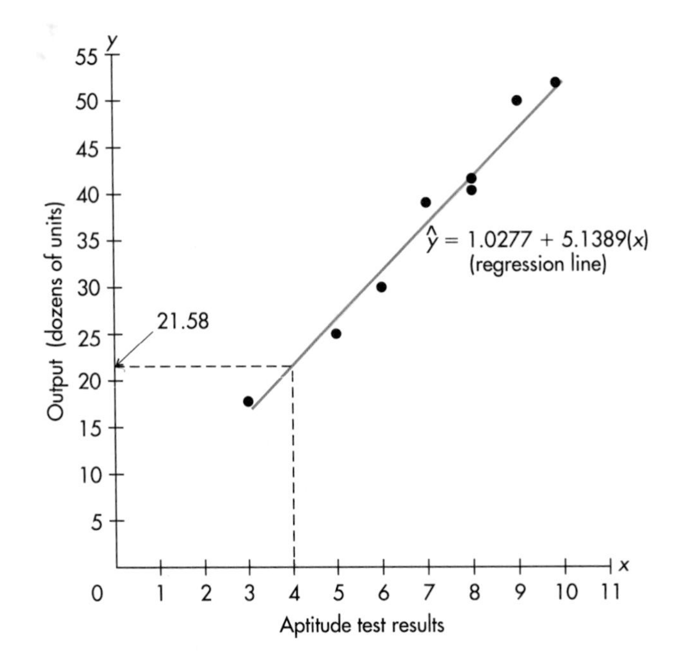

FIGURE 13.6

Toy situation it's ridiculous to assume that an employee's output will increase indefinitely as his or her test results increase. There's obviously some upper limit to aptitude test scores, and regardless of aptitude, the speed of productive equipment and physical endurance also set a limit to productive output. Since the largest value of x we observed is 10, we can place no reliance on estimates that might be made of the output of employees who have test scores of, say, 15 or 16.

The Standard Error of Estimate

When predicting from a regression equation, the question arises: "How dependable is the estimate?" Obviously, an important factor in gauging dependability is the closeness of the relationship between the variables. Let's look at Figure 13.7 and assume that both scatter diagrams have the same scales for the variables and the same regression line. When the points in a scatter diagram are closely spaced around the regression line, as they are in Figure 13.7*a*, it's logical to assume that an estimate based on that relationship will probably be more dependable than an estimate based on a regression line such as that shown in Figure 13.7*b*, where the scatter is much greater. Therefore, if we had a measure of the extent of the spread or scatter of the points around the regression line, we would be in a better position to judge the dependability of estimates made using the line. (You just know we are leading up to something here, don't you?)

You'll not be surprised to learn that *we do have a measure* that indicates the extent of the spread, scatter, or dispersion of the points about the regression line. From an estimating standpoint, *the smaller this measure is, the more dependable the prediction is likely to be.* (The numerical value of this measure for Figure 13.7*a* is smaller than the value for Figure 13.7*b* because the dispersion is smaller in Figure 13.7*a*.) The name of this measure is the **standard error of estimate** (the symbol is $s_{y.x}$), and, as the name implies, it's used to qualify the estimate made with the regression equation by indicating the extent of the possible variation (or error) that may be present.

Computation of $s_{y.x}$. In more precise terms, *the standard error of estimate is a standard deviation that measures the scatter of the observed values around the regression line.* Thus, one formula that may be used to compute the standard error of estimate naturally bears a striking resemblance to the formula used to compute the standard deviation for ungrouped data. The primary difference between the two formulas lies in the fact that the standard deviation is measured from the mean, while the standard error of estimate is measured from the regression line. Both the mean and the regres-

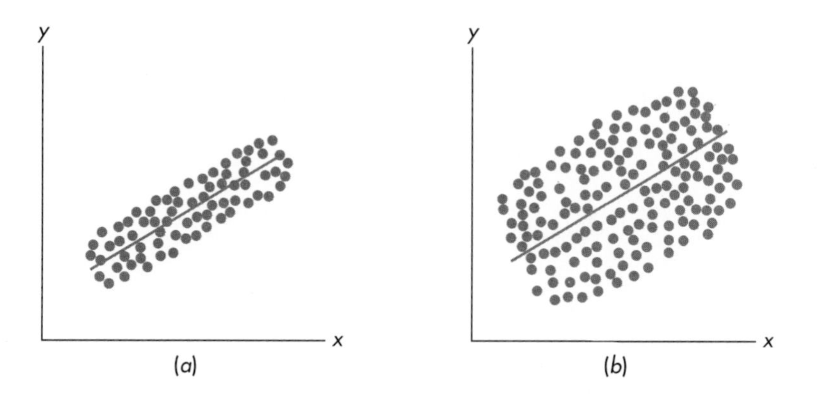

FIGURE 13.7 Varying degrees of spread or scatter.

sion line, of course, indicate central tendency. This formula for the standard error of estimate is:

$$s_{y.x} = \sqrt{\frac{\Sigma (y - \hat{y})^2}{n - 2}} \tag{13.4}$$

□

The $n - 2$ in the denominator is used in this case because our sample data consist of two variables, but we'll omit a detailed explanation of degrees of freedom.

In using formula 13.4, we must compute a value for $\hat{y}$ for each value of x by plugging each x value into the regression equation. We must then compute the difference between these $\hat{y}$ values and the corresponding observed values of y. Table 13.3 shows a work sheet for calculating the standard error of estimate for our Tackey Toy example.

TABLE 13.3

Employee	Aptitude Test Results (x)	Output (Dozens of Units) (y)	$\hat{y}$	$(y - \hat{y})$	$(y - \hat{y})^2$
A	6	30	31.86	−1.86	3.4596
B	9	49	47.28	1.72	2.9584
C	3	18	16.44	1.56	2.4336
D	8	42	42.14	−0.14	0.0196
E	7	39	37.00	2.00	4.0000
F	5	25	26.72	−1.72	2.9584
G	8	41	42.14	−1.14	1.2996
H	10	52	52.42	−0.42	.1764
Total	56	296*	296.00*	0.00	17.3056

*Note that the sums of y and $\hat{y}$ are equal. This must always be true if $\Sigma (y - \hat{y}) = 0$.

The standard error of estimate is calculated as follows:

$$s_{y.x} = \sqrt{\frac{\Sigma (y - \hat{y})^2}{n - 2}} = \sqrt{\frac{17.3056}{6}} = \sqrt{2.884}$$

$= 1.698$ dozens of units of output (The value of $s_{y.x}$ will always be expressed in the units of the y variable.)

Although it's helpful to use formula 13.4 to explain the nature of the $s_{y.x}$, a much easier formula to apply is:

$$s_{y.x} = \sqrt{\frac{\Sigma (y^2) - a(\Sigma y) - b(\Sigma xy)}{n - 2}} \tag{13.5}$$

You'll notice that all the values needed for this formula are available from Table 13.2, which was used to prepare the regression equation. Using the values from Table 13.2:

$$s_{y.x} = \sqrt{\frac{\Sigma (y)^2 - a(\Sigma y) - b(\Sigma xy)}{n-2}} = \sqrt{\frac{(11{,}920) - 1.0277(296) - 5.1389(2{,}257)}{6}}$$

$$= \sqrt{\frac{17.304}{6}} = \sqrt{2.884} = 1.698 \text{ dozens of units of output}$$

(The same result is obtained when using formula 13.4.)

Slip-Sliding Away

Geologists at the U.S. Geological Survey use statistical software to predict damage patterns caused by earthquakes. The geologists try to correlate earthquake variables with the data gathered on landslides and on the distribution of structural damages. Of course, earthquake models are only as good as their input data and assumptions. The January 1994 earthquake that shook the Los Angeles area involved previously undetected fault lines, and so these unknown conditions wouldn't have been included in any such model.

Self-Testing Review 13.2

1–9. Kelly Sanchez is a college senior who is interviewing at several companies for a management position. As she tours the offices at ABC Corporation, she notices that some seem to be much larger than others, and she wonders if there's a relationship between the number of years a person has spent at the company and the width of that person's office space. A random sample produces the following data pairs where x = the number of years a person has spent at the company and y = the width in feet of the person's office space:

x	3	16	7	4	15	7	8	5
y	4	40	16	9	38	16	17	10

1. Find Σx, Σx^2, $(\Sigma x)^2$.
2. Find Σy, Σy^2, $(\Sigma y)^2$.
3. Find Σxy.
4. Find $\bar{x}$ and $\bar{y}$.
5. Find b and a.
6. Write the equation for the line of regression.
7. Predict the width of the office space for a person who has been with the company for 6 years.
8. Predict the width of the office space for a person who has been with the company for 3 years.
9. Calculate the standard error of estimate.

10–18. Mark Z. Papers is a sixth-grade teacher who believes there is a relationship between his student's IQ scores and the hours they spend watching television each week. The following table represents a random sample of Mark's data where x = a sixth grader's IQ score and y = the hours spent watching TV each week:

x	125	116	97	114	85	107	105
y	5	14	30	16	41	25	21

10. Find Σx, Σx^2, $(\Sigma x)^2$.

11. Find Σy, Σy^2, $(\Sigma y)^2$.

12. Find Σxy.

13. Find $\bar{x}$ and $\bar{y}$.

14. Find b and a.

15. Write the equation for the line of regression.

16. Predict the number of hours of TV watching for a sixth grader with an IQ of 91.

17. Predict the number of hours of TV watching for a sixth grader with an IQ of 125.

18. Calculate the standard error of estimate.

19–27. Douglas Michaels, a premed student, wants to see if there's a relationship between the age of a youngster and the number of visits the child makes to a doctor during the year. The following data are collected from a random sample where x = a child's age and y = the number of doctor visits last year:

x	5	6	8	14	15	7	8	12
y	9	2	12	17	9	16	6	15

19. Find Σx, Σx^2, $(\Sigma x)^2$.

20. Find Σy, Σy^2, $(\Sigma y)^2$.

21. Find Σxy.

22. Find $\bar{x}$ and $\bar{y}$.

23. Find b and a.

24. Write the equation for the line of regression.

25. Predict the number of office visits for a 9-year-old child.

26. Predict the number of office visits for a 5-year-old child.

27. Calculate the standard error of estimate.

28–36. Clinical characteristics were measured in a study of autistic children that appeared in *The American Journal of Psychiatry.* A psychologist now wants to see if there's a linear relationship between the x variable and the y variable. The x variable is the total behavioral score made by children on a test of 29 items where each item score ranges from 0 (absence of symptoms) to 116 (maximum severity of symptoms). And the y variable is a language score made on a test that has a scale ranging from 1 to 5 where 1 = normal development and 5 = severe retardation. The psychologist's random sample of autistic children produces the following data pairs:

Behavioral Score	Language Score
27	2
35	4
65	4
67	4
47	3
46	3
63	4
44	4
34	5
51	4
17	2
40	2
41	4
60	5
24	5
48	4
29	2
73	5
60	3
41	4
27	3

28. Find Σx, Σx^2, $(\Sigma x)^2$.

29. Find Σy, Σy^2, $(\Sigma y)^2$.

30. Find Σxy.

31. Find $\bar{x}$ and $\bar{y}$.

32. Find b and a.

33. Write the equation for the line of regression.

34. Predict the language score for a child with a behavioral score of 64.

35. Predict the language score for a child with a behavioral score of 27.

36. Calculate the standard error of estimate.

37–45. Castle Rock Entertainment has produced many movies over the past few years. A vice president wants to see if there's a relationship between the total cost of a film (including production costs, salaries, and marketing expenses) and the gross income produced by the film through ticket sales in American movie theaters. A random sample of films produced the following data pairs (data source: Castle Rock Entertainment):

Cost (Millions of Dollars) (x)	Gross Income (Millions of Dollars) (y)
55	150.5
42	123.0
17	68.0
30	93.0
43	16.0
26	5.0
19	10.0
35	35.0
22	20.0
13	15.0

37. Find Σx, Σx^2, $(\Sigma x)^2$.

38. Find Σy, Σy^2, $(\Sigma y)^2$.

39. Find Σxy.

40. Find $\bar{x}$ and $\bar{y}$.

41. Find b and a.

42. Write the equation for the line of regression.

43. Predict the gross income for a film with a cost of $27 million.

44. Predict the gross income for a film with a cost of $35 million.

45. Calculate the standard error of estimate.

46–54. A U.S. military organization wants to know if there's a relationship between the number of recruiting offices located in particular cities and the total number of persons enlisting in those cities. Total enlistment data are obtained for a 1-month sample period from 10 cities and are given below:

City	Recruiting Offices	Enlistments for Month
Austin	1	20
Pittsburgh	2	40
Chicago	4	60
Los Angeles	3	60
Denver	5	80
Atlanta	4	100
Cleveland	5	80
Louisville	2	50
New Orleans	5	110
Kansas City	1	30

46. Find Σx, Σx^2, $(\Sigma x)^2$.

47. Find Σy, Σy^2, $(\Sigma y)^2$.

48. Find Σxy.

49. Find $\bar{x}$ and $\bar{y}$.

50. Find b and a.

51. Write the equation for the line of regression.

52. Predict the monthly enlistments for a city that has 3 recruiting offices.

53. Predict the monthly enlistments for a city that has 1 recruiting office.

54. Calculate the standard error of estimate.

55–63. The Rip-Off Vending Machine Company operates coffee vending machines in office buildings. The company wants to study the relationship, if any, that exists between the number of cups of coffee sold per day and the number of persons working in each building. Sample data for this study were collected by the company and are presented below:

Number of Persons Working at Location (x)	Number of Cups of Coffee Sold (y)
5	10
6	20
14	30
19	40
15	30
11	20
18	40
22	40
26	50

55. Find Σx, Σx^2, $(\Sigma x)^2$.

56. Find Σy, Σy^2, $(\Sigma y)^2$.

57. Find Σxy.

58. Find $\bar{x}$ and $\bar{y}$.

59. Find b and a.

60. Write the equation for the line of regression.

61. Predict the number of cups of coffee for a building with 17 employees.

62. Predict the number of cups of coffee for a building with 5 employees.

63. Calculate the standard error of estimate.

64–72. A regional development council, composed of members from several local chambers of commerce, gathered the following data from a random sample of firms in the region that produced similar products:

Production Cost (Thousands of Dollars)	Production Output (Thousands of Dollars)
150	40
140	38
160	48
170	56
150	62
162	75
180	70
165	90
190	110
185	120

64. Find Σx, Σx^2, $(\Sigma x)^2$.
65. Find Σy, Σy^2, $(\Sigma y)^2$.
66. Find Σxy.
67. Find $\bar{x}$ and $\bar{y}$.
68. Find b and a.
69. Write the equation for the line of regression.
70. Predict the production output if the production cost is $175,000.
71. Predict the production output if the production cost is $150,000.
72. Calculate the standard error of estimate.

73–81. Katie Dvorak trains employees to use a specific statistical software package. A random sample of trainees have turned in the following performances in recent weeks:

Trainee	Hours of Training	Errors
A	1	6
B	4	3
C	6	2
D	8	1
E	2	5
F	3	4
G	1	7

73. Find Σx, Σx^2, $(\Sigma x)^2$.
74. Find Σy, Σy^2, $(\Sigma y)^2$.
75. Find Σxy.
76. Find $\bar{x}$ and $\bar{y}$.
77. Find b and a.